AF389145

Teaching and Learning of Fluid Mechanics

Teaching and Learning of Fluid Mechanics

Special Issue Editor

Ashwin Vaidya

MDPI • Basel • Beijing • Wuhan • Barcelona • Belgrade • Manchester • Tokyo • Cluj • Tianjin

Special Issue Editor
Ashwin Vaidya
Montclair State University
USA

Editorial Office
MDPI
St. Alban-Anlage 66
4052 Basel, Switzerland

This is a reprint of articles from the Special Issue published online in the open access journal *Fluids* (ISSN 2311-5521) (available at: https://www.mdpi.com/journal/fluids/special_issues/teaching_learning_fluids).

For citation purposes, cite each article independently as indicated on the article page online and as indicated below:

LastName, A.A.; LastName, B.B.; LastName, C.C. Article Title. *Journal Name* **Year**, *Article Number*, Page Range.

ISBN 978-3-03936-443-5 (Pbk)
ISBN 978-3-03936-444-2 (PDF)

Contents

About the Special Issue Editor

Ashwin Vaidya is a faculty member in the Department of Mathematics and Department of Physics & Astronomy at Montclair State University. He also directs the Complex fluids laboratory where students and faculty members work on experimental and theoretical problems pertaining to fluid mechanics. His general scholarly interests, besides fluid mechanics include complex systems, thermodynamics, math and science education and philosophy of science.

Editorial

Teaching and Learning of Fluid Mechanics

Ashwin Vaidya

Department of Mathematics, Montclair State University, Montclair, NJ 07043, USA; vaidyaa@mail.montclair.edu

Received: 8 April 2020; Accepted: 10 April 2020; Published: 13 April 2020

Fluid mechanics is arguably one of the oldest branches of physics, and the literature on this subject is vast and complex. However, this subject has not sufficiently captured the interest of STEM educators like in other subjects such as quantum mechanics [1,2]. The objective of this collection, while not necessarily intended to generate education research, aims at bringing together various ways of teaching and learning about different topics in fluid mechanics.

Fluid mechanics occupies a privileged position in the sciences; it is taught in various science departments including physics, mathematics, environmental sciences and mechanical, chemical and civil engineering, with each highlighting a different aspect or interpretation of the foundation and applications of fluids. While scholarship in fluid mechanics is vast, expanding into the areas of experimental, theoretical and computational, there is little discussion among scientists about the different possible ways of teaching this subject or wide awareness of the how fluid mechanics plays a role in different disciplines. We believe there is much to be learned from an interdisciplinary dialogue about fluids for teachers and students alike.

The terms 'interdisciplinary', 'multidisciplinary' and 'transdisciplinary' have become common parlance in academia, but have been misunderstood and used without distinction [3]. Multidisciplinary teaching refers to diverse parallel viewpoints, with different goals and objectives being presented in the same setting while interdisciplinary or transdiscplinary refer to instances where goals overlap or unify completely [4]. In the context of education, inter- or transdisciplinary instruction allows students get to see the commonalities between different disciplines, thereby allowing students to make new meanings out of old ideas [5]. The education theorist, William Doll [5], articulates this idea very well: "Order emerges from interactions having just the 'right amount' of tension or difference or imbalance among the elements interacting." In a recent paper on education, my co-authors and I have argued that the synergy between different disciplines can result in the emergence of order, which we argue is nothing but creativity [6]. Doll's fluid analogy [5] for this idea is especially relevant to this issue:

"Emergence of creativity from complex flow of knowledge—example of Benard convection pattern as an analogy—dissipation or dispersal of knowledge (complex knowledge) results in emergent structures, i.e., creativity which in the context of education should be thought of as a unique way to arrange information so as to make new meaning of old ideas."

With this philosophy in mind, we have included all kinds of articles in this issue, including research on the pedagogical aspects of fluid mechanics, case studies or lesson plans at the undergraduate or graduate levels, articles on historical aspects of fluids, and novel and interesting experiments or theoretical calculations that can convey complex ideas in creative ways. The current volume includes 14 papers and showcases the work of scientists from different disciplines ranging from mathematics and physics to mechanical, environmental and chemical engineering. It truly is a wonderful collection and provides ideas on theoretical computational and experimental aspects of fluid mechanics that be implemented in a course in fluids in any department. The suitability of these papers ranges from early undergraduate to graduate level.

Overall, this issue contains papers in various, somewhat distinct categories. The articles [7–11] add and *reconsider fundamental ideas in fluid mechanics*. The paper by Brkić and Praks [7] is devoted to the solution of Colebrook's friction equation. This basic equation is used to introduce interesting and sophisticated mathematical tools, such as the fixed point method and Padé approximation, among others. The article by Kariotoglou and Psillos [8] is a synthesis of previous research upon high school and undergraduate students on the teaching of concepts such as pressure in fluids. The two papers by Pal [9,10] are focused on fundamental ideas in thermomechanics; the papers discuss the ways in which important thermodynamic ideas can be introduced and elucidated in a fluid dynamics course through examples such as flow in pipes and flow through packed beds. Vianna et al. [11] take up the concept of *permeability* in porous media flow and discusses new computational concepts that can help convey such complex concepts.

Several collections in this issue deal with the *development of computational tools* to resolve important problems in fluid mechanics [12–14]. Addair and Jaeger [12] consider efficient and effective strategies to convey fundamental concepts in Computational Fluids Dynamics (CFD) to undergraduate students, such as the implementation of the finite volume method to solve Navier–Stokes equations. The paper by Battista [13] provides an open repository of several useful two-dimensional solvers written in MATLAB and Python 3. The contribution by Pawar and San [14] is about CFD Julia, a programming module that teaches the foundations of computational fluid dynamics (CFD). This piece is written for an upper-level undergraduate and early undergraduate course in fluid mechanics and uses the inviscid Burger's equation and the two-dimensional Poisson equation as examples.

Articles [10,15] treat *classical topics in fluid mechanics* in a very thorough and innovative manner and serve to guide the development of lectures on these topics in undergraduate and graduate courses. Medved, Davis and Vasquez [15] deal with the classical problem of a particle motion in a fluid. Pal's contribution [10] cited earlier regarding melding fluid mechanics and thermodynamics in the context of pipeline flow is yet another example of such a contribution.

Articles [16,17] provide *novel and creative new ways of introducing fluid mechanics* to students. They make rich connections between several disciplines and are guaranteed to capture the imagination of students. The paper by Mayer [16] discusses the seemingly simple example of an experiment on the flow of a fluid through a bottle. This seemingly mundane topic is a classic example of the richness and complexity of fluid mechanics, but also its ubiquity. Nita and Ramanathan [17] make a creative connection between fluids and music in their discussion of the physics of the Pan's flute. These are both truly exciting examples that can be introduced in undergraduate or graduate courses on fluids.

Articles [18–20] are more *pedagogically focused*. Huilier's contribution [18] is unique and provides a rich personal history of teaching fluid mechanics for over forty years at the University of Strasbourg. While the subject of fluid mechanics has evolved much in the last several decades, this reflective piece is a guide to all young instructors about how to adapt and evolve one's teaching and pedagogy. The paper by Potter and Wolff [19] discusses the experience and observations of interventions in a second-year fluid mechanics module, and provides insights into the teaching of fluid mechanics at their institution. A second goal of this paper is to conceptualize the use of Legitimation Code Theory (LCT) dimensions towards teaching strategies intended to facilitate improved learning outcomes. The LCT provides educators with a very useful tool to guide their curriculum planning and pedagogy. The article [20] by Valyrakis et al. focuses on the importance of lab-based fluid mechanics instruction. The authors demonstrate the value of using physical models ("floodopoly") to demonstrate complex geophysical processes such as sediment transport and flooding. A survey instrument is used to assess and demonstrate student understanding and perception of the subject.

References

1. Carr, L.D.; McKagan, S.B. Graduate quantum mechanics reform. *Am. J. Phys.* **2009**, *77*, 308–319. [CrossRef]
2. Singh, C. Student understanding of quantum mechanics at the beginning of graduate instruction. *Am. J. Phys.* **2008**, *76*, 277–287. [CrossRef]
3. Lawrence, R.J. Deciphering interdisciplinary and transdisciplinary contributions. *Transdiscipl. J. Eng. Sci.* **2010**, *1*, 125–130. [CrossRef] [PubMed]

4. Park, J.-Y.; Son, J.-B. Transitioning toward transdisciplinary learning in a multidisciplinary environment. *Int. J. Pedagog. Learn.* **2010**, *6*, 82–93. [CrossRef]

5. Doll, W.E., Jr. *A Post-Modern Perspective on Curriculum*; Teachers College Press: New York, NY, USA, 2015.

6. Monahan, C.; Munakata, M.; Vaidya, A. Creativity as an Emergent Property of Complex Educational System. *Northeast J. Complex Syst. (NEJCS)* **2019**, *1*, 4. [CrossRef]

7. Brkić, D.; Praks, P. What Can Students Learn While Solving Colebrook's Flow Friction Equation? *Fluids* **2019**, *4*, 114. [CrossRef]

8. Kariotoglou, P.; Psillos, D. Teaching and Learning Pressure and Fluids. *Fluids* **2019**, *4*, 194. [CrossRef]

9. Pal, R. Teach Second Law of Thermodynamics via Analysis of Flow through Packed Beds and Consolidated Porous Media. *Fluids* **2019**, *4*, 116. [CrossRef]

10. Pal, R. Teaching Fluid Mechanics and Thermodynamics Simultaneously through Pipeline Flow Experiments. *Fluids* **2019**, *4*, 103. [CrossRef]

11. Vianna, R.S.; Cunha, A.M.; Azeredo, R.B.V.; Leiderman, R.; Pereira, A. Computing Effective Permeability of Porous Media with FEM and Micro-CT: An Educational Approach. *Fluids* **2020**, *5*, 16. [CrossRef]

12. Adair, D.; Jaeger, M. An Efficient Strategy to Deliver Understanding of Both Numerical and Practical Aspects When Using Navier-Stokes Equations to Solve Fluid Mechanics Problems. *Fluids* **2019**, *4*, 178. [CrossRef]

13. Battista, N.A. Suite-CFD: An Array of Fluid Solvers Written in MATLAB and Python. *Fluids* **2020**, *5*, 28. [CrossRef]

14. Pawar, S.; San, O. CFD Julia: A Learning Module Structuring an Introductory Course on Computational Fluid Dynamics. *Fluids* **2019**, *4*, 159. [CrossRef]

15. Medved, A.; Davis, R.; Vasquez, P.A. Understanding Fluid Dynamics from Langevin and Fokker–Planck Equations. *Fluids* **2020**, *5*, 40. [CrossRef]

16. Mayer, H.C. Bottle Emptying: A Fluid Mechanics and Measurements Exercise for Engineering Undergraduate Students. *Fluids* **2019**, *4*, 183. [CrossRef]

17. Nita, B.G.; Ramanathan, S. Fluids in Music: The Mathematics of Pan's Flutes. *Fluids* **2019**, *4*, 181. [CrossRef]

18. Huilier, D.G.F. Forty Years' Experience in Teaching Fluid Mechanics at Strasbourg University. *Fluids* **2019**, *4*, 199. [CrossRef]

19. Pott, R.W.M.; Wolff, K. Using Legitimation Code Theory to Conceptualize Learning Opportunities in Fluid Mechanics. *Fluids* **2019**, *4*, 203. [CrossRef]

20. Valyrakis, M.; Latessa, G.; Koursari, E.; Cheng, M. Floodopoly: Enhancing the Learning Experience of Students in Water Engineering Courses. *Fluids* **2020**, *5*, 21. [CrossRef]

Article

Teaching Fluid Mechanics and Thermodynamics Simultaneously through Pipeline Flow Experiments

Rajinder Pal

Department of Chemical Engineering, University of Waterloo, Waterloo, ON N2L 3G1, Canada; rpal@uwaterloo.ca; Tel.: +1-519-888-4567 (ext. 32985)

Received: 12 May 2019; Accepted: 28 May 2019; Published: 1 June 2019

Abstract: Entropy and entropy generation are abstract and illusive concepts for undergraduate students. In general, students find it difficult to visualize entropy generation in real (irreversible) processes, especially at a mechanistic level. Fluid mechanics laboratory can assist students in making the concepts of entropy and entropy generation more tangible. In flow of real fluids, dissipation of mechanical energy takes place due to friction in fluids. The dissipation of mechanical energy in pipeline flow is reflected in loss of pressure of fluid. The degradation of high quality mechanical energy into low quality frictional heat (internal energy) is simultaneously reflected in the generation of entropy. Thus, experiments involving measurements of pressure gradient as a function of flow rate in pipes offer an opportunity for students to visualize and quantify entropy generation in real processes. In this article, the background in fluid mechanics and thermodynamics relevant to the concepts of mechanical energy dissipation, entropy and entropy generation are reviewed briefly. The link between entropy generation and mechanical energy dissipation in pipe flow experiments is demonstrated both theoretically and experimentally. The rate of entropy generation in pipeline flow of Newtonian fluids is quantified through measurements of pressure gradient as a function of flow rate for a number of test fluids. The factors affecting the rate of entropy generation in pipeline flows are discussed.

Keywords: undergraduate education; fluid mechanics; pipeline flow; non-equilibrium thermodynamics; entropy generation; pressure loss; experimental studies

---

1. Introduction

Fluid mechanics, that is, the study of motion of fluids and forces in fluids, is relatively a less abstract subject as compared with thermodynamics. Students find it relatively easy to visualize and understand the motion of fluids and forces in fluids. For example, it is not difficult to convey to the students fluid mechanics concepts such as pressure, pressure distribution, viscosity, velocity distribution, velocity gradient, shear and normal stresses, mechanical energy dissipation and pressure loss in flow of fluids due to friction, etc. Most fluid mechanics quantities are directly measurable. For example, instruments are available to directly measure pressure or pressure drop, flow rate, local velocity, shear and normal stresses, etc. Thermodynamics, on the other hand, is a very abstract subject. Concepts such as entropy and entropy generation in real processes are illusive for students. There are no instruments available which can be used to directly measure entropy and entropy generation in real processes.

It is a well-known fact that students learn and understand concepts better and with relative ease through experiential learning. It is appropriate to quote here a famous Chinese saying credited to the Chinese philosopher Confucius, "I hear and I forget; I see and I remember; I do and I understand". To that end, the undergraduate laboratory experiments can play a very important role in providing experiential learning of concepts and theory to students.

In most undergraduate engineering programs (chemical, mechanical, civil, etc.) students are taught fluid mechanics through in-class instruction and laboratory experiments. In a typical undergraduate fluid mechanics laboratory, the students are required to do pipeline flow experiments involving measurement of pressure loss as a function of flow rate for different diameter pipes. The flow rate is measured with the help of a flowmeter and the pressure drop is measured with the help of pressure transducers. The test fluid is usually water and the experiments are carried out at room temperature. From pressure drop vs. flow rate experimental data, friction factor vs. Reynolds number data are calculated and compared with the available theoretical and empirical relations. What is completely missing in the undergraduate fluid mechanics experiments is the link between the pipeline flow experiments and the second law of thermodynamics, that is, entropy and entropy generation in pipeline flows. In order to appreciate and understand the second law of thermodynamics, it is important for students to be able to relate the directly measureable quantities like pressure loss in pipeline flow to entropy generation in real flows.

The main objectives of this article are: (1) to briefly review the background in fluid mechanics and thermodynamics related to pressure loss, mechanical energy dissipation, and entropy generation in real flows; (2) to demonstrate the link between mechanical energy dissipation and entropy generation in real flows; (3) to carry out experimental work to determine friction factor as a function of Reynolds number and entropy generation rate as a function of fluid velocity in flow of emulsion-type test fluids in different diameter pipelines; and (4) to explain mechanistically the cause of entropy generation in real flows.

2. Background

2.1. Fluid Mechanics

Consider flow of a fluid through a stationary control volume. The macroscopic balance of any entity (mass, momentum, energy) over the control volume can be expressed as:

$$Input - Output + Generation = Accumulation \tag{1}$$

The application of the entity balance equation, Equation (1), to mass gives the following integral equation [1,2]:

$$-\iint \rho\left(\hat{n}\cdot\vec{V}\right)dA = \iiint \frac{\partial \rho}{\partial t}d\vartheta \tag{2}$$

where ρ is the fluid density, $\hat{n}$ is the unit outward normal to the control surface, $\vec{V}$ is the fluid velocity vector, A is the control surface area, t is the time, ϑ is the volume of the control volume, the cyclic double integral is the surface integral over the entire control surface, and the cyclic triple integral is the volume integral over the entire control volume. It should be noted that the surface integral $\iint \rho\left(\hat{n}\cdot\vec{V}\right)dA$ is the net outward flow of mass across the entire control surface. The volume integral $\iiint \frac{\partial \rho}{\partial t}d\vartheta$ is the rate of accumulation of mass within the entire control volume (assumed to be fixed and non-deforming). For a control volume with one inlet and one outlet, the macroscopic or integral mass balance equation, Equation (2), under steady state condition, reduces to:

$$\rho_1 A_1 V_1 = \rho_2 A_2 V_2 = \dot{m} \tag{3}$$

where V is the fluid velocity, A is the cross-section area of inlet or outlet opening, and $\dot{m}$ is the mass flow rate. The subscript 1 indicates inlet variables and the subscript 2 indicates outlet variables.

The application of the entity balance equation, Equation (1), to linear momentum gives the following integral equation [1,2]:

$$-\oiint \rho\vec{V}\left(\hat{n}\cdot\vec{V}\right)dA + \sum \vec{F} = \oiiint \frac{\partial\left(\rho\vec{V}\right)}{\partial t}d\vartheta \tag{4}$$

where $\vec{F}$ is the force vector. Equation (4) is often referred to as *momentum theorem* [1]. The surface integral $\oiint \rho\vec{V}\left(\hat{n}\cdot\vec{V}\right)dA$ is the net outward flow of linear momentum across the entire control surface. The volume integral $\oiiint \frac{\partial\left(\rho\vec{V}\right)}{\partial t}d\vartheta$ is the rate of accumulation of linear momentum within the entire control volume (assumed to be fixed and non-deforming). The momentum balance equation, Equation (4), is a vector equation. In Cartesian coordinates, the vector momentum balance equation can be written as three scalar equations:

$$-\oiint \rho V_x\left(\hat{n}\cdot\vec{V}\right)dA + \sum F_x = \oiiint \frac{\partial(\rho V_x)}{\partial t}d\vartheta \tag{5}$$

$$-\oiint \rho V_y\left(\hat{n}\cdot\vec{V}\right)dA + \sum F_y = \oiiint \frac{\partial(\rho V_y)}{\partial t}d\vartheta \tag{6}$$

$$-\oiint \rho V_z\left(\hat{n}\cdot\vec{V}\right)dA + \sum F_z = \oiiint \frac{\partial(\rho V_z)}{\partial t}d\vartheta \tag{7}$$

where V_x, V_y, V_z and F_x, F_y, F_z are velocity and force components in x, y, z directions. For a control volume with one inlet and one outlet, the macroscopic or integral momentum balance equations, Equations (5)–(7), under steady state condition, reduce to:

$$V_{x,1}(\rho_1 A_1 V_1) - V_{x,2}(\rho_2 A_2 V_2) + \sum F_x = 0 \tag{8}$$

$$V_{y,1}(\rho_1 A_1 V_1) - V_{y,2}(\rho_2 A_2 V_2) + \sum F_y = 0 \tag{9}$$

$$V_{z,1}(\rho_1 A_1 V_1) - V_{z,2}(\rho_2 A_2 V_2) + \sum F_z = 0 \tag{10}$$

Using mass balance, Equation (3), Equations (8)–(10) can be further simplified as:

$$\sum F_x = \dot{m}(V_{x,2} - V_{x,1}) \tag{11}$$

$$\sum F_y = \dot{m}\left(V_{y,2} - V_{y,1}\right) \tag{12}$$

$$\sum F_z = \dot{m}(V_{z,2} - V_{z,1}) \tag{13}$$

Note that we are assuming velocity profiles are uniform at inlet and outlet of the control volume.

The application of the entity balance equation, Equation (1), to mechanical energy gives the following integral equation assuming *frictionless* flow [1,2]:

$$-\oiint \rho\left(\hat{n}\cdot\vec{V}\right)\left(\varphi + KE + \frac{P}{\rho}\right)dA - \dot{W}_{sh} = \oiiint \frac{\partial[\rho(\varphi + KE)]}{\partial t}d\vartheta \tag{14}$$

where φ is the potential energy per unit mass, KE is the kinetic energy per unit mass, P is the pressure, and $\dot{W}_{sh}$ is the rate of shaft work. The surface integral $\oiint \rho\left(\hat{n}\cdot\vec{V}\right)\left(\varphi + KE + \frac{P}{\rho}\right)dA$ is the net outward flow of mechanical energy (including flow work) across the entire control surface. The volume integral $\oiiint \frac{\partial[\rho(\varphi + KE)]}{\partial t}d\vartheta$ is the rate of accumulation of mechanical energy within the entire control volume (assumed to be fixed and non-deforming). For a control volume with one inlet and one outlet, the macroscopic mechanical energy balance equation for frictionless flow, Equation (14), under steady state condition, reduces to:

$$\left(\varphi_1 + KE_1 + \frac{P_1}{\rho_1}\right)(\rho_1 A_1 V_1) - \left(\varphi_2 + KE_2 + \frac{P_2}{\rho_2}\right)(\rho_2 A_2 V_2) - \dot{W}_{sh} = 0 \tag{15}$$

Using mass balance, Equation (3), Equation (15) can be further simplified as:

$$\dot{W}_{sh} + \dot{m}\left[\Delta\varphi + \Delta(KE) + \Delta\left(\frac{P}{\rho}\right)\right] = 0 \tag{16}$$

Note that mechanical energy is conserved only in frictionless flows. In real flows, however, mechanical energy is not conserved as mechanical energy dissipation occurs due to friction in fluids. The mechanical energy loss due to friction in real flows can be treated as negative generation (destruction) in entity balance equation, Equation (1). Thus, Equation (16) can be modified as [2]:

$$\dot{W}_{sh} + \dot{F}_l + \dot{m}\left[\Delta\varphi + \Delta(KE) + \Delta\left(\frac{P}{\rho}\right)\right] = 0 \tag{17}$$

where $\dot{F}_l$ is the rate of mechanical energy dissipation due to friction in fluid. Equation (17) assumes that the velocity profiles are uniform at inlet and outlet of the control volume. Furthermore any work due to viscous effects (shear stresses and viscous normal stresses) at the control surface is assumed to be negligible.

2.2. Thermodynamics

The *first law of thermodynamics* is simply the principle of conservation of energy. When applied to a control volume, it indicates that that the rate of accumulation of total energy inside the control volume is equal to the net rate of total energy addition to the control volume. Equation (1) is applicable to total energy as the entity with no generation. For a stationary control volume (fixed and non-deforming control volume), the first law of thermodynamics can be expressed as [1]:

$$-\iint \rho\left(\hat{n}\cdot\vec{V}\right)\left(e + \frac{P}{\rho}\right)dA + \dot{Q} - \dot{W}_{sh} = \iiint \frac{\partial(\rho e)}{\partial t}d\vartheta \tag{18}$$

where e is the specific total energy (total energy per unit mass) of fluid and $\dot{Q}$ is the rate of heat transfer. The specific total energy e includes internal energy, kinetic energy and potential energy, that is, $e = u + KE + \varphi$, where u is the specific internal energy. The work associated with shear stress and viscous portion of normal stress at the control surface is assumed to be zero [1]. The surface integral $\iint \rho\left(\hat{n}\cdot\vec{V}\right)\left(e + \frac{P}{\rho}\right)dA$ is the net outward flow of total energy (including flow work) across the entire control surface. The volume integral $\iiint \frac{\partial(\rho e)}{\partial t}d\vartheta$ is the rate of accumulation of total energy within the entire control volume.

For a control volume with one inlet and one outlet, the macroscopic total energy balance (first law of thermodynamics), Equation (18), under steady state condition, reduces to:

$$\left(e_1 + \frac{P_1}{\rho_1}\right)(\rho_1 A_1 V_1) - \left(e_2 + \frac{P_2}{\rho_2}\right)(\rho_2 A_2 V_2) + \dot{Q} - \dot{W}_{sh} = 0 \tag{19}$$

Using mass balance, Equation (3), Equation (19) can be further simplified as:

$$\dot{m}\left[\Delta\left(e + \frac{P}{\rho}\right)\right] = \dot{Q} - \dot{W}_{sh} \tag{20}$$

As

$$e + \frac{P}{\rho} = u + V^2/2 + gz + \frac{P}{\rho} = h + V^2/2 + gz \tag{21}$$

where g is acceleration due to gravity, z is elevation and h is specific enthalpy of fluid given as: $h = u + P/\rho$, Equation (20) could also be written as:

$$\dot{m}\left[\Delta h + \Delta\left(V^2/2\right) + g\Delta z\right] = \dot{Q} - \dot{W}_{sh} \tag{22}$$

For reversible (frictionless) flows, it can be readily shown that

$$\dot{Q} = \dot{m}\left[\Delta h - \Delta\left(\frac{P}{\rho}\right)\right] \tag{23}$$

Upon substitution of $\dot{Q}$ from Equation (23) into Equation (22), the mechanical energy balance equation for frictionless flow, Equation (16), is recovered.

The *second law of thermodynamics* states that all irreversible (real) processes are accompanied by entropy generation in the universe [3]. For flow through a stationary control volume, the entropy generation rate ($\dot{S}_G$) in the universe can be expressed as:

$$\dot{S}_{G,universe} = \dot{S}_{G,CV} + \dot{S}_{G,Surr} = \iint \rho s\left(\hat{n}\cdot\vec{V}\right)dA + \iiint \frac{\partial(\rho s)}{\partial t}d\vartheta - \sum \frac{\dot{Q}_i}{T_i} \geq 0 \tag{24}$$

where s is the entropy per unit mass of fluid, $\dot{Q}_i$ is the rate of heat transfer to control volume from *ith* heat reservoir at an absolute temperature of T_i, the subscripts CV and *Surr* refer to control volume and surroundings, respectively. The equality in Equation (24) is valid for any reversible (frictionless) process and the inequality is valid for all irreversible processes. The surface integral $\iint \rho s\left(\hat{n}\cdot\vec{V}\right)dA$ is the net outward flow of entropy across the entire control surface. The volume integral $\iiint \frac{\partial(\rho s)}{\partial t}d\vartheta$ is the rate of accumulation of entropy within the entire control volume (assumed to be fixed and non-deforming).

For a control volume with one inlet and one outlet, Equation (24), under steady state condition, reduces to:

$$\dot{S}_{G,universe} = (s_2)(\rho_2 A_2 V_2) - (s_1)(\rho_1 A_1 V_1) - \sum \frac{\dot{Q}_i}{T_i} \geq 0 \tag{25}$$

Using mass balance, Equation (3), Equation (25) can be further simplified as:

$$\dot{S}_{G,universe} = \dot{m}(\Delta s) - \sum \frac{\dot{Q}_i}{T_i} \geq 0 \tag{26}$$

2.3. Steady Flow in a Pipe

Consider steady flow of an incompressible fluid in a cylindrical pipe of uniform diameter (see Figure 1). From mass balance,

$$\dot{m} = \rho V A = constant \tag{27}$$

As ρ and A are constant, the velocity is constant,

$$V = constant \tag{28}$$

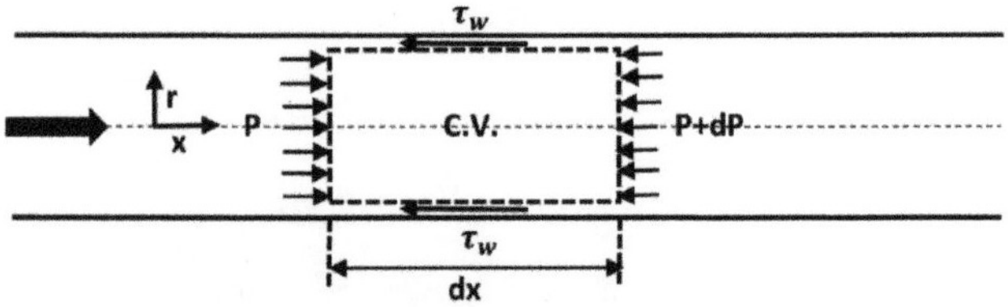

Figure 1. Stresses acting on fluid inside the control volume.

From momentum balance, Equation (11):

$$\sum F_x = \dot{m}(V_{x,2} - V_{x,1}) = 0 \tag{29}$$

Now consider force balance over a differential control volume of length dx as shown in Figure 1:

$$\sum F_x = PA - (P + dP)A - \tau_W(\pi D)dx = \tag{30}$$

where τ_w is the wall shear stress, D is the pipe internal diameter, and $A = \pi D^2/4$. Equation (30) leads to:

$$-(dP/dx) = \frac{\pi D \tau_w}{A} = \frac{4\tau_w}{D} \tag{31}$$

Upon rearrangement, Equation (31) gives

$$\tau_w = -\frac{(dP/dx)D}{4} = -\frac{(dP/dx)R}{2} \tag{32}$$

where R is the pipe radius. Equation (32) can also be applied to any radial position as:

$$\tau = -\frac{(dP/dx)r}{2} \tag{33}$$

where r is any radial position in the pipe. From Equations (32) and (33), it follows that:

$$\frac{\tau}{\tau_w} = \frac{r}{R} \tag{34}$$

Equation (34) describes the variation of shear stress with the radial position. The shear stress varies linearly with the radial position.

Using macroscopic mechanical energy balance, Equation (17), it can be readily shown that for steady incompressible flow in a horizontal pipe of uniform diameter:

$$\dot{F}_l/L = (\dot{m}/\rho)(-dP/dx) \tag{35}$$

where L is the length of the pipe. The rate of mechanical energy loss $\dot{F}_l$ can further be expressed in terms of a friction factor (f) defined as:

$$f = 2\tau_w / \left(\rho \overline{V}^2\right) \tag{36}$$

where $\overline{V}$ is the average velocity in the pipe. Upon substitution of τ_w from Equation (32) into Equation (36), we get

$$f = (-dP/dx)(D)/\left(2\rho\overline{V}^2\right) \tag{37}$$

From Equations (35) and (37), it follows that the mechanical energy loss per unit mass of fluid is:

$$\dot{F}_l/\dot{m} = 4f(L/D)\left(\overline{V}^2/2\right) \tag{38}$$

From Equation (38), it follows that the mechanical energy loss per unit length per unit mass of fluid is:

$$F'_l = \dot{F}_l/(L\dot{m}) = (4f/D)\left(\overline{V}^2/2\right) \tag{39}$$

In order to calculate the mechanical energy loss in pipeline flows, the value of friction factor is required.

In *laminar* flow of Newtonian fluids ($Re \leq 2100$), friction factor is related to Reynolds number through the following theoretical relationship [1,4]:

$$f = 16/Re \tag{40}$$

where the Reynolds number Re is defined as:

$$Re = \rho D \overline{V} / \mu \tag{41}$$

In *turbulent* flow of Newtonian fluids, friction factor is a function of Reynolds number and relative roughness of pipe (ϵ/D). For hydraulically smooth pipes ($\epsilon/D \rightarrow 0$), the friction factor depends only on Re in turbulent regime. The following semi-empirical equation, often referred to as von Karman–Nikuradse equation, describes the f vs. Re turbulent behavior of Newtonian fluids in smooth pipes very well [1]:

$$1/\sqrt{f} = 4 log_{10}\left(Re\sqrt{f}\right) - 0.40 \tag{42}$$

The von Karman–Nikuradse equation is not explicit in friction factor. A number of explicit f vs. Re relations are available in the literature. One of the popular ones is the Blasius friction factor equation for turbulent flow of Newtonian fluids in smooth pipes [4]:

$$f = 0.079/Re^{0.25} \tag{43}$$

Equation (43) is accurate over a Reynolds number range of $3000 \leq Re \leq 100{,}000$. For turbulent flow in rough pipes, the following Colebrook equation [5,6] is widely accepted:

$$1/\sqrt{f} = -4 log_{10}\left(\frac{\epsilon/D}{3.7} + \frac{1.26}{Re\sqrt{f}}\right) \tag{44}$$

This Colebrook equation is implicit in friction factor. A number of explicit f vs. Re relations are available in the literature for turbulent flow of Newtonian fluids in rough pipes [7–9]. An explicit equation which is very accurate for turbulent flow of Newtonian fluids in rough pipes is as follows:

$$1/\sqrt{f} = -4 log_{10}\left[\frac{\epsilon/D}{3.7} - \frac{5.02}{Re} log_{10}\left(\frac{\epsilon/D}{3.7} - \frac{5.02A}{Re}\right)\right] \tag{45}$$

$$A = log_{10}\left(\frac{\epsilon/D}{3.7} + \frac{13}{Re}\right) \tag{46}$$

This equation was originally proposed by Zigrang and Sylvester [10].

2.4. Entropy Generation in Steady Flow in a Pipe

In flow of real fluids, the dissipation of mechanical energy, and hence loss of pressure, is simultaneously reflected in the generation of entropy [11,12]. Consequently, the pipeline flow experiments performed by undergraduate students in the fluid mechanics laboratory can also be used as a tool to teach the second law of thermodynamics which states that all real processes are accompanied by generation of entropy in the universe.

For steady flow in a pipe with no heat transfer, Equation (26) reduces to:

$$\dot{S}_{G,universe} = \dot{S}_{G,CV} = \dot{m}(\Delta s) > 0 \tag{47}$$

There is no entropy generation in the surroundings. All the entropy is generated within the fluid inside the pipe and the rate of entropy generation is the net rate of increase in entropy of the flowing stream.

We can now relate entropy change of the fluid stream to other variables. For pure substances, the relationship between entropy and other state variables is given as [3]:

$$Tds = dh - (dP/\rho) \tag{48}$$

where T is the absolute temperature. From the first law of thermodynamics, Equation (22), the enthalpy change is zero in the absence of heat transfer and shaft work for steady flow in a horizontal pipe of uniform diameter. Consequently, Equation (48) reduces to:

$$Tds = -(dP/\rho) \tag{49}$$

Assuming incompressible flow and constant temperature, Equation (49) upon integration gives:

$$\Delta s = -\frac{\Delta P}{\rho T} \tag{50}$$

Strictly speaking, the temperature is expected to rise somewhat in adiabatic flow due to frictional heating. However, the temperature rise is usually very small in pipeline flow experiments conducted in the undergraduate fluid mechanics laboratory. From Equations (47) and (50), it follows that:

$$\dot{S}_G = \frac{\dot{m}}{\rho}\left(-\frac{\Delta P}{T}\right) > 0 \tag{51}$$

The subscript "CV" has been removed from $\dot{S}_{G,CV}$ for the sake of simplicity. We can also express the rate of entropy generation in a pipe on a unit length basis as:

$$\dot{S}_G' = \dot{m}\left(\frac{F_l'}{T}\right) = \frac{\dot{m}}{\rho T}\left(-\frac{dP}{dx}\right) \tag{52}$$

where $\dot{S}_G'$ is the rate of entropy generation per unit length of the pipe. From Equations (37) and (52), it can be readily shown that:

$$\dot{S}_G' = \left(\frac{\pi}{2T}\right)\left(\rho D \overline{V}^3\right) f \tag{53}$$

In laminar flow, the friction factor f is given by Equation (40). Consequently, Equation (53) yields:

$$\dot{S}_G' = \left(\frac{8\pi}{T}\right)\mu \overline{V}^2 \tag{54}$$

Thus entropy generation rate per unit length of pipe in steady laminar flow of a Newtonian fluid is directly proportional to fluid viscosity and square of average velocity in the pipe.

In turbulent flow of a Newtonian fluid in hydraulically smooth pipe, Equations (43) and (53) give the flowing expression for entropy generation rate per unit length of pipe:

$$\dot{S}_G' = \left(\frac{0.079\pi}{2T}\right)\mu^{0.25}\rho^{0.75}D^{0.75}\overline{V}^{2.75} \tag{55}$$

In turbulent flow, the entropy generation rate per unit length of pipe also depends on pipe diameter and fluid density, in addition to viscosity and fluid velocity dependence. Although the viscosity dependence of $\dot{S}_G'$ in turbulent flow is less severe in comparison with laminar flow, the velocity dependence is stronger in turbulent flow.

Figure 2 shows the plots of $\dot{S}_G'$ vs. $\overline{V}$ on a log-log scale for laminar and turbulent flows generated from Equations (54) and (55), respectively. The fluid properties used in the equations are: $\mu = 1$ mPa·s, $\rho = 1000$ kg/m^3. The temperature used is 298.15 K. A single line of slope 2 is obtained for laminar regime regardless of the pipe diameter. The entropy generation rate per unit length of the pipe increases linearly with the increase in average velocity in the pipe. A family of parallel lines of slope 2.75 is obtained for the turbulent regime. The line shifts upward towards higher entropy generation rate with the increase in the pipe diameter.

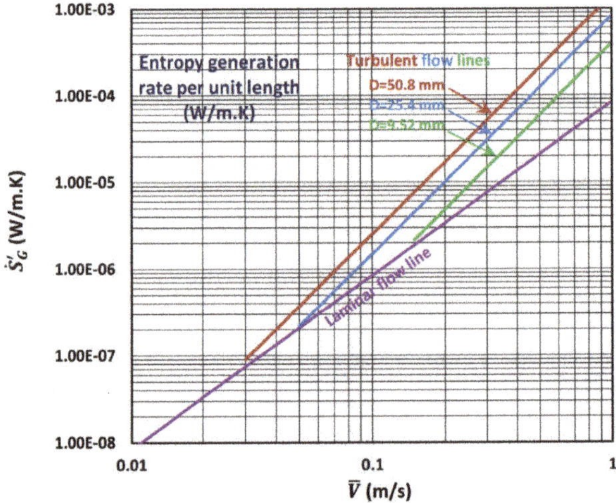

Figure 2. Entropy generation rate per unit length of pipe as a function of average velocity in the pipe.

3. Experimental Work

3.1. Apparatus

A flow rig consisting of five different diameter pipeline test sections (stainless steel seamless tubes, hydraulically smooth) was designed and constructed. The pipelines were installed horizontally. Table 1 gives the dimensions of the test sections. The test fluid was circulated through the pipeline test sections, one at a time, using a centrifugal pump. An electromagnetic flow meter was used to measure the flow rate of a fluid circulated through the test section. The pressure drop in a pipeline test section was measured using pressure transducers covering a broad range of pressure drops. The pressure drop as a function of flow rate was recorded by a computer data acquisition system. The experiments were carried out at a constant temperature of 25 °C.

Table 1. Various dimensions of the pipeline flow test sections.

Pipe Inside Diameter (mm)	Entrance Length (m)	Length of Test Section (m)	Exit Length (m)
7.15	1.07	3.05	0.46
8.89	0.89	3.35	0.48
12.60	1.19	2.74	0.53
15.8	1.65	2.59	0.56
26.54	3.05	1.22	0.67

3.2. Test Fluids

The test fluids used were surfactant-stabilized oil-in-water (O/W) emulsions. The viscosity of the test fluid was increased by increasing the oil concentration of the emulsion. Table 2 summarizes the viscosity and density data of the test fluids at 25 °C.

Table 2. Viscosity and density of test fluids.

Test Fluid Type	Viscosity, mPa·s	Density, kg/m^3
Aqueous surfactant solution	0.935	997.5
16.53% O/W emulsion	1.424	961.56
30.4% O/W emulsion	2.464	931.38
44.4% O/W emulsion	5.216	900.90
49.65% O/W emulsion	7.159	889.51
55.14% O/W emulsion	10.628	877.56

4. Results and Discussion

4.1. Fluid Mechanics Experiments

The experimental data obtained from pipeline flow experiments consist of pressure drop (over a known length of pipe) versus flow rate for different diameter pipes. From pressure drop versus flow rate data, friction factor is calculated from Equation (37), re-written as:

$$f = (-\Delta P/L)(D)/\left(2\rho \overline{V}^2\right) \tag{56}$$

where the average velocity $\overline{V}$ is obtained from:

$$\overline{V} = 4\dot{\vartheta}/\left(\pi D^2\right) \tag{57}$$

Here $\dot{\vartheta}$ is the volumetric flow rate of fluid. The Reynolds is calculated from the defining relation, Equation (41), re-written as:

$$Re = 4\rho\dot{\vartheta}/(\pi D\mu) \tag{58}$$

The f vs. Re data are plotted on a log-log scale and compared with the predictions of available f vs. Re theoretical and empirical relations (Equation (40) for laminar flow and Equation (43) for turbulent flow in hydraulically smooth pipes).

Figures 3–8 show the f vs. Re data for different test fluids obtained from different diameter pipelines. The experimental data follow the existing f vs. Re relations reasonably well in both laminar and turbulent regimes.

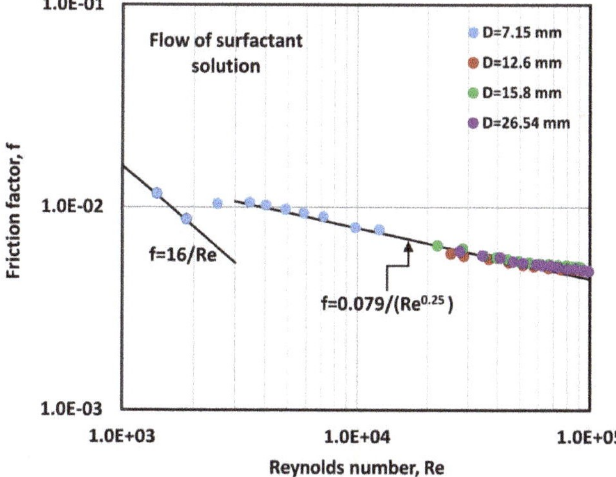

Figure 3. Friction factor versus Reynolds number data for aqueous surfactant solution.

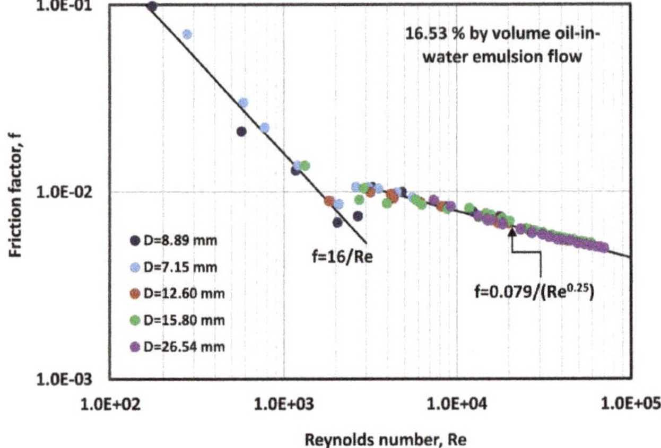

Figure 4. Friction factor versus Reynolds number data for 16.53% by volume oil-in-water emulsion.

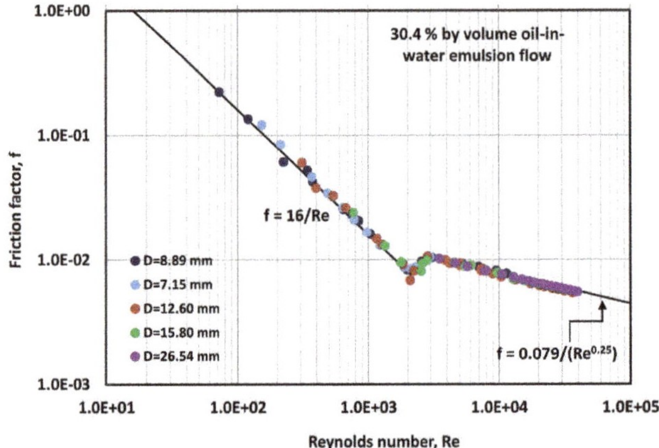

Figure 5. Friction factor versus Reynolds number data for 30.4% by volume oil-in-water emulsion.

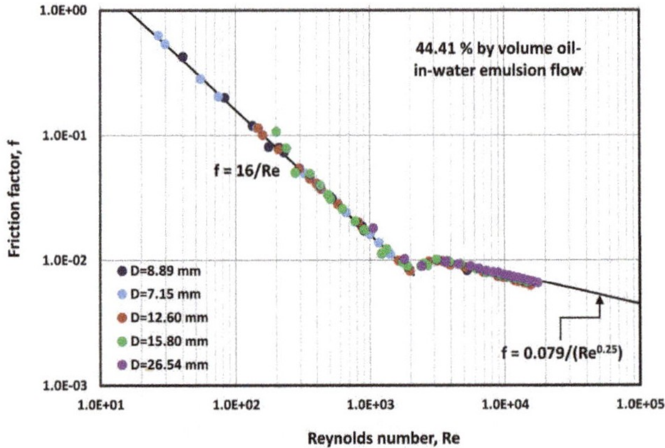

Figure 6. Friction factor versus Reynolds number data for 44.41% by volume oil-in-water emulsion.

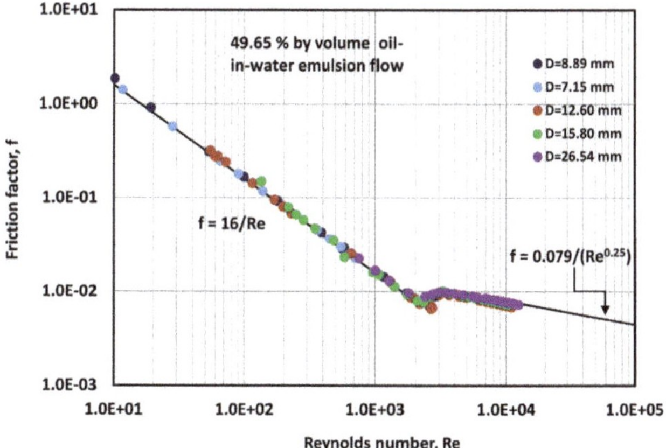

Figure 7. Friction factor versus Reynolds number data for 49.65% by volume oil-in-water emulsion.

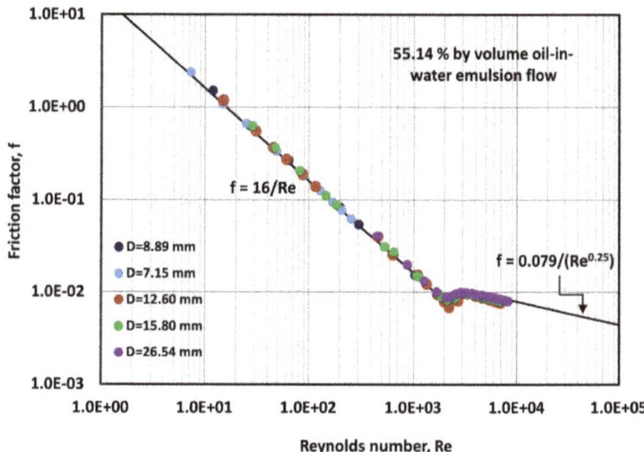

Figure 8. Friction factor versus Reynolds number data for 55.14% by volume oil-in-water emulsion.

4.2. Entropy Generation Results

The entropy generation rate in pipeline flow is calculated from pressure drop versus flow rate measurements using Equation (52), re-written as:

$$\dot{S}'_G = \frac{\dot{\vartheta}}{T}\left(-\frac{\Delta P}{L}\right) \tag{59}$$

where $T = 25 + 273.15 = 298.15K$.

The $\dot{S}'_G$ *vs.* $\overline{V}$ data are plotted on a log-log scale and compared with the predictions of equations developed in Section 2.4, that is, Equation (54) for laminar flows and Equation (55) for turbulent flows in smooth pipes.

Figures 9–14 show the $\dot{S}'_G$ *vs.* $\overline{V}$ data for different test fluids obtained from different diameter pipelines. As expected from Equation (54), the experimental data corresponding to laminar flow is independent of the pipe diameter. Also, the slope of the laminar flow line is 2 as predicted by Equation (54). Equation (54) describes the laminar flow $\dot{S}'_G$ *vs.* $\overline{V}$ data adequately for all the test fluids investigated. According to Equation (55), the turbulent flow $\dot{S}'_G$ *vs.* $\overline{V}$ data for a given diameter pipeline should follow a straight line on a log-log scale with a slope of 2.75. As expected, the experimental data in turbulent regime follow a straight line of slope 2.75. With the increase in pipe diameter, the entropy generation rate increases. The $\dot{S}_G'$ vs. $\overline{V}$ line shifts upward with the increase in pipe diameter but the slope remains the same, that is, 2.75. Thus Equation (55) describes entropy generation in turbulent flows reasonably well.

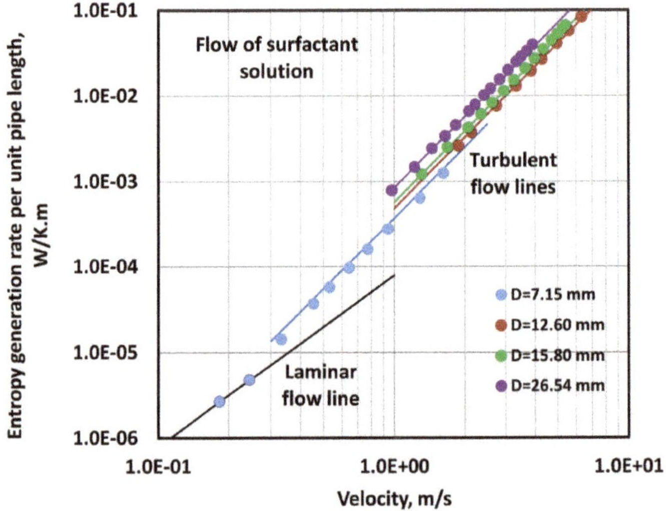

Figure 9. Entropy generation rate per unit length of pipe as a function of average velocity in flow of aqueous surfactant solution.

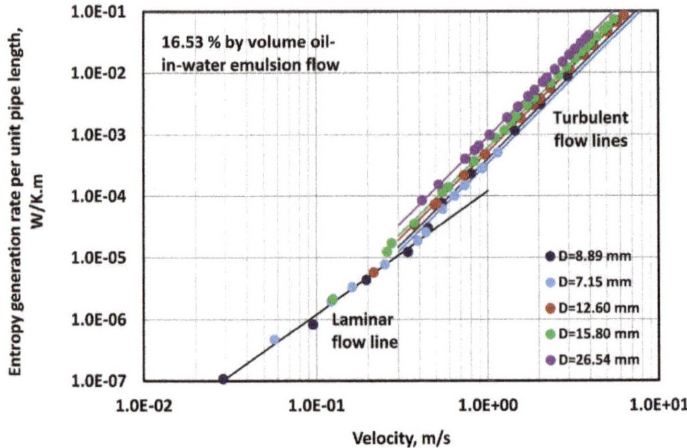

Figure 10. Entropy generation rate per unit length of pipe as a function of average velocity in flow of 16.53% by volume oil-in-water emulsion.

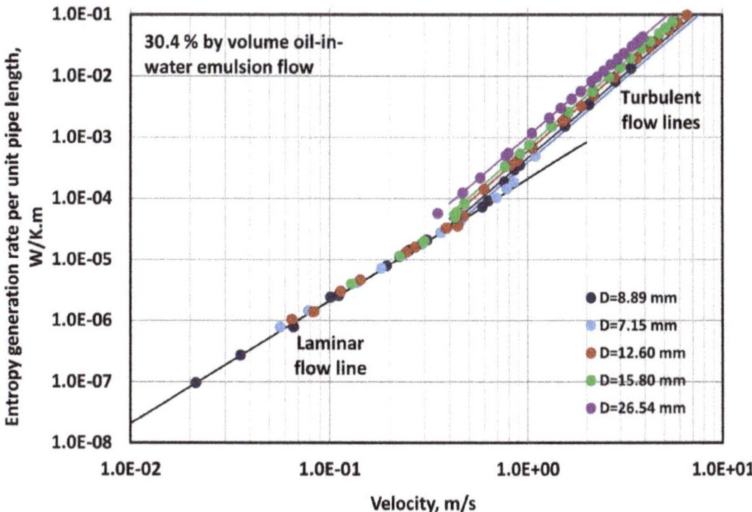

Figure 11. Entropy generation rate per unit length of pipe as a function of average velocity in flow of 30.4% by volume oil-in-water emulsion.

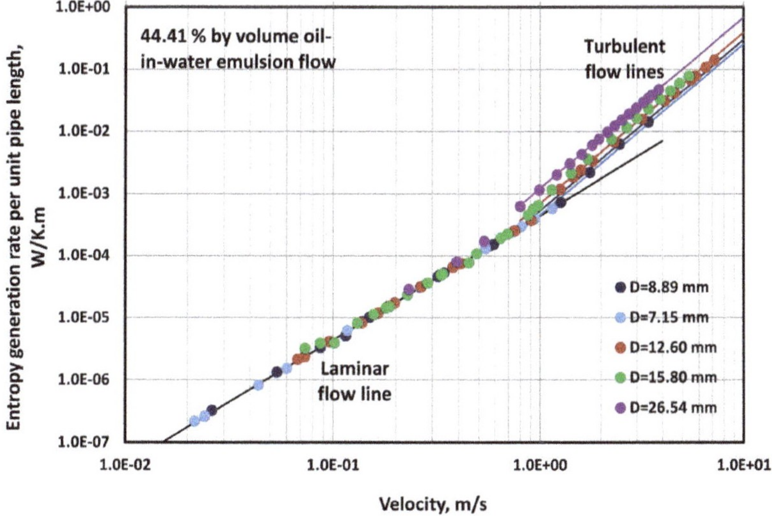

Figure 12. Entropy generation rate per unit length of pipe as a function of average velocity in flow of 44.41% by volume oil-in-water emulsion.

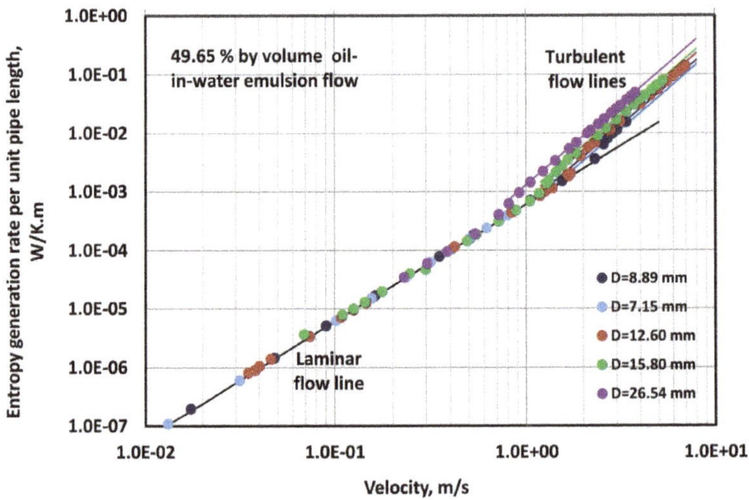

Figure 13. Entropy generation rate per unit length of pipe as a function of average velocity in flow of 49.65% by volume oil-in-water emulsion.

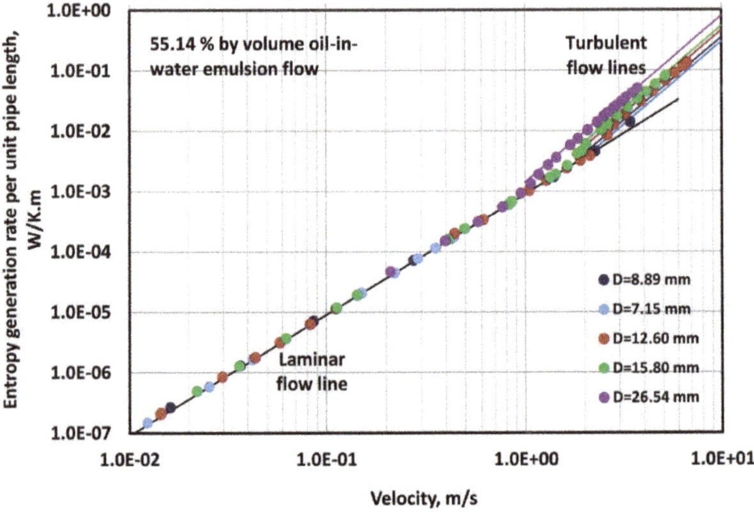

Figure 14. Entropy generation rate per unit length of pipe as a function of average velocity in flow of 55.14% by volume oil-in-water emulsion.

4.3. Mechanism of Entropy Generation in Real Flows

In real flows, entropy generation is a volumetric phenomenon caused by friction (non-zero viscosity) in fluids. Due to non-zero viscosity of real fluids, velocity gradients and viscous stresses are set up when fluid is forced to flow through a pipe. The presence of viscous stresses and velocity gradients in the fluid cause mechanical energy dissipation into frictional heating effect (internal energy). The degradation of highly ordered mechanical energy into disorderly internal energy results in entropy generation.

The entropy generation rate per unit volume of fluid in real flows ($\dot{S}_G'''$) can be expressed as [13]:

$$\dot{S}_G''' = \frac{1}{T}\left(\underset{=}{\tau} : \nabla\vec{V}\right) \tag{60}$$

where $\underset{=}{\tau}$ is the viscous stress tensor and $\nabla\vec{V}$ is the velocity gradient tensor. It should be noted that we are assuming that there are no temperature gradients in the fluid. If temperature is not uniform and temperature gradients are present, then entropy generation can occur due to another mechanism, namely, irreversible transfer of heat.

In general, the local entropy generation rate per unit volume of fluid ($\dot{S}_G'''$) will vary with the position coordinates. As an example, we illustrate the application of Equation (60) to steady laminar flow in cylindrical pipes. For steady laminar flow of a fluid in a uniform diameter pipe, the term $\underset{=}{\tau} : \nabla\vec{V}$ simplifies to:

$$\underset{=}{\tau} : \nabla\vec{V} = \tau_{rx}\left(\frac{\partial V_x}{\partial r}\right) \tag{61}$$

From Equations (60) and (61), it follows that:

$$\dot{S}_G''' = \frac{1}{T}\left(\underset{=}{\tau} : \nabla\vec{V}\right) = \frac{1}{T}\tau_{rx}\left(\frac{\partial V_x}{\partial r}\right) \tag{62}$$

From Newton's law of viscosity,

$$\tau_{rx} = \mu\left(\frac{\partial V_x}{\partial r}\right) \tag{63}$$

Consequently,

$$\dot{S}_G''' = \frac{1}{T}\tau_{rx}\left(\frac{\partial V_x}{\partial r}\right) = \frac{\mu}{T}\left(\frac{\partial V_x}{\partial r}\right)^2 \tag{64}$$

The velocity distribution in steady laminar flow of a Newtonian fluid is given as:

$$V_x = 2\overline{V}\left[1 - \left(\frac{r}{R}\right)^2\right] \tag{65}$$

Thus,

$$\frac{\partial V_x}{\partial r} = -\frac{4r}{R^2}\overline{V} \tag{66}$$

From Equations (64) and (66), we get:

$$\dot{S}_G''' = \frac{\mu}{T}\left(\frac{\partial V_x}{\partial r}\right)^2 = \left(\frac{16\mu}{T}\right)\left(\frac{r^2}{R^4}\right)\overline{V}^2 \tag{67}$$

This equation describes the local rate of entropy generation per unit volume in steady laminar flow of a Newtonian fluid in cylindrical pipe of uniform diameter. Figure 15 shows the plot of $\dot{S}_G'''$ as a function of radial position. The plot is generated using Equation (67). $\dot{S}_G'''$ is zero at the centre of the pipe as velocity gradiant and hence local dissipation of mechanical energy into frictional heating is zero at the centre of the pipe. As we go towards the pipe wall, the local entropy generation increases due to an increase in the velocity gradient and hence frictional heating of the fluid.

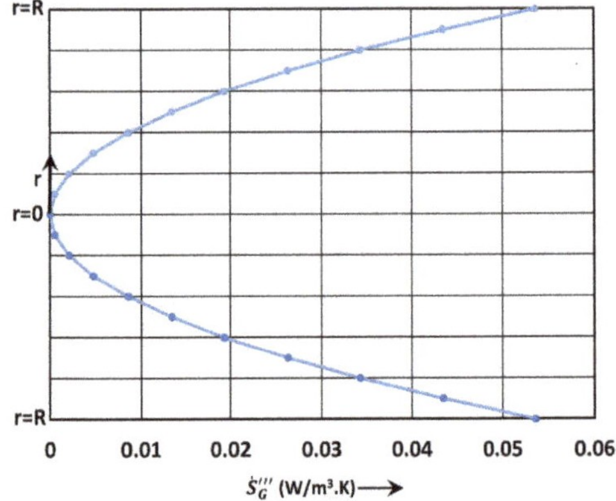

Figure 15. Local entropy generation rate per unit volume ($\dot{S}_G'''$) as a function of radial position in laminar flow of a Newtonian fluid ($R = 10$ mm, $T = 298.15$ K, $\mu = 10$ mPa·s, $\overline{V} = 0.1$ m/s).

Finally it can be readily shown that the global rate of entropy generation (Equation (54)) follows from the integration of the local rate of entropy generation. The global entropy generation rate per unit length $\dot{S}_G'$ can be expressed as:

$$\dot{S}_G' = \frac{1}{T} \int_0^R \left(\dot{S}_G''' \right) (2\pi r)\,dr \tag{68}$$

From Equations (67) and (68), we get:

$$\dot{S}_G' = \frac{32\pi\mu}{TR^4}\overline{V}^2 \int_0^R r^3\,dr = \left(\frac{8\pi}{T} \right)\mu\overline{V}^2 \tag{69}$$

This is the same result that was obtained earlier in Equation (54).

5. Conclusions

In conclusion, a novel approach is described to teach the second law of thermodynamics with the help of an undergraduate fluid mechanics laboratory involving pipeline flow experiments. The relevant background in fluid mechanics and thermodynamics is reviewed briefly. The link between entropy generation and pressure loss in pipeline flow experiments is demonstrated both theoretically and experimentally. Experimental work involving flow of emulsion-type test fluids in different diameter pipes is carried out to determine friction factor versus Reynolds number behavior and entropy generation rates in pipeline flows. Entropy generation in pipeline flows is explained mechanistically considering local entropy generation in the presence of viscous stresses and velocity gradients in flow of real fluids.

Funding: This research received no external funding.

Conflicts of Interest: The author declares no conflict of interest.

References

1. Welty, J.R.; Wicks, C.E.; Wilson, R.E.; Rorrer, G. *Fundamentals of Momentum, Heat, and Mass Transfer*, 4th ed.; Wiley: New York, NY, USA, 2001.
2. Greenkorn, R.A.; Kessler, D.P. *Transfer Operations*; McGraw-Hill: New York, NY, USA, 1972.

3. Smith, J.M.; Van Ness, H.C.; Abbott, M.M. *Introduction to Chemical Engineering Thermodynamics*, 7th ed.; McGraw-Hill: New York, NY, USA, 2005.

4. Bird, R.B.; Stewart, W.E.; Lightfoot, E.N. *Transport Phenomena*, 2nd ed.; Wiley: New York, NY, USA, 2007.

5. Colebrook, C.F. Turbulent flow in pipes with particular reference to the transition region between smooth and rough pipe laws. *J. Inst. Civ. Eng.* **1939**, *11*, 133–156. [CrossRef]

6. Colebrook, C.F.; White, C.M. Experiments with fluid friction in roughened pipes. *Proc. R. Soc. Lond. Ser. A. Maths. Phys. Sci.* **1937**, *161*, 367–381.

7. Asker, M.; Turgut, O.E.; Coban, M.T. A review of non iterative friction factor correlations for the calculation of pressure drop in pipes. *Bitlis Eren Univ. J. Sci. Technol.* **2014**, *4*, 1–8.

8. Brkic, D. Review of explicit approximations to the Colebrook relation for flow friction. *J. Pet. Sci. Eng.* **2011**, *77*, 34–48. [CrossRef]

9. Fang, X.; Xu, Y.; Zhou, Z. New correlations of single-phase friction factor for turbulent pipe flow and evaluation of existing single-phase friction factor correlations. *Nuclear Eng. Des.* **2011**, *241*, 897–902. [CrossRef]

10. Zigrang, D.J.; Sylvester, N.D. Explicit approximations to the solution of Colebrook's friction factor equation. *AIChE J.* **1982**, *28*, 514–515. [CrossRef]

11. Pal, R. Entropy production in pipeline flow of dispersions of water in oil. *Entropy* **2014**, *16*, 4648–4661. [CrossRef]

12. Pal, R. Second law analysis of adiabatic and non-adiabatic pipeline flows of unstable and surfactant-stabilized emulsions. *Entropy* **2016**, *18*, 113. [CrossRef]

13. Pal, R. Entropy generation and exergy destruction in flow of multiphase dispersions of droplets and particles in a polymeric liquid. *Fluids* **2018**, *3*, 19. [CrossRef]

 *fluids*

Article

What Can Students Learn While Solving Colebrook's Flow Friction Equation?

Dejan Brkić [1,2,*,†] and **Pavel Praks** [1,*,†]

[1]   IT4Innovations, VŠB—Technical University of Ostrava, 708 00 Ostrava, Czech Republic
[2]   Research and Development Center "Alfatec", 18000 Niš, Serbia
*   Correspondence: dejanbrkic0611@gmail.com (D.B.); pavel.praks@vsb.cz (P.P.)
†   Both authors contributed equally to this study.

Received: 12 May 2019; Accepted: 25 June 2019; Published: 27 June 2019

Abstract: Even a relatively simple equation such as Colebrook's offers a lot of possibilities to students to increase their computational skills. The Colebrook's equation is implicit in the flow friction factor and, therefore, it needs to be solved iteratively or using explicit approximations, which need to be developed using different approaches. Various procedures can be used for iterative methods, such as single the fixed-point iterative method, Newton–Raphson, and other types of multi-point iterative methods, iterative methods in a combination with Padé polynomials, special functions such as Lambert W, artificial intelligence such as neural networks, etc. In addition, to develop explicit approximations or to improve their accuracy, regression analysis, genetic algorithms, and curve fitting techniques can be used too. In this learning numerical exercise, a few numerical examples will be shown along with the explanation of the estimated pedagogical impact for university students. Students can see what the difference is between the classical vs. floating-point algebra used in computers.

Keywords: Colebrook equation; Lambert W function; Padé polynomials; iterative methods; explicit approximations; learning; teaching strategies; floating-point computations

---

1. Introduction

The Colebrook equation for flow friction is an empirical formula developed by C. F. Colebrook in 1939 [1] and it is based on his joint experiment, which was performed with Prof. C. M. White in 1937 [2]. Colebrook was at this time a PhD student of Prof. White at Imperial College in London, UK [3]. The experiment was performed with a set of pipes of which the inner side was covered with sand of different grain sizes, while one was left completely without sand to be smooth. The sand was fixed on the pipe's inner surface using a sort of glue as adhesive material, while air was used as the fluid in this experiment.

The Colebrook equation follows logarithmic law and, at first sight, it seems to be very simple. However, it is given implicitly, with respect to the unknown flow friction factor f in a way that it can be expressed explicitly only in terms of the Lambert W function, with further difficulties in its evaluation [4–6] (it can be rearranged for gas flow [7]). Otherwise, it needs to be solved iteratively [8–10] or explicitly, using approximate formulas specially developed for such a purpose [10–17]. The unknown flow friction factor f in the Colebrook's equation depends on the known Reynolds number Re and the relative roughness of the inner pipe surface $\frac{\varepsilon}{D}$, all three of which are dimensionless, shown in Equation (1):

$$\frac{1}{\sqrt{f}} = -2 \cdot \log_{10}\left(\frac{2.51}{Re} \cdot \frac{1}{\sqrt{f}} + \frac{\frac{\varepsilon}{D}}{3.71} \right) \tag{1}$$

The relative roughness $\frac{\varepsilon}{D}$ can take values from 0 to 0.05, while the Reynolds number Re is from about 2500 to 10^8 [18]. The Colebrook equation is valid for the turbulent condition of flow [19].

In this learning exercise, students can have numerous numerical tasks that can improve their computational skills. The task can start with a simple trial/error approach that can, in a certain way, put the Colebrook equation in balance, while, furthermore, a simple fixed-point iteration method can be introduced in the curriculum [20]. Then, more complex approaches that require derivatives of the Colebrook equation can lead students to the Newton–Raphson iterative methods [21–23] or to the more complex multi-point iterative methods [8,9]. A first simple explicit approximate formula with inner iterative cycles can be introduced using such approaches [24].

Transcendental functions, such as logarithmic and exponential functions, on computers require the execution of additional floating point operations [25–27]. On the contrary, basic mathematical operations (addition, subtraction, multiplication, and division) are very fast on computers, because they are executed directly. Students can learn how to avoid using these time-consuming functions [27–30]. This simplification process can have a large effect on the students' future jobs, regarding the design of complex pipe networks, which involve extensive computations [31–36]. For example, in Section 2.2.1, students can reuse the value of the logarithm, which is computed only in the first iteration, to all subsequent iterations [30]. The less computationally demanded Padé polynomials are used instead of the logarithmic function in all subsequent iterations [30]. The students in that way need to know how to generate the appropriate polynomial expression (using appropriate software tools).

Regarding special functions, the Lambert W function can be introduced in the curriculum [5,6], as the Lambert W function transforms the Colebrook equation in the explicit form [4]. The widely used explicit form has one fast-growing term that makes calculations impossible in computers, due to an overflow error [37,38]. Students should figure out why such a numerical error appears in computers and then understand how such situations can be avoided. For this reason, the next step involves the series expansion of the Lambert W function, and its cognate, the Wright ω function [39]. In this step, students will learn how to make accurate, but at the same time, efficient approximations of the Colebrook equation using the Wright ω function [27–29,40,41]. Last, but not least, students can use raw numerical data that can efficiently simulate the Colebrook equation and, in that way, learn principles of using artificial intelligence in everyday engineering practice [12,42–45].

Finally, keeping in mind that the Moody diagram represents a graphical interpretation of the Colebrook equation, students should make a sketch of the Moody diagram [18,46], which will add an additional value both from the engineering and mathematical points of view (Figure 1 of this article can be used for this purpose).

2. Tasks for Students

2.1. Iterative Methods

For the purpose of this case study, two numerical examples are given: (1) $Re = 2.3 \times 10^5$ and $\frac{\varepsilon}{D} = 10^{-4}$; (2) $Re = 4.6 \times 10^7$ and $\frac{\varepsilon}{D} = 0.037$. Corresponding solutions for the flow friction factor are (1) $f = 0.016050961$, and (2) $f = 0.062427396$, respectively. Positions of these two examples with respect to the hydraulic regimes (laminar, transition to turbulent, and turbulent, which includes smooth, smooth to rough, and fully turbulent rough regime) are shown in Figure 1. Cases that can capture the nonlinearity of the Colebrook equation are: (a) Smooth pipe at very low, moderate, and very high Re, (b) pipe with moderate roughness at very low, moderate, and very high Re, and (c) pipe with high roughness ($\frac{\varepsilon}{D} \sim 0.05$) and again, low, moderate, and high Re. Example 1 is in the zone of moderate values of the Reynolds number Re, where the proposed value of relative roughness, which is small ($\frac{\varepsilon}{D} \sim 10^{-4}$), is large enough to provide a transition from smooth towards the turbulent regime. On the other hand, Example 2 with higher values of Re and $\frac{\varepsilon}{D}$ is in a fully turbulent zone. Using Figure 1, other different examples can be provided for these teaching exercises.

Example 1 is set to be exactly in the zone where the smooth nature of the viscose zone of the inner pipe surface starts to be too thin and when roughness starts to have influence [47,48].

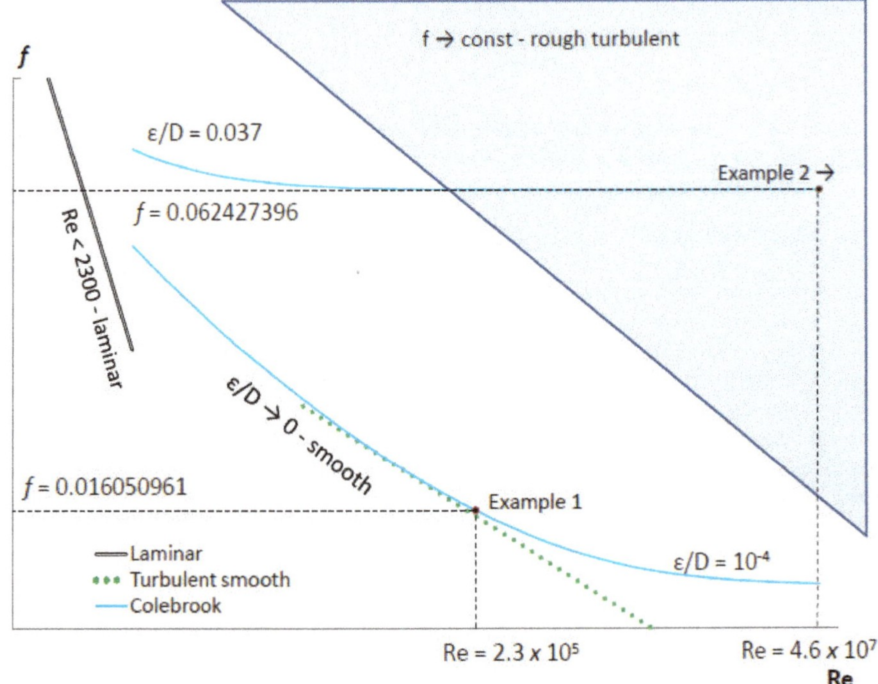

Figure 1. Position of the two presented examples with respect to the hydraulic regimes (laminar, transition to turbulent, and turbulent, which includes smooth, smooth to rough, and fully turbulent rough regime).

From Figure 1, students should see that our numerical Example 1 is set to be in the turbulent zone. More precisely, Example 1 is set at the end of the smooth turbulent regime, where the transition to fully turbulent begins. On the contrary, Example 2 is located in the fully developed rough turbulent zone. Students should understand the physical interpretation of different hydraulic regimes.

For this task, 45 min should be assigned.

2.1.1. Trial/Error Method

To do this task, students should guess somehow the initial value of the friction factor f, then insert it into the Colebrook equation, and finally, calculate a new value of f until an error criterion is fulfilled. Using this iterative procedure, students should show how many iterations will probably be required, as the number of iterations depends on the initial guess and whether such a rudimentary iterative procedure will always lead to the final balanced solution or not. Here will be introduced a concept of convergence, divergence, and oscillations in iterative procedures. Students should understand how significant it is to find the adequate value for the starting point for an iterative procedure [9]. For our Example 1, which uses $Re = 2.3 \times 10^5$ and $\frac{\varepsilon}{D} = 10^{-4}$, the starting point needs to be chosen between −2.469 and 1574003 when Microsoft (MS) Excel is used and students should discuss why [10,20]. Subsequently, students should analyze the starting point for Example 2, in which $Re = 4.6 \times 10^7$ and $\frac{\varepsilon}{D} = 0.037$. Students should explain consequences of the inappropriately chosen initial starting point. Finally, students should understand that the starting point for −2.5, for example, will actually be

$\frac{1}{\sqrt{f_0}} = \frac{\varepsilon/D}{3.71} - 2.5 = \frac{10^{-4}}{3.71} - 2.5 = -2.499973046$ for our Example 1, in which $\frac{\varepsilon}{D} = 10^{-4}$ (the negative starting point does not have physical meaning and it is chosen purely for testing methods and to introduce the concept of "thinking outside of the box", which could contribute to development of creative thinking in general.

For this task, 45 min should be assigned.

2.1.2. Fixed-Point Iterative Procedure

The fixed-point iterative method is suitable for spreadsheet solvers, such as MS Excel [10,20]. To prepare this task, students should rearrange the Colebrook equation in the suitable form $F(x) = 0$; where $x = \frac{1}{\sqrt{f}}$, i.e., $x + 2 \cdot \log_{10}\left(\frac{2.51}{Re} \cdot x + \frac{\varepsilon/D}{3.71}\right) = 0$. The iterative process starts with a selected starting point x_0 and continues by $x_{i+1} = x_i - F(x_i)$ until a stopping criteria is fulfilled, for example, when the results from two subsequent iterations are almost equal ($x_{i+1} - x_{i+1} \sim 0$). From this approach, students will learn that the trial/error method can be used in a more systematic way. This task can improve the ability of students to use spreadsheet solvers, because students should set the solver to allow iterative computations. Moreover, the maximal number of iterations and the stopping criteria represented by the minimal value between two successive iterations should be set in the solver. For example, the MS Excel spreadsheet solver allows a maximal number of iterations of 32,767, and students should figure out why this constraint is set. The answer is that 2^{15} = 32,768 as this represents the maximum it can be set for 16-bit registers in binary logic because one bit is reserved for the sign ($\pm$). In that way, students will learn that for human perception, more common is the decimal system, which is used, for example, for banknotes, \$1, \$10, \$100, while, for computers, the binary system is more suitable, which commonly uses 2, 4, 8, 16, 32, 64, etc.

In our case, (1) $Re = 2.3 \times 10^5$ and $\frac{\varepsilon}{D} = 10^{-4}$, for $\frac{1}{\sqrt{f_0}} = \frac{\varepsilon/D}{3.71} - 2.4 \approx -2.4$ and for $\frac{1}{\sqrt{f_0}} = \frac{\varepsilon/D}{3.71} + 1{,}574{,}003 \approx 1{,}574{,}003$, the number of required iterations for nine decimal accuracy will be 11 in both cases, i.e., the accurate solution will be $\frac{1}{\sqrt{f_{11}}} = 7.8931$. Students should make further tests with the same example, but with different starting points and, in addition, perform a similar test for Example 2.

Using the spreadsheet solver, students should first solve the Colebrook equation iteratively, using only one cell of the Colebrook equation for smooth conditions, when $\frac{\varepsilon}{D}$ is zero. Students should realize how important the role of the initial starting point is, because, in this case, software will declare a division by zero error. In other conditions, when $\frac{\varepsilon}{D} \neq 0$, the initial starting point will be automatically set as $f_0 = \log_{10}\left(\frac{\varepsilon/D}{3.71}\right)$.

Here, students may try to make a sketch of the function, i.e., to make a two-dimensional (2D) diagram of the Colebrook equation. Here, two possibilities should be analyzed: $\frac{1}{\sqrt{f}} = F\left(Re \cdot \sqrt{f}, \frac{\varepsilon}{D}\right)$ as suggested by Rose [46], or $f = F\left(Re, \frac{\varepsilon}{D}\right)$ as suggested by Moody [18] (approach from Figure 1 of this article can be used). In addition, students can analyze whether a simple transformation $x = \frac{1}{\sqrt{f}}$ accelerates solving of the Colebrook equation.

For this task, 45 min should be assigned.

2.1.3. Newton–Raphson Iterative Procedure

In this part, students will learn how to find the first derivative of the function with respect to the variable flow friction factor f, where $x = \frac{1}{\sqrt{f}}$ and $F(x) = x + 2\log_{10}\left(\frac{2.51}{Re}\cdot x + \frac{\varepsilon}{3.71}\right)$, as in Equation (2):

$$\frac{\partial(F(x))}{\partial x} = F'(x) = F'\left(x + 2\log_{10}\left(\frac{2.51}{Re}\cdot x + \frac{\varepsilon}{3.71}\right)\right) = -\frac{1}{2}\left(\frac{1}{\sqrt{|f_i|}}\right)^3\left(1 + 2\cdot\frac{2.51}{Re\cdot\ln(10)\cdot\left(\frac{2.51}{Re}\cdot\frac{1}{\sqrt{|f_i|}} + \frac{\varepsilon}{3.71}\right)}\right), \tag{2}$$

The procedure for finding the first derivative can be done manually, but also for Equation (2), it can be generated using appropriate software, e.g., Matlab by the "diff" command.

The first derivative is used in the Newton–Raphson procedure as $f_{i+1} = f_i - \frac{F(f_i)}{F'(f_i)}$ and students may test whether the Newton–Raphson procedure is faster or not, compared with the simple fixed-point method of the Colebrook equation, where $F(f_{i+1}) = \frac{1}{\sqrt{f}} + 2\log_{10}\left(\frac{2.51}{Re}\cdot\frac{1}{\sqrt{f}} + \frac{\varepsilon}{3.71}\right)$.

Students should realize that the previous example with the fixed point method is set for solving $x = \frac{1}{\sqrt{f}}$, while here, the Newton–Raphson method is set for solving $F(x) = 0$. For Example 1, in which $Re = 2.3 \times 10^5$ and $\frac{\varepsilon}{D} = 10^{-4}$, we can use the already analyzed starting point $\frac{1}{\sqrt{f_0}} = \frac{\varepsilon}{3.71} + 1{,}574{,}003 \approx 1{,}574{,}003 \rightarrow f_0 = 4.04 \times 10^{-13}$, which will give accurate results. Unfortunatelly, this iteration process requires 27 iterations, $f_{27} = 0.016050961$, whereas $\frac{1}{\sqrt{f_0}} = \frac{\varepsilon}{3.71} - 2.4 \approx -2.4 \rightarrow f_0 = 0.173611$ will report a division by zero error in the fourteenth iteration and the calculation will stop without obtaining the precise solution. Here, students should understand that even a more powerful iterative procedure can fail when the problem is not set in an appropriate way and when the role of a good initial starting point is not fully understood.

The Newton–Raphson method, in which $F'(f_i) = 1$ is assumed, is equivalent to the fixed-point method in the form $f_{i+1} = f_i - F(f_i)$. However, this formulation shows very bad convergence properties. Students should understand why the iterative process $x_{i+1} = \frac{1}{\sqrt{f_{i+1}}} = x_i - F(x_i)$ gives the fast solution, while the iterative process $f_{i+1} = f_i - F(f_i)$ is slow. Students should again analyze coordinates in diagrams from Rose and Moody, and discuss which formulation is better [18].

Here, it should be noted that the function $F(x) = x + 2\log_{10}\left(\frac{2.51}{Re}\cdot x + \frac{\varepsilon}{3.71}\right)$ in form $F\left(\frac{1}{\sqrt{f}}\right) = \left(2\log_{10}\left(\frac{2.51}{Re}\cdot\frac{1}{\sqrt{f}} + \frac{\varepsilon}{3.71}\right)\right)^{-2} + \frac{1}{\sqrt{f}}$ shows very slow convergence when it is used in the iterative process $f_{i+1} = f_i - \frac{F(f_i)}{F'(f_i)}$. Students should discover why. Students should investigate how the initial starting point affects the required number of iterations.

For this task, 90 min should be assigned (including repetition and clarifications from the previous tasks).

2.1.4. Multi-Point Iterative Procedures

The shown Newton–Raphson method belongs to the two-point iterative methods and, in this task, students should find more powerful iterative methods. For example, the proposed methods for this exercise are three-point iterative methods, such as Halley or Schröder or even more powerful multi-point methods [8,9]. Students should figure out whether the number of required iterations to find the accurate solution decreases with the increased complexity of the chosen method. If the number of required iterations does not decrease, students should analyze why. In addition, it needs to be found how important it is to find the adequate initial starting point. In this task, students should find the second derivative of the Colebrook equation, which will increase their computing skills.

In this exercise, students should develop the ability to make a balance between complexity, speed, and required accuracy. Students obtain an experience for choosing the appropriate iterative method.

Here, we will not repeat the Newton–Raphson calculations in the form of Equation (3), which gives fast accurate results, but we will show the Halley method: Equation (4) for which the first derivative $F'(x)$ and the second $F''(x)$, with respect to variable $x = \frac{1}{\sqrt{f_i}}$, are given by Equations (5) and (6), respectively.

$$x_{i+1} = x_i - \frac{F(x_i)}{F'(x_i)}, \tag{3}$$

$$x_{i+1} = x_i - \frac{2 \cdot F(x_i) \cdot F'(x_i)}{2 \cdot F'(x_i)^2 - F(x_i) \cdot F''(x_i)}, \tag{4}$$

$$F'(x_i) = \frac{9287 \cdot \ln(10) \cdot x_i + 1000 \cdot \frac{\varepsilon}{D} \cdot Re \cdot \ln(10) + 18574}{\ln(10) \cdot \left(9287 \cdot x_i + 1000 \cdot \frac{\varepsilon}{D} \cdot Re\right)}, \tag{5}$$

$$F''(x_i) = -\frac{172496738}{\ln(10) \cdot \left(9287 \cdot x_i + 1000 \cdot \frac{\varepsilon}{D} \cdot Re\right)^2}, \tag{6}$$

For the first derivatives, [8,9,21–23] should be consulted.

For Example 1, in which $Re = 2.3 \times 10^5$ and $\frac{\varepsilon}{D} = 10^{-4}$, for initial starting point $x_0 = \frac{1}{\sqrt{f_0}} = \frac{\frac{\varepsilon}{D}}{3.71} + 1574003 \approx 1574003$, the final balanced solution will be achieved after 27 iterations; $x_{27} = \frac{1}{\sqrt{f_{27}}} = 7.89313398624169 \rightarrow f_{27} = 0.016050961$. However, for the initial starting point $x_0 = \frac{1}{\sqrt{f_0}} = \frac{\frac{\varepsilon}{D}}{3.71} - 2.4 \approx -2.4$, the procedure in MS Excel will report a division by zero error in the third step, so it will be a good opportunity for students to reexamine the importance of the appropriate chosen starting point. Thus, $x_0 = \frac{1}{\sqrt{f_0}}$, which in this case can be selected from -1.172 to $1,583,778$. Of course, negative starting points do not represent physical meaning. A similar exercise should be done for Example 2, in which $Re = 4.6 \times 10^7$ and $\frac{\varepsilon}{D} = 0.037$.

For this task, at least 90 min should be assigned, as multi-point iterative methods are more complex than two-point methods.

2.2. Special Iterative Methods

While simple operations, such as adding, subtracting, multiplication, and division, require minimal system resources of computers, more complex operations, such as logarithmic, non-integer power, and similar function calls, require multiple floating point operations to be executed in the Central Processing Unit (CPU), which represents a time- and energy-consuming task [15,25–28]. For example, students should understand that the execution of one non-integer power operation in a computer goes through execution of one logarithmic and one exponential function. Some numbers are too big or too small to be placed in registers of computers, which have 32 or 64 bits [37,38].

2.2.1. Padé Approximant

As noted, one iterative cycle involves the evaluation of one logarithmic function when solving the Colebrook equation. The Colebrook's law is in its nature logarithmic, but here, students will learn how to avoid multiple evaluations of the logarithmic function by replacing it with a Padé approximant in all iterations, with the exception of the first one [30]. A Padé approximant is the "best" approximation of the given function by a rational function of given order. For example, $\log_{10}(95)$ can be accurately approximated by its Padé approximant at the expansion point $x = 1$, keeping in mind that $\log_{10}(100) = 2$ and using the fact that $\log_{10}(95) = \log_{10}\left(\frac{100}{\frac{100}{95}}\right) = \log_{10}(100) - \log_{10}\left(\frac{100}{95}\right)$ where

$\frac{100}{95} \approx 1.0526$ is close to 1. The Padé approximant will be very accurate in this case, because 1.0526 is very close to the expansion point $x = 1$.

Evaluation of the decadic logarithm can be expressed by the natural logarithm $\log_{10}(z) = \frac{\ln(z)}{\ln(10)}$, where $\ln(10) \approx 2.302585093$ is constant. The first task for students would be to make research related to Padé approximants. For students with interest in programming, the additional task would be to develop a code in appropriate software, such as Matlab. Padé approximants of different orders can be used for approximation of $\ln(z)$, for arguments z close to 1, i.e., $z \approx 1$. As the expansion point $z_0 = 1$ is a root of $\ln(z)$, the accuracy of the Padé approximant decreases. Setting the "OrderMode" option in the Matlab Padé command to relative compensates for the decreased accuracy. Thus, here, the Padé approximant of the (m,n) order uses the form $\ln(z) \approx \frac{(z-z_0)\cdot(a_0+a_1\cdot(z-z_0)+...+a_m(z-z_0)^m)}{1+\beta_1\cdot(z-z_0)+...+\beta_n(z-z_0)^n}$, where α and β are coefficients (the coefficients of the polynomials need not be rational numbers). The Horner algorithm transforms polynomials into a computationally efficient form [30]. One possible procedure for solving the Colebrook equation using Padé approximants and the simple fixed-point iterative method is shown in Figure 2. Detailed step-by-step calculation using a table can be found in [30].

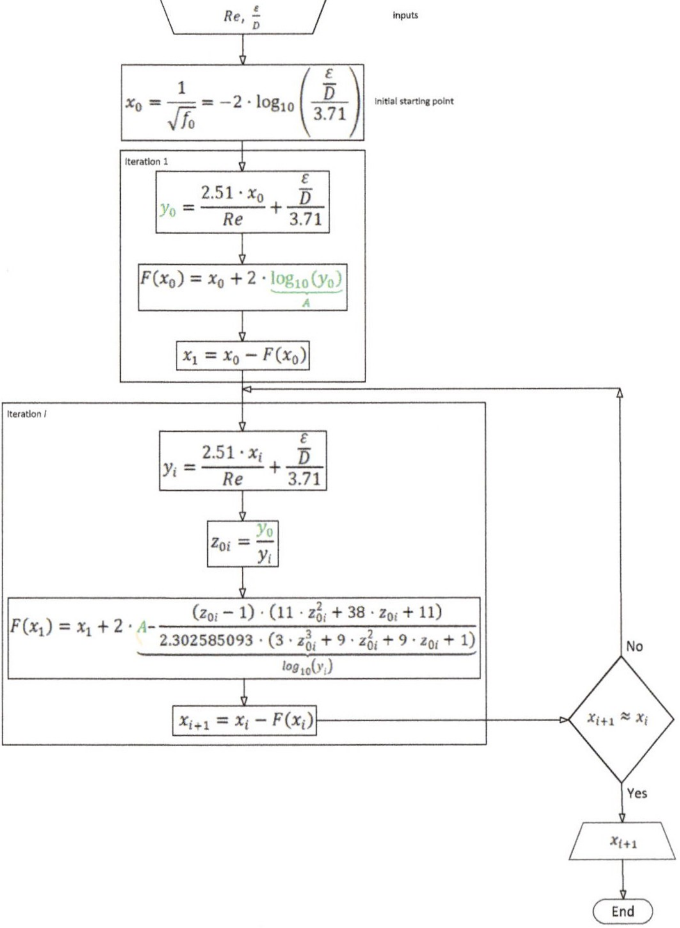

Figure 2. A possible procedure for solving the Colebrook equation using Padé approximants and the simple fixed-point iterative method.

Students interested in programming can use tic and toc functions of Matlab and GNU/Octave [49,50] to measure the speed of calculations using the classical fixed-point method vs. the fixed-point method combined with Padé approximants (let us remind readers that GNU/Octave software is available for free [49]). Students can combine the Newton–Raphson method with Padé approximants and check the accuracy of Padé approximants of different orders [30].

For a general understanding of the concept of Padé approximants, 90 min is required (45 min for an introduction and a further 45 min for numerical tests). Then, that additional 90 min would be required for the application of Padé polynomials for solving the Colebrook equation (again, 45 min for an introduction and 45 min for numerical examples). Students with an interest in programming should have an additional 45-min exercise (a presentation of complexity vs. accuracy of Padé approximants).

2.2.2. Approximations by Multi-Point Methods with Internal Cycles

Some iterative methods belong to the group of multi-point methods [8] that are very powerful and that can reach the accurate solution of the Colebrook equation in the first iteration. An example of an approximation of the Colebrook equation based on internal cycles is given by Serghides [51]. Serghides' methods belong to Steffensen types of methods, which do not require derivatives (other Steffensen types of methods can be used in addition [52,53]).

In order to develop an approximation with the internal cycle, students can choose one initial fixed point f_0, for example $f_0 = \frac{f_{min} + f_{max}}{2} = \frac{0.0064 + 0.077}{2} = 0.0417$, and then calculate $\frac{1}{\sqrt{f_0}} \approx 4.9$, which will give $\frac{1}{\sqrt{f}} = -2 \cdot \log_{10}\left(\frac{2.51 \cdot 4.9}{Re} + \frac{\frac{\varepsilon}{D}}{3.71}\right)$ and after one additional cycle $\frac{1}{\sqrt{f}} \approx -2 \cdot \log_{10}\left(\left(\frac{\frac{\varepsilon}{D}}{3.71} - \frac{5.02}{Re} \cdot \log_{10}\left(\frac{12.3}{Re} + \frac{\frac{\varepsilon}{D}}{3.71}\right)\right)\right)$. Using this exercise, students should see how important constant 12.3 is. Students should evaluate the relative error over the domain of applicability of the Colebrook's equation, then replace the constant 12.3 with a more appropriate value. Students can compare their chosen constant with values shown by Shacham [54] and by Zigrang and Sylvester [24]. For Example 1, in which $Re = 2.3 \cdot 10^5$ and $\frac{\varepsilon}{D} = 10^{-4}$, the approximate value is $\frac{1}{\sqrt{f}} \approx -2 \cdot \log_{10}\left(\left(\frac{10^{-4}}{3.71} - \frac{5.02}{2.3 \cdot 10^5} \cdot \log_{10}\left(\frac{12.3}{2.3 \cdot 10^5} + \frac{10^{-4}}{3.71}\right)\right)\right)$. Students should calculate this value and evaluate the introduced error compared with the accurate iterative solution [20]. An additional task could be to find additional approximations with internal cycles, such as Romeo et al. [55] or Offor and Alabi [56] and to discuss what additional strategies were used in these cases.

For this task, 45 min is required.

2.3. Special Functions

The Colebrook equation is implicit with respect to the unknown flow friction factor. However, the Colebrook equation can be rearranged in the explicit form using the Lambert W function, where further, this special function can be evaluated only approximately [5]. Some numerical constraints in using the Lambert W function, which can occur in our case, will be detected [37] and a way to mitigate this inconvenience will be shown [27–29].

For tasks related to special functions, students need 45 min for a general introduction and an additional 90 min to work on examples.

2.3.1. Lambert W Function

Interested readers may find more information on the Lambert W function through [5,6,57,58]. Then, students should understand why some functions, such as logarithmic, square root, or exponential, are available in pocket calculators and spreadsheet solvers, while the Lambert W function is not [6]. In this task, students should understand how computers use a series expansion to interpret functions. Moreover, they need to understand how to use Visual Basic for Applications (VBA) modules in order to introduce new functions in MS Excel [59] (related to this task, students interested in programming should take at least 90 min).

The Lambert W function is used in mathematics to solve equations in which the unknown appears both outside and inside an exponential function or a logarithm, such as $3x + 2 = e^x$ or $x = \ln(4 \cdot x)$. According to Fukushima [60], the Lambert W function is defined as the solution of a transcendental equation with $\kappa = W(\kappa) \cdot e^{W(\kappa)}$. Such equations cannot be, except in special cases, solved explicitly in terms of the finite number of arithmetic operations, nor in terms of exponentials or logarithms. In the case of the Colebrook equation, students can consult scientific literature, while here we will show formulation from [28,29], Equation (7), while a similar formulation can be found in [27]:

$$\left. \begin{array}{c} \frac{1}{\sqrt{f}} = \frac{2}{\ln(10)} \cdot \left(\ln\left(\frac{Re}{2.51} \cdot \frac{\ln(10)}{2} \right) + W(e^x) - x \right) \\ x = \ln\left(\frac{Re}{2.51} \cdot \frac{\ln(10)}{2} \right) + \frac{Re \cdot \frac{\varepsilon}{D}}{2.51 \cdot 3.71} \cdot \frac{\ln(10)}{2} \end{array} \right\}, \tag{7}$$

Our carefully chosen examples, $Re = 2.3 \times 10^5$ and $\frac{\varepsilon}{D} = 10^{-4}$, (2) $Re = 4.6 \times 10^7$ and $\frac{\varepsilon}{D} = 0.037$, respectively, are chosen to show that using floating-point arithmetic, Equation (7), gives the accurate solution for Example 1, while for Example 2, an arithmetic overflow error will occur. It is because $W(e^x)$ becomes too large for ordinary computer registers.

2.3.2. Wright ω Function and Related Approximations

As shown in Biberg [27] and Brkić and Praks [28,29], the problem with too large numbers in computer registers can be solved by introducing a cognate of the Lambert W function—the Wright ω function. The Wright ω function $y = \omega(x) = W(e^x)$ is a solution to the equation $y + \ln(y) = x$. Here, students should reformulate Equation (7) through the Wright ω function [50].

For Example 2, which uses $Re = 4.6 \times 10^7$ and $\frac{\varepsilon}{D} = 0.037$, x is approximately 210441.7. In Matlab, the command $y = wrightOmega(210441.7)$ returns the value $y = 210429.44$.

In Matlab, but also in the GNU/Octave software [49], students can use the free WrightOmegaq library [50]. In this case, the command is $y = wrightOmegaq(210441.7)$.

Students can easily verify that the solution y satisfies the relation $y + \ln(y) = x$. Let us reiterate that for Example 2, the value of x is approximately 210441.7.

In the following task, students should explore how to estimate the Wright ω function without Matlab or GNU/Octave. The simplification $W(e^x) - x \approx \ln(x) \cdot \left(\frac{1}{x} - 1 \right)$ gives an accurate approximation suitable for engineering practice—Equation (8) [28,29]:

$$\left. \begin{array}{c} \frac{1}{\sqrt{f}} \approx 0.8686 \cdot \left[B + \ln(B + A) \cdot \left(\frac{1}{B+A} - 1 \right) \right] \\ A \approx \frac{Re \cdot \frac{\varepsilon}{D}}{8.0878} \\ B \approx \ln\left(\frac{Re}{2.18} \right) \approx \ln(Re) - 0.7794 \end{array} \right\}, \tag{8}$$

where $\frac{2 \cdot 2.51}{\ln(10)} \approx 2.18$ and $2.18 \cdot 3.71 \approx 8.0878$.

Students should evaluate the relative error of the approximation, Equation (8), and develop a more accurate approximation using the enhanced model $W(e^x) - x \approx c + \ln(x) \cdot \left(\frac{1}{x} - 1 \right)$, where c is a constant. Of course, if $c = 0$, the enhanced model is equivalent to the model (8). For the general case, the constant c can be found by the MS Excel Solver, in order to minimize the relative error of the model with the iterative solution of the Colebrook equation provided by MS Excel. Students should do this optimization task with randomly selected input values of the Colebrook model $\left(Re, \frac{\varepsilon}{D} \right)$, which are sampled from the domain of applicability of the Colebrook equation. Finally, students can compare the relative error of this enhanced model with the relative error of the original model.

2.3.3. Tania Function and Related Approximations

Students should examine the Tania function, which is defined as $Tania(x) = \omega(x + 1)$, $Tania(x) = W(e^x)$, all with reference to [61] (the Euler T function should be an appropriate task to be further examined [39]).

For this task, 45–90 min should be assigned, during which students need to try to develop ability accepting new functions with new concepts [62].

3. Conclusions

Here we present how the relatively simple Colebrook equation, widely known by all students of hydraulics, can be used for introducing in curricula various iterative methods, special functions such as Lambert W, polynomials such as Padé approximants, optimization methods, etc.

The proposed exercises are suitable for students of hydraulics (civil, mechanical, chemical, or petroleum engineering curriculum, including the exploitation of oil and gas), but can be used as practical examples in teaching of numerical methods for students of mathematics. The proposed exercises can be extended for the teaching of pipe network analysis [31–36]. These methods can be combined with recent experimental works [63]. Furthermore, instead of pipes, a modified version of the Colebrook equation can be used for fuel cells [64,65].

Author Contributions: The paper is a product of the joint efforts of the authors, who worked together on models of natural gas distribution networks. P.P. has a scientific background in applied mathematics and programming, while D.B. has a background in control and applied computing in mechanical and petroleum engineering. They used the experience gained from their previous papers related to the Colebrook equation to develop proposed teaching exercises. D.B. wrote a draft of this article and prepared figures.

Funding: This work has been partially funded by the Technology Agency of the Czech Republic, partially by the National Centre for Energy TN01000007 and partially by the project "Energy System for Grids" TK02030039. Work of D.B. has been also supported by the Ministry of Education, Youth and Sports from the National Programme of Sustainability (NPS II) project "IT4Innovations excellence in science - LQ1602" and by the Ministry of Education, Science, and Technological Development of the Republic of Serbia through the project "Development of new information and communication technologies, based on advanced mathematical methods, with applications in medicine, telecommunications, power systems, protection of national heritage and education" iii44006.

Conflicts of Interest: The authors declare no conflict of interest.

References

1. Colebrook, C.F. Turbulent flow in pipes with particular reference to the transition region between the smooth and rough pipe laws. *J. Inst. Civ. Eng.* **1939**, *11*, 133–156. [CrossRef]
2. Colebrook, C.F.; White, C.M. Experiments with fluid friction in roughened pipes. *Proc. R. Soc. Lond. Ser. A-Math. Phys. Sci.* **1937**, *161*, 367–381. [CrossRef]
3. Hager, W.H. Cedric Masey White and his solution to the pipe flow problem. *Proc. Inst. Civ. Eng. Water Manag.* **2010**, *163*, 529–537. [CrossRef]
4. Keady, G. Colebrook-White formula for pipe flows. *J. Hydraul. Eng.* **1998**, *124*, 96–97. [CrossRef]
5. Corless, R.M.; Gonnet, G.H.; Hare, D.E.; Jeffrey, D.J.; Knuth, D.E. On the LambertW function. *Adv. Comput. Math.* **1996**, *5*, 329–359. [CrossRef]
6. Hayes, B. Why W? On the Lambert W function, a candidate for a new elementary function in mathematics. *Am. Sci.* **2005**, *93*, 104–108. [CrossRef]
7. More, A.A. Analytical solutions for the Colebrook and White equation and for pressure drop in ideal gas flow in pipes. *Chem. Eng. Sci.* **2006**, *61*, 5515–5519. [CrossRef]
8. Praks, P.; Brkić, D. Choosing the optimal multi-point iterative method for the Colebrook flow friction equation. *Processes* **2018**, *6*, 130. [CrossRef]
9. Praks, P.; Brkić, D. Advanced iterative procedures for solving the implicit Colebrook equation for fluid flow friction. *Adv. Civ. Eng.* **2018**, *2018*, 5451034. [CrossRef]
10. Brkić, D. Determining friction factors in turbulent pipe flow. *Chem. Eng. N. Y.* **2012**, *119*, 34–39.
11. Brkić, D. Review of explicit approximations to the Colebrook relation for flow friction. *J. Pet. Sci. Eng.* **2011**, *77*, 34–48. [CrossRef]
12. Brkić, D.; Ćojbašić, Ž. Evolutionary optimization of Colebrook's turbulent flow friction approximations. *Fluids* **2017**, *2*, 15. [CrossRef]
13. Brkić, D. New explicit correlations for turbulent flow friction factor. *Nucl. Eng. Des.* **2011**, *241*, 4055–4059. [CrossRef]

14. Brkić, D. An explicit approximation of Colebrook's equation for fluid flow friction factor. *Pet. Sci. Technol.* **2011**, *29*, 1596–1602. [CrossRef]

15. Winning, H.K.; Coole, T. Improved method of determining friction factor in pipes. *Int. J. Numer. Methods Heat Fluid Flow* **2015**, *25*, 941–949. [CrossRef]

16. Pimenta, B.D.; Robaina, A.D.; Peiter, M.X.; Mezzomo, W.; Kirchner, J.H.; Ben, L.H. Performance of explicit approximations of the coefficient of head loss for pressurized conduits. *Rev. Bras. Eng. Agrícola Ambient.* **2018**, *22*, 301–307. [CrossRef]

17. Vatankhah, A.R. Approximate analytical solutions for the Colebrook equation. *J. Hydraul. Eng.* **2018**, *144*, 06018007. [CrossRef]

18. Moody, L.F. Friction factors for pipe flow. *Trans. ASME* **1944**, *66*, 671–684.

19. Brkić, D.; Praks, P. Unified friction formulation from laminar to fully rough turbulent flow. *Appl. Sci.* **2018**, *8*, 2036. [CrossRef]

20. Brkić, D. Solution of the implicit Colebrook equation for flow friction using Excel. *Spreadsheets Educ.* **2017**, *10*, 2.

21. Santos-Ruiz, I.; Bermúdez, J.R.; López-Estrada, F.R.; Puig, V.; Torres, L. Estimación experimental de la rugosidad y del factor de fricción en una tubería. In Proceedings of the Memorias del Congreso Nacional de Control Automático, San Luis Potosí, Mexico, 10–12 October 2018. Available online: https://www.researchgate.net/publication/328332798 (accessed on 19 April 2019). (In Spanish)

22. Olivares, A.; Guerra, R.; Alfaro, M.; Notte-Cuello, E.; Puentes, L. Experimental evaluation of correlations used to calculate friction factor for turbulent flow in cylindrical pipes. *Rev. Int. Métodos Numér. Cálc. Diseño Ing.* **2019**, *35*, 15. (In Spanish) [CrossRef]

23. Olivares Gallardo, A.P.; Guerra Rojas, R.A.; Alfaro Guerra, M.A. Evaluación experimental de la solución analítica exacta de la ecuación de Colebrook-White. *Ing. Investig. Tecnol.* **2019**, *20*, 1–11. (In Spanish) [CrossRef]

24. Zigrang, D.J.; Sylvester, N.D. Explicit approximations to the solution of Colebrook's friction factor equation. *AIChE J.* **1982**, *28*, 514–515. [CrossRef]

25. Winning, H.K.; Coole, T. Explicit friction factor accuracy and computational efficiency for turbulent flow in pipes. *Flow Turbul. Combust.* **2013**, *90*, 1–27. [CrossRef]

26. Clamond, D. Efficient resolution of the Colebrook equation. *Ind. Eng. Chem. Res.* **2009**, *48*, 3665–3671. [CrossRef]

27. Biberg, D. Fast and accurate approximations for the Colebrook equation. *J. Fluids Eng.* **2017**, *139*, 031401. [CrossRef]

28. Brkić, D.; Praks, P. Accurate and efficient explicit approximations of the Colebrook flow friction equation based on the Wright ω-function. *Mathematics* **2019**, *7*, 34. [CrossRef]

29. Brkić, D.; Praks, P. Accurate and efficient explicit approximations of the Colebrook flow friction equation based on the Wright ω-function: Reply to discussion. *Mathematics* **2019**, *7*, 410. [CrossRef]

30. Praks, P.; Brkić, D. One-log call iterative solution of the Colebrook equation for flow friction based on Padé polynomials. *Energies* **2018**, *11*, 1825. [CrossRef]

31. Brkić, D.; Praks, P. An efficient iterative method for looped pipe network hydraulics free of flow-corrections. *Fluids* **2019**, *4*, 73. [CrossRef]

32. Brkić, D.; Praks, P. Short overview of early developments of the Hardy Cross type methods for computation of flow distribution in pipe networks. *Appl. Sci.* **2019**, *9*, 2019. [CrossRef]

33. Brkić, D. Spreadsheet-based pipe networks analysis for teaching and learning purpose. *Spreadsheets Educ.* **2016**, *9*, 4.

34. Brkić, D. A gas distribution network hydraulic problem from practice. *Pet. Sci.Technol.* **2011**, *29*, 366–377. [CrossRef]

35. Brkić, D. Iterative methods for looped network pipeline calculation. *Water Resour. Manag.* **2011**, *25*, 2951–2987. [CrossRef]

36. Brkić, D. An improvement of Hardy Cross method applied on looped spatial natural gas distribution networks. *Appl. Energy* **2009**, *86*, 1290–1300. [CrossRef]

37. Sonnad, J.R.; Goudar, C.T. Constraints for using Lambert W function-based explicit Colebrook–White equation. *J. Hydraul. Eng.* **2004**, *130*, 929–931. [CrossRef]

38. Brkić, D. Comparison of the Lambert W-function based solutions to the Colebrook equation. *Eng. Comput.* **2012**, *29*, 617–630. [CrossRef]

39. Belkić, D. The Euler T and Lambert W functions in mechanistic radiobiological models with chemical kinetics for repair of irradiated cells. *J. Math. Chem.* **2018**, *56*, 2133–2193. [CrossRef]

40. Brkić, D. Lambert W function in hydraulic problems. *Math. Balk.* **2012**, *26*, 285–292.

41. Brkić, D. W solutions of the CW equation for flow friction. *Appl. Math. Lett.* **2011**, *24*, 1379–1383. [CrossRef]

42. Ćojbašić, Ž.; Brkić, D. Very accurate explicit approximations for calculation of the Colebrook friction factor. *Int. J. Mech. Sci.* **2013**, *67*, 10–13. [CrossRef]

43. Brkić, D.; Ćojbašić, Ž. Intelligent flow friction estimation. *Comput. Intell. Neurosci.* **2016**, *2016*, 5242596. [CrossRef] [PubMed]

44. Brkić, D. Discussion of "Gene expression programming analysis of implicit Colebrook–White equation in turbulent flow friction factor calculation" by Saeed Samadianfard [J. Pet. Sci. Eng. 92–93 (2012) 48–55]. *J. Pet. Sci. Eng.* **2014**, *124*, 399–401. [CrossRef]

45. Praks, P.; Brkić, D. Symbolic regression-based genetic approximations of the Colebrook equation for flow friction. *Water* **2018**, *10*, 1175. [CrossRef]

46. Ettema, R. Hunter Rouse—His work in retrospect. *J. Hydraul. Eng.* **2006**, *132*, 1248–1258. [CrossRef]

47. Brkić, D. Can pipes be actually really that smooth? *Int. J. Refrig.* **2012**, *35*, 209–215. [CrossRef]

48. Brkić, D. A note on explicit approximations to Colebrook's friction factor in rough pipes under highly turbulent cases. *Int. J. Heat Mass Transf.* **2016**, *93*, 513–515. [CrossRef]

49. Eaton, J.W. GNU Octave Manual, U.K., Bristol: Network Theory Ltd. 1997. Available online: Ftp://ftp.eeng.dcu.ie/pub/ee454/cygwin/usr/share/doc/octave-2.1.73/pdf/Octave-FAQ.pdf (accessed on 13 June 2019).

50. Horchler, A.D. WrightOmegaq: Complex Double-Precision Evaluation of the Wright Omega Function, a Solution of W + LOG (W) = Z. Version 1.0, 3-12-13. Available online: https://github.com/horchler/wrightOmegaq (accessed on 13 June 2019).

51. Serghides, T.K. Estimate friction factor accurately. *Chem. Eng. N. Y.* **1984**, *91*, 63–64.

52. Khdhr, F.W.; Saeed, R.K.; Soleymani, F. Improving the computational efficiency of a variant of Steffensen's method for nonlinear equations. *Mathematics* **2019**, *7*, 306. [CrossRef]

53. Behl, R.; Salimi, M.; Ferrara, M.; Sharifi, S.; Alharbi, S.K. Some real-life applications of a newly constructed derivative free iterative scheme. *Symmetry* **2019**, *11*, 239. [CrossRef]

54. Shacham, M. Comments on: "An explicit equation for friction factor in pipe". *Ind. Eng. Chem. Fundam.* **1980**, *19*, 228–229. [CrossRef]

55. Romeo, E.; Royo, C.; Monzón, A. Improved explicit equations for estimation of the friction factor in rough and smooth pipes. *Chem. Eng. J.* **2002**, *86*, 369–374. [CrossRef]

56. Offor, U.H.; Alabi, S.B. An accurate and computationally efficient explicit friction factor model. *Adv. Chem. Eng. Sci.* **2016**, *6*, 66711. [CrossRef]

57. Barry, D.A.; Parlange, J.Y.; Li, L.; Prommer, H.; Cunningham, C.J.; Stagnitti, F. Analytical approximations for real values of the Lambert W-function. *Math. Comput. Simul.* **2000**, *53*, 95–103. [CrossRef]

58. Boyd, J.P. Global approximations to the principal real-valued branch of the Lambert W-function. *Appl. Math. Lett.* **1998**, *11*, 27–31. [CrossRef]

59. Demir, S.; Duman, S.; Demir, N.M.; Karadeniz, A. An MS Excel add-in for calculating Darcy friction factor. *Spreadsheets Educ.* **2018**, *10*, 2. Available online: https://sie.scholasticahq.com/article/4669 (accessed on 12 May 2019).

60. Fukushima, T. Precise and fast computation of Lambert W-functions without transcendental function evaluations. *J. Comput. Appl. Math.* **2013**, *244*, 77–89. [CrossRef]

61. Kouznetsov, D. Doya Function and Tania Function. Available online: http://www.ils.uec.ac.jp/~{}dima/PAPERS/2013taniadoya.pdf (accessed on 12 May 2019).

62. Stewart, S. A new elementary function for our curricula? *Aust. Sr. Math. J.* **2005**, *19*, 8–26. Available online: https://files.eric.ed.gov/fulltext/EJ720055.pdf (accessed on 12 May 2019).

63. Pal, R. Teaching Fluid Mechanics and Thermodynamics Simultaneously through Pipeline Flow Experiments. *Fluids* **2019**, *4*, 103. [CrossRef]

64. Barreras, F.; López, A.M.; Lozano, A.; Barranco, J.E. Experimental study of the pressure drop in the cathode side of air-forced open-cathode proton exchange membrane fuel cells. *Int. J. Hydrog. Energy* **2011**, *36*, 7612–7620. [CrossRef]
65. Brkić, D. Comments on "Experimental study of the pressure drop in the cathode side of air-forced open-cathode proton exchange membrane fuel cells" by Barreras et al. *Int. J. Hydrog. Energy* **2012**, *37*, 10963–10964. [CrossRef]

Article

Teach Second Law of Thermodynamics via Analysis of Flow through Packed Beds and Consolidated Porous Media

Rajinder Pal

Department of Chemical Engineering, University of Waterloo, Waterloo, ON N2L 3G1, Canada; rpal@uwaterloo.ca; Tel.: +1-519-888-4567 (ext. 32985)

Received: 12 June 2019; Accepted: 24 June 2019; Published: 27 June 2019

Abstract: The second law of thermodynamics is indispensable in engineering applications. It allows us to determine if a given process is feasible or not, and if the given process is feasible, how efficient or inefficient is the process. Thus, the second law plays a key role in the design and operation of engineering processes, such as steam power plants and refrigeration processes. Nevertheless students often find the second law and its applications most difficult to comprehend. The second law revolves around the concepts of entropy and entropy generation. The feasibility of a process and its efficiency are directly related to entropy generation in the process. As entropy generation occurs in all flow processes due to friction in fluids, fluid mechanics can be used as a tool to teach the second law of thermodynamics and related concepts to students. In this article, flow through packed beds and consolidated porous media is analyzed in terms of entropy generation. The link between entropy generation and mechanical energy dissipation is established in such flows in terms of the directly measurable quantities such as pressure drop. Equations are developed to predict the entropy generation rates in terms of superficial fluid velocity, porous medium characteristics, and fluid properties. The predictions of the proposed equations are presented and discussed. Factors affecting the rate of entropy generation in flow through packed beds and consolidated porous media are identified and explained.

Keywords: undergraduate education; applications of fluids; fluid mechanics; packed bed; porous media; non-equilibrium thermodynamics; entropy generation; pressure loss; Ergun equation; Forchheimer equation

---

1. Introduction

Thermodynamics is a difficult subject to learn and teach. It is considered to be one of the most abstract disciplines of the physical sciences [1]. Students all over the world face difficulties in learning thermodynamics. More specifically, it is the second law of thermodynamics dealing with entropy and entropy production that is difficult for students to fully comprehend. The second law of thermodynamics simply states that all real (irreversible) processes are accompanied by the production of entropy in the universe. However, the students of thermodynamics often find it difficult to visualize entropy generation in real processes at a mechanistic level. The quantification of entropy generation in real processes is an equally problematic issue for students. What causes entropy generation and how do we quantify entropy generation? These are some of the fundamental questions faced by students. Moreover, there are no instruments which can be used to directly measure entropy and entropy generation in real processes.

Entropy generation occurs in flow of all real (viscous) fluids [2]. When a viscous fluid is forced to flow through any geometry, such as a pipe, viscous stresses and velocity gradients are established. For fluid to remain in motion, work has to be done against the viscous stresses which oppose its motion.

Consequently, part of the mechanical energy of fluid is dissipated into frictional heat (internal energy) during motion. The dissipation of highly ordered mechanical energy into disorderly internal energy is reflected in entropy generation. Thus, the analysis of fluid mechanics problems and measurement of the appropriate flow variables such as pressure loss could be used as a tool to demonstrate and quantify entropy generation in real processes.

The main objectives of this article are: (a) to analyze the flow of viscous fluids through packed beds of discrete particles and through consolidated porous media in terms of entropy generation; (b) to quantify entropy generation in flow through packed beds and consolidated porous media in terms of directly measureable quantities such as pressure loss as a function of flow rate of fluid; and (c) to discuss various factors which affect the rate of entropy generation in flow through packed beds and consolidated porous media.

2. Brief Review of the Second Law of Thermodynamics

The second law of thermodynamics could be stated in several different but equivalent ways. The classical statements of the second law are the Kelvin–Planck statement and the Clausius statement.

The Kelvin–Planck statement says that "It is impossible for a system operating in a cycle and connected to a single heat reservoir to produce a positive amount of work in the surroundings [3]". According to the Kelvin–Planck statement, it is impossible to build a heat engine shown schematically in Figure 1 where heat absorbed from a heat reservoir is completely converted into work without altering the properties of the system. Mathematically, the Kelvin–Planck statement can be expressed as [4]:

$$W_{cycle} \leq 0 \text{ (single heat reservoir)} \tag{1}$$

where the system communicates thermally only with a single heat reservoir. The sign convention for work (W) used in this article is W = positive, if it is produced by the system (flows out of the system to surroundings) and W = negative, if it is absorbed by the system (flows into the system from surroundings). Thus, no cyclic process is possible where $W_{cycle} > 0$ using a single heat reservoir. The equality in Equation (1) is valid for a reversible process and the inequality is valid for an irreversible process. For a reversible process, no work is produced or destroyed, that is, $W_{cycle} = 0$ whereas work is destroyed when the process is irreversible, that is, $W_{cycle} < 0$ (assuming that the system communicates thermally only with a single heat reservoir).

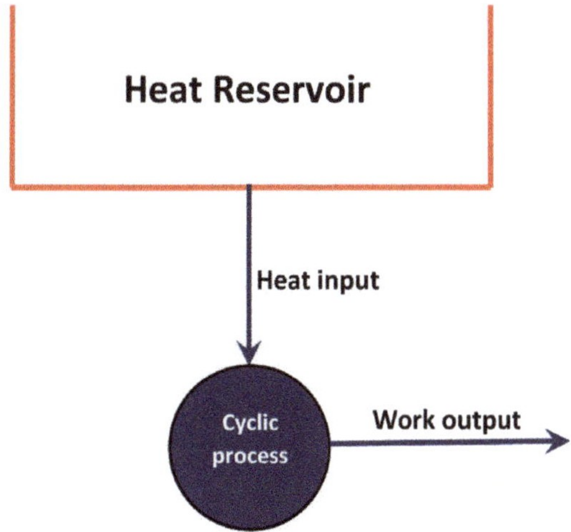

Figure 1. Impossible heat engine.

The Clausius statement of the second law [3] says that "It is impossible for a system operating in a cycle to have its *sole effect* the transfer of heat from a low temperature heat reservoir to a high temperature heat reservoir". According to the Clausius statement, it is impossible to construct a device based on the scheme shown schematically in Figure 2 where the sole result of the process is the transfer of heat from a cooler body at low temperature (T_L) to a hotter body at high temperature (T_H).

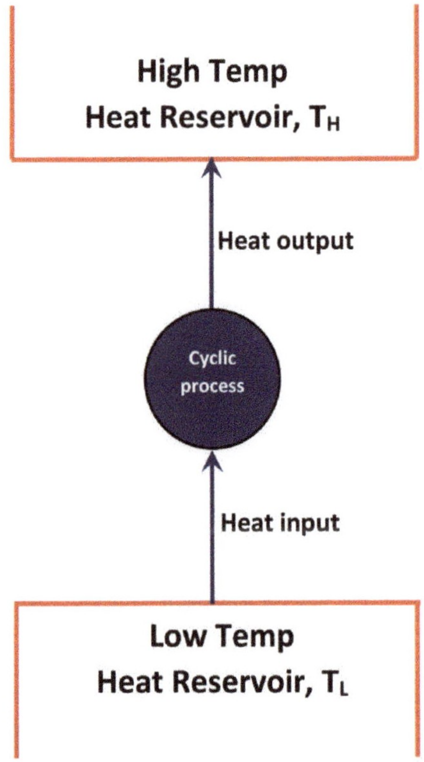

Figure 2. Impossible scheme.

Another powerful statement of the second law of thermodynamics is that all irreversible (real) processes are accompanied by entropy generation in the universe [5], that is:

$$S_{G,univ} = \Delta S_{sys} + \Delta S_{surr} \geq 0 \tag{2}$$

where $S_{G,univ}$ is the total amount of entropy generated in the universe (system + surroundings), ΔS_{sys} is the entropy change of the system and ΔS_{surr} is the entropy change of the surroundings. The equality in Equation (2) is valid for a reversible process and the inequality is valid for an irreversible process. According to this statement of the second law, no process is possible for which $S_{G,univ} < 0$.

Entropy is a measure of disorderliness of a system. According to the Boltzmann entropy equation, the entropy of a system can be expressed as [3]:

$$S = k_B ln\Omega \tag{3}$$

where Ω is the number of possible configurations of the system and k_B is the Boltzmann constant. The larger the number of possible configurations of the system, greater is the disorderliness of the system and higher is the entropy. Therefore, we can interpret the second law of thermodynamics in yet another way, that is, "Only those processes are possible processes which lead to an increase in the disorderliness of the universe".

The scheme shown in Figure 1 is impossible as it decreases the disorder of the universe, that is, $S_{G,univ} < 0$. Here $\Delta S_{sys} = 0$ but $\Delta S_{surr} < 0$ as the surroundings heat reservoir loses heat. For this scheme, Equation (2) gives:

$$S_{G,univ} = -\frac{Q_{cycle}}{T} \geq 0 \tag{4}$$

where T is the absolute temperature of the heat reservoir and Q is the heat transferred. As $Q_{cycle} = W_{cycle}$ from the first law of thermodynamics, Equation (4) reduces to Equation (1), that is, $W_{cycle} \leq 0$ when there is only one heat reservoir involved. Similarly, the scheme shown in Figure 2 is impossible as it decreases the disorder of the universe:

$$S_{G,univ} = \Delta S_{cold-body} + \Delta S_{hot-body} < 0 \tag{5}$$

Note that $\Delta S_{cold-body} < 0$ as it loses heat whereas $\Delta S_{hot-body} > 0$ as it gains heat. However, due to different temperatures of the cold and hot bodies, $\left|\Delta S_{cold-body}\right| > \left|\Delta S_{hot-body}\right|$.

Thus, the Kelvin–Planck and Clausius statements of the second law of thermodynamics are special cases of the statement of the second law expressed in the form of Equation (2).

For a flow process (see Figure 3), the second law of thermodynamics can be written as [2]:

$$\dot{S}_{G,univ} = \dot{S}_{G,CV} + \dot{S}_{G,Surr} = \iint \rho s(\hat{n}\cdot\vec{V})dA + \iiint \frac{\partial(\rho s)}{\partial t}d\vartheta - \sum \frac{\dot{Q}_i}{T_i} \geq 0 \tag{6}$$

where $\dot{S}_G$ is the rate of entropy generation, s is the entropy per unit mass of fluid, ρ is the fluid density, $\hat{n}$ is unit outward normal to the control surface, $\vec{V}$ is fluid velocity vector, A is the control surface area, t is time, ϑ is the volume of the control volume, $\dot{Q}_i$ is the rate of heat transfer to control volume from ith heat reservoir at an absolute temperature of T_i, the subscripts CV and $Surr$ refer to control volume and surroundings, respectively. As noted earlier, the equality in Equation (6) is valid for a reversible (frictionless) process and the inequality is valid for an irreversible process. The surface integral $\iint \rho s(\hat{n}\cdot\vec{V})dA$ is the net outward flow of entropy (associated with mass) across the entire control surface. The volume integral $\iiint \frac{\partial(\rho s)}{\partial t}d\vartheta$ is the rate of accumulation of entropy within the entire control volume (assumed to be fixed and non-deforming). For a control volume with one inlet and one outlet, Equation (6), under steady state condition, reduces to:

$$\dot{S}_{G,universe} = \dot{m}(\Delta s) - \sum \frac{\dot{Q}_i}{T_i} \geq 0 \tag{7}$$

where $\dot{m}$ is the mass flow rate.

The quantification of entropy generation in real processes is important from a practical point of view as entropy generation is directly related to the efficiency of the process. Higher the rate of entropy generation in a process, lower is the thermodynamic efficiency of the process. According to the Gouy–Stodola theorem [6] of thermodynamics, the loss of power or work potential in a real process, due to irreversibilities in the process, is directly proportional to the total rate of entropy generation. Thus:

$$\dot{W}_{lost} \propto \dot{S}_{G,univ} \tag{8}$$

where $\dot{W}_{lost}$ is the rate of work lost (wasted) as a result of irreversibilities in the process.

As an example of a flow process, consider flow through a control volume shown in Figure 3. The first law of thermodynamics for open systems under steady state condition gives:

$$\dot{m}\left[\Delta h + \Delta\left(V^2/2\right) + g\Delta z\right] = \dot{Q} - \dot{W}_{sh} \tag{9}$$

where h is the specific enthalpy of fluid, V is the fluid velocity, g is the acceleration due to gravity, z is the elevation, $\dot{Q}$ is the rate of heat transfer, and $\dot{W}_{sh}$ is the rate of shaft work. Neglecting kinetic and potential energy changes, Equation (9) simplifies to:

$$\dot{W}_{sh} = \dot{Q} - \dot{m}\Delta h = \dot{m}(Q - \Delta h) \tag{10}$$

where Q is the heat transfer per unit mass of fluid. The second law, Equation (7), can be written as:

$$\dot{S}_{G,univ} = \dot{m}(\Delta s) - \frac{\dot{Q}}{T_0} \geq 0 \tag{11}$$

where T_0 is the absolute temperature of the heat reservoir (surroundings). For the flow process to be reversible:

$$\dot{S}_{G,univ} = \dot{m}\left[(\Delta s) - \frac{Q_{rev}}{T_0}\right] = 0 \tag{12}$$

$$Q_{rev} = T_0(\Delta s) \tag{13}$$

where Q_{rev} is the heat transfer per unit mass of fluid for a reversible process. From Equations (10) and (13):

$$\dot{W}_{sh,rev} = \dot{m}[T_0(\Delta s) - \Delta h] \tag{14}$$

The lost work, $\dot{W}_{lost}$, is defined as:

$$\dot{W}_{lost} = \dot{W}_{sh,rev} - \dot{W}_{sh} \tag{15}$$

From Equations (10), (14), and (15), we get:

$$\dot{W}_{lost} = \dot{m}[T_0(\Delta s) - Q] = \dot{m}T_0\left[(\Delta s) - \frac{Q}{T_0}\right] \tag{16}$$

Using Equation (11), Equation (16) can be re-written as:

$$\dot{W}_{lost} = T_0\dot{S}_{G,univ} \geq 0 \tag{17}$$

Equation (17) is the Gouy–Stodola theorem [6] of thermodynamics. Thus, the rate of work lost due to irreversibilities is directly proportional to the rate of entropy generation. The thermodynamic efficiency ϵ of a flow process can be defined as:

$$\epsilon = \frac{\dot{W}_{sh}}{\dot{W}_{sh,rev}} = 1 - \frac{\dot{W}_{lost}}{\dot{W}_{sh,rev}} \leq 1.0 \tag{18}$$

In a reversible process, $\dot{W}_{lost} = 0$ and $\epsilon = 1$. In any irreversible process, $\dot{W}_{lost} > 0$ and $\epsilon < 1$. If no work is produced in the process, that is, actual $\dot{W}_{sh}$ is zero, then all the work potential is lost due to irreversibilities in the process and consequently, $\epsilon = 0$.

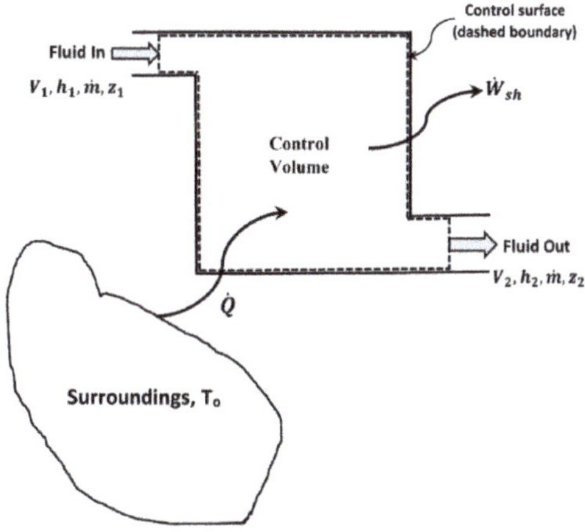

Figure 3. Typical flow process.

3. Flow through Unconsolidated and Consolidated Porous Media

Porous medium is a composite material in that it consists of two phases, namely pores (voids, free space pockets) and solid-phase. The pores may be occupied by a fluid (gas, oil, water, etc.). A large variety of natural and synthetic materials are porous in nature. Examples include: underground oil reservoirs, ceramics, solid foams, sand filters, wood, and packed beds of particles used widely in chemical engineering applications. The pores of a porous medium usually form a three-dimensional inter-connected network and, therefore, fluids can flow through the porous medium. If all pores of a porous medium are inter-connected, then the porosity (ε) of the porous medium is simply the fraction of the total volume of the medium that is occupied by the pores. Thus, the fraction of the total volume that is occupied by the solid phase is $(1-\varepsilon)$. When some pores are isolated or disconnected or have dead ends, then the effective porosity, defined as the ratio of connected void volume to total volume of the medium, is lower than the total porosity.

Porous media could be classified as consolidated or unconsolidated. In consolidated porous medium, the solid phase is basically a single piece of material or the grains of the solid phase are cemented or fused together to form a single piece of solid phase. In unconsolidated porous media, on the other hand, the grains or the particles of the solid phase are not cemented together and, therefore, the porous medium is a multi-particle system like a packed bed of individual (un-cemented, un-glued) particles. Flow of single-phase Newtonian fluids (gas, water, oil) through packed beds and consolidated porous media is important from a practical point of view.

3.1. Analysis of Flow through Packed Beds (Unconsolidated Porous Media)

3.1.1. Pressure-Loss in Flow through Packed Beds

Consider flow of a Newtonian fluid through a horizontal cylindrical packed bed of particles with bed diameter D_b and bed length L_b. As flow through a packed bed is quite complex, a rigorous theoretical derivation of pressure drop-flow rate relationship is not possible. Only approximate models have been developed. In one approach, used widely to model flow through packed bed of particles, the packed bed is visualized as a bundle of identical capillary tubes [7–9]. In its simplest form, the capillary tube bundle model assumes that the capillary tubes are straight, cylindrical of constant cross-section (uniform radius), and parallel (all oriented in the same direction), as shown in

Figure 4. The average velocity in any capillary tube is the same as that in the packed bed. The capillary diameter is equal to the hydraulic diameter of the bed, d_H, defined as:

$$d_H = 4\left(\frac{volume\ of\ voids}{total\ wetted\ surface\ area}\right) = \left(\frac{4\varepsilon}{1-\epsilon}\right)\left(\frac{V_p}{S_p}\right) \tag{19}$$

where ε is the void fraction or porosity, V_p is the volume of a single particle, and S_p is the surface area of a single particle. In writing the above expression for d_H, it is assumed that all the particles are identical and that the wetted surface of the cylindrical container walls of the bed is negligible as compared with the total wetted surface area of the particles. For a bed of identical spherical particles, $V_p/S_p = d_p/6$, where d_p is the particle diameter. Thus, the hydraulic diameter becomes:

$$d_H = \left(\frac{2}{3}\right)\left(\frac{\varepsilon}{1-\epsilon}\right)d_p \tag{20}$$

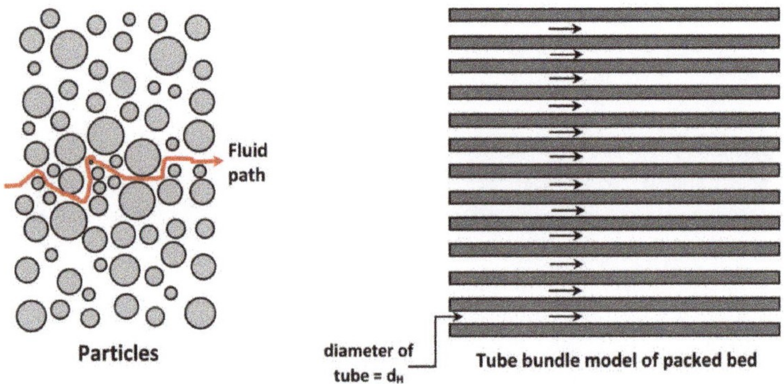

Figure 4. Path taken by fluid in a bed of particles and tube bundle model of packed bed.

The superficial velocity of fluid (V_s) is defined is:

$$V_s = \frac{\dot{\vartheta}}{A_b} \tag{21}$$

where $\dot{\vartheta}$ is the volumetric flow rate of fluid and A_b is the total cross-sectional area of the bed. Thus, V_s is the velocity of fluid in the bed if no particles were present in the bed. The average velocity $\overline{V}$ of fluid in the bed, also called interstitial velocity, is defined as:

$$\overline{V} = \frac{\dot{\vartheta}}{A_{flow}} \tag{22}$$

where A_{flow} is the cross-sectional area of bed through which the fluid flows. The porosity of the bed ε is given as:

$$\varepsilon = \frac{A_{flow}}{A_b} \tag{23}$$

Therefore, the average or interstitial velocity of fluid in the bed is:

$$\overline{V} = \frac{V_s}{\varepsilon} \tag{24}$$

This expression of average velocity does not consider the tortuous path taken by fluid in the bed. Due to tortuosity of the bed, the actual average velocity of fluid in the bed is larger than that given by Equation (24). The tortuosity τ of a bed is defined as:

$$\tau = \frac{L_e}{L_b} \tag{25}$$

where L_e is average length of the tortuous path taken by the fluid and L_b is the straight length of the bed. To account for the tortuosity, the actual average velocity of the fluid in the bed can be expressed as follows:

$$\overline{V} = \tau \frac{V_s}{\varepsilon} \tag{26}$$

In laminar flow of a Newtonian fluid through a cylindrical tube, the pressure gradient in the direction of flow is given as:

$$-\left(\frac{dP}{dx}\right) = 32\left(\frac{\mu \overline{V}}{D^2}\right) \tag{27}$$

where μ is the fluid viscosity and D is the tube diameter. Replacing D with hydraulic diameter d_H and the constant factor of 32 with C, the pressure-gradient in a capillary tube model of the packed-bed model can be expressed as:

$$-\left(\frac{dP}{dx}\right) = C\left(\frac{\mu \overline{V}}{d_H^2}\right) \tag{28}$$

Note that flow in the bed is assumed to be laminar here. The constant C is expected to be larger than 32, as the path of fluid in the bed is not straight. The fluid follows a tortuous path in the bed and consequently, the pressure drop over a certain straight length is expected to be more than that observed over the same length if the fluid path was non-tortuous. Therefore, the constant C is taken as $32\,\tau$ where τ is the tortuosity. Upon substitution of the expressions for d_H and $\overline{V}$, and taking $C = 32\,\tau$, Equation (28) gives:

$$-\left(\frac{dP}{dx}\right) = 72\,\tau^2\left(\frac{\mu V_s}{d_p^2}\right)\frac{(1-\varepsilon)^2}{\varepsilon^3} \tag{29}$$

Equation (29) assumes that the cross-section of the representative flow passage (capillary tube) in the porous medium is circular and constant. In reality the fluid moves through converging-diverging flow passages of non-uniform cross-sections. In converging-diverging flows, fluid experiences stretching or extensional deformation. The elongation or stretching of fluid elements results in additional dissipation of energy and pressure drop. Thus, this equation needs to be modified further as follows:

$$-\left(\frac{dP}{dx}\right) = 72\,\tau^2 K\left(\frac{\mu V_s}{d_p^2}\right)\frac{(1-\varepsilon)^2}{\varepsilon^3} \tag{30}$$

where K is an empirical factor that takes into account the influence of non-constant cross-section of flow passage on pressure-gradient. The tortuosity τ is often taken to be $\sqrt{2}$ for random packing of uniform spheres [9]. It is generally a function of the porosity of the porous medium [10]. For example, the following model, based on the Maxwell equation for electrical conductivity of composite, is often used to describe the relationship between tortuosity and porosity [10]:

$$\tau = 1.5 - 0.5\varepsilon \tag{31}$$

When $\tau = \sqrt{2}$ and $K = 25/24$, the following well-known Blake–Kozeny equation for laminar flow through randomly packed bed of uniform spheres is obtained [7]:

$$-\left(\frac{dP}{dx}\right) = 150\left(\frac{\mu V_s}{d_p^2}\right)\frac{(1-\varepsilon)^2}{\varepsilon^3} \tag{32}$$

When $\tau = \sqrt{2}$ and $K = 5/4$, the following Carman–Kozeny equation, another well-known equation for laminar flow through randomly packed bed of uniform spheres, is obtained [11]:

$$-\left(\frac{dP}{dx}\right) = 180\left(\frac{\mu V_s}{d_p^2}\right)\frac{(1-\varepsilon)^2}{\varepsilon^3} \tag{33}$$

Both Blake—Kozeny and Carman–Kozeny equations are popular in the literature although they use different values of the factor K. The observed difference in K values is probably related to differences in particle-shape, surface roughness of particles, and porosity of bed. Note that these equations are restricted to only laminar flow through bed of nearly uniform spheres. The packed-bed Reynolds number Re_b, defined below, should be less than 10 for flow to be laminar in the bed.

$$Re_b = \left(\frac{1}{1-\varepsilon}\right)\frac{\rho d_p V_s}{\mu} \tag{34}$$

For turbulent flow through packed beds $(Re_b > 1000)$, the following Burke–Plummer equation is often used [12]:

$$-\left(\frac{dP}{dx}\right) = 1.75\left(\frac{\rho V_s^2}{d_p}\right)\frac{(1-\varepsilon)}{\varepsilon^3} \tag{35}$$

In the transition region, the following equation, obtained by superposition of Blake–Kozeny and Burke–Plummer equations, proposed by Ergun [13] is widely used:

$$-\left(\frac{dP}{dx}\right) = 150\left(\frac{\mu V_s}{d_p^2}\right)\frac{(1-\varepsilon)^2}{\varepsilon^3} + 1.75\left(\frac{\rho V_s^2}{d_p}\right)\frac{(1-\varepsilon)}{\varepsilon^3} \tag{36}$$

As the Ergun equation, Equation (36), is obtained by the superposition of laminar and turbulent expressions, it is valid over the full range of the packed-bed Reynolds number Re_b. The Ergun equation could also be re-cast in the following form [14]:

$$f_b = \frac{150}{Re_b} + 1.75 \tag{37}$$

where f_b is the packed-bed friction factor defined as:

$$f_b = \frac{-\left(\frac{dP}{dx}\right)}{\left(\frac{\rho V_s^2}{d_p}\right)\frac{(1-\varepsilon)}{\varepsilon^3}} \tag{38}$$

The Ergun equation, Equation (36) or (37), is used extensively in the literature to describe pressure loss in packed beds. However, the following points should be kept in mind when using the Ergun equation: (a) it is applicable to unconsolidated beds of nearly uniform-size spherical particles with appreciable roughness. For smooth spheres, it tends to over-predict f_b in the high Re_b range $(Re_b > 700)$ [14–16]; (b) if the bed particles are non-uniform in sizes, the Sauter mean diameter of particles should be used in the application of the Ergun equation; (c) when particle shape deviates significantly from a sphere, the Ergun equation tends to under-predict f_b [17]; and (d) wall effects can be important when $D_b/d_p < 10$. When wall effects are important, the experimental data show deviation from the predictions of the Ergun equation [18].

Figure 5 shows the prediction of the packed bed friction factor as a function of packed bed Reynolds number using the Ergun equation, Equation (37). In the limits of low Re_b and high Re_b, the Ergun equation predictions overlap with the predictions of the Blake–Kozeny equation (low Re_b) and the Burke–Plummer equation (high Re_b), as expected.

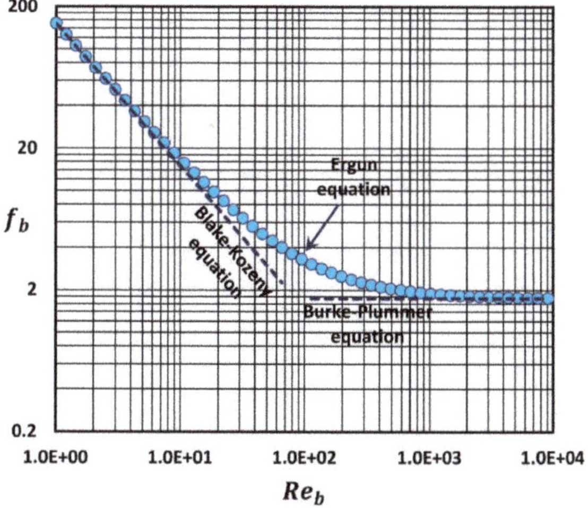

Figure 5. Prediction of packed bed friction factor as a function of packed bed Reynolds number. The data points are generated using the Ergun equation (Equation (37)).

3.1.2. Entropy Generation in Flow through Packed Beds

Consider steady flow of an incompressible fluid through a bed of particles (see Figure 6). We can apply the following macroscopic mechanical energy balance between the inlet and outlet of the bed [2]:

$$\dot{W}_{sh} + \dot{F}_l + \dot{m}\left[\Delta\varphi + \Delta(KE) + \frac{1}{\rho}(\Delta P)\right] = 0 \tag{39}$$

where $\dot{F}_l$ is the rate of mechanical energy dissipation due to friction in fluid, $\Delta\varphi$ is the potential energy change per unit mass of fluid, $\Delta(KE)$ is the kinetic energy change per unit mass of fluid, and ΔP is the pressure change of the fluid. Neglecting potential and kinetic energy changes and taking $\dot{W}_{sh} = 0$, the mechanical energy balance gives:

$$\dot{F}_l = -\frac{\dot{m}}{\rho}(\Delta P) \tag{40}$$

The rate of mechanical energy dissipation per unit length of packed bed ($\dot{F}_l'$) can be expressed as:

$$\dot{F}_l' = \frac{\dot{m}}{\rho}\left(-\frac{dP}{dx}\right) \tag{41}$$

From the second law of thermodynamics, Equation (11), we get:

$$\dot{S}_{G,univ} = \dot{m}(\Delta s) > 0 \tag{42}$$

Note that we are assuming flow to be adiabatic with negligible heat transfer. There is no entropy generation in the surroundings. All the entropy is generated within the fluid inside the packed bed and the rate of entropy generation is the net rate of increase in entropy of the flowing stream.

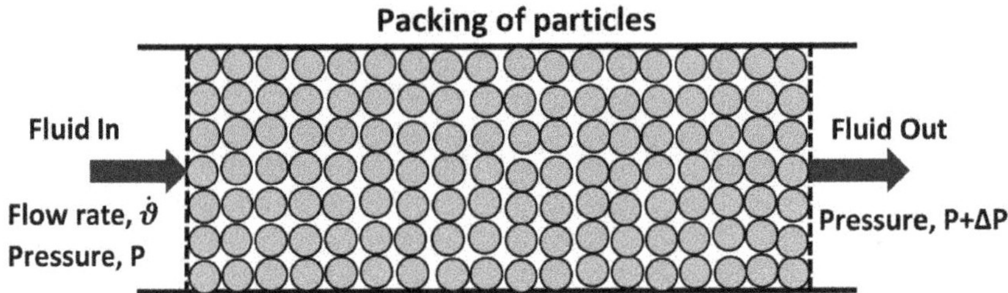

Figure 6. Flow through a packed bed

We can now relate entropy change of the fluid stream to pressure change. For pure substances, the relationship between entropy and other state variables is given as [5]:

$$Tds = dh - (dP/\rho) \tag{43}$$

where T is the absolute temperature. From the first law of thermodynamics, Equation (9), the enthalpy change is zero in the absence of heat transfer and shaft work for steady flow in a horizontal bed. Consequently, Equation (43) reduces to:

$$Tds = -(dP/\rho) \tag{44}$$

Assuming incompressible flow and constant temperature, Equation (44) upon integration gives:

$$\Delta s = -\frac{\Delta P}{\rho T} \tag{45}$$

From Equations (42) and (45), it follows that:

$$\dot{S}_G = \frac{\dot{m}}{\rho}\left(-\frac{\Delta P}{T}\right) > 0 \tag{46}$$

The subscript "*univ*" has been removed from $\dot{S}_{G,univ}$ as $\dot{S}_{G,univ}$ is simply the rate of entropy generation within the fluid inside a packed bed. From Equations (41) and (46), we can also express the rate of entropy generation in a packed bed on a unit volume basis as:

$$\dot{S}_G''' = \frac{\dot{F}_l'}{TA_b} = \frac{V_s}{T}\left(-\frac{dP}{dx}\right) \tag{47}$$

where A_b is the total cross-sectional area of the bed and $\dot{S}_G'''$ is the rate of entropy generation per unit volume of the bed. From Equations (38) and (47), we get:

$$\dot{S}_G''' = \left(\frac{1}{T}\right)\left(\frac{1-\varepsilon}{\varepsilon^3}\right)\left(\frac{\rho V_s^3}{d_p}\right)f_b \tag{48}$$

The packed bed friction factor f_b for laminar flow ($Re_b < 10$) is given as:

$$f_b = \frac{150}{Re_b} \tag{49}$$

Consequently, Equation (48) reduces to:

$$\dot{S}_G''' = \left(\frac{1}{T}\right)\left(\frac{150(1-\varepsilon)^2}{d_p^2\varepsilon^3}\right)\mu V_s^2 \tag{50}$$

Thus, entropy generation rate per unit volume of bed in steady laminar flow of a Newtonian fluid is directly proportional to the fluid viscosity and to the square of the superficial fluid velocity in the bed. The entropy generation rate also depends on the particle diameter, the bed porosity ε, and the temperature.

The packed bed friction factor in turbulent flow ($Re_b > 1000$) is given as [7–9]:

$$f_b = 1.75 \tag{51}$$

Consequently, Equation (48) reduces to:

$$\dot{S}_G''' = \left(\frac{1}{T}\right)\left(\frac{1.75(1-\varepsilon)}{d_p\varepsilon^3}\right)\left(\rho V_s^3\right) \tag{52}$$

Thus, entropy generation rate per unit volume of bed in steady turbulent flow of a Newtonian fluid is independent of the fluid viscosity and is directly proportional to the cube of the superficial fluid velocity in the bed. The entropy generation rate also depends on the fluid density, the particle diameter, the bed porosity ε, and the temperature.

To cover the full range of packed bed Reynolds number Re_b, we substitute the Ergun equation, Equation (37), into Equation (48) to obtain the following equation valid over the full range of packed bed Reynolds number Re_b:

$$\dot{S}_G''' = \frac{1}{T}\left[\left(\frac{150(1-\varepsilon)^2}{d_p^2\varepsilon^3}\right)\mu V_s^2 + \left(\frac{1.75(1-\varepsilon)}{d_p\varepsilon^3}\right)\left(\rho V_s^3\right)\right] \tag{53}$$

Note that Equation (53) is simply the superposition of Equations (50) and (52).

3.2. Analysis of Flow through Consolidated Porous Media

3.2.1. Pressure-Loss in Flow through Consolidated Porous Media

Laminar flow in consolidated porous media (see Figure 7) is usually described by Darcy's law given as follows [19]:

$$-\frac{dP}{dx} = \frac{\mu}{k}V_s \tag{54}$$

where k is the permeability of the porous medium. Permeability is a measure of the ability of porous medium to conduct fluid. Higher k means lower resistance to flow and consequently, larger flow rate for the same pressure gradient. k has the dimensions of (length)^2. It is usually expressed in darcies. For example, 1 darcy = 1 (cm/s). centipoise/(atm/cm) = 9.87×10^{-13} m^2.

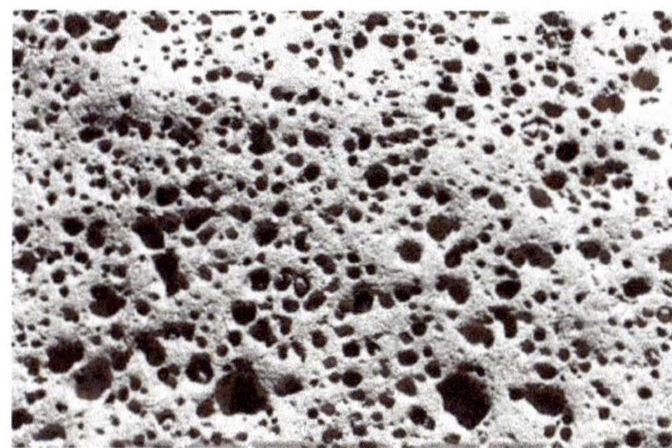

Figure 7. Consolidated porous medium (adapted from www.fotosearch.com.au).

Darcy's law is equally applicable to unconsolidated porous medium such as packed bed of free (un-cemented/un-fused) particles. Thus, the Blake–Kozeny equation or the Carman–Kozeny equation are special cases of the Darcy law. Upon comparison of the Darcy law with Blake–Kozeny/Carman–Kozeny equations, the permeability of a packed bed of uniform size spherical particles can be expressed as follows:

$$k = \frac{d_p^2 \varepsilon^3}{150(1-\varepsilon)^2} \tag{55}$$

$$k = \frac{d_p^2 \varepsilon^3}{180(1-\varepsilon)^2} \tag{56}$$

Equation (55) gives permeability based on the Blake–Kozeny equation whereas Equation (56) gives permeability based on the Carman–Kozeny equation.

Darcy's law is restricted to flows where viscous forces dominate over the inertial forces. At high flow rates, the flow in the porous medium becomes turbulent and Darcy's law is no longer valid. The pressure-gradient in the turbulent regime is much higher compared with that in the laminar regime at the same superficial velocity V_s. In order to extend the Darcy law to turbulent regime, Forchheimer [19] modified the Darcy law by adding a quadratic term as follows:

$$-\frac{dP}{dx} = \frac{\mu}{k} V_s + \beta \rho V_s^2 \tag{57}$$

where the first term on the right side of the equation reflects the viscous effects and the second term reflects the inertial effects. β is called the non-Darcy flow coefficient with a dimension of 1/length. This equation could be re-written in dimensionless form as:

$$f_{pm} = \frac{-\frac{dP}{dx}}{\beta \rho V_s^2} = 1 + \frac{1}{Re_{pm}} \tag{58}$$

where f_{pm} is the friction factor for flow in consolidated porous medium and Re_{pm} is the porous-medium Reynolds number, also referred to as Forchheimer number, defined as:

$$Re_{pm} = \frac{\rho V_s \beta k}{\mu} \tag{59}$$

At low Reynolds number ($Re_{pm} < 0.1$), the first term on the right side of the dimensionless Forchheimer equation, Equation (58), can be neglected and it reduces to Darcy's law. At high Reynolds number ($Re_{pm} > 10$), the second term on the right side of the dimensionless Forchheimer equation, Equation (58), can be neglected and it reduces to $f_{pm} = 1$. Figure 8 shows the prediction of f_{pm} as a function of Re_{pm} using the Forchheimer equation, Equation (58).

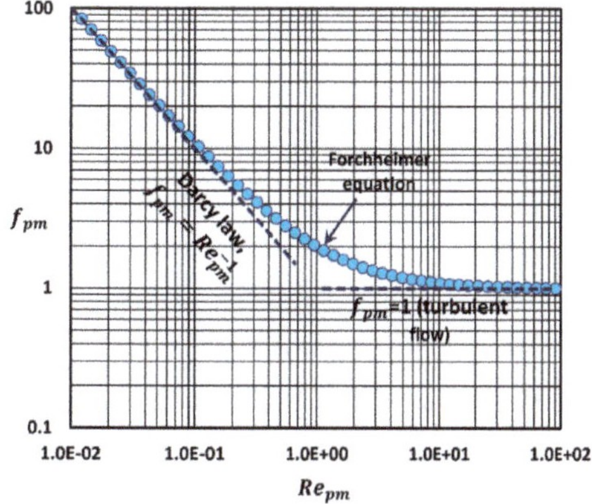

Figure 8. Prediction of porous medium friction factor as a function of porous medium Reynolds number. The data points are generated using the Forchheimer equation (Equation (58)).

In order to apply the Forchheimer equation, the value of non-Darcy flow coefficient is needed. It can be determined experimentally. The experimental procedure to determine β involves two steps: In the first step, the permeability k of the porous sample is determined from Darcy's law by restricting the experiments to low flow rates, and in the second step, the experiments are conducted at high flow rates and β is evaluated directly from the Forchheimer equation from the knowledge of pressure drop versus flow rate. The non-Darcy flow coefficient decreases with the increase in the permeability of the porous medium [20].

The Forchheimer equation is equally applicable to unconsolidated porous medium such as a packed bed of particles. Ergun equation is a special case of the Forchheimer equation. The Forchheimer equation, Equation (57) reduces to the Ergun equation, Equation (36), when permeability k and non-Darcy flow coefficient β are replaced by the following expressions:

$$k = \frac{d_p^2 \varepsilon^3}{150(1-\varepsilon)^2} \text{ and } \beta = \frac{7}{4}\left(\frac{1-\varepsilon}{d_p \varepsilon^3}\right) \tag{60}$$

The relationships between f_{pm} and f_b, and Re_{pm} and Re_b are as follows:

$$f_b = 1.75 \, f_{pm} \tag{61}$$

$$Re_b = \left(\frac{600}{7}\right)Re_{pm} \tag{62}$$

3.2.2. Entropy Generation in Flow through Consolidated Porous Media

Consider one-dimensional flow of an incompressible Newtonian fluid in consolidated porous medium, as shown in Figure 9. The rate of entropy generation in consolidated porous medium per unit volume is given by Equation (47), re-written as:

$$\dot{S}_G''' = \frac{V_s}{T}\left(-\frac{dP}{dx}\right) = \left(\frac{1}{T}\right)\left(\beta \rho V_s^3\right)f_{pm} \tag{63}$$

The porous medium friction factor f_{pm} for laminar flow ($Re_{pm} < 0.1$) is given as:

$$f_{pm} = \frac{1}{Re_{pm}} \tag{64}$$

Consequently, Equation (63) reduces to:

$$\dot{S}_G''' = \left(\frac{1}{T}\right)\left(\frac{\mu}{k}\right)V_s^2 \tag{65}$$

Thus, entropy generation rate per unit volume of the porous medium in steady laminar flow of a Newtonian fluid is directly proportional to the fluid viscosity, inversely proportional to the porous medium permeability, and directly proportional to the square of the superficial fluid velocity in the medium. The entropy generation rate also depends on the temperature.

Figure 9. One-dimensional flow in consolidated porous medium.

The porous medium friction factor in turbulent flow ($Re_{pm} > 10$) is given as:

$$f_{pm} = 1 \tag{66}$$

Consequently, Equation (63) reduces to:

$$\dot{S}_G''' = \left(\frac{1}{T}\right)(\rho\beta)V_s^3 \tag{67}$$

Thus, entropy generation rate per unit volume of the porous medium in steady turbulent flow of a Newtonian fluid is independent of the fluid viscosity and is directly proportional to the cube of the superficial fluid velocity in the bed. The entropy generation rate also depends on the fluid density, the non-Darcy flow coefficient β, and the temperature.

To cover the full range of porous medium Reynolds number Re_{pm}, we substitute the Forchheimer equation, Equation (58), into Equation (63) to obtain the following equation valid over the full range of porous medium Reynolds number Re_{pm}:

$$\dot{S}_G''' = \frac{1}{T}\left[\left(\frac{\mu}{k}\right)V_s^2 + (\rho\beta)V_s^3\right] \tag{68}$$

Equation (68) is simply the superposition of Equations (65) and (67).

4. Simulation Results and Discussion

4.1. Entropy Generation in Flow through Packed Beds

4.1.1. Laminar Flow

The entropy generation rate in laminar flow through packed beds, per unit volume of the bed, as a function of superficial bed velocity is shown in Figure 10 for the following conditions: $T = 298.15$ K, $d_p = 1$ mm, $\mu = 18.5$ μPa·s, and $\rho = 1.184$ kg/m^3. Equation (50) is used to generate the plots. For a given value of the bed porosity ε, the entropy generation rate $\dot{S}_G'''$ increases with the increase in the superficial velocity V_s of the fluid in the bed.

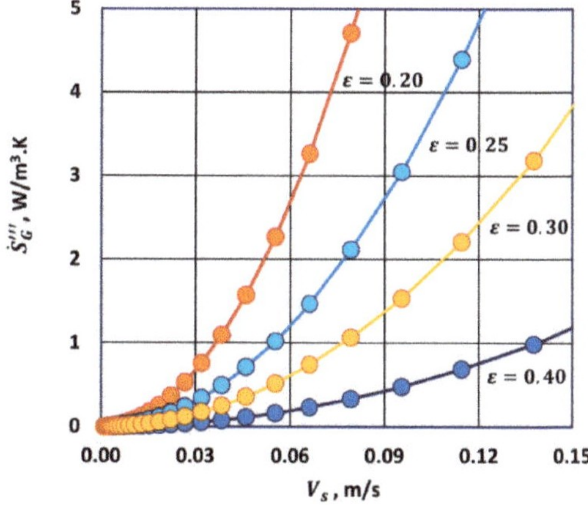

Figure 10. Effect of bed porosity on the rate of entropy generation per unit volume of the packed bed in laminar flow (T = 298.15 K, d_p = 1 mm, μ = 18.5 µPa·s, ρ = 1.184 kg/m³). The data points are generated using Equation (50).

The increase in $\dot{S}_G'''$ with the increase in V_s is larger if the bed porosity ε is small. Additionally, the entropy generation rate falls sharply with the increase in the bed porosity. These results are as expected. With the increase in the fluid velocity, the rate of mechanical energy dissipation into frictional heat increases due to increases in viscous stresses and velocity gradients in the fluid. Consequently, there occurs an increase in the rate of entropy generation. When the bed porosity is increased, the bed structure becomes more open to fluid flow resulting in a decrease in the resistance to fluid motion and mechanical energy dissipation and, hence, a decrease in the rate of entropy generation.

Figure 11 shows the effect of fluid viscosity on the rate of entropy generation in laminar flow. The plots are generated using Equation (50). With the increase in fluid viscosity, the resistance to fluid motion increases due to an increase in the viscous stresses. With the increase in viscous stresses, the rate of mechanical energy dissipation into frictional heat (internal energy) increases. The conversion of highly ordered mechanical energy into disorderly internal energy is reflected in an increase in the rate of entropy generation.

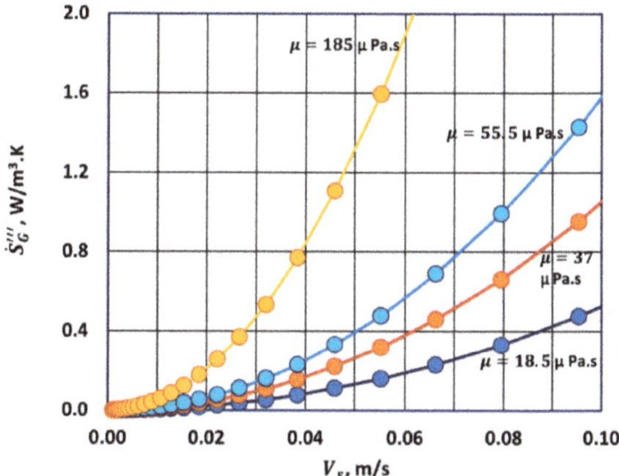

Figure 11. Effect of fluid viscosity on the rate of entropy generation per unit volume of the packed bed in laminar flow (T = 298.15 K, d_p = 1 mm, ε = 0.40, ρ = 1.184 kg/m³). The data points are generated using Equation (50).

The effect of particle diameter on the rate of entropy generation in laminar flow through packed beds is shown in Figure 12. A sharp reduction in the rate of entropy generation occurs with the increase in the particle diameter. With the increase in the particle diameter, the number density of particles is decreased for the same bed porosity ε. It can be readily shown that the number of particles per unit volume of bed n_p and the fluid-solids contact area per unit volume of bed A_{solids} are:

$$n_p = \frac{6(1-\varepsilon)}{\pi d_p^3}; \ A_{solids} = \pi d_p^2 n_p = \frac{6(1-\varepsilon)}{d_p} \tag{69}$$

Equation (69) assumes that the particles are uniform spheres of diameter d_p. The decrease in the number density of particles with the increase in particle diameter results in a decrease in contact area between the fluid and solids resulting in a decrease in the resistance to fluid motion. Consequently the rate of mechanical energy dissipation into internal energy and hence the rate of entropy generation decrease with the increase in particle diameter.

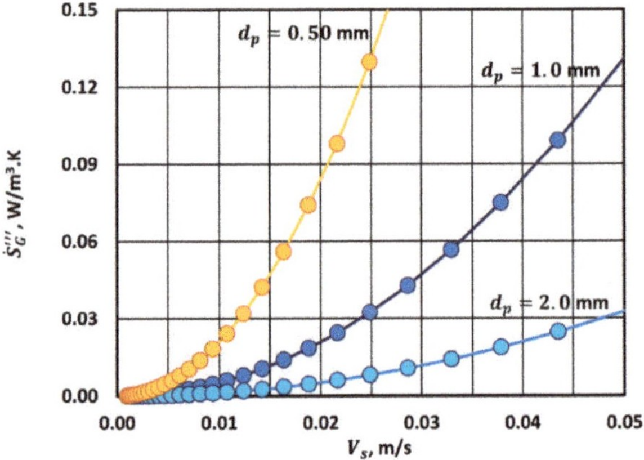

Figure 12. Effect of particle diameter on the rate of entropy generation per unit volume of the packed bed in laminar flow (T = 298.15 K, ε = 0.40, μ = 18.5 µPa·s, ρ = 1.184 kg/m³). The data points are generated using Equation (50).

Figure 13 shows the effect of temperature on the rate of entropy generation in laminar flow through packed beds. The rate of entropy generation decreases with the increase in temperature keeping other factors (fluid properties, bed characteristics) constant. At a given fluid velocity V_s, the rate of mechanical energy dissipation into internal energy is the same at different temperatures as the fluid properties are kept constant. Then why does the rate of entropy generation decrease with the increase in temperature? It so happens that the increase in entropy with the increase in internal energy is sensitive to temperature. For incompressible materials, the rate of change of entropy with respect to internal energy, that is, the derivative dS/dU, is given as:

$$\frac{dS}{dU} = \frac{1}{T} \tag{70}$$

Higher the temperature, smaller is the increase in entropy for the same increase in internal energy.

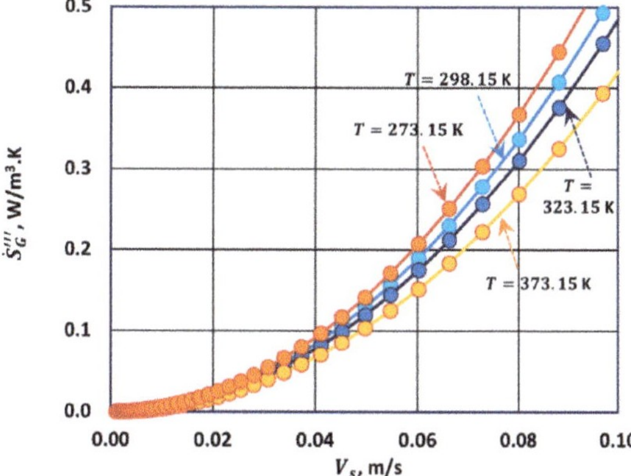

Figure 13. Effect of temperature on the rate of entropy generation per unit volume of the packed bed in laminar flow ($\varepsilon = 0.40$, $d_p = 1$ mm, $\mu = 18.5$ µPa·s, $\rho = 1.184$ kg/m³). The data points are generated using Equation (50).

4.1.2. Turbulent Flow

Figures 14–16 show the entropy generation rates for turbulent flow through packed beds under different conditions. Due to broad ranges of entropy generation rates and superficial fluid velocities, the plots are drawn using log-log scale. The plots on log-log scale are linear with a slope of 3 as expected from Equation (52). With the increases in bed porosity and particle diameter, the entropy generation rates decrease due to a decrease in the resistance to fluid motion and mechanical energy dissipation. There is no dependence of entropy generation on fluid viscosity in turbulent flow. However, fluid density now plays a role. With the increase in fluid density, the rate of entropy generation in turbulent flow increases due to an increase in the resistance to fluid motion and mechanical energy dissipation caused by inertial effects.

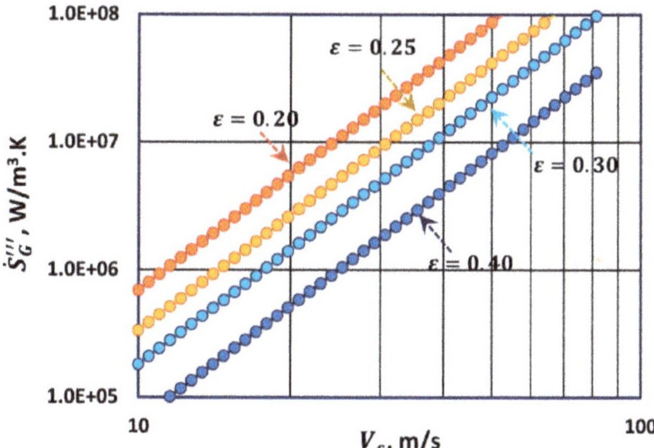

Figure 14. Effect of bed porosity on the rate of entropy generation per unit volume of the packed bed in turbulent flow ($T = 298.15$ K, $d_p = 1$ mm, $\mu = 18.5$ µPa·s, $\rho = 1.184$ kg/m³). The data points are generated using Equation (52).

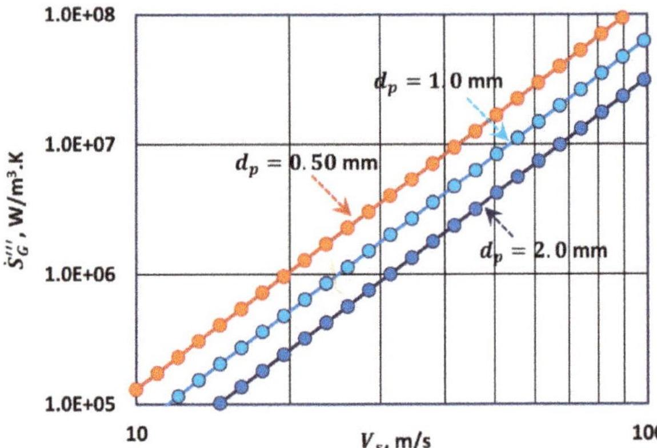

Figure 15. Effect of particle diameter on the rate of entropy generation per unit volume of the packed bed in turbulent flow (T = 298.15 K, ε = 0.40, μ = 18.5 µPa·s, ρ = 1.184 kg/m³). The data points are generated using Equation (52).

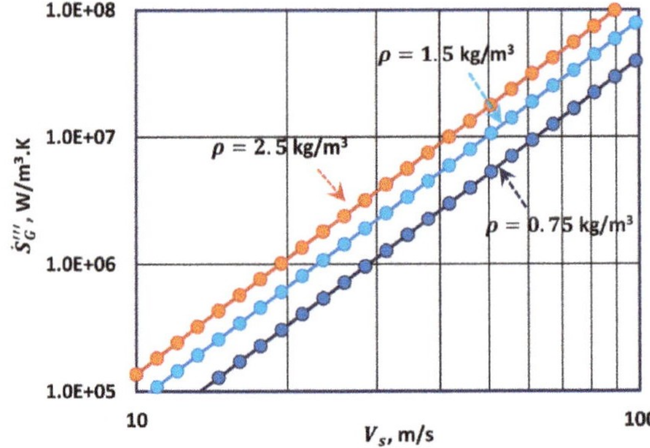

Figure 16. Effect of fluid density on the rate of entropy generation per unit volume of the packed bed in turbulent flow (T = 298.15 K, ε = 0.40, d_p = 1 mm, μ = 18.5 µPa·s). The data points are generated using Equation (52).

Figure 17 shows the entropy generation rate in a packed bed over a broad range of fluid superficial velocity covering the full range of packed bed Reynolds number. The plot is generated from Equation (53) under the following conditions: T = 298.15 K, ε = 0.40, d_p = 1 mm, μ = 18.5 µPa·s, and ρ = 1.184 kg/m³. The predictions of Equation (53) overlap with the two asymptotes, Equation (50) for low velocities and Equation (52) for high velocities.

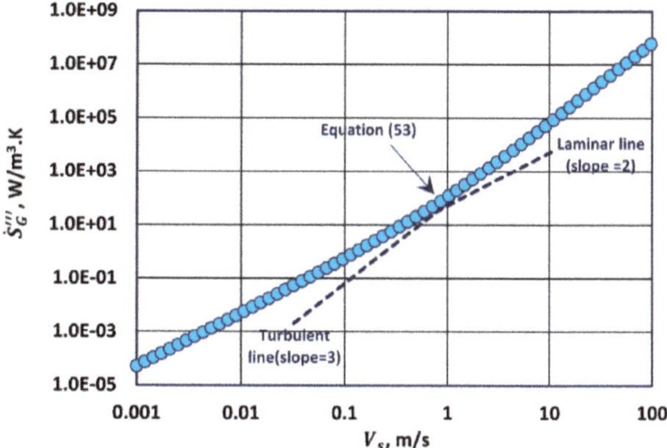

Figure 17. Prediction of $\dot{S}_G'''$ for packed beds over a broad range of fluid superficial velocity under the conditions: $T = 298.15$ K, $\varepsilon = 0.40$, $d_p = 1$ mm, $\mu = 18.5$ μPa·s, and $\rho = 1.184$ kg/m^3. The data points are generated using Equation (53).

4.2. Entropy Generation in Flow through Consolidated Porous Media

The entropy generation rate in laminar flow through consolidated porous medium is given by Equation (65). The key factor affecting the $\dot{S}_G'''$ vs. V_s behavior is the ratio of fluid viscosity to permeability, μ/k. Figure 18 shows the plots of $\dot{S}_G'''$ vs. V_s for different values of μ/k. The plots on a log-log scale are linear with slopes of 2. With the increase in μ/k ratio, the entropy generation rate increases at any given superficial velocity V_s. When the fluid viscosity is increased (keeping k the same), the entropy generation rate increases due to an increase in mechanical energy dissipation caused by viscous stresses. When the permeability k of the porous medium is decreased (keeping μ the same), the porous medium becomes less permeable to fluid flow resulting in larger resistance to fluid motion and hence larger rates of mechanical energy dissipation and entropy generation. Note that k is typically in the range of 10^{-15} to 10^{-12} m^2 for porous sandstones [20].

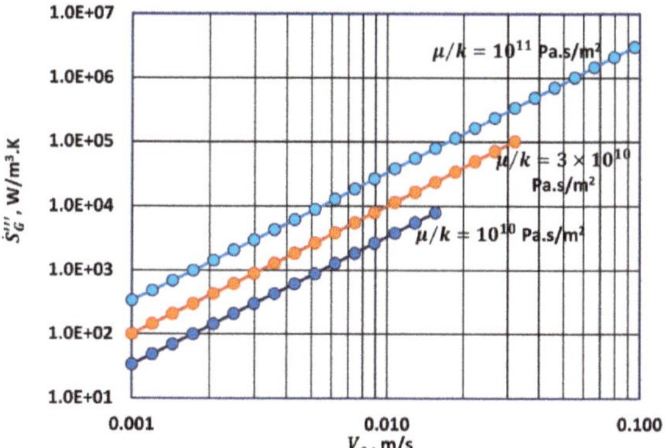

Figure 18. Effect of viscosity to permeability ratio (μ/k) on the rate of entropy generation per unit volume of consolidated porous medium in laminar flow ($T = 298.15$ K, $\beta = 10^8$ m^{-1}, $\rho = 10^3$ kg/m^3). The data points are generated using Equation (65).

In turbulent flow through consolidated porous medium, the key factor governing the $\dot{S}_G'''$ vs. V_s behavior is $\rho\beta$ where β is the non-Darcy flow coefficient. With the increase in $\rho\beta$, the entropy generation rate increases at any given V_s, as shown in Figure 19. Note that the non-Darcy flow coefficient is inversely related to the porous medium permeability [20]. For sandstones, β is typically in the range of 10^8 to 10^{12} m^{-1}. With the increase in β, the porous medium becomes less permeable to fluid flow resulting in an increase in the resistance to fluid motion. Consequently, the rates of mechanical energy dissipation and entropy generation increase with the increase in $\rho\beta$. The fluid density has a similar effect on entropy generation rate in the turbulent regime.

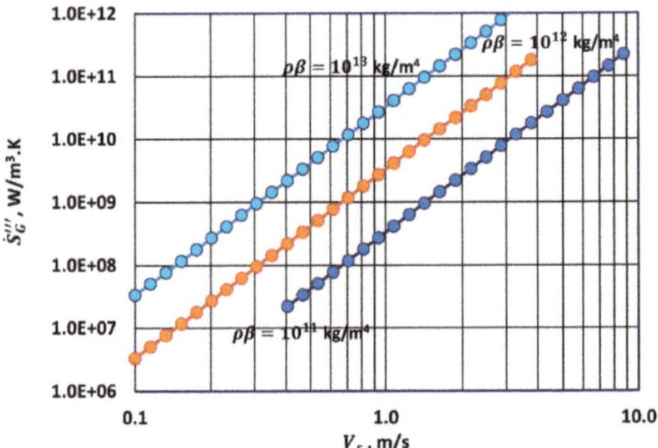

Figure 19. Effect of $\rho\beta$ on the rate of entropy generation per unit volume of consolidated porous medium in turbulent regime (T = 298.15 K, μ = 10 mPa·s, $k = 10^{-12}$ m^2). The data points are generated using Equation (67).

Figure 20 shows the entropy generation rate in a consolidated porous medium over a broad range of fluid superficial velocity ($10^{-4} \leq V_s \leq 10$ m/s) covering laminar, transition, and turbulent flow regimes. The plot is generated from Equation (68) under the following conditions: T = 298.15 K, μ = 1 mPa·s, ρ = 1000 kg/m^3, $k = 10^{-12}$ m^2, $\beta = 10^8$ m^{-1}. The predictions of Equation (68) overlap with the limiting low Re_{pm} asymptote (Equation (65)) at low superficial velocities and with the high Re_{pm} asymptote (Equation (67)) at high superficial velocities.

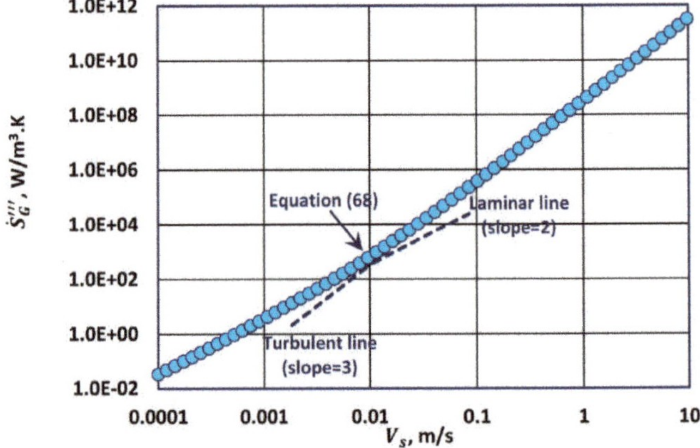

Figure 20. Prediction of $\dot{S}_G'''$ for consolidated porous medium over a broad range of fluid superficial velocity under the conditions: T = 298.15 K, μ = 1 mPa·s, ρ = 1000 kg/m^3, $k = 10^{-12}$ m^2, $\beta = 10^8$ m^{-1}. The data points are generated using Equation (68).

5. Conclusions

In conclusion, a novel approach is described to teach the second law of thermodynamics via the analysis of flow through packed beds and consolidated porous media. The second law of thermodynamics and the relevant background in fluid mechanics are reviewed briefly. The link between entropy generation and mechanical energy dissipation in flow through packed beds and consolidated porous media is established in terms of the directly measureable pressure loss. Equations are developed to predict the entropy generation rates in a porous medium in terms of the flow variables, fluid properties, and structural properties of the medium. Simulation results dealing with entropy generation in a porous medium are presented and discussed. The proposed approach can be implemented at an undergraduate level either as an experimental exercise dealing with pressure loss measurements in flow through a porous medium such as a packed bed or as a simulation exercise. The material presented in the article is suited for third year engineering students after they have completed introductory courses in fluid mechanics and thermodynamics.

Funding: This research received no external funding.

Conflicts of Interest: The author declares no conflict of interest.

References

1. Pal, R. Conceptual issues related to reversibility and reversible work in closed and open flow systems. *Educ. Chem. Eng.* **2017**, *19*, 29–37. [CrossRef]
2. Pal, R. Teaching fluid mechanics and thermodynamics simultaneously through pipeline flow experiments. *Fluids* **2019**, *4*, 103. [CrossRef]
3. Castellan, G.W. *Physical Chemistry*, 3rd ed.; Benjamin/Cummings Pub. Co: Menlo Park, CA, USA, 1983.
4. Bejan, A.; Tsatsaronis, G.; Moran, M. *Thermal Design and Optimization*; Wiley: New York, NY, USA, 1996.
5. Smith, J.M.; Van Ness, H.C.; Abbott, M.M. *Introduction to Chemical Engineering Thermodynamics*, 7th ed.; McGraw-Hill: New York, NY, USA, 2005.
6. Pal, R. On the Gouy–Stodola theorem of thermodynamics for open systems. *Int. J. Mech. Eng. Educ.* **2017**, *45*, 194–206. [CrossRef]
7. Bird, R.B.; Stewart, W.E.; Lightfoot, E.N. *Transport Phenomena*, 2nd ed.; Wiley: New York, NY, USA, 2007.
8. Wilkes, J.O. *Fluid Mechanics for Chemical Engineers*, 3rd ed.; Prentice Hall: Boston, MA, USA, 2018.
9. Chhabra, R.P.; Richardson, J.F. *Non-Newtonian Flow in Process Industries*; Butterworth Heinemann: Oxford, UK, 1999.
10. Barrande, M.; Bouchet, R.; Denoyel, R. Tortuosity of porous particles. *Anal. Chem.* **2007**, *79*, 9115–9121. [CrossRef] [PubMed]
11. Kruczek, B. Carman–Kozeny equation. In *Encyclopedia of Membranes*; Droli, E., Giorno, L., Eds.; Springer: Berlin, Germany, 2014.
12. Burke, S.P.; Plummer, W.B. Gas flow through packed columns. *Ind. Eng. Chem.* **1928**, *20*, 1196–1200. [CrossRef]
13. Ergun, S. Flow through packed columns. *Chem. Eng. Prog.* **1952**, *48*, 89–94.
14. Allen, K.G.; Von Backstrom, T.W.; Kroger, D.G. Packed bed pressure drop dependence on particle shape, size distribution, packing arrangement and roughness. *Powder Technol.* **2013**, *246*, 590–600. [CrossRef]
15. Tallmadge, J.A. Packed bed pressure drop—An extension to higher Reynolds numbers. *AIChE J.* **1970**, *16*, 1092–1093. [CrossRef]
16. Wentz, C.A.; Thodos, G. Pressure drops in the flow of gases through packed and distended beds of spherical particles. *AIChE J.* **1963**, *9*, 81–84. [CrossRef]
17. Mayerhofer, M.; Govaerts, J.; Parmentier, N.; Jeanmart, H.; Helsen, L. Experimental investigation of pressure drop in packed beds of irregular shaped wood particles. *Powder Technol.* **2011**, *205*, 30–35. [CrossRef]
18. Pesic, R.; Kaluderovic-Radoicic, T.K.; Boskovic-Vragolovic, N.; Arsenijevic, Z.; Grbavcic, Z. Pressure drop in packed beds of spherical particles at ambient and elevated air temperatures. *Chem. Ind. Chem. Eng.* **2015**, *21*, 419–427. [CrossRef]

19. De Schampheleire, S.; De Kerpel, K.; Ameel, B.; De Jaeger, P.; Bagci, O.; De Paepe, M. A discussion on the interpretation of the Darcy equation in case of open-cell metal foam based on numerical simulations. *Materials* **2016**, *9*, 409. [CrossRef] [PubMed]
20. Choi, C.S.; Song, J.J. Estimation of the Non-Darcy coefficient using supercritical CO_2 and various sandstones. *JCR Solid Earth* **2019**, *124*, 442–455.

 *fluids*

Article

CFD Julia: A Learning Module Structuring an Introductory Course on Computational Fluid Dynamics

Suraj Pawar and Omer San *

School of Mechanical and Aerospace Engineering, Oklahoma State University, Stillwater, OK 74078, USA
* Correspondence: osan@okstate.edu; Tel.: +1-405-744-2457; Fax: +1-405-744-7873

Received: 1 July 2019; Accepted: 19 August 2019; Published: 23 August 2019

Abstract: CFD Julia is a programming module developed for senior undergraduate or graduate-level coursework which teaches the foundations of computational fluid dynamics (CFD). The module comprises several programs written in general-purpose programming language Julia designed for high-performance numerical analysis and computational science. The paper explains various concepts related to spatial and temporal discretization, explicit and implicit numerical schemes, multi-step numerical schemes, higher-order shock-capturing numerical methods, and iterative solvers in CFD. These concepts are illustrated using the linear convection equation, the inviscid Burgers equation, and the two-dimensional Poisson equation. The paper covers finite difference implementation for equations in both conservative and non-conservative form. The paper also includes the development of one-dimensional solver for Euler equations and demonstrate it for the Sod shock tube problem. We show the application of finite difference schemes for developing two-dimensional incompressible Navier-Stokes solvers with different boundary conditions applied to the lid-driven cavity and vortex-merger problems. At the end of this paper, we develop hybrid Arakawa-spectral solver and pseudo-spectral solver for two-dimensional incompressible Navier-Stokes equations. Additionally, we compare the computational performance of these minimalist fashion Navier-Stokes solvers written in Julia and Python.

Keywords: CFD; Julia; numerical analysis; finite difference; spectral methods; multigrid

---

1. Introduction

Coursework in computational fluid dynamics (CFD) usually starts with an introduction to discretization techniques of partial differential equations (PDEs), stability analysis of numerical methods, the order of convergence, iterative methods for elliptic equations, shock-capturing methods for compressible flows, and development of incompressible flow solver. A lot of emphases is put on the theory of discretization, stability analysis, and different formulations of governing equations for compressible and incompressible flows. Students can gain a better understanding of CFD subject with actual hands-on programming of numerical solutions to mathematical models that present different fluid behavior. To develop a flow solver from scratch is a fairly complex task. Therefore, almost all developmental CFD courses begin with one-dimensional problems to explain theory and fundamentals and then build on to complicated problems like compressible or incompressible flow solvers.

CFD Julia is a programming module that contains several codes for problems ranging from one-dimensional heat equation to two-dimensional Navier-Stokes incompressible flow solver. Some of these problems can be included as part of programming assignments or coursework projects. We include different types of problems, different techniques to solve the same problem and try to introduce CFD using a practical approach. There are a couple of programming modules available online related to CFD. CFD Python learning module is a set of Jupyter notebooks that tries to teach

CFD in twelve steps [1]. This module also contains bonus modules for numerical stability analysis, and advanced programming in Python using Numpy [2]. It starts with an introduction to Python and ends with the development of incompressible flow solver for lid-driven cavity problem and channel flow. There is an online module available for a course on numerical methods covering finite difference methods [3]. The material for this coursework was created using IPython notebooks. HyperPython is a set of IPython notebooks created to teach hyperbolic conservation laws [4]. HyperPython module covers high-resolution methods for compressible flows and also teaches how one can use the PyClaw package to solve compressible flow problems. PyClaw is a hyperbolic PDE solver in 1D, 2D, and 3D and it has several features like parallel implementation and choice of higher-order accurate numerical methods [5].

In this work, we use Julia programming language [6] as a tool to develop codes for fundamental CFD problems. Julia is a new programming language which first appeared in the year 2012 and it has gained popularity in the scientific community in recent years. There is a number of programming languages that are already available such as Fortran [7], Python [8], C/C++ [9], Matlab [10], etc. Many of the state of the art CFD codes are written in Fortran and C/C++ because of their high-performance characteristics. Python and Matlab programming languages have simple syntax and students usually use these languages with ease in their coursework. Since the developer's language is different from the user's language, there will always be a hindrance to developing numerical codes. Many times a user has to resort to developer's language like C/C++ for getting the fast performance or have to use other packages to exchange information between codes written in different languages. Julia tries to solve this two-language problem. Julia language was designed to have the syntax that is easy to understand and will also have fast computational performance. Julia shows that one can have machine performance without sacrificing human convenience. Julia tries to achieve the Fortran performance through the automatic translation of formulas into efficient executable code. It allows the user to write clear, high-level, generic code that resembles mathematical formulas. Julia's ability to combine high-performance with productivity makes it the perfect choice for scientists working in different scientific domains.

In this paper, we use finite difference discretization for different types of PDEs encountered in CFD. For all problems, we provide the main script in Julia so that reader will get familiar with the Julia syntax. We start with the heat equation as the prototype of parabolic PDE. We present an explicit forward in time numerical scheme and implicit Crank-Nicolson numerical scheme for the heat equation. We also cover the multistage Runge-Kutta numerical approach to show how we can get higher temporal accuracy by taking multiple steps between two time intervals. We derive an implicit compact Pade scheme for the heat equation. This will give the reader an understanding of high-resolution numerical methods which are widely used in the direct numerical simulation (DNS) and large eddy simulation (LES). All fundamental concepts related to finite difference numerical methods are covered in Section 2 with the help of heat equation. This will lay the foundation of CFD and will familiarize the reader with basic concepts of discretization methods which will be used further for more complex problems.

We then present the hyperbolic equation in Section 3 using the inviscid Burgers equation. Hyperbolic equations allow having a discontinuous solution. Therefore, higher-order numerical methods have been developed that can capture discontinuities such as shocks. We present weighted essentially non-oscillatory (WENO) formulation for finite difference schemes and show their capability to capture shock using one-dimensional case. We show the mathematical formulation of WENO schemes, their implementation in Julia for two different boundary conditions. We use Dirichlet boundary condition and periodic boundary condition for the inviscid Burgers equation. This will help the reader to understand how to treat different boundary conditions in CFD problems. We also present a compact reconstruction of WENO scheme which has better accuracy and resolution characteristic than the WENO scheme with the price of solving a tridigonal system. Furthermore, we present discretization of the inviscid Burgers equation in its conservative form using different finite difference grid arrangement. This type of formulation is more applicable to finite volume method. We present a method of flux splitting and constructing flux at the interface using a Riemann solver approach in Section 4. We use a Riemann solver approach to solve the one-dimensional Euler equation and is presented in Section 5. The Euler equation solver is demonstrated for Sod shock tube problem using three different Riemann solvers: Rusanov scheme, Roe scheme, and Harten-Lax-van Leer Contact (HLLC) scheme.

We present different methods for solving elliptic partial differential equations using the Poisson equation as an example in Section 6. The Poisson equation is a boundary value problem and is encountered in solution to incompressible flow problems. We show different methods for solving the Poisson equation and their implementation in Julia. We demonstrate direct solver using fast Fourier transform (FFT) for periodic boundary condition problem and fast sine transform (FST) for Dirichlet boundary condition. We also present different iterative methods for solving elliptic equations. We demonstrate the implementation of Gauss-Seidel iterative method and the conjugate gradient method in Julia. The readers will learn the application of both stationary and non-stationary iterative methods. We also show the implementation of a V-cycle multigrid solver to accelerate the convergence of iterative methods.

We show solution to incompressible Navier-Stokes equation with vorticity-streamfunction formulation in Section 7. This is one of the simplest formulations to Navier-Stokes equation as there is no coupling between the velocity and pressure. This allows us to use collocated grid and not a staggered grid for the Navier-Stokes equation. We present a periodic boundary condition case using the vortex-merger problem. The Dirichlet boundary condition case is demonstrated using the lid-driven cavity problem. We use the direct solver to solve the elliptic equation giving the relation between vorticity and streamfunction. However, any of the iterative methods can also be used. We use second-order Arakawa numerical scheme which has better energy conservation properties for the discretization of the Jacobian term. The reader will get familiar with different types of numerical methods apart from standard finite difference schemes with these examples. This type of formulation can also be used for different types of problems such as two-dimensional decaying turbulence, and Taylor green vortex problem. Additionally, we present hybrid Arakawa-spectral solver for two-dimensional incompresible Navier-Stokes equations in Section 8. The flow solver is hybridized using explicit and implicit scheme for time integration. Also, the Navier-Stokes equations are solved in its Fourier space with the nonlinear term computed in physical space using Arakawa finite difference scheme and converted to Fourier space. Section 9 details the implementation fully pseudo-spectral solver for two-dimensional Navier-Stokes equations. We show two approaches to reduce aliasing error: 3/2 padding and 2/3 padding rules. We also show the comparison between CPU time for codes written in Julia and Python for solving the Navier-Stokes equations.

CFD Julia module covers several topics pertinent to both compressible and incompressible flows. This module provides different finite difference codes that will help students learn not just the fundamentals of CFD but also the implementation of numerical methods to solve CFD problems. Using these examples, students can go about developing their solvers for more challenging problems for their coursework or research. We make all these codes available on Github and are accessible to everyone (https://github.com/surajp92/CFD_Julia). Table 1 outlines all codes available in the CFD Julia module with their Github index. Table 1 presents a summary of topics covered and numerical schemes employed for problems included in the CFD Julia module. We write results for all problems into a text file and then use separate plotting scripts to plot results. We provide installation instruction for Julia and required packages, and plotting scripts in Appendices A and B.

Table 1. Summary of different problems included in the CFD Julia module and the description of topics covered in these selected problems. Source codes are accessible to everyone at https://github.com/surajp92/CFD_Julia.

Github Index	Description
01	1D heat equation: Forward time central space (FTCS) scheme
02	1D heat equation: Third-order Runge-Kutta (RK3) scheme
03	1D heat equation: Crank-Nicolson (CN) scheme
04	1D heat equation: Implicit compact Pade (ICP) scheme
05	1D inviscid Burgers equation: WENO-5 with Dirichlet and periodic boundary condition
06	1D inviscid Burgers equation: CRWENO-5 with Dirichlet and periodic boundary conditions
07	1D inviscid Burgers equation: Flux-splitting approach with WENO-5
08	1D inviscid Burgers equation: Riemann solver approach with WENO-5 using Rusanov solver
09	1D Euler equations: Roe solver, WENO-5, RK3 for time integration
10	1D Euler equations: HLLC solver, WENO-5, RK3 for time integration
11	1D Euler equations: Rusanov solver, WENO-5, RK3 for time integration
12	2D Poisson equation: Finite difference fast Fourier transform (FFT) based direct solver
13	2D Poisson equation: Spectral fast Fourier transform (FFT) based direct solver
14	2D Poisson equation: Fast sine transform (FST) based direct solver for Dirichlet boundary
15	2D Poisson equation: Gauss-Seidel iterative method
16	2D Poisson equation: Conjugate gradient iterative method
17	2D Poisson equation: V-cycle multigrid iterative method
18	2D incompressible Navier-Stokes equations (cavity flow): Arakawa, FST, RK3 schemes
19 *	2D incompressible Navier-Stokes equations (vortex merging): Arakawa, FFT, RK3 schemes
20	2D incompressible Navier-Stokes equations (vortex merging): Hybrid RK3/CN approach
21 *	2D incompressible Navier-Stokes equations (vortex merging): Hybrid pseudo-spectral 3/2 padding approach
22	2D incompressible Navier-Stokes equations (vortex merging): Hybrid pseudo-spectral 2/3 padding approach

* The codes for these problems are provided in Python in addition to Julia language.

2. Heat Equation

We start with the one-dimensional heat equation, which is one of the simplest and most basic models to introduce the concept of finite difference methods. The one-dimensional heat equation is given as

$$\frac{\partial u}{\partial t} = \alpha \frac{\partial^2 u}{\partial x^2}, \tag{1}$$

where u is the field variable, t is the time variable, and α is the diffusivity of the medium. The heat equation describes the evolution of the field variable over time in a medium.

There are different methods to discretize the heat equation, such as, finite volume method [11], finite difference method [12], finite element method [13], spectral methods [14], etc. In this paper, we will study the finite difference method to discretize all partial differential equations (PDEs). The finite difference method is comparatively simpler and is easy to understand in the initial stage of learning CFD. The finite difference method converts the linear or nonlinear PDEs into a set of linear or nonlinear ordinary differential equations (ODEs). These ODEs can be solved using matrix algebra techniques.

Before starting with the derivation of finite difference scheme for the heat equation, we define a grid of points in the (t, x) plane. Let Δx and Δt be the grid spacing in space and time direction, respectively. Therefore, $(t_n, x_i) = (n\Delta t, i\Delta x)$ for arbitrary integers n and i as shown in Figure 1. We use the notation $u_i^{(n)}$ for $u(t_n, x_i)$.

The finite difference method uses Taylor series expansion to derive the discrete approximation for PDEs. Taylor series gives the series expansion of a function f about a point. Taylor series expansion for a function f about the point x_0 is given by:

$$f(x_0 + \Delta x) = f(x_0) + \frac{\Delta x}{1!}\frac{\partial f(x_0)}{\partial x} + \frac{\Delta x^2}{2!}\frac{\partial^2 f(x_0)}{\partial x^2} + \cdots + \frac{\Delta x^n}{n!}\frac{\partial^n f(x_0)}{\partial x^n} + \mathcal{T}_n(x), \tag{2}$$

where $n!$ denotes the factorial of n, and $\mathcal{T}_n$ is a truncation term, giving the difference between the Taylor series expansion of degree n and the original function. The difference between the exact solution to the original differential equation and finite difference approximation is termed as the discretization or truncation error. The leading term in the discretization error determines the accuracy of the finite difference scheme. The finite difference scheme for any PDE must have two properties: consistency and stability to ensure the convergence. We would like to suggest a textbook "Finite difference schemes and partial differential equations" [15] which explains numerical behavior of finite difference scheme in detail. In this paper, we focus on the application of the finite difference method to study problems encountered in CFD.

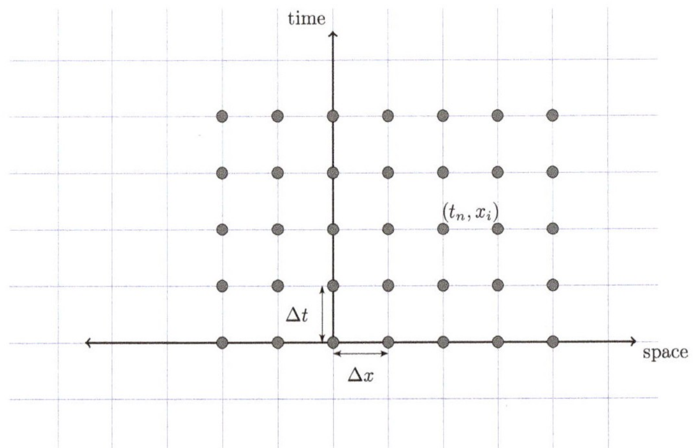

Figure 1. Finite difference grid for one-dimensional problems.

2.1. Forward Time Central Space (FTCS) Scheme

We derive the forward time central space (FTCS) numerical scheme for approximating the heat equation. Using the Taylor series expansion in time for the discrete field $u_i^{(n)}$, we get

$$u_i^{(n+1)} = u_i^{(n)} + \frac{\partial u_i^{(n)}}{\partial t}\Delta t + \mathcal{O}(\Delta t^2), \tag{3}$$

$$\frac{\partial u_i^{(n)}}{\partial t} = \frac{u_i^{(n+1)} - u_i^{(n)}}{\Delta t} + \mathcal{O}(\Delta t). \tag{4}$$

Equation (4) is the first-order accurate expression for the time derivative term in the heat equation. To obtain the finite difference formula for second-derivative term in the heat equation, we write the Taylor series expansion for $u_{i+1}^{(n)}$ and $u_{i-1}^{(n)}$ about $u_i^{(n)}$

$$u_{i+1}^{(n)} = u_i^{(n)} + \frac{\partial u_i^{(n)}}{\partial x}\Delta x + \frac{\partial^2 u_i^{(n)}}{\partial x^2}\frac{\Delta x^2}{2!} + \frac{\partial^3 u_i^{(n)}}{\partial x^2}\frac{\Delta x^3}{3!} + \mathcal{O}(\Delta x^4), \tag{5}$$

$$u_{i-1}^{(n)} = u_i^{(n)} - \frac{\partial u_i^{(n)}}{\partial x}\Delta x + \frac{\partial^2 u_i^{(n)}}{\partial x^2}\frac{\Delta x^2}{2!} - \frac{\partial^3 u_i^{(n)}}{\partial x^2}\frac{\Delta x^3}{3!} + \mathcal{O}(\Delta x^4). \tag{6}$$

Adding Equations (5) and (6) and dividing by Δx^2, we get the second-order accurate formula for second-derivative term in heat equation

$$\frac{\partial^2 u_i^{(n)}}{\partial x^2} = \frac{u_{i+1}^{(n)} - 2u_i^{(n)} + u_{i-1}^{(n)}}{\Delta x^2} + \mathcal{O}(\Delta x^2). \tag{7}$$

Substituting Equations (4) and (7) in Equation (1), we get the FTCS numerical scheme for the heat equation as given below

$$\frac{u_i^{(n+1)} - u_i^{(n)}}{\Delta t} = \alpha \frac{u_{i+1}^{(n)} - 2u_i^{(n)} + u_{i-1}^{(n)}}{\Delta x^2},\tag{8}$$

and we can re-write Equation (8) as an explicit update formula

$$u_i^{(n+1)} = u_i^{(n)} + \frac{\alpha \Delta t}{\Delta x^2}\left(u_{i+1}^{(n)} - 2u_i^{(n)} + u_{i-1}^{(n)}\right).\tag{9}$$

We denote the above formula as $\mathcal{O}(\Delta t, \Delta x^2)$ formula meaning that the leading term of the discretization error is first-order accurate in time and second-order accurate in space. We can obtain the higher-order formula using the larger stencil to get better approximations. If the initial solution to the heat equation is given, then we can proceed in time to get the solution at future times using Equation (9). We can observe from Equation (9), that the solution at the next time step depends only on the solution at previous time steps. These numerical schemes are called an explicit numerical scheme. They are simple to code. However, explicit numerical schemes are restricted by some stability criteria. The stability criteria restricts the maximum step size we can use in spatial and temporal direction and hence explicit numerical schemes are computationally expensive.

We will now demonstrate how one can go about finding the solution at the final time step given an initial condition. We use the computational domain $x \in [-1, 1]$ and $\alpha = 1/\pi^2$. The initial condition for our problem is $u(t = 0, x) = -\sin(\pi x)$. We use $\Delta x = 0.025$ and $\Delta t = 0.0025$ for spatial and temporal discretization. First, we initialize the array u and assign each element of the array using an initial condition. The solution after one time step is calculated using Equation (9) and we will store the solution in first column of matrix un[k,i] (variable in Julia) with the shape $(N_x + 1) \times (N_t + 1)$, where N_x is the number of divisions of the domain and N_t is the number of time steps between initial and final time. There is no need to store the solution at every time step, and we can just store the solution at intermediate time steps in a temporary array. Here, we store the solution at every time step to see how the field u evolves. The main script of the code which computes and stores the numerical solution at every time step is given in Listing 1.

We compare the solution at final time $t = 1$ using the analytical solution to the one-dimensional heat equation. The analytical solution for Equation (1) is given by

$$u(t, x) = -e^{-t}\sin(\pi x).\tag{10}$$

We also compute the absolute error at $t = 1$ using the below formula

$$\epsilon(x_i) = |u_i^{exact} - u_i^{numerical}|.\tag{11}$$

Listing 1. Implementation of FTCS numerical scheme for heat equation in Julia.

```julia
# nx: total number of grid points
# nt: total number of time steps between initial and final time
# dt, dx: time step and grid size
# beta: (alpha*dt)/(dx*dx)
# un: matrix for storing solution at every time step
for k = 2:nt+1
for i = 2:nx
un[k,i] = un[k-1,i] + beta*(un[k-1,i+1] - 2.0*un[k-1,i] + un[k-1,i-1])
end
un[k,1] = 0.0 # boundary condition at x = -1
un[k,nx+1] = 0.0 # boundary condition at x = 1
end
```

The comparison of the exact and numerical solution using the FTCS scheme and absolute error plot are shown in Figure 2. We see that we get a very good agreement between the exact and numerical solution due to very small grid size in both space and temporal direction. If we use a larger grid size, then we will see deprecation in accuracy.

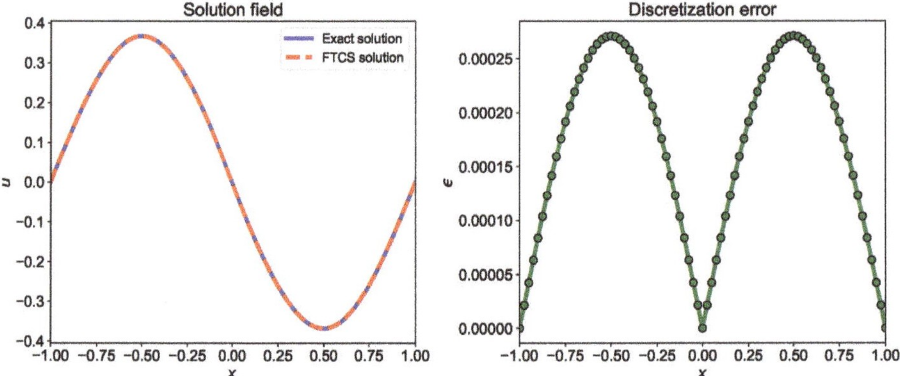

Figure 2. Comparison of exact solution and numerical solution computed using FTCS numerical scheme. The numerical solution is computed using $\Delta t = 0.0025$ and $\Delta x = 0.025$.

2.2. Runge-Kutta Numerical Scheme

We saw in Section 2.1 that the accuracy of the numerical scheme depends upon the number of terms included from the Taylor series expansion. One of the ways to increase accuracy is to include more terms. The additional term involves partial derivative of $f(x,t)$ which provides additional information on f at $t = t_n$. For the time stepping, we only have information about the data at one or more time steps and the analytical form of f is not known between two time steps. Runge-Kutta methods tries to improve the accuracy of temporal term by evaluating f at intermediate points between t_n and t_{n+1}. The additional steps lead to an increase in computational time, but the temporal accuracy is increased.

We use a third-order accurate Runge-Kutta scheme [16] for the discretization of the temporal term in the heat equation. We use the same second-order central difference scheme for the spatial term. The truncation error of this numerical approximation of heat equation is $\mathcal{O}(\Delta t^3, \Delta x^2)$. We move from time step t_n to t_{n+1} using three steps. The time integration of the heat equation using third-order Runge-Kutta scheme is given below:

$$u_i^{(1)} = u_i^{(n)} + \frac{\alpha \Delta t}{\Delta x^2} \left(u_{i+1}^{(n)} - 2u_i^{(n)} + u_{i-1}^{(n)} \right), \tag{12}$$

$$u_i^{(2)} = \frac{3}{4}u_i^{(n)} + \frac{1}{4}u_i^{(1)} + \frac{\alpha \Delta t}{4\Delta x^2} \left(u_{i+1}^{(n)} - 2u_i^{(n)} + u_{i-1}^{(n)} \right), \tag{13}$$

$$u_i^{(n+1)} = \frac{1}{3}u_i^{(n)} + \frac{2}{3}u_i^{(2)} + \frac{2\alpha \Delta t}{3\Delta x^2} \left(u_{i+1}^{(n)} - 2u_i^{(n)} + u_{i-1}^{(n)} \right). \tag{14}$$

We can also construct the higher-order Runge-Kutta numerical scheme using more intermediate steps. The implementation of the third-order Runge-Kutta numerical scheme in Julia is given in Listing 2. The comparison of the exact and numerical solutions using the Runge-Kutta scheme and absolute error plot is shown in Figure 3. The discretization error is less for Runge-Kutta numerical scheme compared to the FTCS numerical scheme due to a higher-order approximation of temporal term.

Listing 2. Implementation of third-order Runge-Kutta numerical scheme for heat equation in Julia.

```julia
# nx: total number of grid points
# nt: total number of time steps between initial and final time
# dt, dx: time step and grid size
# un: matrix for storing solution at every time step
for j = 2:nt+1
rhs(nx,dx,un,r,alpha)
for i = 2:nx
ut[i] = un[i] + dt*r[i] # 1st step
end
rhs(nx,dx,ut,r,alpha)
for i = 2:nx
ut[i] = 0.75*un[i] + 0.25*ut[i] + 0.25*dt*r[i] # 2nd step
end
rhs(nx,dx,ut,r,alpha)
for i = 2:nx
un[i] = (1.0/3.0)*un[i] + (2.0/3.0)*ut[i] + (2.0/3.0)*dt*r[i] # 3rd step
end
k = k+1 # index for solution storage at every time step
u[:,k] = un[:]
end

# function to compute the right hand side of heat equation
function rhs(nx,dx,u,r,alpha)
for i = 2:nx
r[i] = alpha*(u[i+1] - 2.0*u[i] + u[i-1])/(dx*dx)
end
end
```

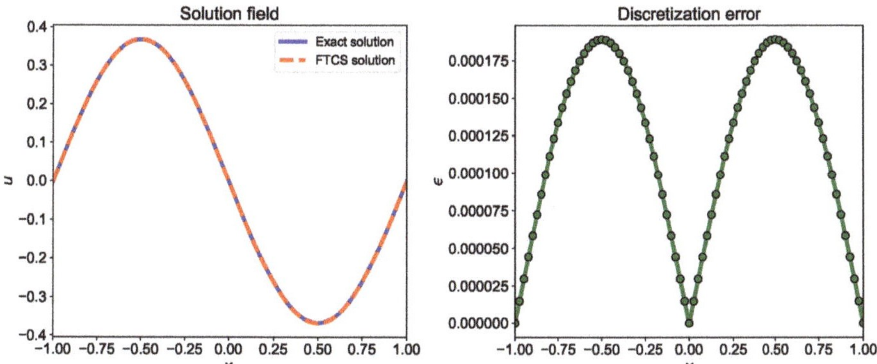

Figure 3. Comparison of exact solution and numerical solution computed using third-order Runge-Kutta numerical scheme. The numerical solution is computed using $\Delta t = 0.0025$ and $\Delta x = 0.025$.

2.3. Crank-Nicolson Scheme

The Crank-Nicolson scheme is a second-order accurate in time numerical scheme. The Crank-Nicolson scheme is derived by combining forward in time method at (n) and the backward in time method at $(n+1)$. Similar to Equation (3), we can write the Taylor series in time for the discrete field $u_i^{(n+1)}$ as

$$u_i^{(n)} = u_i^{(n+1)} - \frac{\partial u_i^{(n+1)}}{\partial t} \Delta t + \mathcal{O}(\Delta t^2), \tag{15}$$

$$\frac{\partial u_i^{(n+1)}}{\partial t} = \frac{u_i^{(n+1)} - u_i^{(n)}}{\Delta t} + \mathcal{O}(\Delta t). \tag{16}$$

The second-derivative term in heat equation at $(n+1)^{th}$ time step can be approximated using the second-order central difference scheme similar to the way we derived in Section 2.1

$$\frac{\partial^2 u_i^{(n+1)}}{\partial x^2} = \frac{u_{i+1}^{(n+1)} - 2u_i^{(n+1)} + u_{i-1}^{(n+1)}}{\Delta x^2} + \mathcal{O}(\Delta x^2). \tag{17}$$

Substituting finite difference approximation derived in Equations (16) and (17) in the heat equation, we get

$$\frac{u_i^{(n+1)} - u_i^{(n)}}{\Delta t} = \alpha \frac{u_{i+1}^{(n+1)} - 2u_i^{(n+1)} + u_{i-1}^{(n+1)}}{\Delta x^2}. \tag{18}$$

If we add Equations (8) and (18), the leading term in discretization error cancels each other due to their opposite sign and same magnitude. Hence, we get the second-order accurate in time numerical scheme. The Crank-Nicolson scheme is $\mathcal{O}(\Delta t^2, \Delta x^2)$ numerical scheme. The Crank-Nicolson scheme is given as

$$\frac{u_i^{(n+1)} - u_i^{(n)}}{\Delta t} = \frac{\alpha}{2} \left[\frac{u_{i+1}^{(n)} - 2u_i^{(n)} + u_{i-1}^{(n)}}{\Delta x^2} + \frac{u_{i+1}^{(n+1)} - 2u_i^{(n+1)} + u_{i-1}^{(n+1)}}{\Delta x^2} \right]. \tag{19}$$

We observe from the above equation that the solution at $(n+1)^{th}$ time step depends on the solution at previous time step $(n)^{th}$ and the solution at $(n+1)^{th}$ time step. Such numerical schemes are called implicit numerical schemes. For implicit numerical schemes, the system of algebraic equations has to be solved by inverting the matrix. For one-dimensional problems, the tridiagonal matrix is formed and can be solved efficiently using the Thomas algorithm, which gives a fast $\mathcal{O}(N)$ direct solution as opposed to the usual $\mathcal{O}(N^3)$ for a full matrix, where N is the size of the square tridiagonal matrix. The main advantage of an implicit numerical scheme is that they are unconditionally stable. This allows the use of larger time step and grid size without affecting the convergence of the numerical scheme. An illustration of a tridiagonal system is given in Equation (20). The Thomas algorithm [17] used for solving the tridiagonal matrix is outlined in Algorithm 1.

$$\underbrace{\begin{bmatrix} b_1 & c_1 & 0 & \cdots & 0 \\ a_2 & b_2 & c_2 & \cdots & 0 \\ \vdots & \vdots & & \ddots & \vdots \\ 0 & \cdots & a_{N-1} & b_{N-1} & c_{N-1} \\ 0 & 0 & \cdots & a_N & b_N \end{bmatrix}}_{A} \underbrace{\begin{bmatrix} u_1 \\ u_2 \\ \vdots \\ u_{N-1} \\ u_N \end{bmatrix}}_{u} = \underbrace{\begin{bmatrix} d_1 \\ d_2 \\ \vdots \\ d_{N-1} \\ d_N \end{bmatrix}}_{d}. \tag{20}$$

Algorithm 1 Thomas algorithm

1: Given a, b, c, d $\triangleright\ Au = d$
2: Allocate q $\triangleright$ Storage of superdiagonal array
3: $u_1 = d_1/b_1$
4: **for** $i = 2$ to N **do** $\triangleright$ Forward elimination
5:      $q_i = c_{i-1}/b_{i-1}$
6:      $b_i = b_i - q_i a_i$
7:      $u_i = (d_i - a_i u_{i-1})/b_i$
8: **end for**
9: **for** $i = N - 1$ to 1 **do** $\triangleright$ Backward substitution
10:      $u_i = (u_i - q_{i+1} u_{i+1})$
11: **end for**

The implementation of the Crank-Nicolson numerical scheme for the heat equation is presented in Listing 3 and Thomas algorithm in Listing 4. The comparison of exact and numerical solution computed using the Crank-Nicolson scheme, and absolute error plot are displayed in Figure 4. The absolute error between the exact and numerical solution is less for the Crank-Nicolson scheme than the FTCS numerical scheme due to smaller truncation error.

Listing 3. Implementation of Crank-Nicolson numerical scheme for heat equation in Julia.

```julia
# nx: total number of grid points
# nt: total number of time steps between initial and final time
# dt, dx: time step and grid size
# beta: (alpha*dt)/(dx*dx)
# p: temporary storage variable
# un: matrix for storing solution at every time step
s, e = 1, nx
for k = 2:nt+1
i = 1 # left boundary
a[i], b[i], c[i], d[i] = 0.0, 1.0, 0.0, 0.0
for i = 2:nx # interior points
a[i], b[i], c[i] = -beta, 1.0+2.0*beta, -beta
d[i] = beta*un[k-1,i+1] + (1.0-2.0*beta)*un[k-1,i] + beta*un[k-1,i-1]
end
i = nx+1 # right boundary
a[i], b[i], c[i], d[i] = 0.0, 1.0, 0.0, 0.0
tdms(a,b,c,r,p,s,e) # call Thomas algorithm function to calculate p
un[k,:] = p
end
```

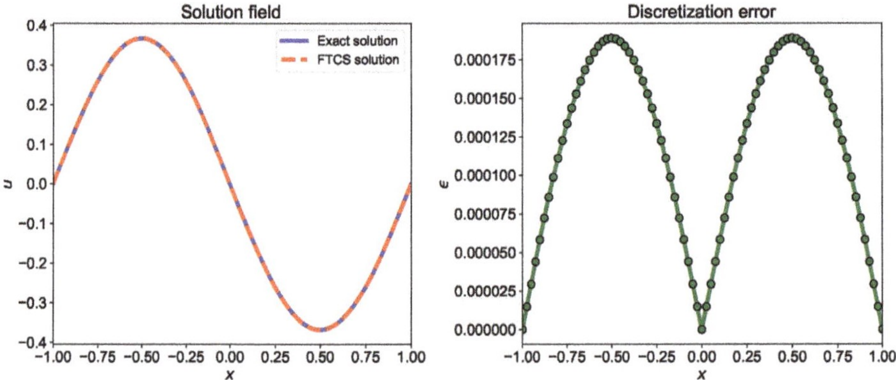

Figure 4. Comparison of exact solution and numerical solution computed using Crank-Nicolson numerical scheme. The numerical solution is computed using $\Delta t = 0.0025$ and $\Delta x = 0.025$.

Listing 4. Implementation of Thomas algorithm to solve tridiagonal system of equations in Julia.

```julia
# a, b, c: subdiagonal, diagonal, superdiagonal entries of tridiagonal matrix
# r: right hand side of tridiagonal system
# x: solution to the tridiagonal system
# s,e: start and end index of the matrix
function tdms(a,b,c,r,x,s,e)
gam = Array{Float64}(undef, e)
bet = b[s]
x[s] = r[s]/bet

for i = s+1:e
gam[i] = c[i-1]/bet
bet = b[i] - a[i]*gam[i]
x[i] = (r[i] - a[i]*x[i-1])/bet
end

for i = e-1:-1:s
x[i] = x[i] - gam[i+1]*x[i+1]
end
return x
end
```

2.4. Implicit Compact Pade (ICP) Scheme

We usually use the order of accuracy as an indicator of the ability of finite difference schemes to approximate the exact solution as it tells us how the discretization error will reduce with mesh refinement. Another way of measuring the order of accuracy is the modified wavenumber approach. In this approach, we see how much the modified wave number is different from the true wave number. The solution of a nonlinear partial differential equation usually contains several frequencies and the modified wavenumber approach provides a way to assess how well the different components of the solution are represented. The modified wavenumber varies for every finite difference scheme and can be found using Fourier analysis of the differencing scheme [12].

Compact finite difference schemes have very good resolution characteristics and can be used for capturing high-frequency waves [18]. In compact formulation, we express the derivative of a function as a linear combination of values of function defined on a set of nodes. The compact method tries to mimic global dependence and hence has good resolution characteristics. Lele [18] investigated the behavior of compact finite difference schemes for approximating a first and second derivative. The fourth-order accurate approximation to the second derivative is given by

$$\frac{1}{12}\frac{\partial^2 u}{\partial x^2}\bigg|_{i-1} + \frac{10}{12}\frac{\partial^2 u}{\partial x^2}\bigg|_{i} + \frac{1}{12}\frac{\partial^2 u}{\partial x^2}\bigg|_{i+1} = \frac{u_{i-1} - 2u_i + u_{i+1}}{\Delta x^2}. \tag{21}$$

Taking the linear combination of heat equation and using the same coefficients as the above equation, we get

$$\frac{1}{12}\frac{\partial u}{\partial t}\bigg|_{i-1} + \frac{10}{12}\frac{\partial u}{\partial t}\bigg|_{i} + \frac{1}{12}\frac{\partial u}{\partial t}\bigg|_{i+1} = \frac{u_{i-1} - 2u_i + u_{i+1}}{\Delta x^2}. \tag{22}$$

We then use Crank-Nicolson numerical scheme for time discretization and we arrive to below equation

$$\frac{u_{i-1}^{(n+1)} - u_{i-1}^{(n)}}{12\Delta t} + 10\frac{u_{i}^{(n+1)} - u_{i}^{(n)}}{12\Delta t} + \frac{u_{i+1}^{(n+1)} - u_{i+1}^{(n)}}{12\Delta t} = \frac{u_{i-1}^{(n+1)} - 2u_{i}^{(n+1)} + u_{i+1}^{(n+1)} + u_{i-1}^{(n)} - 2u_{i}^{(n)} + u_{i+1}^{(n)}}{2\Delta x^2}. \tag{23}$$

Simplifying above equation, we get the implicit compact Pade (ICP) scheme for heat equation

$$a_i u_{i-1}^{(n+1)} + b_i u_{i}^{(n+1)} + c_i u_{i+1}^{(n+1)} = r_i^{(n)}, \tag{24}$$

where

$$a_i = \frac{12}{\Delta x^2} - \frac{2}{\alpha \Delta t}, \tag{25}$$

$$b_i = \frac{-24}{\Delta x^2} - \frac{20}{\alpha \Delta t}, \tag{26}$$

$$c_i = \frac{12}{\Delta x^2} - \frac{2}{\alpha \Delta t}, \tag{27}$$

$$r_i = \frac{-2}{\alpha \Delta t}\left(u_{i-1}^{(n+1)} - 2u_{i}^{(n+1)} + u_{i+1}^{(n+1)}\right) - \frac{12}{\Delta x^2}\left(u_{i-1}^{(n)} - 2u_{i}^{(n)} + u_{i+1}^{(n)}\right). \tag{28}$$

The ICP numerical scheme forms the tridiagonal matrix which can be solved using the Thomas algorithm to get the solution. The implementation of the ICP numerical scheme in Julia is given in Listing 5. Figure 5 shows the comparison of the exact and numerical solution and discretization error for ICP scheme. The ICP scheme is $\mathcal{O}(\Delta t^2, \Delta x^4)$ accurate scheme and hence, the discretization error is very small.

Listing 5. Implementation of implicit compact scheme Pade numerical scheme for heat equation in Julia.

```julia
# nx: total number of grid points
# nt: total number of time steps between initial and final time
# dt, dx: time step and grid size
# un: matrix for storing solution at every time step
s, e = 1, nx
for k = 2:nt+1
i = 1 # left boundary
a[i], b[i], c[i], d[i] = 0.0, 1.0, 0.0, 0.0
for i = 2:nx # interior points
a[i] = 12.0/(dx*dx)-2.0/(alpha*dt)
b[i] = -24.0/(dx*dx) - 20.0/(alpha*dt)
c[i] = 12.0/(dx*dx)-2.0/(alpha*dt)
d[i] = -2.0/(alpha*dt)*(un[k-1,i+1] + 10.0*un[k-1,i] + un[k-1,i-1])
-12.0/(dx*dx)*(un[k-1,i+1] -2.0*un[k-1,i] + un[k-1,i-1])
end
```

```
i = nx+1 # right boundary
a[i], b[i], c[i], d[i] = 0.0, 1.0, 0.0, 0.0
un[k,:] = tdma(a,b,c,d,s,e) # thomas algorithm function
end
```

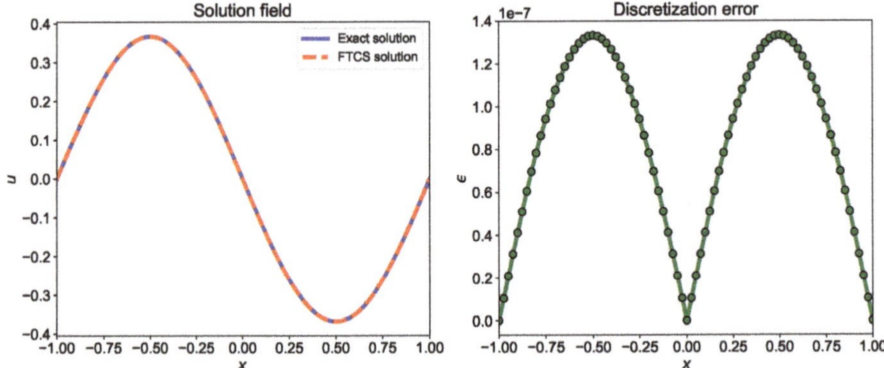

Figure 5. Comparison of exact solution and numerical solution computed using implicit compact Pade numerical scheme. The numerical solution is computed using $\Delta t = 0.0025$ and $\Delta x = 0.025$.

3. Inviscid Burgers Equation: Non-Conservative Form

In this section, we discuss the solution of the inviscid Burgers equation which is a nonlinear hyperbolic partial differential equation. The hyperbolic equations admit discontinuities, and the numerical schemes used for solving hyperbolic PDEs need to be higher-order accurate for smooth solutions, and non-oscillatory for discontinuous solutions [19]. The inviscid Burgers equation is given below

$$\frac{\partial u}{\partial t} + u\frac{\partial u}{\partial x} = 0. \tag{29}$$

First we use FTCS numerical scheme for discretization of the inviscid Burgers equation as follow

$$\frac{u_i^{(n+1)} - u_i^{(n)}}{\Delta t} + u_i^{(n)}\frac{u_{i+1}^{(n)} - u_{i-1}^{(n)}}{2\Delta x} = 0. \tag{30}$$

We use the shock generated by a sine wave to demonstrate the capability of FTCS numerical scheme to compute the numerical solution. The initial condition is $u_0 = \sin(2\pi x)$. We integrate the solution from time $t = 0$ to $t = 0.25$ with $\Delta t = 0.0001$ and divide the computational domain into 200 grids. The numerical solution computed by FTCS at 10 equally spaced time steps between initial and final time is shown in Figure 6. It can be observed that the FTCS scheme does not capture the shock formed at $x = 0.5$ and produce oscillations in the solution. It can be demonstrated that FTCS scheme is numerically unstable for advection equations.

There are several ways in which the discontinuity in the solution (e.g., shock) can be handled. There are classical shock-capturing methods like MacCormack method, Lax-Wanderoff method, and Beam-Warming method as well as flux limiting methods [20,21]. We can also use higher-order central difference schemes with artificial dissipation [22]. Low pass filtering might also help to diminish the growth of the instability modes [23]. There is also a class of modern shock capturing methods called as high-resolution schemes. These high-resolution schemes are generally upwind biased and take the direction of the flow into account. Here, in particular, we discuss higher-order weighted essentially non-oscillatory (WENO) schemes [24] that are used widely as shock-capturing numerical methods.

We also use compact reconstruction weighted essentially non-oscillatory (CRWENO) schemes [25] which have lower truncation error compared to WENO schemes.

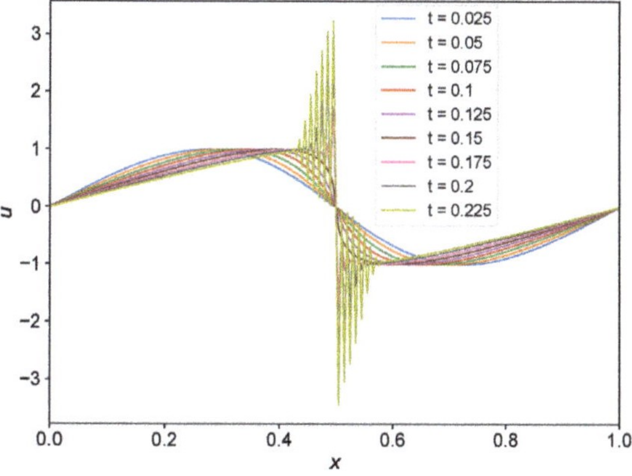

Figure 6. Evolution of shock with time for the initial condition $u_0 = \sin(2\pi x)$ using FTCS scheme.

3.1. WENO-5 Scheme

We refer readers to a text by Shu [26] for a comprehensive overview of the development of WENO schemes, their mathematical formulation, and the application of WENO schemes for convection dominated problems. We will use WENO-5 scheme for one-dimensional inviscid Burgers equation with the shock developed by a sine wave. We will discuss how one can go about applying WENO reconstruction for finite difference schemes. The finite difference approximation to the inviscid Burgers equation in non-conserved form can be written as

$$\frac{\partial u_i}{\partial t} + u_i \frac{u^L_{i+\frac{1}{2}} - u^L_{i-\frac{1}{2}}}{\Delta x}, \quad \text{if} \quad u_i > 0 \tag{31}$$

$$\frac{\partial u_i}{\partial t} + u_i \frac{u^R_{i+\frac{1}{2}} - u^R_{i-\frac{1}{2}}}{\Delta x}, \quad \text{otherwise,} \tag{32}$$

which can be combined into a more compact upwind/downwind notation as follows

$$\frac{\partial u_i}{\partial t} + u_i^+ \frac{u^L_{i+\frac{1}{2}} - u^L_{i-\frac{1}{2}}}{\Delta x} + u_i^- \frac{u^R_{i+\frac{1}{2}} - u^R_{i-\frac{1}{2}}}{\Delta x} = 0, \tag{33}$$

where we define $u_i^+ = \max(u_i, 0)$ and $u_i^- = \min(u_i, 0)$. the reconstruction of $u^L_{i+\frac{1}{2}}$ uses a biased stencil with one more point to the left, and that for $u^R_{i+\frac{1}{2}}$ uses a biased stencil with one more point to the right. The stencil used for reconstruction of left and right side state (i.e., $u^L_{i+1/2}$ and $u^R_{i+1/2}$) is shown in Figure 7. The WENO-5 reconstruction for left and right side reconstructed values is given as [24]

$$u^L_{i+\frac{1}{2}} = w^L_0 \left(\frac{1}{3}u_{i-2} - \frac{7}{6}u_{i-1} + \frac{11}{6}u_i \right) + w^L_1 \left(-\frac{1}{6}u_{i-1} + \frac{5}{6}u_i + \frac{1}{3}u_{i+1} \right) + w^L_2 \left(\frac{1}{3}u_i + \frac{5}{6}u_{i+1} - \frac{1}{6}u_{i+2} \right), \tag{34}$$

$$u^R_{i-\frac{1}{2}} = w^R_0 \left(-\frac{1}{6}u_{i-2} + \frac{5}{6}u_{i-1} + \frac{1}{3}u_i \right) + w^R_1 \left(\frac{1}{3}u_{i-1} + \frac{5}{6}u_i - \frac{1}{6}u_{i+1} \right) + w^R_2 \left(\frac{11}{6}u_i - \frac{7}{6}u_{i+1} + \frac{1}{3}u_{i+2} \right), \tag{35}$$

where nonlinear weights are defined as

$$w^L_k = \frac{\alpha_k}{\alpha_0 + \alpha_1 + \alpha_2}, \quad \alpha_k = \frac{d^L_k}{(\beta_k + \epsilon)^2}, \quad k = 0, 1, 2 \tag{36}$$

$$w_k^R = \frac{\alpha_k}{\alpha_0 + \alpha_1 + \alpha_2}, \quad \alpha_k = \frac{d_k^R}{(\beta_k + \epsilon)^2}, \quad k = 0, 1, 2 \tag{37}$$

in which the smoothness indicators are defined as

$$\beta_0 = \frac{13}{12}(u_{i-2} - 2u_{i-1} + u_i)^2 + \frac{1}{4}(u_{i-2} - 4u_{i-1} + 3u_i)^2, \tag{38}$$

$$\beta_1 = \frac{13}{12}(u_{i-1} - 2u_i + u_{i+1})^2 + \frac{1}{4}(u_{i-1} - u_{i+1})^2, \tag{39}$$

$$\beta_2 = \frac{13}{12}(u_i - 2u_{i+1} + u_{i+2})^2 + \frac{1}{4}(3u_i - 4u_{i+1} + 3u_{i+2})^2. \tag{40}$$

We use $d_0^L = 1/10$, $d_1^L = 3/5$, and $d_2^L = 3/10$ in Equation (36) to compute nonlinear weights for calculation of left side state using Equation (34) and $d_0^R = 3/10$, $d_1^R = 3/5$, and $d_2^R = 1/10$ in Equation (37) to compute nonlinear weights for calculation of right side state using Equation (35). We set $\epsilon = 1 \times 10^{-6}$ to avoid division by zero. To avoid repetition, we provide only the implementation of Julia function to compute the left side state in Listing 6. The right side state is computed in a similar way with coefficients for d_k^R in Equation (37).

We perform the time integration using third-order Runge-Kutta numerical scheme as discussed in Section 2.2 and its implementation is similar to Listing 2. The right hand side term of the inviscid Burgers equation is computed using the point value u_i and the derivative term is evaluated using the left and right side states at the interface. Based on the direction of flow i.e., the sign of u_i, we either use left side state or right side state to compute the derivative. The implementation of right hand side state calculation in Julia for the inviscid Burgers equation is given in Listing 7.

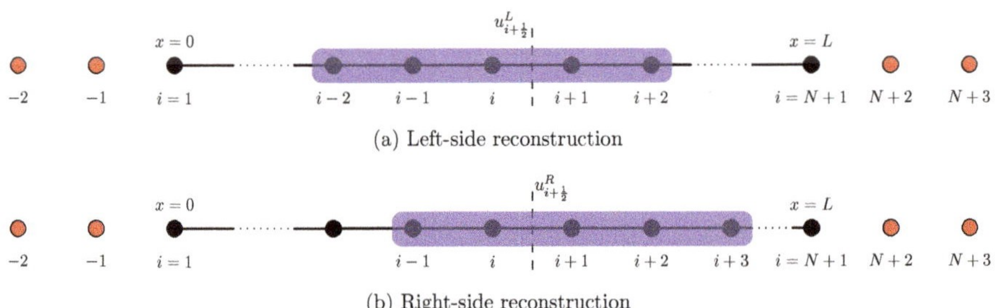

Figure 7. Finite difference grid for one-dimensional problems where the the solution is stored at the interface of the cells. The first and last points lie on the left and right boundary of the domain. The stencil required for reconstruction of left and right side state is highlighted with blue rectangle. Ghost points are shown by red color.

Listing 6. Implementation of left side state calculation function in Julia.

```julia
# v1,v2,v3,v4,v5: point values of v in the stencil (v3 = v[i])
# f: left side numerical reconstruction at (i+1/2) interface
function wcL(v1,v2,v3,v4,v5)
eps = 1.0e-6

# smoothness indicators
s1 = (13.0/12.0)*(v1-2.0*v2+v3)^2 + 0.25*(v1-4.0*v2+3.0*v3)^2
s2 = (13.0/12.0)*(v2-2.0*v3+v4)^2 + 0.25*(v2-v4)^2
s3 = (13.0/12.0)*(v3-2.0*v4+v5)^2 + 0.25*(3.0*v3-4.0*v4+v5)^2

# computing nonlinear weights w1,w2,w3
c1 = 1.0e-1/((eps+s1)^2)
c2 = 6.0e-1/((eps+s2)^2)
```

```julia
c3 = 3.0e-1/((eps+s3)^2)

w1 = c1/(c1+c2+c3)
w2 = c2/(c1+c2+c3)
w3 = c3/(c1+c2+c3)

# candiate stencils
q1 = v1/3.0 - 7.0/6.0*v2 + 11.0/6.0*v3
q2 =-v2/6.0 + 5.0/6.0*v3 + v4/3.0
q3 = v3/3.0 + 5.0/6.0*v4 - v5/6.0

# reconstructed value at interface
f = (w1*q1 + w2*q2 + w3*q3)

return f
end
```

Listing 7. Implementation of right hand side calculation for the inviscid Burgers equation in Julia.

```julia
# nx: number of grid points in domain
# dx:  grid spacing
# u: point values of solution field
# r: right hand side of Burgers equation r = -udu/dx
function rhs(nx,dx,u,r)
uL = Array{Float64}(undef, nx) # left side state
uR = Array{Float64}(undef, nx+1) # right side~state

wenoL(nx,u,uL) # compute left side state at the interface
wenoR(nx,u,uR) # compute right side state at the~interface

for i = 2:nx
if (u[i] >= 0.0)
r[i] = -u[i]*(uL[i] - uL[i-1])/dx
else
r[i] = -u[i]*(uR[i+1] - uR[i])/dx
end
end
end
```

We demonstrate two types of boundary conditions for the inviscid Burgers equation. The first one is the Dirichlet boundary condition. For Dirichlet boundary condition, we use $u(x = 0) = 0$ and $u(x = 1) = 0$. For computing numerical state at the interface, we need the information of five neighboring grid points (see Figure 7). When we compute the left side state at $i + \frac{1}{2}$ interface, we use three points on left side (i.e., u_{i-2}, u_{i-1}, u_i) and two points on right side (i.e., u_{i+1}, u_{i+2}). Therefore, we use two ghost points on the left side of the domain and one ghost point on right side of the domain for computing left side state (i.e., u^L at $x = 3/2$ and $x = N - 1/2$ respectively). Similarly, we use one ghost points on the left side of the domain and two ghost point on right side of the domain for computing right side state (i.e., u^R at $x = 3/2$ and $x = N - 1/2$ respectively). We use linear interpolation to compute the value of discrete field u at ghost points. The computation of u at ghost points is given below

$$u_{-2} = 3u_1 - 2u_2, \qquad u_{-1} = 2u_1 - u_2, \tag{41}$$

$$u_{N+3} = 3u_{N+1} - 2u_N, \quad u_{N+2} = 2u_{N+1} - u_N, \tag{42}$$

where u is stored from 1 to $N+1$ from left ($x = 0$) to right boundary ($x = L$) on the computational domain respectively (The default indexing in Julia for an Array starts with 1). The Julia function to select the stencil for left side state calculation is given in Listing 8. A similar function is also used for selecting the stencil for right side state calculation.

Listing 8. Implementation of Julia function to select the stencil for computing left side numerical state at the interface for Dirichlet boundary condition.

```julia
# n: number of grid points in domain
# u: point variable stored at the center of cells
# f: array of numerical state at the interface between two cells
function wenoL(n,u,f)
a = Array{Float64}(undef, n)
b = Array{Float64}(undef, n)
c = Array{Float64}(undef, n)
r = Array{Float64}(undef, n)

i = 1
v1 = 3.0*u[i] - 2.0*u[i+1]
v2 = 2.0*u[i] - u[i+1]
v3, v4, v5 = u[i], u[i+1], u[i+2]
f[i] = wcL(v1,v2,v3,v4,v5)

i = 2
v1 = 2.0*u[i-1] - u[i]
v2, v3, v4, v5 = u[i-1], u[i], u[i+1], u[i+2]
f[i] = wcL(v1,v2,v3,v4,v5)

for i = 3:n-1
v1, v2, v3, v4, v5 = u[i-2], u[i-1], u[i], u[i+1], u[i+2]
f[i] = wcL(v1,v2,v3,v4,v5)
end

i = n
v1, v2, v3, v4 = u[i-2], u[i-1], u[i], u[i+1]
v5 = 2.0*u[i+1]-u[i]
f[i] = wcL(v1,v2,v3,v4,v5)
end
```

We also use the periodic boundary condition for the same problem. For periodic boundary condition, we do not need to use any interpolation formula to compute the variable at ghost points. The periodic boundary condition for two left and right side points outside the domain is given below

$$u_{-2} = u_{N-1}, \quad u_{-1} = u_N, \quad u_{N+2} = u_2, \quad u_{N+3} = u_3, \tag{43}$$

where u is stored from 1 to $N+1$ from left ($x = 0$) to right ($x = L$) boundary on the computational domain, respectively.

The evolution of shock generated from sine wave with time is shown in Figure 8 for both Dirichlet and periodic boundary conditions. We use the domain between $x \in [0, 1]$ and divide it into 200 grids. We use the time step $\Delta t = 0.0001$ and integrate the solution from $t = 0$ to $t = 0.25$. We save snapshots of the solution field at different time steps to see the evolution of the shock. It can be seen from Figure 8 that the WENO-5 scheme is able to capture the shock formed at $x = 0.5$ at final time $t = 0.25$.

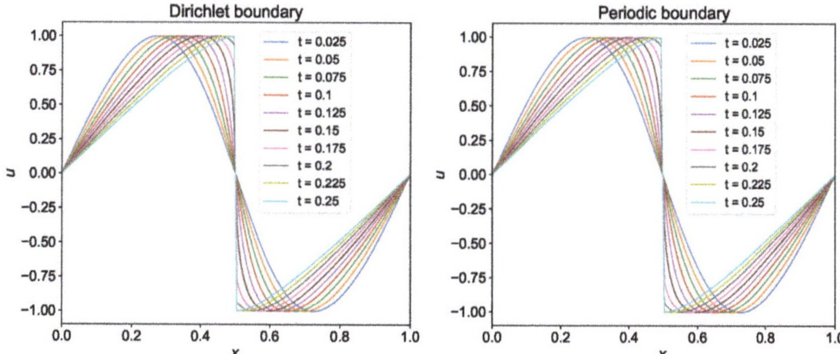

Figure 8. Evolution of shock with time for the initial condition $u_0 = \sin(2\pi x)$ using WENO-5 scheme.

3.2. Compact Reconstruction WENO-5 Scheme

The main drawback of the WENO-5 scheme is that we have to increase the stencil size to get more accuracy. Compact reconstruction of WENO-5 scheme has been developed that uses smaller stencil without reducing the accuracy of the solution [25]. CRWENO-5 scheme uses compact stencils as their basis to calculate the left and right side state at the interface. The procedure for CRWENO-5 is similar to the WENO-5 scheme. However, its candidate stencils are implicit and hence smaller stencils can be used to get the same accuracy. The implicit system used to compute the left and right side state is given below

$$\left(\frac{2}{3}w_1^L + \frac{1}{3}w_2^L\right)u_{i-\frac{1}{2}}^L + \left[\frac{1}{3}w_1^L + \frac{2}{3}(w_2^L + w_3^L)\right]u_{i+\frac{1}{2}}^L + \frac{1}{3}w_3^L u_{i+\frac{3}{2}}^L = \tag{44}$$
$$\frac{w_1^L}{6}u_{i-1} + \frac{5(w_1^L + w_2^L) + w_3^L}{6}u_i + \frac{w_2^L + 5w_3^L}{6}u_{i+1},$$

$$\frac{1}{3}w_3^R u_{i-\frac{1}{2}}^R + \left[\frac{1}{3}w_1^R + \frac{2}{3}(w_2^R + w_3^R)\right]u_{i+\frac{1}{2}}^R + \left(\frac{2}{3}w_1^R + \frac{1}{3}w_2^R\right)u_{i+\frac{3}{2}}^R = \tag{45}$$
$$\frac{w_1^R}{6}u_{i-1} + \frac{5(w_1^R + w_2^R) + w_3^R}{6}u_i + \frac{w_2^R + 5w_3^R}{6}u_{i+1},$$

where the nonlinear weights are calculated using Equations (36) and (37), in which the linear weights are given by $d_0^L = 1/5$, $d_1^L = 1/2$, and $d_2^L = 3/10$ in Equation (36), and $d_0^R = 3/10$, $d_1^R = 1/2$, and $d_2^R = 1/5$ in Equation (37). The smoothness indicators are computed using Equations (38)–(40). The tridiagonal system formed using Equations (44) and (45) can be solved using the Thomas algorithm with $O(N)$ complexity, where N is the total number of grid points. In our implementation of CRWENO-5 scheme, we write a function that takes u in the candidate stencil (i.e., $u_{i-2}, u_{i-1}, u_i, u_{i+1}, u_{i+2}$) and computes all coefficients in Equations (44) and (45). The implementation of Julia function to compute the coefficients of tridiagonal system for reconstruction of left side state is given in Listing 9. A similar function can be implemented for right side state also.

Listing 9. Implementation of Julia function to compute coefficients used for forming the implicit system in computation of left side numerical state at the interface.

```julia
# v1,v2,v3,v4,v5: point values of v in the stencil (v3 = v[i])
# a1, a2, a3: coefficients of CRWENO-5 scheme that forms the tridiagonal system
# b1, b2, b3: coefficients to compute right hand side term of tridiagonal system
function crwcL(v1,v2,v3,v4,v5)
eps = 1.0e-6

s1 = (13.0/12.0)*(v1-2.0*v2+v3)^2 + 0.25*(v1-4.0*v2+3.0*v3)^2
s2 = (13.0/12.0)*(v2-2.0*v3+v4)^2 + 0.25*(v2-v4)^2
```

```julia
s3 = (13.0/12.0)*(v3-2.0*v4+v5)^2 + 0.25*(3.0*v3-4.0*v4+v5)^2

c1 = 2.0e-1/((eps+s1)^2)
c2 = 5.0e-1/((eps+s2)^2)
c3 = 3.0e-1/((eps+s3)^2)

w1 = c1/(c1+c2+c3)
w2 = c2/(c1+c2+c3)
w3 = c3/(c1+c2+c3)

a1 = (2.0*w1 + w2)/3.0
a2 = (w1 + 2.0*w2 + 2.0*w3)/3.0
a3 = w3/3.0

b1 = w1/6.0
b2 = (5.0*w1 + 5.0*w2 + w3)/6.0
b3 = (w2 + 5.0*w3)/6.0

return a1,a2,a3,b1,b2,b3
end
```

We demonstrate CRWENO-5 scheme for both Dirichlet and periodic boundary condition. The Julia function to form the tridiagonal system for computing left side state at the interface in CRWENO-5 is given in Listing 10. We use a similar function for computing right side state.

Listing 10. Implementation of Julia function to form the tridiagonal system for computing left side numerical state at the interface for periodic boundary condition.

```julia
# n: number of grid points in domain
# u: point variable stored at the center of cells
# f: array of numerical state at the interface between two cells
function crwenoL(n,u,f)
a = Array{Float64}(undef, n) # subdiagonal array of tridiagonal matrix
b = Array{Float64}(undef, n) # diagonal array of tridiagonal matrix
c = Array{Float64}(undef, n) # superdiagonal array of tridiagonal matrix
r = Array{Float64}(undef, n) # right hand~side

i = 1
v1 = u[n-1]
v2 = u[n]
v3, v4, v5 = u[i], u[i+1], u[i+2]
a1,a2,a3,b1,b2,b3 = wcL(v1,v2,v3,v4,v5)
a[i], b[i], c[i] = a1, a2, a3
r[i] = b1*u[n] + b2*u[i] + b3*u[i+1]

i = 2
v1 = u[n]
v2, v3, v4, v5 = u[i-1], u[i], u[i+1], u[i+2]
a1,a2,a3,b1,b2,b3 = wcL(v1,v2,v3,v4,v5)
a[i], b[i], c[i] = a1, a2, a3
r[i] = b1*u[i-1] + b2*u[i] + b3*u[i+1]

for i = 3:n-1
v1, v2, v3, v4, v5 = u[i-2], u[i-1], u[i], u[i+1], u[i+2]
a1,a2,a3,b1,b2,b3 = wcL(v1,v2,v3,v4,v5)
```

```
a[i], b[i], c[i] = a1, a2, a3
r[i] = b1*u[i-1] + b2*u[i] + b3*u[i+1]
end


i = n
v1, v2, v3, v4 = u[i-2], u[i-1], u[i], u[i+1]
v5 = u[2]
a1,a2,a3,b1,b2,b3 = wcL(v1,v2,v3,v4,v5)
a[i], b[i], c[i] = a1, a2, a3
r[i] = b1*u[i-1] + b2*u[i] + b3*u[i+1]


alpha = c[n]
beta = a[1]


ctdms(a,b,c,alpha,beta,r,f,1,n) # call cyclic Thomas algorithm function
end
```

For periodic boundary condition, we use the relation given by Equation (43) for ghost points. The tridiagonal system formed for the periodic boundary condition is not purely tridiagonal and it has the following form

$$
\underbrace{\begin{bmatrix}
b_1 & c_1 & 0 & \cdots & \beta \\
a_2 & b_2 & c_2 & \cdots & 0 \\
\vdots & \vdots & & \ddots & \vdots \\
0 & \cdots & a_{N-1} & b_{N-1} & c_{N-1} \\
\alpha & 0 & \cdots & a_N & b_N
\end{bmatrix}}_{\hat{A}}
\underbrace{\begin{bmatrix}
u_1 \\ u_2 \\ \vdots \\ u_{N-1} \\ u_N
\end{bmatrix}}_{u}
=
\underbrace{\begin{bmatrix}
d_1 \\ d_2 \\ \vdots \\ d_{N-1} \\ d_N
\end{bmatrix}}_{d},
\tag{46}
$$

where $\hat{A}$ is the tridiagonal system, u is the solution field and d is the right hand side of the tridiagonal system. We use Sherman-Morrison formula [17] to solve the periodic tridiagonal system given in Equation (46). The Sherman-Morrison formula can be used to solve a sparse linear system of the equation if there is a faster way of calculating A^{-1}, where A is the modified matrix without α and β. We can compute the A^{-1} using the regular Thomas algorithm explained in Section 2.3. We can write Equation (46) as given below

$$
(A + p \otimes q)u = d,
\tag{47}
$$

where

$$
p = \begin{bmatrix} \gamma \\ 0 \\ \vdots \\ 0 \\ \alpha \end{bmatrix}, \quad
q = \begin{bmatrix} 1 \\ 0 \\ \vdots \\ 0 \\ \beta/\gamma \end{bmatrix},
\tag{48}
$$

where γ is arbitrary and the matrix A is the tridiagonal part of the matrix $\hat{A}$, and $\otimes$ represents the outer product of two vectors. Then the solution to Equation (46) can be obtained using below formulas

$$
A \cdot y = d, \quad A \cdot z = p,
\tag{49}
$$

$$
u = y - \left[\frac{q \cdot y}{1 + q \cdot z} \right] z.
\tag{50}
$$

The cyclic Thomas algorithm for solving the periodic tridiagonal system is given in Algorithm 2.

Algorithm 2 Cyclic Thomas algorithm

1: Given $a, b, c, d, \alpha, \beta$ $\triangleright \hat{A}u = d$
2: $\gamma = -b_1$ $\triangleright$ Avoid subtraction error in forming bb_0
3: $bb_1 = b_1 - \gamma, bb_N = b_N - \alpha\beta/\gamma$ $\triangleright$ Set up diagonal of the modified tridiagonal system
4: **for** $i = 2$ to $N - 1$ **do**
5:       $bb_i = b_i$
6: **end for**
7: Thomas(a, bb, c, d, y) $\triangleright$ Solve $A \cdot y = d$
8: $u_1 = \gamma, u_N = \alpha$ $\triangleright$ Set up the vector u
9: **for** $i = 2$ to $N - 1$ **do**
10:       $u_i = 0.0$
11: **end for**
12: Thomas(a, bb, c, u, z) $\triangleright$ Solve $A \cdot z = p$
13: $f = (y_1 + \beta y_N/\gamma)/(1.0 + z_1 + \beta z_N/\gamma)$ $\triangleright$ Form $q \cdot y/(1 + q \cdot z)$
14: **for** $i = 1$ to N **do**
15:       $u_i = y_i - f z_i$ $\triangleright$ Calculate the solution vector u
16: **end for**

Once the left and right side states (u^L, u^R) are available we compute the right hand side term of the Burgers equation using Listing 7. We use third-order Runge-Kutta numerical scheme for time integration. We solve the tridiagonal system formed with periodic boundary condition using cyclic Thomas algorithm and for Dirichlet boundary condition, we use regular Thomas algorithm. The implementation of cyclic Thomas algorithm in Julia is given in Listing 11. The implementation of regular Thomas algorithm is given in Listing 4.

Listing 11. Implementation of cyclic Thomas algorithm in Julia for periodic boundary condition domain.

```julia
# a, b, c: subdiagonal, diagonal, superdiagonal entries of tridiagonal matrix
# alpha: bottom-left entry in tridiagonal matrix (first value for N-th equation)
# beta: top-right entry in tridiagonal matrix (last value for 1-st equation)
# r: right hand side of tridiagonal system
# x: solution to the tridiagonal system
# s,e: stat and end index of the matrix
function ctdms(a,b,c,alpha,beta,r,x,s,e)
bb = Array{Float64}(undef, e)
u = Array{Float64}(undef, e)
z = Array{Float64}(undef, e)
gamma = -b[s]
bb[s] = b[s] -gamma
bb[e] = b[e] - alpha*beta/gamma

for i = s+1:e-1
bb[i] = b[i]
end

tdms(a,bb,c,r,x,s,e) # call regular Thomas~algorithm

u[s] = gamma
u[e] = alpha
for i = s+1:e-1
u[i] = 0.0
end

tdms(a,bb,c,u,z,s,e) # call regular Thomas~algorithm
```

```
fact = (x[s] + beta*x[e]/gamma)/(1.0 + z[s] + beta*z[e]/gamma)

for i = s:e
x[i] = x[i] - fact*z[i]
end
end
```

We test the CRWENO-5 method for the same problem as the WENO-5 scheme. The evolution of shock generated from sine wave with time is shown in Figure 9 for both Dirichlet and periodic boundary conditions computed using CRWENO-5 method. We use the same parameter as the WENO-5 scheme. We store snapshots of the solution field at different time steps to see the evolution of the shock. It can be seen from Figure 9 that the CRWENO-5 scheme captures the shock formed at final time $t = 0.25$ accurately.

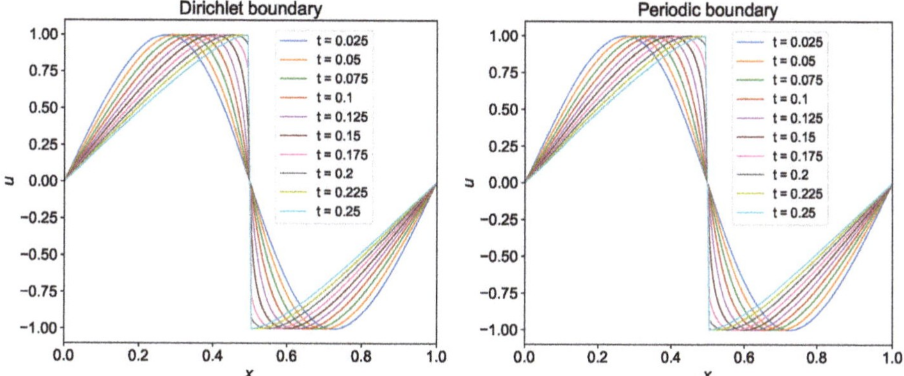

Figure 9. Evolution of shock with time for the initial condition $u_0 = \sin(2\pi x)$ using CRWENO scheme.

4. Inviscid Burgers Equation: Conservative Form

In this section, we explain two approaches for solving the inviscid Burgers equation in its conservative form. The inviscid Burgers equation can be represented in the conservative form as below

$$\frac{\partial u}{\partial t} + \frac{\partial f}{\partial x} = 0, \qquad \text{where} \quad f = \left(\frac{u^2}{2}\right). \tag{51}$$

The f is termed as the flux. We need to modify the grid arrangement to solve the inviscid Burgers equation in its conservative form. The modified grid for one-dimensional problem is shown in Figure 10. Here, $\Delta x = x_{i+1/2} - x_{i-1/2}$. The solution field is stored at the center of each cell and this type of grid is primarily used for finite volume discretization. We explain two approaches to compute the nonlinear term in the inviscid Burgers equation.

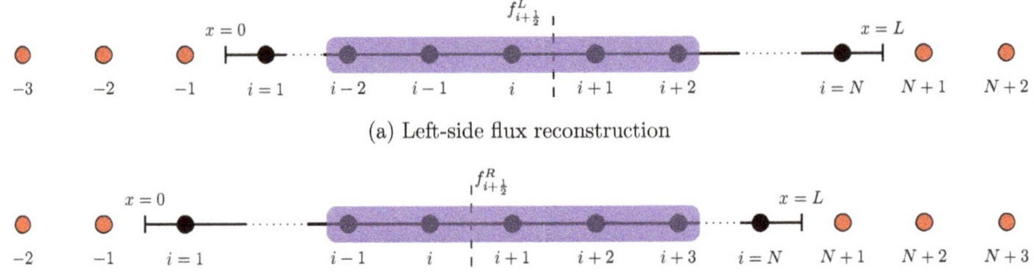

(a) Left-side flux reconstruction

(b) Right-side flux reconstruction

Figure 10. Finite difference grid for one-dimensional problems where the the solution is stored at the center of the cells. The flux is also computed at left and right boundary. The stencil required for reconstruction of left and right side flux is highlighted with blue rectangle. Ghost points are shown by red color.

4.1. Flux Splitting

The conservative form of the inviscid Burgers equation allows us to use the flux splitting method and WENO reconstruction to compute the flux at the interface. In this method, we first compute the flux f at the nodal points. The nonlinear advection term of the inviscid Burgers equation is computed as

$$\left.\frac{\partial f}{\partial x}\right|_{x=x_i} = \frac{f_{i+1/2} - f_{i-1/2}}{\Delta x}, \tag{52}$$

$$\left.\frac{\partial f}{\partial x}\right|_{x=x_i} = \frac{f^L_{i+1/2} - f^L_{i-1/2}}{\Delta x} + \frac{f^R_{i+1/2} - f^R_{i-1/2}}{\Delta x}, \tag{53}$$

at a particular node i. The superscripts L and R refer to the positive and negative flux component at the interface and the subscripts $i - 1/2$ and $i + 1/2$ stands for the left and right side interface of the nodal point i. We use WENO-5 reconstruction discussed in Section 3.1 to compute the left and right side flux at the interface. First, we compute the flux at nodal points and then split it into positive and negative components. This process is called as Lax-Friedrichs flux splitting and the positive and negative component of the flux at a nodal location is calculated as

$$f_i^+ = \frac{1}{2}(f_i + \alpha u_i), \quad f_i^- = \frac{1}{2}(f_i - \alpha u_i), \tag{54}$$

where α is the absolute value of the flux Jacobian ($\alpha = |\frac{\partial f}{\partial u}|$). We chose the values of α as the maximum value of u_i over the stencil used for WENO-5 reconstruction, i.e.,

$$\alpha = \max(|u_{i-2}|, |u_{i-1}|, |u_i|, |u_{i+1}|, |u_{i+2}|). \tag{55}$$

Once we split the nodal flux into its positive and negative component we reconstruct the left and right side fluxes at the interface using WENO-5 scheme using below formulas [24]

$$f^L_{i+\frac{1}{2}} = w^L_0\left(\frac{1}{3}f^+_{i-2} - \frac{7}{6}f^+_{i-1} + \frac{11}{6}f^+_i\right) + w^L_1\left(-\frac{1}{6}f^+_{i-1} + \frac{5}{6}f^+_i + \frac{1}{3}f^+_{i+1}\right) + w^L_2\left(\frac{1}{3}f^+_i + \frac{5}{6}f^+_{i+1} - \frac{1}{6}f^+_{i+2}\right), \tag{56}$$

$$f^R_{i-\frac{1}{2}} = w^R_0\left(-\frac{1}{6}f^-_{i-2} + \frac{5}{6}f^-_{i-1} + \frac{1}{3}f^-_i\right) + w^R_1\left(\frac{1}{3}f^-_{i-1} + \frac{5}{6}f^-_i - \frac{1}{6}f^-_{i+1}\right) + w^R_2\left(\frac{11}{6}f^-_i - \frac{7}{6}f^-_{i+1} + \frac{1}{3}f^-_{i+2}\right), \tag{57}$$

where the nonlinear weights are computed using Equations (36) and (37) . Once we have left and right side fluxes at the interface we use Equation (53) to compute the flux derivative. We demonstrate the flux splitting method for a shock generated by sine wave similar to Section 3. We use the third-order

Runge-Kutta numerical scheme for time integration and periodic boundary condition for the domain. For the conservative form, we need three ghost points on left and right sides, since we compute the flux at the left and right boundary of the domain also. The periodic boundary condition for fluxes at ghost points is given by

$$f_{-3} = f_{N-2}, \quad f_{-2} = f_{N-1}, \quad f_{-1} = f_N, \quad f_{N+1} = f_1, \quad f_{N+2} = f_2, \quad f_{N+3} = f_3. \tag{58}$$

The Julia function to compute the right hand side of the inviscid Burgers equation is detailed in Listing 12. The implementation of flux reconstruction and weight computation in Julia is similar to Listings 6 and 8.

Listing 12. Implementation of Julia function to calculate the right hand side of inviscid Burgers equation using flux splitting method.

```julia
# nx: number of grid points
# dx: grid spacing
# u: solution field at discrete nodal points
# r: right hand side of inviscid Burgers equation
function rhs(nx,dx,u,r)
f = Array{Float64}(undef,nx) # flux at the nodal points
fP = Array{Float64}(undef,nx) # positive part of flux at the nodal points
fN = Array{Float64}(undef,nx) # negative part of flux at the nodal~points

ps = Array{Float64}(undef,nx)   # wave speed at nodal~points

fL = Array{Float64}(undef,nx+1) # left side flux at the interface
fR = Array{Float64}(undef,nx+1) # right side flux at the~interface

for i = 1:nx
f[i] = 0.5*u[i]*u[i] # compute flux
end

wavespeed(nx,u,ps) # call function to compute the~Jacobian

for i = 1:nx
fP[i] = 0.5*(f[i] + ps[i]*u[i]) # split the flux
fN[i] = 0.5*(f[i] - ps[i]*u[i])
end

# compute upwind reconstruction for positive flux (left to right)
fL = wenoL(nx,fP)
# compute downwind reconstruction for negative flux (right to left)
fR = wenoR(nx,fN)

# compute RHS using flux splitting
for i = 1:nx
r[i] = -(fL[i+1] - fL[i])/dx - (fR[i+1] - fR[i])/dx
end
end

# function to compute the propagation speed (Jacobian)
function wavespeed(n,u,ps)
for i = 3:n-2
ps[i] = max(abs(u[i-2]), abs(u[i-1]), abs(u[i]), abs(u[i+1]), abs(u[i+2]))
end
```

```
# periodicity
i = 1
ps[i] = max(abs(u[n-1]), abs(u[n]), abs(u[i]), abs(u[i+1]), abs(u[i+2]))
i = 2
ps[i] = max(abs(u[n]), abs(u[i-1]), abs(u[i]), abs(u[i+1]), abs(u[i+2]))
i = n-1
ps[i] = max(abs(u[i-2]), abs(u[i-1]), abs(u[i]), abs(u[i+1]), abs(u[1]))
i = n
ps[i] = max(abs(u[i-2]), abs(u[i-1]), abs(u[i]), abs(u[1]), abs(u[2]))
end
```

4.2. Riemann Solver: Rusanov Scheme

Another approach to computing the nonlinear term in the inviscid Burgers equation is to use Riemann solvers. Riemann solvers are used for accurate and efficient simulations of Euler equations along with higher-order WENO schemes [27,28]. In this method, first, we reconstruct the left and right side fluxes at the interface similar to the inviscid Burgers equation in non-conservative form. Once we have $f^L_{i+1/2}$ and $f^R_{i+1/2}$ reconstructed, we use Riemann solvers to compute the fluxes at the interface. We follow the same procedure discussed in Section 3.1 to reconstruct the fluxes at the interface. We can also use CRWENO-5 scheme presented in Section 3.2 to determine the fluxes.

We use a simple Rusanov scheme as the Riemann solver [29]. The Rusanov scheme uses maximum local wave propagation speed to compute the flux as follows

$$f_{i+1/2} = \frac{1}{2}\left(f^L_{i+1/2} + f^R_{i+1/2}\right) - \frac{c_{i+1/2}}{2}\left(u^L_{i+1/2} - u^R_{i+1/2}\right), \tag{59}$$

where f^L is the flux component using the right reconstructed state $f^L_{i+1/2} = f(u^L_{i+1/2})$, and f^R is the flux component using the right reconstructed state $f^R_{i+1/2} = f(u^R_{i+1/2})$. Here, $c_{i+1/2}$ is the local wave propagation speed which is obtained by taking the maximum absolute value of the eigenvalues corresponding to the Jacobian, $\frac{\partial f}{\partial u} = u$, between cells i and $i+1$ can be obtained as

$$c_{i+1/2} = \max(|u_i|, |u_{i-1}|), \tag{60}$$

or in a wider stencil as shown in Equation (55). The implementation of Riemann solver with Rusanov scheme for inviscid Burgers equation is given in Listing 13. The implementation of state reconstruction and weight computation in Julia is similar to Listings 6 and 8. Figure 11 shows the evolution of shock for the sine wave. We use the same parameters for the spatial and temporal discretization of the inviscid Burgers equation as used for the non-conservative form in Section 3.

Listing 13. Implementation of Julia function to calculate the right hand side of inviscid Burgers equation using Riemann solver based on Rusanov scheme.

```
# nx: number of grid points
# dx: grid spacing
# u: solution field at discrete nodal points
# r: right hand side of inviscid Burgers equation
function rhs(nx,dx,u,r)
uL = Array{Float64}(undef,nx+1)
uR = Array{Float64}(undef,nx+1)
fL = Array{Float64}(undef,nx+1)
fR = Array{Float64}(undef,nx+1)
f = Array{Float64}(undef,nx+1)

uL = wenoL(nx,u) # construct left state
```

```julia
uR = wenoR(nx,u) # construct right~state

fluxes(nx,uL,fL) # compute flux for left state
fluxes(nx,uR,fR) # compute flux for right~state

rusanov(nx,u,uL,uR,f,fL,fR) # compute Riemann solver (flux at interface)

for i = 1:nx
r[i] = -(f[i+1] - f[i])/dx
end
end

function rusanov(nx,u,uL,uR,f,fL,fR)
ps = Array{Float64}(undef,nx+1) # propagation speed
for i = 2:nx
ps[i] = max(abs(u[i]), abs(u[i-1]))
end
ps[1] = max(abs(u[1]), abs(u[nx])) # left boundary
ps[nx+1] = max(abs(u[1]), abs(u[nx])) # right~boundary

# Interface fluxes (Rusanov)
for i = 1:nx+1
f[i] = 0.5*(fR[i]+fL[i]) - 0.5*ps[i]*(uR[i]-uL[i])
end
end

function fluxes(nx,u,f)
for i = 1:nx+1
f[i] = 0.5*u[i]*u[i]
end
end
```

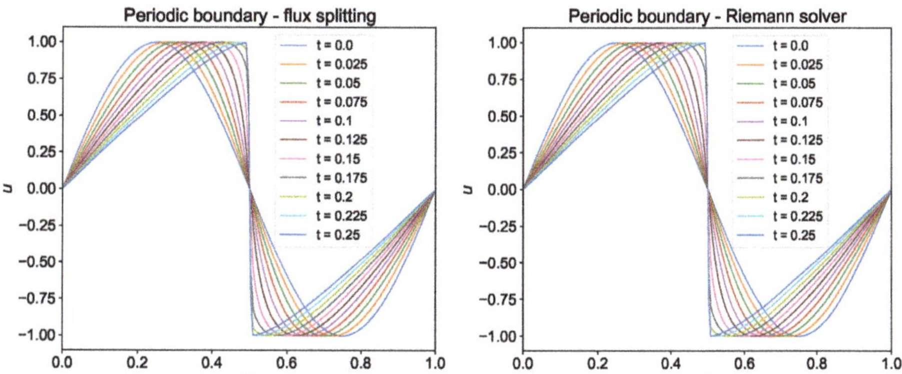

Figure 11. Evolution of shock from sine wave ($u_0 = \sin(2\pi x)$) computed using conservative form of inviscid Burgers equation. The reconstruction of the field is done using WENO-5 scheme, and flux splitting and Riemann solver approach is used to compute the nonlinear term.

5. One-Dimensional Euler Solver

In this section, we present a solution to one-dimensional Euler equations. Euler equations contain a set of nonlinear hyperbolic conservation laws that govern the dynamics of compressible fluids neglecting the effect of body forces, and viscous stress [30]. The one-dimensional Euler equations in its conservative form can be written as

$$\frac{\partial q}{\partial t} + \frac{\partial F}{\partial x} = 0,$$
(61)

where

$$q = \begin{pmatrix} \rho \\ \rho u \\ \rho e \end{pmatrix}, \quad F = \begin{pmatrix} \rho u \\ \rho u^2 + p \\ \rho u h \end{pmatrix},$$

in which

$$h = e + \frac{p}{\rho}, \quad p = \rho(\gamma - 1)\left(e - \frac{1}{2}u^2\right).$$
(62)

Here, ρ and p are the density, and pressure respectively; u is the horizontal component of velocity; e and h stand for the internal energy and static enthalpy, and γ is the ratio of specific heats. Equation (61) can also be written as

$$\frac{\partial q}{\partial t} + A\frac{\partial q}{\partial x} = 0,$$
(63)

where $A = \frac{\partial F}{\partial q}$ is the convective flux Jacobian matrix. The matrix A is given as

$$A = \frac{\partial F}{\partial q} = \begin{pmatrix} 0 & 1 & 0 \\ \phi^2 - u^2 & (3 - \gamma)u & \gamma - 1 \\ (\phi^2 - h)u & h - (\gamma - 1)u^2 & \gamma u \end{pmatrix},$$
(64)

where $\phi^2 = \frac{1}{2}(\gamma - 1)u^2$. We can write the matrix A as $A = R\Lambda L$ in which Λ is the diagonal matrix of real eigenvalues of A, i.e., $\Lambda = \text{diag}[u, u + a, u - a]$. Here, L and R are the matrices of which the columns are the left and right eigenvectors of A; $L = R^{-1}$. The matrices L and R are given below [31]

$$L = \begin{pmatrix} 1 - \frac{\phi^2}{a^2} & (\gamma - 1)\frac{u}{a^2} & -\frac{(\gamma-1)}{a^2} \\ \phi^2 - ua & a - (\gamma - 1)u & \gamma - 1 \\ \phi^2 + ua & -a - (\gamma - 1)u & \gamma - 1 \end{pmatrix}, \quad R = \begin{pmatrix} 1 & \beta & \beta \\ u & \beta(u + a) & \beta(u - a) \\ \frac{\phi^2}{(\gamma-1)} & \beta(h + ua) & \beta(h - ua) \end{pmatrix},$$
(65)

where a is the speed of sound and is given by $a^2 = \gamma p/\rho$, and $\beta = 1/(2a^2)$.

We use the grid similar to the one used for the conservative form of the inviscid Burgers equation and is shown in Figure 10. The semi-discrete form of the Euler equations can be written as

$$\frac{\partial q_i}{\partial t} + \frac{F_{i+1/2} - F_{i-1/2}}{\Delta x} = 0,$$
(66)

where q_i is the cell-centered values stored at nodal points and $F_{i\pm1/2}$ are the fluxes at left and right cell interface. We use third-order Runge-Kutta numerical method for time integration. We use WENO-5 reconstruction to compute the left and sight states of the fluxes at the interface. We include three different Riemann solvers to compute the flux $F_{i\pm1/2}$ at interfaces to be used in Equation (66).

5.1. Roe's Riemann Solver

First, we present the Roe solver for one-dimensional Euler equations. According to the Gudanov theorem [32], for a hyperbolic system of equations, if the Jacobian matrix of the flux vector is constant (i.e., $A = \frac{\partial F}{\partial q} = \text{constant}$), the exact values of fluxes at the interfaces can be calculated as

$$F_{i+1/2} = \frac{1}{2}\left(F^R_{i+1/2} + F^L_{i+1/2}\right) - \frac{1}{2}R|\Lambda|L\left(q^R_{i+1/2} - q^L_{i+1/2}\right),$$
(67)

where $|\Lambda|$ is the diagonal matrix consisting of absolute values eigenvalues of the Jacobian matrix and the matrices L and R are given in Equation (65). However, Jacobian matrix is not constant (i.e., $F = F(q)$). The Roe solver [33] is an approximate Riemann solver and it uses below formulation to find the numerical fluxes at the interface

$$F_{i+1/2} = \frac{1}{2}\left(F^R_{i+1/2} + F^L_{i+1/2}\right) - \frac{1}{2}\bar{R}|\bar{\Lambda}|\bar{L}\left(q^R_{i+1/2} - q^L_{i+1/2}\right),\tag{68}$$

where the bar represents the Roe average between the left and right states. In order to derive $\bar{A}$ we need to know $\bar{u}$, $\bar{h}$, and $\bar{a}$ and they are computed using Roe averaging procedure. In order to derive $\bar{u}$, $\bar{h}$, and $\bar{a}$ Roe used Taylor series expansion of F around q^L and q^R points.

$$F(q) = F(q^L) + A(q^L)(q - q^L),\tag{69}$$
$$F(q) = F(q^R) + A(q^R)(q - q^R).\tag{70}$$

Neglecting higher-order terms and equation above two equations

$$F(q^L) - F(q^R) = A(q^L - q^R) + (A(q^R) - A(q^L))q.\tag{71}$$

Roe [20] derived approximate matrix $\bar{A}$ such that it satisfies $\bar{A}(q^L, q^R) \to A(q)$ as $\bar{q} \to q$. Therefore, we get the following set of equations (written in vector form)

$$F(q^L) - F(q^R) = \bar{A}(q^L - q^R).\tag{72}$$

Using Equation (72) we can compute $\bar{u}$, $\bar{h}$, and $\bar{a}$. First we reconstruct the left and right states of q at the interface (i.e., $q^R_{i+1/2}$ and $q^L_{i+1/2}$) using WENO-5 scheme. Then, we can compute the left and right state of the fluxes (i.e., F^L and F^R) using Equation (61). The Roe averaging formulas to compute approximate values for constructing $\bar{R}$ and $\bar{L}$ are given below

$$\bar{u} = \frac{u_R\sqrt{\rho_R} + u_L\sqrt{\rho_L}}{\sqrt{\rho_R} + \sqrt{\rho_L}},\tag{73}$$

$$\bar{h} = \frac{h_R\sqrt{\rho_R} + h_L\sqrt{\rho_L}}{\sqrt{\rho_R} + \sqrt{\rho_L}},\tag{74}$$

where the left and right states are computed using WENO-5 reconstruction. The speed of the sound is computed using the below equation

$$\bar{a} = \sqrt{(\gamma - 1)\left[\bar{h} - \frac{1}{2}\bar{u}^2\right]}.\tag{75}$$

The eigenvalues of the Jacobian matrix are $\lambda_1 = \bar{u}$, $\lambda_2 = \bar{u} + a$, and $\lambda_3 = \bar{u} - a$. The implementation of Roe's Riemann solver is given in Listing 14.

Listing 14. Implementation of Julia function to calculate the right hand side of Euler equations using Roe's Riemann solver.

```
# nx: number of grid points
# dx: grid spacing
# q: solution field at discret nodal points
# r: right hand side of Euler equations
function rhs(nx,dx,gamma,q,r)
qL = Array{Float64}(undef,nx+1,3)
qR = Array{Float64}(undef,nx+1,3)
fL = Array{Float64}(undef,nx+1,3)
fR = Array{Float64}(undef,nx+1,3)
```

```julia
f = Array{Float64}(undef,nx+1,3)

qL = wenoL(nx,q) # compute left state of q using WENO-5
qR = wenoR(nx,q) # compute right state of q using~WENO-5

fluxes(nx,gamma,qL,fL) # compute fluxes for left state at the interface
fluxes(nx,gamma,qR,fR) # compute fluxes for right state at the~interface

roe(nx,gamma,q,qL,qR,f,fL,fR) # call Rusanov scheme~function

for i = 1:nx for m = 1:3
r[i,m] = -(f[i+1,m] - f[i,m])/dx # compute right hand side term (-dF/dx)
end end
end

# flux computation function
function fluxes(nx,gamma,q,f)
for i = 1:nx+1
p = (gamma-1.0)*(q[i,3]-0.5*q[i,2]*q[i,2]/q[i,1])
f[i,1] = q[i,2]
f[i,2] = q[i,2]*q[i,2]/q[i,1] + p
f[i,3] = q[i,2]*q[i,3]/q[i,1] + p*q[i,2]/q[i,1]
end
end
```

The implementation of WENO-5 reconstruction is similar to Listings 8 and 9. We implemented all Riemann solvers for Euler equations in such a way that we compute smoothness indicators for all equations separately. We do not incur much computational expense since our problem is one-dimensional. However, for higher-dimension usually smoothness indicator is computed only for equation (usually momentum or energy equation) and the same smoothness indicators are used for all equations. The computation of flux using Roe averaging is detailed in Listing 15.

Listing 15. Implementation of Julia function to calculate the interface flux using Riemann solver based on Roe averaging.

```julia
# nx: number of grid points
# gamma: ratios of specific heats
# qL, qR: left and right reconstructed states at the interface
# fL, fR: left and right side flux for reconstructed state
# f: flux at the interface computed using Rusanov scheme
function roe(nx,gamma,uL,uR,f,fL,fR)
dd = Array{Float64}(undef,3)
dF = Array{Float64}(undef,3)
V = Array{Float64}(undef,3)
gm = gamma-1.0

for i = 1:nx+1
#Left and right states:
rhLL = uL[i,1]
uuLL = uL[i,2]/rhLL
eeLL = uL[i,3]/rhLL
ppLL = gm*(eeLL*rhLL - 0.5*rhLL*(uuLL*uuLL))
hhLL = eeLL + ppLL/rhLL

rhRR = uR[i,1]
```

```
uuRR = uR[i,2]/rhRR
eeRR = uR[i,3]/rhRR
ppRR = gm*(eeRR*rhRR - 0.5*rhRR*(uuRR*uuRR))
hhRR = eeRR + ppRR/rhRR

alpha = 1.0/(sqrt(abs(rhLL)) + sqrt(abs(rhRR)))

uu = (sqrt(abs(rhLL))*uuLL + sqrt(abs(rhRR))*uuRR)*alpha
hh = (sqrt(abs(rhLL))*hhLL + sqrt(abs(rhRR))*hhRR)*alpha
aa = sqrt(abs(gm*(hh-0.5*uu*uu)))

D11 = abs(uu)
D22 = abs(uu + aa)
D33 = abs(uu - aa)

beta = 0.5/(aa*aa)
phi2 = 0.5*gm*uu*uu

R11, R21, R31 = 1.0, uu, phi2/gm
R12, R22, R32 = beta, beta*(uu + aa), beta*(hh + uu*aa)
R13, R23, R33 = beta, beta*(uu - aa), beta*(hh - uu*aa)

L11, L21, L31 = 1.0-phi2/(aa*aa), phi2 - uu*aa, phi2 + uu*aa
L12, L22, L32 = gm*uu/(aa*aa), aa - gm*uu, -aa - gm*uu
L13, L23, L33 = -gm/(aa*aa), gm, gm

for m = 1:3
V[m] = 0.5*(uR[i,m]-uL[i,m])
end

dd[1] = D11*(L11*V[1] + L12*V[2] + L13*V[3])
dd[2] = D22*(L21*V[1] + L22*V[2] + L23*V[3])
dd[3] = D33*(L31*V[1] + L32*V[2] + L33*V[3])

dF[1] = R11*dd[1] + R12*dd[2] + R13*dd[3]
dF[2] = R21*dd[1] + R22*dd[2] + R23*dd[3]
dF[3] = R31*dd[1] + R32*dd[2] + R33*dd[3]

for m = 1:3
f[i,m] = 0.5*(fR[i,m]+fL[i,m]) - dF[m]
end
end
end
```

We test the Riemann solver using Sod shock tube problem [34]. The time evolution of Sod shock tube problem is governed buy Euler equations. This problem is used for testing numerical schemes to study how well they can capture and resolve shocks and discontinuities and their ability to produce correct density profile. The initial conditions for the Sod shock tube problem are

$$\begin{pmatrix} \rho_L \\ p_L \\ u_L \end{pmatrix} = \begin{pmatrix} 1.0 \\ 1.0 \\ 0.0 \end{pmatrix} \quad \text{when } x < 0.5; \qquad \begin{pmatrix} \rho_R \\ p_R \\ u_R \end{pmatrix} = \begin{pmatrix} 0.125 \\ 0.1 \\ 0.0 \end{pmatrix} \quad \text{when } x > 0.5. \tag{76}$$

We use third-order Runge-Kutta numerical scheme for time integration from $t = 0$ to $t = 0.2$. We use two grid resolutions to see the capability of Riemann solver to capture the shock. The Sod shock tube problem has Dirichlet boundary condition at the left and right boundary. We use linear interpolation to find the values of q at ghost points. Figure 12 shows the density, velocity, pressure, and energy at final time $t = 0.2$ using Roe solver. We consider the solution obtained with high-grid resolution as the true solution and compare it with the low-resolution results. From Figure 12, we observe that there are small oscillations near discontinuities at low-resolution. We use WENO reconstruction to compute the flux at interface and WENO scheme is not strictly non-oscillatory. However, WENO scheme does not allow oscillations to amplify to large values. To remove these oscillations further, for example, we can use characteristic-wise reconstructions [35].

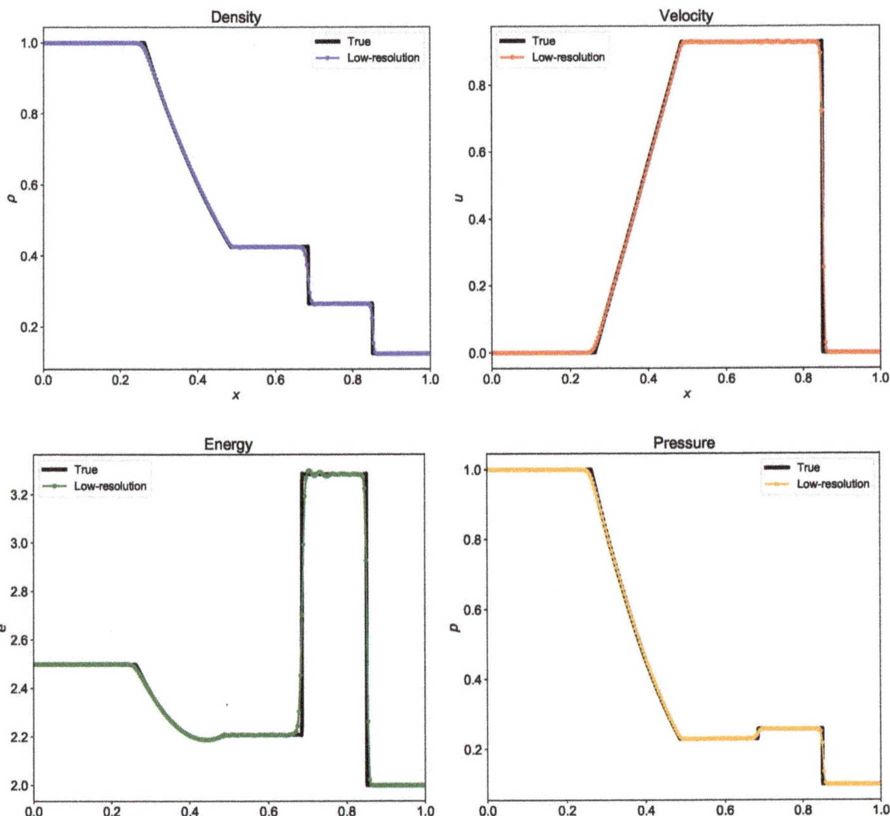

Figure 12. Evolution of density, velocity, energy, and pressure at $t = 0.2$ for the shock tube problem computed using Riemann solver based on Roe averaging. The true solution is calculated with $N = 8192$ grid resolution with $\Delta t = 0.00005$ and the low-resolution results are for $N = 256$ grid resolution with $\Delta t = 0.0001$.

5.2. HLLC Riemann Solver

We present one of the most widely used approximate Riemann solver based on HLLC scheme [30,36]. They used lower and upper bounds on the characteristics speeds in the solution of Riemann problem involving left and right states. These bounds are approximated as

$$S_L = \min(u_L, u_R) - \max(a_L, a_R), \tag{77}$$

$$S_R = \min(u_L, u_R) + \max(a_L, a_R), \tag{78}$$

where S_L and S_R are the lower and upper bound for the left and right state characteristics speed. For HLLC scheme we also include middle wave of speed S_* given by

$$S_* = \frac{p_R - p_L + \rho_L u_L(S_L - u_L)\rho_R u_R(S_R - uR)}{\rho_L(S_L - u_L) - \rho_R(S_R - u_R)}. \tag{79}$$

The mean pressure is given by

$$P_{LR} = \frac{1}{2}\left[p_L + p_R + \rho_L(S_L - u_L)(S_* - u_l) + \rho_R(S_R - u_R)(S_* - u_R)\right]. \tag{80}$$

The fluxes are computed as

$$F_{i+1/2} = \begin{cases} F^L, & \text{if } S_L \geq 0 \\ F^R, & \text{if } S_R \leq 0 \\ \frac{S_*(S_L u_L F^L) + S_L P_{LR} D_*}{S_L - S_*}, & \text{if } S_L \leq 0 \text{ and } S_* \geq 0 \\ \frac{S_*(S_R u_R F^R) + S_R P_{LR} D_*}{S_R - S_*}, & \text{if } S_R \geq 0 \text{ and } S_* \leq 0 \end{cases} \tag{81}$$

where $D_* = [0, 1, S_*]$. Once the flux is available at the interface, we can integrate the solution in time using Equation (66). The implementation of HLLC scheme is given in Listing 16. We perform the time integration using third-order Runge-Kutta numerical scheme. Figure 13 shows the density, velocity, pressure, and energy at final time $t = 0.2$ computed using Riemann solver based on HLLC scheme. The oscillations in the numerical solution calculated by the HLLC scheme is slightly less than those calculated by the Roe's Riemann solver. The true solution is computed using $N = 8192$ grid points and $\Delta t = 0.00005$ and is compared with the low-resolution results for $N = 256$ grid points and $\Delta t = 0.0001$.

Listing 16. Implementation of Julia function to calculate the interface flux using Riemann solver based on HLLC scheme.

```julia
# nx: number of grid points
# gamma: ratios of specific heats
# uL, uR: left and right reconstructed states at the interface
# fL, fR: left and right side flux for reconstructed state
# f: flux at the interface computed using HLLC scheme
function hllc(nx,gamma,uL,uR,f,fL,fR)
gm = gamma-1.0
Ds = Array{Float64}(undef,3)
Ds[1], Ds[2] = 0.0, 1.0

for i = 1:nx+1
# left state
rhLL = uL[i,1]
uuLL = uL[i,2]/rhLL
eeLL = uL[i,3]/rhLL
ppLL = gm*(eeLL*rhLL - 0.5*rhLL*(uuLL*uuLL))
```

```
aaLL = sqrt(abs(gamma*ppLL/rhLL))

# right state
rhRR = uR[i,1]
uuRR = uR[i,2]/rhRR
eeRR = uR[i,3]/rhRR
ppRR = gm*(eeRR*rhRR - 0.5*rhRR*(uuRR*uuRR))
aaRR = sqrt(abs(gamma*ppRR/rhRR))

# compute SL and Sr
SL = min(uuLL,uuRR) - max(aaLL,aaRR)
SR = max(uuLL,uuRR) + max(aaLL,aaRR)

# compute compound speed
SP = (ppRR - ppLL + rhLL*uuLL*(SL-uuLL) - rhRR*uuRR*(SR-uuRR))/
(rhLL*(SL-uuLL) - rhRR*(SR-uuRR)) #never get~zero

# compute compound pressure
PLR = 0.5*(ppLL + ppRR + rhLL*(SL-uuLL)*(SP-uuLL)+
rhRR*(SR-uuRR)*(SP-uuRR))

Ds[3] = SP # compute~D

if (SL >= 0.0)
for m = 1:3
f[i,m] = fL[i,m]
end
elseif (SR <= 0.0)
for m =1:3
f[i,m] = fR[i,m]
end
elseif ((SP >=0.0) & (SL <= 0.0))
for m = 1:3
f[i,m] = (SP*(SL*uL[i,m]-fL[i,m]) + SL*PLR*Ds[m])/(SL-SP)
end
elseif ((SP <= 0.0) & (SR >= 0.0))
for m = 1:3
f[i,m] = (SP*(SR*uR[i,m]-fR[i,m]) + SR*PLR*Ds[m])/(SR-SP)
end
end
end
end
end
```

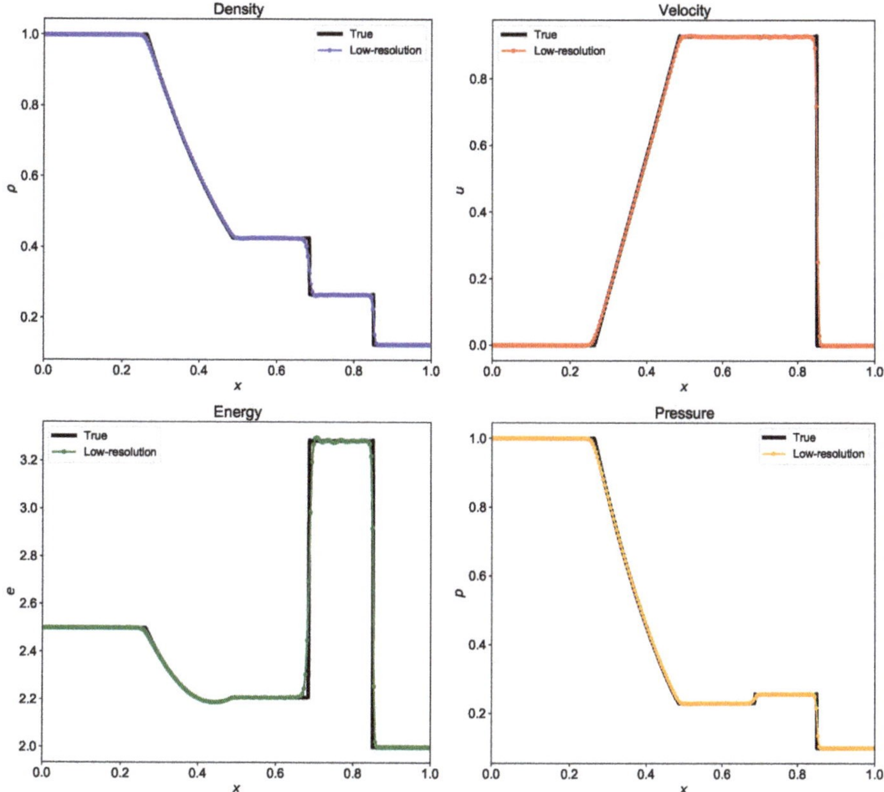

Figure 13. Evolution of density, velocity, energy, and pressure at $t = 0.2$ for the shock tube problem computed using Riemann solver based on Harten-Lax-van Leer Contact (HLLC) scheme. The true solution is calculated with $N = 8192$ grid resolution with $\Delta t = 0.00005$ and the low-resolution results are for $N = 256$ grid resolution with $\Delta t = 0.0001$.

5.3. Rusanov's Riemann Solver

We show the implementation of Riemann solver using Rusanov scheme. This is similar to what we discussed in Section 4.2. The Riemann solver based on Rusanov scheme is simple compared to Roe's Riemann solver and HLLC based Riemann solver. For Euler equations, instead of solving one equation (as in inviscid Burgers equation), now we have to follow the procedure for three equations (i.e., density, velocity, and energy). We need to approximate wave propagation speed at the interface $c_{i+1/2}$ to compute flux at the interface. For Rusanov scheme, we simply use maximum eigenvalue of the Jacobian matrix as the wave propagation speed. We have $c_{i+1/2} = \max(\bar{u}, |\bar{u} + \bar{a}|, |\bar{u} - \bar{a}|)$, where $\bar{u}$ and $\bar{a}$ are computed using Roe averaging. Figure 14 shows the density, velocity, pressure, and energy at final time $t = 0.2$ computed using Riemann solver based on Rusanov scheme. The oscillations in the numerical solution calculated by the Rusanov scheme is more than those calculated by the Roe's Riemann solver or HLLC scheme based Riemann solver. One of the advantage of Rusanov scheme is that it is simple to implement and is computationally faster. The true solution is computed using $N = 8192$ grid points and $\Delta t = 0.00005$ and is compared with the low-resolution results for $N = 256$ grid points and $\Delta t = 0.0001$. The implementation of Riemann solver based on Rusanov's scheme is given in Listing 17.

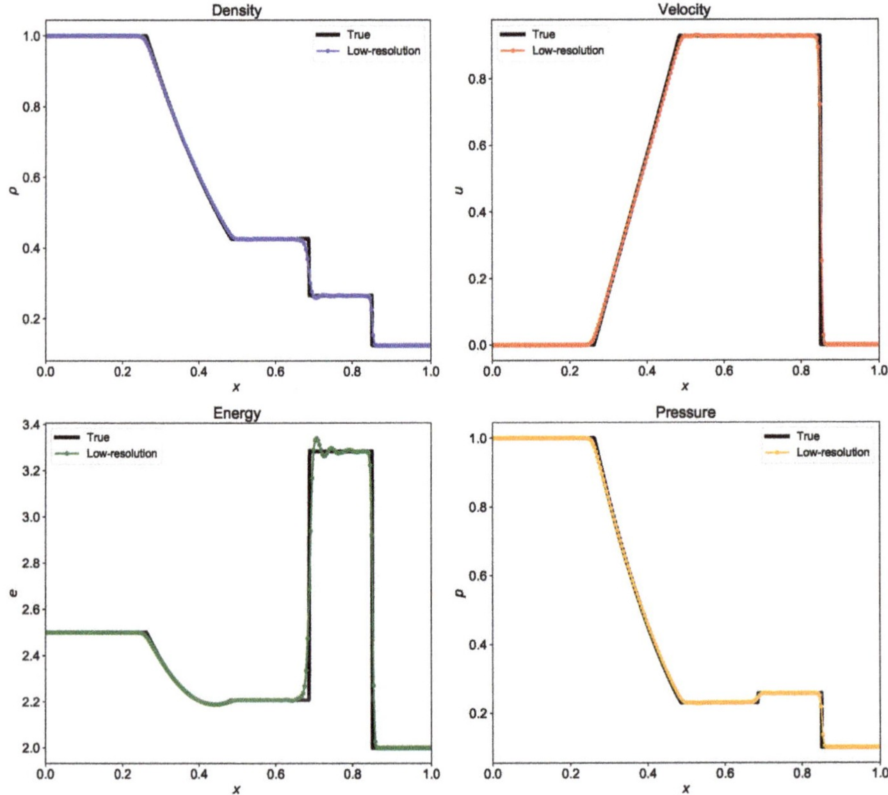

Figure 14. Evolution of density, velocity, energy, and pressure at $t = 0.2$ for the shock tube problem computed using Riemann solver based on Rusanov scheme. The true solution is calculated with $N = 8192$ grid resolution with $\Delta t = 0.00005$ and the low-resolution results are for $N = 256$ grid resolution with $\Delta t = 0.0001$.

Listing 17. Implementation of Julia function to calculate the interface flux using Riemann solver based on Rusanov scheme.

```julia
# nx: number of grid points
# gamma: ratios of specific heats
# qL, qR: left and right reconstructed states at the interface
# fL, fR: left and right side flux for reconstructed state
# f: flux at the interface computed using Rusanov scheme
function rusanov(nx,gamma,qL,qR,f,fL,fR)
ps = Array{Float64}(undef,nx+1)

wavespeed(nx,gamma,qL,qR,ps)
# Interface fluxes (Rusanov scheme)
for i = 1:nx+1 for m = 1:3
f[i,m] = 0.5*(fR[i,m]+fL[i,m]) - 0.5*ps[i]*(qR[i,m]-qL[i,m])
end end
end

function wavespeed(nx,gamma,uL,uR,ps)
```

```
gm = gamma-1.0
for i = 1:nx+1
#Left and right states:
rhLL = uL[i,1]
uuLL = uL[i,2]/rhLL
eeLL = uL[i,3]/rhLL
ppLL = gm*(eeLL*rhLL - 0.5*rhLL*(uuLL*uuLL))
hhLL = eeLL + ppLL/rhLL

rhRR = uR[i,1]
uuRR = uR[i,2]/rhRR
eeRR = uR[i,3]/rhRR
ppRR = gm*(eeRR*rhRR - 0.5*rhRR*(uuRR*uuRR))
hhRR = eeRR + ppRR/rhRR

alpha = 1.0/(sqrt(abs(rhLL)) + sqrt(abs(rhRR)))

uu = (sqrt(abs(rhLL))*uuLL + sqrt(abs(rhRR))*uuRR)*alpha
hh = (sqrt(abs(rhLL))*hhLL + sqrt(abs(rhRR))*hhRR)*alpha
aa = sqrt(abs(gm*(hh-0.5*uu*uu)))

ps[i] = abs(aa + uu)
end
end
```

6. Two-Dimensional Poisson Equation

In this section, we explain different methods to solve the Poisson equation which is encountered in solution to the incompressible flows. The Poisson equation is solved at every iteration step in the solution to the incompressible Navier-Stokes equation due to the continuity constraint. Therefore, many studies have been done to accelerate the solution to the Poisson equation using higher-order numerical methods [37], and developing fast parallel algorithms [38]. The Poisson equation is a second-order elliptic equation and can be represented as

$$\frac{\partial^2 u}{\partial x^2} + \frac{\partial^2 u}{\partial y^2} = f. \tag{82}$$

Using the second-order central difference formula for discretization of the Poisson equation, we get

$$\frac{u_{i+1,j} - 2u_{i,j} + u_{i-1,j}}{\Delta x^2} + \frac{u_{i,j+1} - 2u_{i,j} + u_{i,j-1}}{\Delta y^2} = f_{i,j}, \tag{83}$$

where Δx and Δy is the grid spacing in the x and y directions, and $f_{i,j}$ is the source term at discrete grid locations. If we write Equation (83) at each grid point, we get a system of linear equations. For the Dirichlet boundary condition, we assume that the values of $u_{i,j}$ are available when (x_i, y_j) is a boundary point. These equations can be written in the standard matrix notation as

$$
\underbrace{\begin{bmatrix}
D_{xy} & D_y & 0 & \cdots & D_x & \cdots & \cdots & 0 \\
D_y & D_{xy} & D_y & \ddots & & \ddots & & \vdots \\
0 & D_y & D_{xy} & D_y & \ddots & & \ddots & \vdots \\
\vdots & \ddots & \ddots & \ddots & \ddots & \ddots & & D_x \\
D_x & & \ddots & \ddots & \ddots & \ddots & \ddots & \vdots \\
\vdots & \ddots & & \ddots & D_y & D_{xy} & D_y & 0 \\
\vdots & & \ddots & & \ddots & D_y & D_{xy} & D_y \\
0 & \cdots & \cdots & D_x & \cdots & 0 & D_y & D_{xy}
\end{bmatrix}}_{A}
\underbrace{\begin{bmatrix}
u_{1,1} \\ u_{1,2} \\ \vdots \\ \vdots \\ u_{2,1} \\ \vdots \\ \vdots \\ u_{N_x,N_y}
\end{bmatrix}}_{u}
=
\underbrace{\begin{bmatrix}
f_{1,1} \\ f_{1,2} \\ \vdots \\ \vdots \\ f_{2,1} \\ \vdots \\ \vdots \\ f_{N_x,N_y}
\end{bmatrix}}_{b},
\tag{84}
$$

where

$$
D_{xy} = \frac{-2}{\Delta x^2} + \frac{-2}{\Delta y^2}; \quad D_x = \frac{1}{\Delta x^2}; \quad D_y = \frac{1}{\Delta y^2}.
$$

The boundary points of the domain are incorporated in the source term vector b in Equation (84). We can solve Equation (84) by standard methods for systems of linear equations, such as Gaussian elimination [39]. However, the matrix A is very sparse and the standard methods are computationally expensive for large size of A. In this paper, we discuss direct methods based on fast Fourier transform (FFT) and fast sine transform (FST) for periodic and Dirichlet boundary condition. We also explain iterative methods to solve Equation (84). We will further present a multigrid framework which scales linearly with a number of discrete grid points in the domain.

6.1. Direct Solver

Figure 15 shows the finite difference grid for two-dimensional CFD problems. We use two different boundary conditions for direct Poisson solver: periodic and Dirichlet boundary condition. For the Dirichlet boundary condition, the nodal values of a solution are already known and do not change, and hence, we do not do any calculation for the boundary points. In case of periodic boundary condition, we do calculation for grid points (i,j) between $[1, N_x] \times [1, N_y]$. The solution at right $(i = N_x + 1)$ and top $(j = N_y + 1)$ boundary is obtained from left $(i = 1)$ and bottom $(j = 1)$ boundary, respectively.

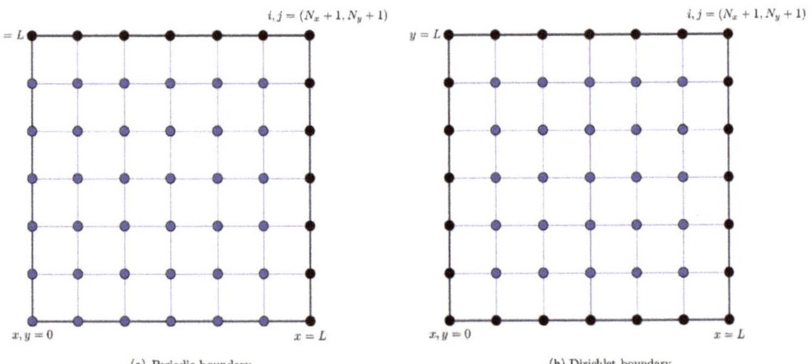

Figure 15. Finite difference grid for two-dimensional problems. The calculation is done only for points shown by blue color. The solution at black points is extended from left and bottom boundary for periodic boundary condition. The solution is already available at all four boundaries for Dirichlet boundary condition.

We perform the assessment of direct solver using the method of manufactured solution. We assume certain field u and compute the source term f at each grid location. We then solve the Poisson equation for this source term f and compare the numerically calculated field u with the exact solution field u. The exact field u and the corresponding source term f used for direct Poisson solver are given below

$$u(x,y) = \sin(2\pi x)\sin(2\pi y) + \frac{1}{16^2}\sin(32\pi x)\sin(32\pi y), \tag{85}$$

$$f(x,y) = -8\pi^2\sin(2\pi x)\sin(2\pi y) - 8\pi^2\sin(32\pi x)\sin(32\pi y). \tag{86}$$

This problem can have both periodic and Dirichlet boundary conditions. The computational domain is square in shape and is divided into 512×512 grid in x and y directions.

6.1.1. Fast Fourier Transform

There are two different ways to implement fast Poisson solver for the periodic domain. One way is to perform FFTs directly on the Poisson equation, which will give us the spectral accuracy. The second method is to discretize the Poisson equation first and then apply FFTs on the discretized equation. The second approach will give us the same spatial order of accuracy as the numerical scheme used for discretization. We use the second-order central difference scheme given in Equation (83) for developing a direct Poisson solver.

The Fourier transform decomposes a spatial function into its sine and cosine components. The output of the Fourier transform is the function in its frequency domain. We can recover the function from its frequency domain using the inverse Fourier transform. We use both the function and its Fourier transform in the discretized domain which is called the discrete Fourier transform (DFT). Cooley-Tukey proposed an FFT algorithm that reduces the complexity of computing DFT from $O(N^2)$ to $O(N\log N)$, where N is the data size [17]. In two-dimensions, the DFT for a square domain discretized equally in both directions is defined as

$$\tilde{u}_{m,n} = \sum_{i=0}^{N_x-1}\sum_{j=0}^{N_y-1} u_{i,j}e^{-\iota 2\pi\left(\frac{mi}{N_x} + \frac{nj}{N_y}\right)}, \tag{87}$$

where $u_{i,j}$ is the function in the spatial domain and the exponential term is the basis function corresponding to each point $\tilde{u}_{m,n}$ in the Fourier space, m and n are the wavenumbers in Fourier space in x and y directions, respectively. This equation can be interpreted as the value in the frequency domain $\tilde{u}_{m,n}$ can be obtained by multiplying the spatial function with the corresponding basis function and summing the result. The basis functions are sine and cosine waves with increasing frequencies. Here, $\tilde{u}_{0,0}$ represents the component of the function with the lowest wavenumber and $\tilde{u}_{N_x-1,N_y-1}$ represents the component with the highest wavenumber. Similarly, the function in Fourier space can be transformed to the spatial domain. The inverse discrete Fourier transform (IDFT) is given by

$$u_{i,j} = \frac{1}{N_x}\frac{1}{N_y}\sum_{m=-N_x/2}^{N_x/2-1}\sum_{n=-N_y/2}^{N_y/2-1} \tilde{u}_{m,n}e^{\iota 2\pi\left(\frac{mi}{N_x} + \frac{nj}{N_y}\right)}, \tag{88}$$

where $1/(N_x N_y)$ is the normalization term. The normalization can also be applied to forward transform, but it should be used only once. If we use Equation (88) in Equation (83), we get

$$\tilde{u}_{m,n}\left(\frac{e^{\iota\frac{2\pi m}{N_x}} - 2 + e^{\iota\frac{-2\pi m}{N_x}}}{\Delta x^2} + \frac{e^{\iota\frac{2\pi n}{N_y}} - 2 + e^{\iota\frac{-2\pi n}{N_y}}}{\Delta y^2}\right) = \tilde{f}_{m,n}, \tag{89}$$

in which we use the definition of cosine to yield

$$\tilde{u}_{m,n} \left(\frac{2\cos\left(\frac{2\pi m}{N_x}\right) - 2}{\Delta x^2} + \frac{2\cos\left(\frac{2\pi n}{N_y}\right) - 2}{\Delta y^2} \right) = \tilde{f}_{m,n}. \tag{90}$$

If we take the forward DFT of Equation (90), we get a spatial function $u_{i,j}$. The three step procedure to develop a Poisson solver using FFT is given below

- Apply forward FFT to find the Fourier coefficients of the source term in the Poisson equation ($\tilde{f}_{m,n}$ from the grid values $f_{i,j}$)
- Get the Fourier coefficients for the solution $\tilde{u}_{m,n}$ using below relation

$$\tilde{u}_{m,n} = \frac{\tilde{f}_{m,n}}{\left(\frac{2}{\Delta x^2}\cos\left(\frac{2\pi m}{N_x}\right) + \frac{2}{\Delta y^2}\cos\left(\frac{2\pi n}{N_y}\right) - \frac{2}{\Delta x^2} - \frac{2}{\Delta y^2} \right)}. \tag{91}$$

- Apply inverse FFT to get the grid values $u_{i,j}$ from the Fourier coefficients $\tilde{u}_{m,n}$.

Implementation of Poisson solver using FFT is given in Listing 18. Figure 16 shows the exact and numerical solution for for the test problem.

Listing 18. Implementation of Julia function for the Poisson solver using fast Fourier transform (FFT) for the domain with periodic boundary condition.

```julia
# nx,ny: number of grid points in x and y direction
# dx,dy: grid spacing in x and y direction
# f: source term of the Poisson equation
function ps_fft(nx,ny,dx,dy,f)
eps = 1.0e-6
kx = Array{Float64}(undef,nx)
ky = Array{Float64}(undef,ny)
data = Array{Complex{Float64}}(undef,nx,ny)
data1 = Array{Complex{Float64}}(undef,nx,ny)
e = Array{Complex{Float64}}(undef,nx,ny)
u = Array{Complex{Float64}}(undef,nx,ny)

aa = -2.0/(dx*dx) - 2.0/(dy*dy)
bb = 2.0/(dx*dx)
cc = 2.0/(dy*dy)

hx = 2.0*pi/nx #wavenumber~indexing

for i = 1:Int64(nx/2)
kx[i] = hx*(i-1.0)
kx[i+Int64(nx/2)] = hx*(i-Int64(nx/2)-1)
end
kx[1] = eps
ky = kx

for i = 1:nx
for j = 1:ny
data[i,j] = complex(f[i,j],0.0)
end
end

e = fft(data) # FFT of the data
```

```julia
e[1,1] = 0.0
for i = 1:nx
for j = 1:ny
data1[i,j] = e[i,j]/(aa + bb*cos(kx[i]) + cc*cos(ky[j]))
end
end

u = real(ifft(data1)) # inverse FFT of the data1
return u
end
```

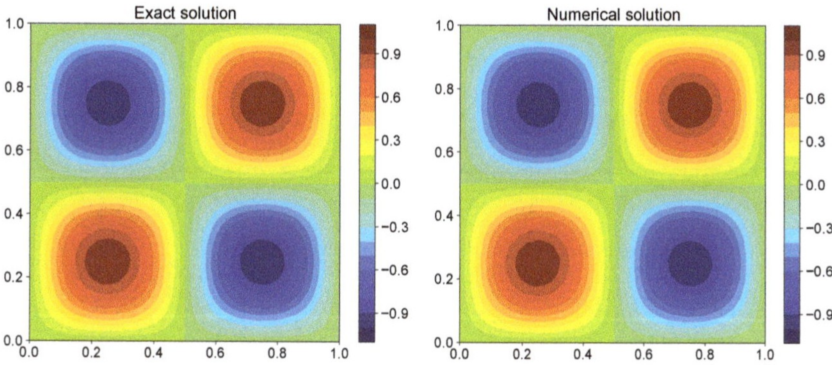

Figure 16. Comparison of exact solution and numerical solution computed using fast Fourier transform method for the Poisson equation with periodic boundary condition.

We develop the Poisson solver using a second-order central difference scheme for discretization of the Poisson equation. Instead of using the discretized equation, we can directly use Fourier analysis for the Poisson equation. This will compute the exact solution and the solution will not depend upon the grid size. The only change will be the change in Equation (91). We can easily derive the equation for Fourier coefficients in spectral solver by performing forward FFT of Equation (82). The Fourier coefficients for spectral solver can be written as

$$\tilde{u}_{m,n} = -\frac{\tilde{f}_{m,n}}{m^2 + n^2}. \tag{92}$$

We can get the solution by performing inverse FFT of the above equation. The implementation of spectral Poisson solver in Julia is outlined in Listing 19.

Listing 19. Implementation of Julia function for the spectral Poisson solver using fast Fourier transform (FFT) for the domain with periodic boundary condition.

```julia
# nx,ny: number of grid points in x and y direction
# dx,dy: grid spacing in x and y direction
# f: source term of the Poisson equation
function ps_spectral(nx,ny,dx,dy,f)
eps = 1.0e-6
kx = Array{Float64}(undef,nx)
ky = Array{Float64}(undef,ny)
data = Array{Complex{Float64}}(undef,nx,ny)
data1 = Array{Complex{Float64}}(undef,nx,ny)
e = Array{Complex{Float64}}(undef,nx,ny)
```

```julia
u = Array{Complex{Float64}}(undef,nx,ny)

hx = 2.0*pi/(nx*dx) #wavenumber indexing
for i = 1:Int64(nx/2)
kx[i] = hx*(i-1.0)
kx[i+Int64(nx/2)] = hx*(i-Int64(nx/2)-1)
end
kx[1] = eps
ky = kx

for i = 1:nx
for j = 1:ny
data[i,j] = complex(f[i,j],0.0)
end
end

e = fft(data)
e[1,1] = 0.0
for i = 1:nx
for j = 1:ny
data1[i,j] = e[i,j]/(-kx[i]^2 -ky[j]^2)
end
end

u = real(ifft(data1))
return u
end
```

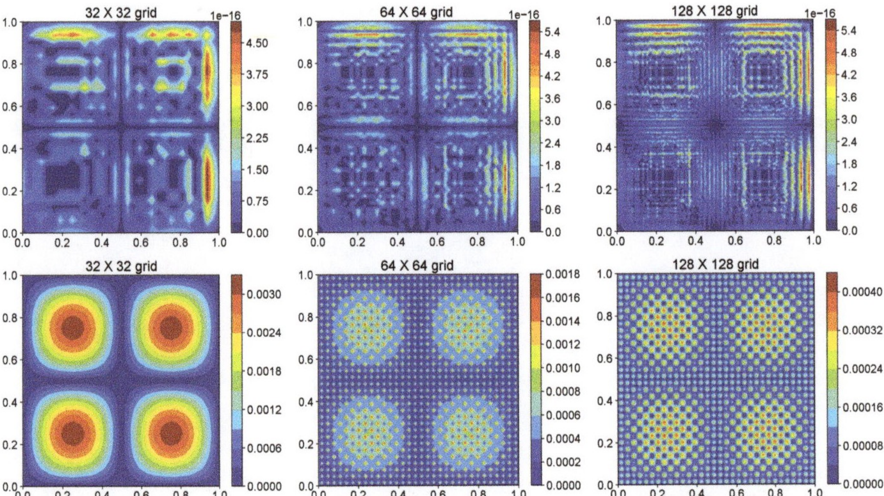

Figure 17. Comparison of error for spectral method (Top) and second-order CDS (bottom) for three different grid resolutions. The error is defined as $\epsilon(x,y) = |u^{exact} - u^{numerical}|$.

The spectral Poisson solver gives the exact solution and the error between exact and numerical solution is machine zero error. We demonstrate it by using spectral Poisson solver for three different grid resolutions. We also use the central-difference scheme (CDS) Poisson solver at these grid resolutions and compare the discretization errors for two methods. We systematically change the grid spacing by a factor of two in both x and y direction and compute the L_2 norm of the discretization error. Figure 17 shows the contour plot for the error at three grid resolutions for spectral and second-order CDS. We can notice that the error for the spectral method does not depend on the grid resolution and the value of error is of the same order as machine zero error. On the other hand, for CDS the error reduces as we increase the grid resolution. This is because the truncation error goes down with the decrease in grid spacing. Figure 18 plots the L_2 norm of the error for spectral method and CDS. The slope of the discretization error is two for CDS showing that the scheme has second-order of accuracy. This method is used in CFD to validate the code and numerical methods. For the spectral method, the L_2 norm of the error has the machine zero value for all three grid resolutions.

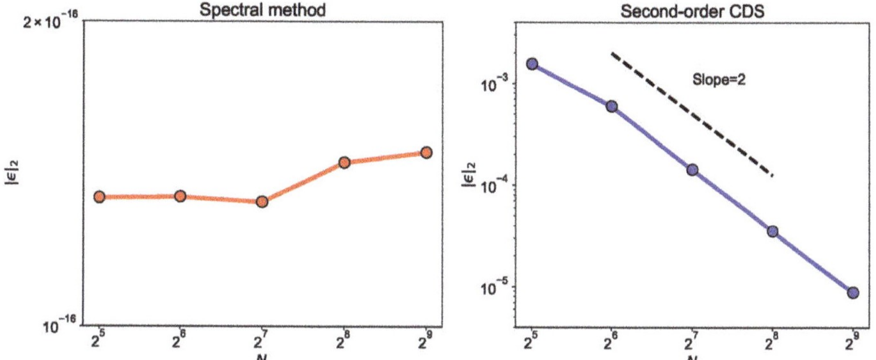

Figure 18. Order of accuracy for spectral method and second-order CDS. We calculate discretization error at five grid resolutions and change the grid size by a factor of 2 in both x and y direction. The CDS Poisson solver give second-order accuracy. Spectral method give exact solution and error is of the same order as machine zero.

6.1.2. Fast Sine Transform

The procedure described in Section 6.1.1 is valid only when the problem has periodic boundary condition, i.e., the solution that satisfies $u_{i,j} = u_{i+N_x,j} = u_{i,j+N_y}$. If we have homogeneous Dirichlet boundary condition for the square domain problem, we need to use the sine transform given by

$$\tilde{u}_{m,n} = \sum_{i=0}^{N_x-1} \sum_{j=0}^{N_x-1} u_{i,j} \sin\frac{\pi m i}{N_x} \sin\frac{\pi n j}{N_y}, \tag{93}$$

where m and n are the wavenumbers. This satisfies the homogeneous Dirichlet boundary conditions that $u = 0$ at $i = 0, N_x$ and $j = 0, N_y$. If we substitute Equation (93) in Equation (83) and use trigonometric identity $(\sin(A \pm B) = \sin(A)\cos(B) \pm \sin(B)\cos(A))$ we get

$$\tilde{u}_{m,n} \left(\frac{2\cos\left(\frac{\pi m}{N_x}\right) - 2}{\Delta x^2} + \frac{2\cos\left(\frac{\pi n}{N_y}\right) - 2}{\Delta y^2} \right) = \tilde{f}_{m,n}. \tag{94}$$

We can compute the spatial field using inverse sine transform given by

$$u_{i,j} = \frac{2}{N_x} \frac{2}{N_y} \sum_{m=0}^{N_x-1} \sum_{n=0}^{N_y-1} \tilde{u}_{m,n} \sin\frac{\pi m i}{N_x} \sin\frac{\pi n j}{N_y}. \tag{95}$$

The three-step procedure to develop Poisson solver using sine transform can be listed as

- Apply forward sine transform to find the Fourier coefficients of the source term in the Poisson equation ($\tilde{f}_{m,n}$ from the grid values $f_{i,j}$)
- Get the Fourier coefficients for the solution $\tilde{u}_{m,n}$ using below relation

$$\tilde{u}_{m,n} = \frac{\tilde{f}_{m,n}}{\left(\frac{2}{\Delta x^2}\cos\left(\frac{\pi m}{N_x}\right) + \frac{2}{\Delta y^2}\cos\left(\frac{\pi n}{N_y}\right) - \frac{2}{\Delta x^2} - \frac{2}{\Delta y^2} \right)}. \tag{96}$$

- Apply inverse sine transform to get the grid values $u_{i,j}$ from the Fourier coefficients $\tilde{u}_{m,n}$.

The implementation of Poisson solver for problems with homogeneous Dirichlet boundary condition using the sine transform is given in Listing 20. Figure 19 displays the comparison of exact and numerical solution for the test problem computed using sine transform.

Listing 20. Implementation of Julia function for the Poisson solver using fast sine transform (FST) for the domain with Dirichlet boundary condition.

```julia
# nx,ny: number of grid points in x and y direction
# dx,dy: grid spacing in x and y direction
# f: source term of the Poisson equation
function ps_fst(nx,ny,dx,dy,f)
data = Array{Complex{Float64}}(undef,nx-1,ny-1)
data1 = Array{Complex{Float64}}(undef,nx-1,ny-1)
e = Array{Complex{Float64}}(undef,nx-1,ny-1)

u = Array{Complex{Float64}}(undef,nx-1,ny-1)

for i = 1:nx-1
for j = 1:ny-1
data[i,j] = f[i+1,j+1]
end
end

e = FFTW.r2r(data,FFTW.RODFT00)

for i = 1:nx-1
for j = 1:ny-1
alpha = (2.0/(dx*dx))*(cos(pi*i/nx) - 1.0) +
(2.0/(dy*dy))*(cos(pi*j/ny) - 1.0)
data1[i,j] = e[i,j]/alpha
end
end

u = FFTW.r2r(data1,FFTW.RODFT00)/((2*nx)*(2*ny))
return u
end
```

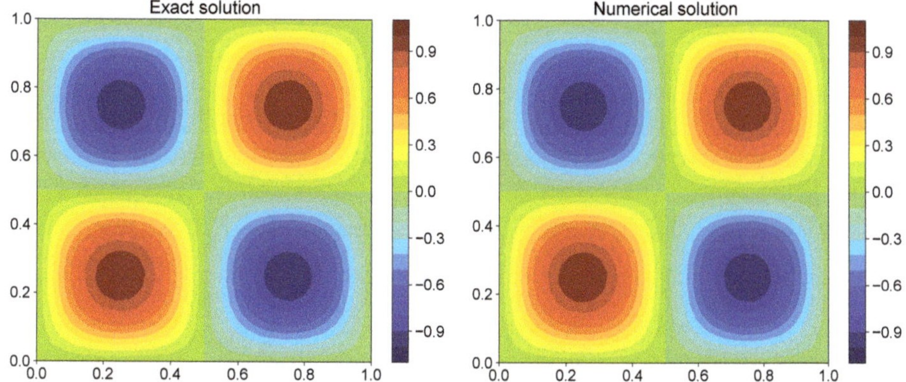

Figure 19. Comparison of exact solution and numerical solution computed using fast sine transform method for the Poisson equation with Dirichlet boundary condition.

6.2. Iterative Solver

Iterative methods use successive approximations to obtain the most accurate solution to a linear system at every iteration step. These methods start with an initial guess and proceed to generate a sequence of approximations, in which the k-th approximation is derived from the previous ones. There are two main types of iterative methods. Stationary methods that are easy to understand and are simple to implement, but are not very effective. Non-stationary methods which are based on the idea of the sequence of orthogonal vectors. We would like to recommend a text by Barrett et al. [40] which provides a good discussion on the implementation of iterative methods for solving a linear system of equations.

We show only Dirichlet boundary condition implementation for iterative methods. For iterative methods, we stop iterations when the L_2 norm of the residual for Equation (83) goes below 1×10^{-10}. We test the performance of all iterative methods using the method of manufactured solution. The exact field u and the corresponding source term f used for the method of manufactured solution for iterative methods are given below [12]

$$u(x,y) = (x^2 - 1)(y^2 - 1), \tag{97}$$
$$f(x,y) = -2(2 - x^2 - y^2). \tag{98}$$

6.2.1. Stationary Methods: Gauss-Seidel

The iterative methods work by splitting the matrix A into

$$A = M - P. \tag{99}$$

Equation (84) becomes

$$Mu = Pu + b. \tag{100}$$

The solution at $(k+1)$ iteration is given by

$$Mu^{(k+1)} = Pu^{(k)} + b. \tag{101}$$

If we take the difference between the above two equations, we get the evolution of error as $\epsilon^{(k+1)} = M^{-1}P\epsilon^{(k)}$. For the solution to converge to the exact solution and the error to go to zero, the largest eigenvalue of iteration matrix $M^{-1}P$ should be less than 1. The approximate number of iterations required for the error ϵ to go below some specified tolerance δ is given by

$$Q = \frac{\ln(\delta)}{\ln(\lambda_1)}, \tag{102}$$

where Q is the number of iterations and λ_1 is the maximum eigenvalue of iteration matrix $M^{-1}P$. If the convergence tolerance is specified to be 1×10^{-5} then the number of iterations for convergence will be $Q = 1146$ and $Q = 23,020$ for $\lambda_1 = 0.99$ and $\lambda_1 = 0.9995$ respectively. Therefore, the maximum eigenvalue of the iteration matrix should be less for faster convergence. When we implement iterative methods, we do not form the matrix M and P. Equation (101) is just the matrix representation of the iterative method. In matrix terms, the Gauss-Seidel method can be expressed as

$$M = D - L, \tag{103}$$

$$u^{(k+1)} = (D - L)^{-1}\big(Uu^{(k)} + b\big), \tag{104}$$

where D, L and U represent the diagonal, the strictly lower-triangular, and the strictly upper-triangular parts of A, respectively. We do not explicitly construct the matrices D, L, and U, instead we use the update formula based on data vectors (i.e., we obtain a vector once a matrix operates on a vector).

The update formulas for Gauss-Seidel iterations if we iterate from left to right and from bottom to top can be written as

$$u_{i,j}^{(k+1)} = u_{i,j}^{(k)} - \left(\frac{2}{\Delta x^2} + \frac{2}{\Delta y^2}\right)r_{i,j}, \tag{105}$$

where

$$r_{i,j} = f_{i,j}^{(k)} - \frac{u_{i+1,j}^{(k)} - 2u_{i,j}^{(k)} + u_{i-1,j}^{(k+1)}}{\Delta x^2} - \frac{u_{i,j+1}^{(k)} - 2u_{i,j}^{(k)} + u_{i,j-1}^{(k+1)}}{\Delta y^2}. \tag{106}$$

The implementation of the Gauss-Seidel method for Poisson equation in Julia is given in Listing 21. The pseudo algorithm is given by Algorithm 3, where we define an operator $\mathcal{A}$ such that

$$\mathcal{A}u_{i,j} = \frac{u_{i+1,j} - 2u_{i,j} + u_{i-1,j}}{\Delta x^2} + \frac{u_{i,j+1} - 2u_{i,j} + u_{i,j-1}}{\Delta y^2}, \tag{107}$$

gives same results as matrix-vector multiplication Au for $(i,j)^{th}$ grid point. It can be noticed that the matrix M and P are not formed, and the solution is computed using Equation (105) for every point on the grid. Since we are traversing bottom to top and left to right, the updated values on left $(u_{i-1,j})$ and bottom $(u_{i,j-1})$ will be used for updating the solution at $(i,j)^{th}$ grid point. This will help in accelerating the convergence compared to the Jacobi method which uses values at the previous iteration step for all its neighboring points.

Figure 20 shows the contour plot for exact and numerical solution for the Gauss-Seidel solver. As seen in Figure 24, the residual reaches 10^{-10} values after more than 400,000 iterations for Gauss-Seidel method. The maximum eigenvalue of the iteration matrix is large for the Gauss-Seidel method and hence, the rate of convergence is slow. The rate of convergence is higher in the beginning while the high wavenumber errors are still present. Once the high wavenumber errors are resolved, the rate of convergence for low wavenumber errors drops. There are ways to accelerate the convergence such as using successive overrelaxation methods [41]. The residual is computed from Equation (83) as the difference between the source term and the left hand side term computed based on the central-difference scheme. We would like to point out that the residual is different from the discretization error. The discretization error is the difference between numerical and exact solution.

Algorithm 3 Gauss-Seidel iterative method

1: Given b ▷ source term
2: $u = u^{(0)}$ ▷ initialize the solution
3: $d = -\left(\frac{2}{\Delta x^2} + \frac{2}{\Delta y^2}\right)$
4: **while** tolerance met **do**
5:      **for** $i = 1$ to N_x **do**
6:          **for** $j = 1$ to N_y **do**
7:             $r_{i,j} = b_{i,j} - \mathcal{A}u_{i,j}$ ▷ compute the residual
8:             $u_{i,j} = u_{i,j} + r_{i,j}/d$ ▷ update the solution
9:          **end for**
10:      **end for**
11:      check convergence criteria, continue if necessary
12: **end while**

Listing 21. Implementation of Gauss-Seidel iterative method for Poisson equation in Julia.

```julia
# nx, ny: number of grid points in x and y directions
# nx, ny: grid spacing in x and y directions
# un: numerical solution matrix for the Poisson equation
# r: residual matrix
# max_iter: maximum number if iterations
# init_rms: L-2 norm of the residual at the start of iterations
den = -2.0/dx^2 - 2.0/dy^2
k = 0

for k = 1:max_iter
# calculate numerical solution at k-th step using Gauss-Seidel iterative method
for j = 2:ny for i = 2:nx
d2udx2 = (un[i+1,j] - 2*un[i,j] + un[i-1,j])/(dx^2)
d2udy2 = (un[i,j+1] - 2*un[i,j] + un[i,j-1])/(dy^2)
r[i,j] = f[i,j] - d2udx2 - d2udy2
un[i,j] = un[i,j] + r[i,j]/den # update the solution at every step
end~end

# compute the residual at k-th step using the new solution
for j = 2:ny for i = 2:nx
d2udx2 = (un[i+1,j] - 2*un[i,j] + un[i-1,j])/(dx^2)
d2udy2 = (un[i,j+1] - 2*un[i,j] + un[i,j-1])/(dy^2)
r[i,j] = f[i,j]  - d2udx2 - d2udy2
end~end

# calculate the L-2 norm of the residual vector
rms = 0.0
for j = 2:ny for i = 2:nx
rms = rms + r[i,j]^2
end end
rms = sqrt(rms/((nx-1)*(ny-1)))

# if the convergence critera (rms<tolerance) is satidfied, stop the iteration
if (rms/init_rms) <= 1e-12
break
end
end
```

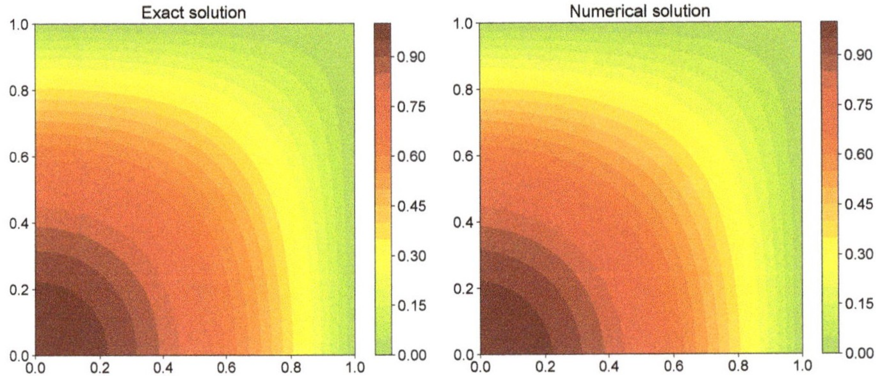

Figure 20. Comparison of exact solution and numerical solution computed using Gauss-Seidel iterative method for the Poisson equation.

6.2.2. Non-Stationary Methods: Conjugate Gradient Algorithm

Non-stationary methods differ from stationary methods in that the iterative matrix changes at every iteration. These methods work by forming a basis of a sequence of matrix powers times the initial residual. The basis is called as the Krylov subspace and mathematically given by $\mathcal{K}_n(A, b) = \text{span}\{b, Ab, A^2b, \ldots, A^{n-1}b\}$. The approximate solution to the linear system is found by minimizing the residual over the subspace formed. In this paper, we discuss the conjugate gradient method which is one of the most effective methods for symmetric positive definite systems.

The conjugate gradient method proceeds by calculating the vector sequence of successive approximate solution, residual corresponding the approximate solution, and search direction used in updating the solution and residuals. The approximate solution $u^{(k)}$ is updated at every iteration by a scalar multiple α_k of the search direction vector $p^{(k)}$:

$$u^{(k+1)} = u^{(k)} + \alpha_k p^{(k)}. \tag{108}$$

Correspondingly, the residuals $r^{(k)}$ are updated as

$$r^{(k+1)} = r^{(k)} - \alpha_k q^{(k)} \quad \text{where} \quad q^{(k)} = \mathcal{A}p^{(k)}. \tag{109}$$

The search directions are then updated using the residuals

$$p^{(k+1)} = r^{(k+1)} + \beta_k p^{(k)}, \tag{110}$$

where the choice $\beta_k = r^{(k)^T} r^{(k)} / r^{(k-1)^T} r^{(k-1)}$ ensures that $p^{(k+1)}$ and $r^{(k+1)}$ are orthogonal to all previous $\mathcal{A}p^{(k)}$ and $r^{(k)}$ respectively. A detailed derivation of the conjugate gradient method can be found in [42]. The complete procedure for the conjugate gradient method is given in Algorithm 4. The implementation of conjugate gradient method in Julia is given in Listing 22.

The comparison of the exact and numerical solution is shown in Figure 21. The residual history for the conjugate gradient method is displayed in Figure 24. It can be seen that the rate of convergence is slower at the start of iterations. As the conjugate gradient algorithm finds the correct direction for residual, the rate of convergence increases. There are several types of iterative methods which are not discussed in this paper and for further reading we refer to a book "Iterative Methods for Sparse Linear Systems" [43].

Listing 22. Implementation of conjugate gradient method for Poisson equation in Julia.

```julia
# nx, ny: number of grid points in x and y directions
# nx, ny: grid spacing in x and y directions
# un: numerical solution matrix for the Poisson equation
# r: residual matrix
# max_iter: maximum number if iterations
# init_rms: L-2 norm of the residual at the start of iterations
p = r # assign initial residual to the conjugate vector
for k = 1:max_iter
# conjugate gradient algorithm
for j = 2:ny for i = 2:nx
q[i,j] = (p[i+1,j] - 2.0*p[i,j] + p[i-1,j])/(dx^2) +
(p[i,j+1] - 2.0*p[i,j] + p[i,j-1])/(dy^2)
end~end

aa, bb = 0.0, 0.0
for j = 2:ny for i = 2:nx
aa = aa + r[i,j]*r[i,j] # <r,r>
bb = bb + q[i,j]*p[i,j] # <q,p>
end end
cc = aa/(bb + tiny) #alpha

for j = 2:ny for i = 2:nx
un[i,j] = un[i,j] + cc*p[i,j] # update the numerical solution
end~end

bb, aa = aa, 0.0
for j = 2:ny for i = 2:nx
r[i,j] = r[i,j] - cc*q[i,j] # update the residual
aa = aa + r[i,j]*r[i,j]
end end
cc = aa/(bb+tiny) # beta

for j = 1:ny for i = 1:nx
p[i,j] = r[i,j] + cc*p[i,j] # update the conjugate vector
end~end

# compute the residual at k-th step using the new solution
for j = 2:ny for i = 2:nx
d2udx2 = (un[i+1,j] - 2*un[i,j] + un[i-1,j])/(dx^2)
d2udy2 = (un[i,j+1] - 2*un[i,j] + un[i,j-1])/(dy^2)
r[i,j] = f[i,j]  - d2udx2 - d2udy2
end end
# calculate the L-2 norm of the residual vector
rms = 0.0
for j = 2:ny for i = 2:nx
rms = rms + r[i,j]^2
end end
rms = sqrt(rms/((nx-1)*(ny-1)))
# if the convergence critera (rms<tolerance) is satidfied, stop the iteration
if (rms/initial_rms) <= 1e-12
break
end
end
```

Algorithm 4 Conjugate gradient algorithm

1: Given $\boldsymbol{b}$ ▷ Given source term in space, i.e., $\boldsymbol{b} = f(x,y)$ in Equation (82)
2: Given matrix operator $\mathcal{A}$ ▷ Given discretization by Equation (107)
3: $\boldsymbol{u}^{(0)} = \boldsymbol{b}$ ▷ Initialize the solution
4: $\boldsymbol{r}^{(0)} = \boldsymbol{b} - \mathcal{A}\boldsymbol{u}^{(0)}$ ▷ Initialize the residual
5: $\boldsymbol{p}^{(0)} = \boldsymbol{r}^{(0)}$ ▷ Initialize the conjugate
6: $k = 0$
7: $\rho_0 = \boldsymbol{r}^{(0)T}\boldsymbol{r}^{(0)}$
8: **while** tolerance met (or $k < \mathcal{N}$) **do**
9:      $\boldsymbol{q}^{(k)} = \mathcal{A}\boldsymbol{p}^{(k)}$
10:      $\alpha_k = \rho_k / \boldsymbol{p}^{(k)T}\boldsymbol{q}^{(k)}$
11:      $\boldsymbol{u}^{(k+1)} = \boldsymbol{u}^{(k)} + \alpha_k \boldsymbol{p}^{(k)}$ ▷ Update the solution
12:      $\boldsymbol{r}^{(k+1)} = \boldsymbol{r}^{(k)} - \alpha_k \boldsymbol{q}^{(k)}$
13:      $\rho_{k+1} = \boldsymbol{r}^{(k+1)T}\boldsymbol{r}^{(k+1)}$
14:      $\beta_k = \rho_{k+1} / \rho_k$
15:      $\boldsymbol{p}^{(k+1)} = \boldsymbol{r}^{(k+1)} + \beta_k \boldsymbol{p}^{(k)}$ ▷ Update the residual
16:      check convergence; continue if necessary
17:      $k \leftarrow k + 1$
18: **end while**

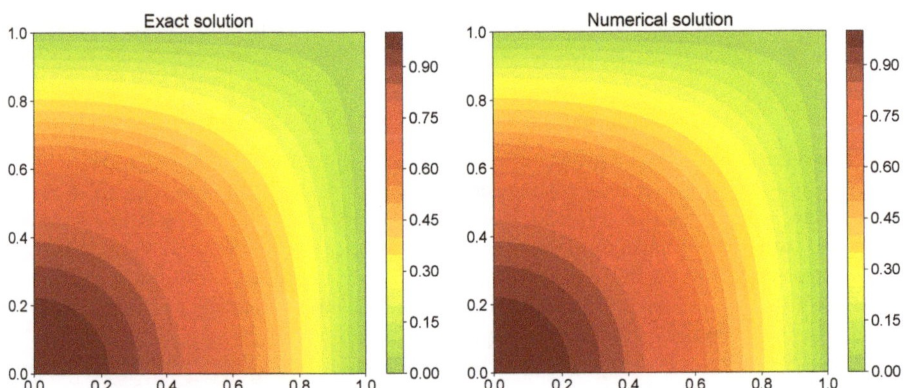

Figure 21. Comparison of exact solution and numerical solution computed using conjugate gradient algorithm for the Poisson equation.

6.3. Multigrid Framework

We saw in Section 6.2 that the rate of convergence for iterative methods depends upon the iteration matrix $M^{-1}N$. Point iterative methods like Jacobi and Gauss-Seidel methods have large eigenvalue and hence the slow convergence. As the grid becomes finer, the maximum eigenvalue of the iteration matrix becomes close to 1. Therefore, for very high-resolution simulation, these iterative methods are not feasible due to the large computational time required for residuals to go below some specified tolerance.

The multigrid framework is one of the most efficient iterative algorithm to solve the linear system of equations arising due to the discretization of the Poisson equation. The multigrid framework works on the principle that low wavenumber errors on fine grid behave like a high wavenumber error on a coarse grid. In the multigrid framework, we restrict the residuals on the fine grid to the coarser grid. The restricted residual is then relaxed to resolve the low wavenumber errors and the correction to the solution is prolongated back to the fine grid. We can use any of the iterative methods like Jacobi,

Gauss-Seidel method for relaxation. The algorithm can be implemented recursively on the hierarchy of grids to get faster convergence.

Let $\tilde{u}^{\Delta x}$ denote the approximate solution after applying few steps of iterations of a relaxation method. The fine grid solution will be

$$A(\tilde{u}^{\Delta x} + e^{\Delta x}) = b^{\Delta x}, \tag{111}$$

$$Ae^{\Delta x} = b^{\Delta x} - A\tilde{u}^{\Delta x}, \tag{112}$$

where $e^{\Delta x}$ is the error at fine grid and the operator A is defined in Equation (84) for the Cartesian grid. We define the residual vector $r^{\Delta x} = Ae^{\Delta x}$. This residual vector contains mostly low wavenumber errors. The correction at the fine grid is obtained by restricting the residual vector to a coarse grid ($2\Delta x$ grid spacing) and solving an equivalent system to obtain the correction. The equivalent system on a coarse grid can be written as

$$A^{2\Delta x}\tilde{e}^{2\Delta x} = \mathcal{R}(r^{\Delta x}), \tag{113}$$

where $A^{2\Delta x}$ is the elliptic operator on coarse grid, and $\mathcal{R}$ is the restriction operator. The discretized form of Equation (113) can be written as

$$\frac{\tilde{e}^{2\Delta x}_{i+1,j} - 2\tilde{e}^{2\Delta x}_{i,j} + \tilde{e}^{2\Delta x}_{i-1,j}}{2\Delta x^2} + \frac{\tilde{e}^{2\Delta x}_{i,j+1} - 2\tilde{e}^{2\Delta x}_{i,j} + \tilde{e}^{2\Delta x}_{i,j-1}}{2\Delta y^2} = \mathcal{R}(r^{\Delta x}_{i,j}). \tag{114}$$

We compute the approximate error at an intermediate grid using some relaxation operation. Once the approximation to $\tilde{e}^{2\Delta x}$ is obtained, it is prolongated back to the fine grid using

$$\hat{e}^{\Delta x} = \mathcal{P}(\tilde{e}^{2\Delta x}), \tag{115}$$

where $\mathcal{P}$ is the prolongation operator. The approximate solution at fine grid $\tilde{u}^{\Delta x}$ is corrected with $\hat{e}^{\Delta x}$. The approximate solution is relaxed for a specific number of iterations and convergence criteria is checked. If the residuals are below specified tolerance we stop or we repeat the steps. The procedure can be easily extended to more number of levels. The illustration of the V-cycle multigrid framework for two levels is shown in Figure 22. In this study, we use two iterations of the Gauss-Seidel method for relaxation at every grid level. The number of iterations can be set to a different number or we can also set the residual tolerance instead of a fixed number of iterations.

The residuals corresponding to the fine grid can be projected on to the coarse grid either using full-weighting. The full-weighting restriction operator is given by

$$r^{2\Delta x}_{i,j} = \frac{4r^{\Delta x}_{2i-1,2j-1} + 2(r^{\Delta x}_{2i-1,2j} + r^{\Delta x}_{2i-1,2j-2} + r^{\Delta x}_{2i,2j-1} + r^{\Delta x}_{2i-2,2j-1}) + r^{\Delta x}_{2i,2j} + r^{\Delta x}_{2i,2j-2} + r^{\Delta x}_{2i-2,2j} + r^{\Delta x}_{2i-2,2j-2}}{16}. \tag{116}$$

For boundary points, the residuals are directly injected from fine grid to coarse grid. The implementation of restriction operator in Julia is given in Listing 23.

The prolongation operator transfers the error from the coarse grid to the fine grid. We use direct injection for points which are common to both coarse and fine grid and bilinear interpolation for points which are on the fine grid but not on the coarse grid. The prolongation operations are given by below equations

$$e^{\Delta x}_{2i-1,2j-1} = e^{2\Delta x}_{i,j},\tag{117}$$

$$e^{\Delta x}_{2i-1,2j-1+1} = \frac{e^{2\Delta x}_{i,j} + e^{2\Delta x}_{i,j+1}}{2},\tag{118}$$

$$e^{\Delta x}_{2i-1+1,2j-1} = \frac{e^{2\Delta x}_{i,j} + e^{2\Delta x}_{i+1,j}}{2},\tag{119}$$

$$e^{\Delta x}_{2i-1+1,2j-1+1} = \frac{e^{2\Delta x}_{i,j} + e^{2\Delta x}_{i,j+1} + e^{2\Delta x}_{i+1,j} + e^{2\Delta x}_{i+1,j+1}}{2}.\tag{120}$$

The implementation of prolongation operation in Julia is given in Listing 24.

Listing 23. Implementation of restriction operation in multigrid framework.

```julia
# nxf, nyf: number of grid points in x and y directions on fine grid
# nxc, nyc: number of grid points in x and y directions on coarse grid
# (nxc = nxf/2, nyc = nyf/2)
# r: residual matrix for the Poisson equation on fine grid
# ec: error correction on coarse grid
function restriction(nxf, nyf, nxc, nyc, r, ec)
for j = 2:nyc for i = 2:nxc
# grid index for fine grid for the same coarse point
center = 4.0*r[2*i-1, 2*j-1]
# E, W, N, S with respect to coarse grid point in fine grid
grid = 2.0*(r[2*i-1, 2*j-1+1] + r[2*i-1, 2*j-1-1] +
r[2*i-1+1, 2*j-1] + r[2*i-1-1, 2*j-1])
# NE, NW, SE, SW with respect to coarse grid point in fine grid
corner = 1.0*(r[2*i-1+1, 2*j-1+1] + r[2*i-1+1, 2*j-1-1] +
r[2*i-1-1, 2*j-1+1] + r[2*i-1-1, 2*j-1-1])
# restriction using trapezoidal rule
ec[i,j] = (center + grid + corner)/16.0
end end
end
```

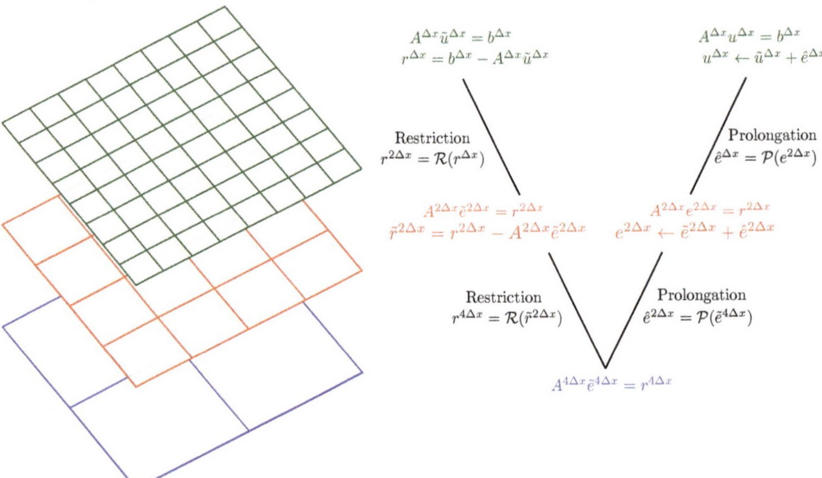

Figure 22. Illustration of multigrid V-cycle framework with three levels between fine and coarse grid. For full multigrid cycle, the coarsest mesh has 2×2 grid resolution and is solved directly.

Listing 24. Implementation of prolongation operation in multigrid framework.

```julia
# nxf, nyf: number of grid points in x and y directions on fine grid
# nxc, nyc: number of grid points in x and y directions on coarse grid
# (nxc = nxf/2, nyc = nyf/2)
# unc: solution to the error correction on coarse grid
# ef: error correction on fine grid
function prolongation(nxc, nyc, nxf, nyf, unc, ef)
for j = 1:nyc for i = 1:nxc
# direct injection at center point
ef[2*i-1, 2*j-1] = unc[i,j]
# east neighnour on fine grid corresponding to coarse grid point
ef[2*i-1, 2*j-1+1] = 0.5*(unc[i,j] + unc[i,j+1])
# north neighbout on fine grid corresponding to coarse grid point
ef[2*i-1+1, 2*j-1] = 0.5*(unc[i,j] + unc[i+1,j])
# north-east neighbour on fine grid corresponding to coarse grid point
ef[2*i-1+1, 2*j-1+1] = 0.25*(unc[i,j] + unc[i,j+1] +
unc[i+1,j] + unc[i+1,j+1])
end end
end
```

The relaxation of the solution is done using the Gauss-Seidel method for each grid level as formulated in Section 6.2.1 for a fixed number of iterations. The Julia implementation of relaxation operation is provided in Listing 25. The pseudocode for V-cycle multigrid framework for three levels if provided in Algorithm 5. The implementation of a complete multigrid framework in Julia for two levels is given in Listing 26.

Listing 25. Implementation of relaxation operation (using Gauss-Seidel iterative method) in multigrid framework.

```julia
# nxf, nyf: number of grid points in x and y directions on fine grid
# nx, ny: number of grid points in x and y directions
# dx, dy: grid spacing in x and y directions
# un: relaxation solution after fixed number of iterations
# f: source term
# v: number of iterations
function gauss_seidel_mg(nx, ny, dx, dy, f, un, V)
rt = zeros(Float64, nx+1, ny+1) # temporary variable
den = -2.0/dx^2 - 2.0/dy^2

for iteration_count = 1:V
for j = 2:nx for i = 2:ny
rt[i,j] = f[i,j] - (un[i+1,j] - 2.0*un[i,j] + un[i-1,j])/dx^2
-(un[i,j+1] - 2.0*un[i,j] + un[i,j-1])/dy^2
un[i,j] = un[i,j] + rt[i,j]/den
end end
end
end
```

Algorithm 5 V-cycle multigrid framework

1: Given operator $\mathcal{A}$
2: Given v1 $\triangleright$ number of relaxation operation during restriction
3: Given v2 $\triangleright$ number of relaxation operation during prolongation
4: Given v3 $\triangleright$ number of relaxation operation on coarsest grid
5: **while** tolerance met (or $k < \mathcal{N}$) **do**
6:      **for** v1 iterations **do**
7:          $A^{\Delta x} \tilde{u}^{\Delta x} = b^{\Delta x}$ $\triangleright$ relaxation for v1 iterations
8:      **end for**
9:      check convergence; continue if necessary
10:     $r^{\Delta x} = b^{\Delta x} - A^{\Delta x} \tilde{u}^{\Delta x}$ $\triangleright$ compute residual for fine grid
11:     $r^{2\Delta x} = \mathcal{R}(r^{\Delta x})$ $\triangleright$ restrict the residual from fine grid to intermediate grid
12:     **for** v1 iterations **do**
13:         $A^{2\Delta x} \tilde{e}^{2\Delta x} = r^{2\Delta x}$ $\triangleright$ relaxation of error for v1 iterations
14:     **end for**
15:     $\tilde{r}^{2\Delta x} = r^{2\Delta x} - A^{2\Delta x} \tilde{e}^{2\Delta x}$ $\triangleright$ compute residual using error at intermediate grid
16:     $r^{4\Delta x} = \mathcal{R}(\tilde{r}^{2\Delta x})$ $\triangleright$ restrict the residual for error to coarsest grid
17:     **for** v3 iterations **do** $\triangleright$ can be solved directly instead of v3 iterations
18:         $A^{4\Delta x} \tilde{e}^{4\Delta x} = r^{4\Delta x}$ $\triangleright$ relaxation of error for v3 iterations
19:     **end for**
20:     $\hat{e}^{2\Delta x} = \mathcal{P}(\tilde{e}^{4\Delta x})$ $\triangleright$ prolongate error from coarsest grid to intermediate level
21:     $e^{2\Delta x} \leftarrow \tilde{e}^{2\Delta x} + \hat{e}^{2\Delta x}$ $\triangleright$ add correction to the error at intermediate grid
22:     **for** v2 iterations **do**
23:         $A^{2\Delta x} e^{2\Delta x} = r^{2\Delta x}$ $\triangleright$ relaxation of error for v2 iterations
24:     **end for**
25:     $\hat{e}^{\Delta x} = \mathcal{P}(e^{2\Delta x})$ $\triangleright$ prolongate error from intermediate grid to coarse grid
26:     $u^{\Delta x} \leftarrow \tilde{u}^{\Delta x} + \hat{e}^{\Delta x}$ $\triangleright$ add correction to the solution at fine level
27:     **for** v2 iterations **do**
28:         $A^{\Delta x} u^{\Delta x} = b^{\Delta x}$ $\triangleright$ relaxation of solution for v2 iterations
29:     **end for**
30: **end while**

Listing 26. Implementation of multigrid framework in Julia with two levels in V-cycle.

```julia
# nxf, nyf: number of grid points in x and y directions on fine grid
# dx,dy: grid spacing in x and y direction for fine grid
# nxc, nyc: number of grid points in x and y directions on coarse grid
# (nxc = nxf/2, nyc = nyf/2)
# un: numerical solution on fine grid (required solution)
# unc: solution to the error correction on coarse grid
# v1, v2, v3: number of iterations in relaxation step for different grid levels
# ef: error correction on fine~grid
for k = 1:max_iter
# call relaxation on fine grid and compute the numerical solution
gauss_seidel_mg(nxf, nyf, dx, dy, f, un, v1)

# compute the residual and L2 norm
compute_residual(nxf, nyf, dx, dy, f, un, r)
rms = compute_l2norm(nxf, nyf, r)

if (rms/init_rms) <= 1e-12
break
end

# restrict the residual from fine level to coarse level
```

```julia
restriction(nxf, nyf, nxc, nyc, r, ec)

# set solution zero on coarse grid
unc[:,:] = zeros(nxc+1, nyc+1)

# solve on the coarsest level and relax V3 times
gauss_seidel_mg(nxc, nyc, 2.0*dx, 2.0*dy, fc, unc, v3)

# prolongate solution from coarse level to fine level
prolongation(nxc, nyc, nxf, nyf, unc, ef)

# correct the solution on fine level
for j = 2:nyf for i = 2:nxf
un[i,j] = un[i,j] + ef[i,j]
end~end

# relax v2 times
gauss_seidel_mg(nxf, nyf, dx, dy, f, un, v2)
end
end
```

Figure 23 displays the comparison of exact solution and numerical solution for full V-cycle multigrid framework. We use 512×512 grid resolution and hence we use nine levels from fine mesh to coarsest mesh. When the coarsest mesh has 2×2 grid resolution (i.e., since boundaries are set zero, only one grid point value is unknown and that can be solved algebraically), then it is usually referred to as the full multigrid framework. Figure 24 shows that the residual drops below tolerance in just nine iterations. This does not consist of a number of iterations for relaxation operations done for each intermediate grid levels. Although we report the outer iteration counter for multigrid results, we highlight that the workload in multigrid is more than the outer iteration counts. Equation (121) gives the total iteration count for a typical V-cycle multigrid method:

$$\text{V-cycle multigrid cost} = V\left((v_1 + v_2) \sum_{d=0}^{L-2} \frac{1}{2^d} + v_3 \frac{1}{2^{L-1}} \right), \tag{121}$$

where V is the outer iteration count, v_1 and v_2 are the number of iterations during prolongation and restriction, v_3 is the number of iterations for coarsest grid, and L is total number of levels in full multigrid cycle. In a typical application, the values of $v_1 = 2$, $v_2 = 2$, and $v_3 = 1$ can work effectively (i.e., a higher value of v_3 can be used if the coarsest resolution is larger than 2×2 and intended to be solved by the same relaxation method). Table 2 provides the comparison of different iterative solvers for the Poisson equation.

Table 2. Comparison of different iterative methods in terms of iteration count, residual and CPU time for solving the Poisson equation on a resolution of $N_x = 512$ and $N_y = 512$.

Iterative Solver	Iteration Count	Residual	CPU Time (s)
Gauss-Seidel	416,946	9.99×10^{-11}	1662.08
Conjugate-Gradient	1687	9.88×10^{-11}	4.30
V-cycle Multigrid	9	5.04×10^{-11}	0.55

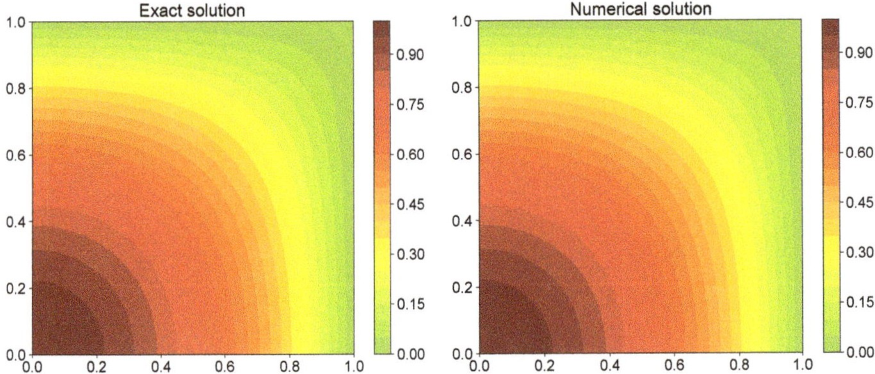

Figure 23. Comparison of exact solution and numerical solution computed using multigrid framework for the Poisson equation.

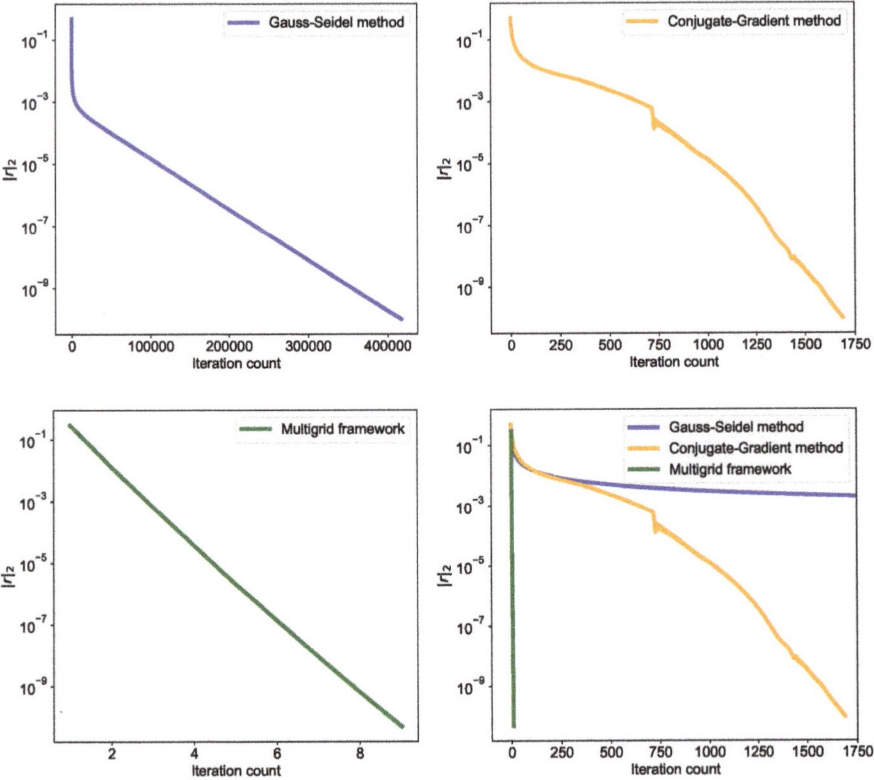

Figure 24. Residual history comparison for different iterative methods for Poisson equation with Dirichlet boundary condition. The stopping criteria for all iterative methods is that the residual should reduce below 10^{-10}. The grid resolution for all cases is $N_x = 512$ and $N_y = 512$.

7. Incompressible Two-Dimensional Navier-Stokes Equation

The Navier-Stokes equations for incompressible flow can be written as

$$\nabla \cdot \mathbf{u} = 0, \tag{122}$$

$$\frac{\partial \mathbf{u}}{\partial t} + \mathbf{u} \cdot \nabla \mathbf{u} = -\frac{1}{\rho}\nabla p + \nu\nabla^2\mathbf{u}, \tag{123}$$

where $\mathbf{u} = [u, v]$ is the velocity vector, t is the time, ρ is the density, p is pressure, and ν is the kinematic viscosity. By taking curl of Equation (123) and using $\nabla \times \mathbf{u} = \omega$, we can derive the vorticity equation

$$\nabla \times \frac{\partial \mathbf{u}}{\partial t} + \nabla \times (\mathbf{u} \cdot \nabla \mathbf{u}) = -\nabla \times \left(\frac{1}{\rho}\nabla p\right) + \nabla \times \nu\nabla^2\mathbf{u}. \tag{124}$$

The first term on the left side and last term on the right side becomes

$$\nabla \times \frac{\partial \mathbf{u}}{\partial t} = \frac{\partial \omega}{\partial t}; \quad \nabla \times \nu\nabla^2\mathbf{u} = \nu\nabla^2\omega. \tag{125}$$

Applying the identity $\nabla \times \nabla f = 0$ for any scalar function f, the pressure term vanishes for the incompressible flow since the density is constant

$$\nabla \times \nabla\left(\frac{p}{\rho}\right) = 0. \tag{126}$$

The second term in Equation (123) can be writtern as

$$\mathbf{u} \cdot \nabla \mathbf{u} = \frac{1}{2}\nabla(\mathbf{u} \cdot \mathbf{u}) - \mathbf{u} \times (\nabla \times \mathbf{u}) = \nabla\left(\frac{\mathbf{u}^2}{2}\right) - \mathbf{u} \times \omega \text{ where } \mathbf{u}^2 = \mathbf{u} \cdot \mathbf{u}. \tag{127}$$

The second term in Equation (124) becomes

$$\begin{aligned}
\nabla \times (\mathbf{u} \cdot \nabla \mathbf{u}) &= \nabla \times \nabla\left(\frac{\mathbf{u}^2}{2}\right) - \nabla \times \mathbf{u} \times \omega = \nabla \times (\omega \times \mathbf{u}) \\
&= (\mathbf{u} \cdot \nabla)\omega - \underbrace{(\omega \cdot \nabla)\mathbf{u}}_{\text{(vortex stretching)}} + \underbrace{\omega(\nabla \cdot \mathbf{u})}_{\nabla\cdot\mathbf{u}=0 \text{ (incompressible flow)}} + \underbrace{\mathbf{u}(\nabla \cdot \omega)}_{\nabla\cdot(\nabla \times \mathbf{u}) = 0},
\end{aligned} \tag{128}$$

and the vortex stretching term vanishes in two-dimensional flows (i.e., $\omega_x = 0$, $\omega_y = 0$, $\omega_z = \omega$)

$$(\omega \cdot \nabla)\mathbf{u} = \left(\underbrace{\omega_x}_{0}\frac{\partial}{\partial x} + \underbrace{\omega_y}_{0}\frac{\partial}{\partial y} + \underbrace{\omega_z}_{0}\frac{\partial}{\partial z}\right) = 0. \tag{129}$$

We can further use below relations for two-dimensional flows

$$(\mathbf{u} \cdot \nabla)\omega = u\frac{\partial \omega}{\partial x} + v\frac{\partial \omega}{\partial y}, \tag{130}$$

and introducing a streamfunction with the following definitions:

$$u = \frac{\partial \psi}{\partial y}, \text{ and } v = -\frac{\partial \psi}{\partial x}, \tag{131}$$

then the advection term becomes (i.e., sometimes called nonlinear Jacobian)

$$(\mathbf{u} \cdot \nabla)\omega = \frac{\partial \psi}{\partial y}\frac{\partial \omega}{\partial x} - \frac{\partial \psi}{\partial x}\frac{\partial \omega}{\partial y}. \tag{132}$$

The vorticity equation for two-dimensional incompressible flow becomes

$$\frac{\partial \omega}{\partial t} + \frac{\partial \psi}{\partial y}\frac{\partial \omega}{\partial x} - \frac{\partial \psi}{\partial x}\frac{\partial \omega}{\partial y} = \nu\left(\frac{\partial^2 \omega}{\partial x^2} + \frac{\partial^2 \omega}{\partial y^2}\right). \tag{133}$$

The kinematic relationship between streamfunction and vorticity is given by a Poisson equation

$$\frac{\partial^2 \psi}{\partial x^2} + \frac{\partial^2 \psi}{\partial y^2} = -\omega. \tag{134}$$

The vorticity-streamfunction formulation has several advantages over solving Equations (122) and (123). It eliminates the pressure term from the momentum equation and hence, there is no odd-even coupling between the pressure and velocity. We can use vorticity-streamfunction formulation directly on the collocated grid instead of using staggered grid. The number of equations to be solved in the vorticity-streamfunction formulation is also less than primitive variable formulation as it satisfies the divergence-free condition.

We use third-order Runge-Kutta numerical scheme (as discussed in Section 2.2) for the time integration. The right hand side terms in Equation (133) is discretized using the second-order central difference scheme similar to the diffusion term in heat equation (refer to Section 2.1). The nonlinear terms in Equation (133) is defined as the Jacobian

$$J(\omega, \psi) = \frac{\partial \psi}{\partial y}\frac{\partial \omega}{\partial x} - \frac{\partial \psi}{\partial x}\frac{\partial \omega}{\partial y}. \tag{135}$$

We can use any discretization method like explicit finite difference scheme (refer to Section 2.1) or compact finite difference scheme (refer to Section 2.4) for each term in Equation (135). Here, we use a numerical scheme introduced by Arakawa [44] for the discretization of Equation (135). This numerical scheme has conservation of energy, enstrophy and skew symmetry property and avoids computational instabilities arising from nonlinear interactions. The second-order Arakawa scheme for Equation (135) is given below

$$J(\omega, \psi) = \frac{J_1(\omega, \psi) + J_2(\omega, \psi) + J_3(\omega, \psi)}{3}, \tag{136}$$

where the discrete parts of the Jacobian are

$$J_1(\omega, \psi) = \frac{(\omega_{i+1,j} - \omega_{i-1,j})(\psi_{i,j+1} - \psi_{i,j-1}) - (\omega_{i,j+1} - \omega_{i,j-1})(\psi_{i+1,j} - \psi_{i-1,j})}{4\Delta x \Delta y}, \tag{137}$$

$$J_2(\omega, \psi) = \frac{1}{4\Delta x \Delta y}[\omega_{i+1,j}(\psi_{i+1,j+1} - \psi_{i+1,j-1}) - \omega_{i-1,j}(\psi_{i-1,j+1} - \psi_{i-1,j-1})$$
$$-\omega_{i,j+1}(\psi_{i+1,j+1} - \psi_{i-1,j+1}) + \omega_{i,j-1}(\psi_{i+1,j-1} - \psi_{i-1,j-1})], \tag{138}$$

$$J_3(\omega, \psi) = \frac{1}{4\Delta x \Delta y}[\omega_{i+1,j+1}(\psi_{i,j+1} - \psi_{i+1,j}) - \omega_{i-1,j-1}(\psi_{i-1,j} - \psi_{i,j-1})$$
$$-\omega_{i-1,j+1}(\psi_{i,j+1} - \psi_{i-1,j}) + \omega_{i+1,j-1}(\psi_{i+1,j} - \psi_{i,j-1})]. \tag{139}$$

7.1. Lid-Driven Cavity Problem

We test our two-dimensional Navier-Stokes solver using the lid-driven cavity benchmark problem for viscous incompressible flow [45]. The problem deals with a square cavity consisting of three rigid walls with no-slip conditions and a lid moving with a tangential unit velocity. The density of the fluid is taken to be unity. Therefore, we get $\nu = 1/\text{Re}$, where Re is the Reynolds number of flow. The vorticity equation for lid-driven cavity problem can be written as

$$\frac{\partial \omega}{\partial t} = -\left(\frac{\partial \psi}{\partial y}\frac{\partial \omega}{\partial x} - \frac{\partial \psi}{\partial x}\frac{\partial \omega}{\partial y} \right) + \frac{1}{\text{Re}}\left(\frac{\partial^2 \omega}{\partial x^2} + \frac{\partial^2 \omega}{\partial y^2} \right). \tag{140}$$

The computational domain is square in shape with $(x,y) \in [0,1] \times [0,1]$. We divide the computational domain into 64×64 grid resolution. All the walls have Dirichlet boundary conditions. The time integration implementation in Julia for two-dimensional Navier-Stokes solver is similar to Listing 2. We perform time integration from time $t = 0$ to $t = 10$ to make sure that steady state is reached and the residual reaches below 10^{-6}. The residual is defined as the L_2 norm of the difference between two consecutive solutions. At each step of the Runge-Kutta numerical scheme, we also update the boundary condition for vorticity (i.e., further details can be found in [46]) and solve Equation (134) to update streamfunction. Any of the Poisson solvers mentioned in Section 6 can be used to solve this equation. We use fast sine transform Poisson solver discussed in Section 6.1.2 to get streamfunction from vorticity field as we have Dirichlet boundary conditions for all four walls. The implementation of FST Poisson solver in Julia is given in Listing 20. The functions to compute the right-hand side term for Runge-Kutta numerical scheme and to update boundary conditions are given in Listings 27 and 28 respectively. Figure 25 shows the vorticity and streamfunction field for the lid driven cavity problem.

Listing 27. Computation of right-hand side term in Equation (140) in Julia for two-dimensional incompressible Navier-Stokes equations.

```julia
# nx,ny: total number of grids in x and y direction
# dx,dy: grid spacing in x and y direction
# re: Reynolds number of the flow
# w: vorticity field
# s: streamfunction
# r: right hand side of the Runge-Kutta scheme (-Jacobian+Laplacian terms)
function rhs(nx,ny,dx,dy,re,w,s,r)
aa = 1.0/(re*dx*dx)
bb = 1.0/(re*dy*dy)
gg = 1.0/(4.0*dx*dy)
hh = 1.0/3.0


for i = 2:nx for j = 2:ny
# Arakawa numerical scheme for Jacobian
j1 = gg*((w[i+1,j]-w[i-1,j])*(s[i,j+1]-s[i,j-1]) -
(w[i,j+1]-w[i,j-1])*(s[i+1,j]-s[i-1,j]))

j2 = gg*(w[i+1,j]*(s[i+1,j+1]-s[i+1,j-1]) -
w[i-1,j]*(s[i-1,j+1]-s[i-1,j-1]) -
w[i,j+1]*(s[i+1,j+1]-s[i-1,j+1]) +
w[i,j-1]*(s[i+1,j-1]-s[i-1,j-1]))

j3 = gg*(w[i+1,j+1]*(s[i,j+1]-s[i+1,j]) -
w[i-1,j-1]*(s[i-1,j]-s[i,j-1]) -
w[i-1,j+1]*(s[i,j+1]-s[i-1,j]) +
w[i+1,j-1]*(s[i+1,j]-s[i,j-1]))

jac = (j1+j2+j3)*hh

#Central difference for Laplacian
r[i,j] = -jac + aa*(w[i+1,j]-2.0*w[i,j]+w[i-1,j]) +
bb*(w[i,j+1]-2.0*w[i,j]+w[i,j-1])
```

```
end  end
end
```

Listing 28. Boundary condition update function in Julia for the lid-driven cavity problem.

```julia
# nx,ny: total number of grids in x and y direction
# dx,dy: grid spacing in x and y direction
# w: vorticity field
# s: streamfunction
function bc(nx,ny,dx,dy,w,s)
# boundary condition for vorticity (Jensen) left and right
for j = 1:ny+1
w[1,j] = (-4.0*s[2,j]+0.5*s[3,j])/(dx*dx)
w[nx+1,j]= (-4.0*s[nx,j]+0.5*s[nx-1,j])/(dx*dx)
end
# boundary condition for vorticity (Jensen) bottom and top
for i = 1:nx+1
w[i,1] = (-4.0*s[i,2]+0.5*s[i,3])/(dy*dy)
w[i,ny+1]= (-4.0*s[i,ny]+0.5*s[i,ny-1])/(dy*dy) - 3.0/dy
end
end
```

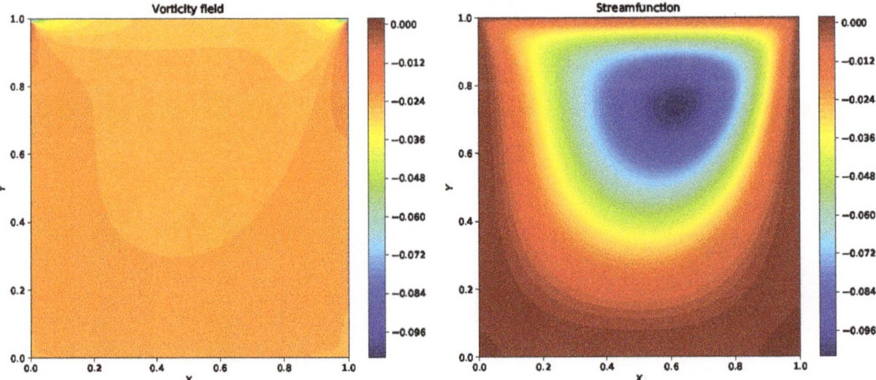

Figure 25. Vorticity field and streamfunction field for the lid-driven cavity benchmark problem.

7.2. Vortex-Merger Problem

We demonstrate the two-dimensional Navier-Stokes solver for the domain with periodic boundary condition using vortex-merger problem. The merging process occurs when two vortices of the same sign with parallel axes are within a certain critical distance from each other. The two vortices merge to form a single vortex. It is a two-dimensional process and is one of the fundamental processes of fluid motion and it plays a key role in a variety of simulations, such as decaying two-dimensional turbulence, three-dimensional turbulence, and mixing layers [47–49]. This phenomenon also occurs in other fields such as astrophysics, meteorology, and geophysics [50].

We use the Cartesian computational domain $(x, y) \in [0, 2\pi] \times [0, 2\pi]$ and divide it into 128×128 grid resolution. The vorticity equation for the vortex-merger problem is same as the lid-driven cavity problem and is given in Equation (140). We use Re $= 2000$ and integrate the solution from time $t = 0$ to $t = 20$ with $\Delta t = 0.01$. We use third-order Runge-Kutta method for time integration similar to the lid-driven cavity problem. The Julia function to compute the right hand side term of vorticity equation is the same as the lid-driven cavity problem and is given in Listing 27. The Julia function to assign the initial condition for vortex-merger problem is detailed in Listing 29.

Listing 29. Initial condition assignment function in Julia for vortex-merger problem.

```julia
# nx,ny: total number of grids in x and y direction
# x,y: array of grid locations of the domain
# w: vorticity field
function vm_ic(nx,ny,x,y,w)
xc1 = pi-pi/4.0 # horizontal location of left vortex
yc1 = pi        # vertical location of left vortex
xc2 = pi+pi/4.0 # horizontal location of right vortex
yc2 = pi        # vertical location of right-vortex

# initial vorticity field
for i = 2:nx+2 for j = 2:ny+2
w[i,j] = exp(-pi*((x[i-1]-xc1)^2 + (y[j-1]-yc1)^2)) +
exp(-pi*((x[i-1]-xc2)^2 + (y[j-1]-yc2)^2))
end end
end
```

The vortex-merger problem has the periodic domain and hence we use fast Fourier transform (FFT) Poisson solver discussed in Section 6.1.1 to get the streamfunction from vorticity field at every time step. The Julia function for FFT Poisson solver is outlined in Listing 18. The evolution of the merging process for two vortices into a single vortex is displayed in Figure 26. If we let the simulation run for some more time, we will see a single vortex getting formed.

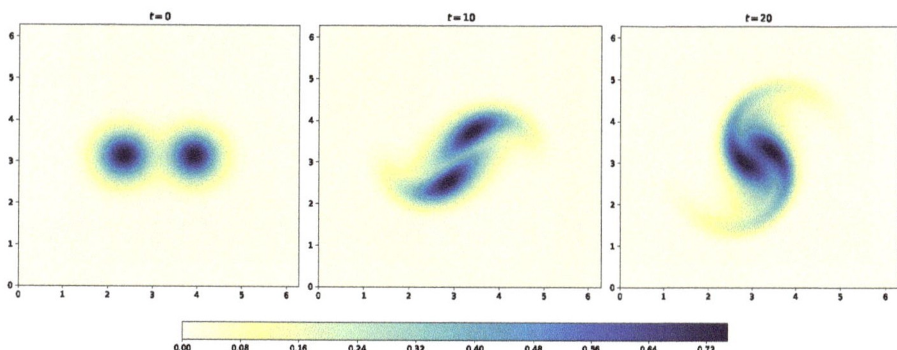

Figure 26. Evolution of the vorticity field at three time steps during the co-rotating vortex interaction. The solution is computed using second-order explicit solver.

We compare the computational time required for two-dimensional incompressible Navier-Stokes solver in Julia with Python code. We use two versions of Python code. In the first version, we use for loop similar to the way we write Julia code. The Python code for computing the right hand side term using Arakawa numerical scheme is given in Listing 30. The other version is the vectorized version which speeds up the code without using any loops and is given in Listing 31. In vectorization, several operations such as multiplication, addition, division are performed over a vector at the same time. From Table 3, we can easily see the benefit of Julia over Python for computationally intensive problems like the ones encountered in CFD applications.

Table 3. Comparison of CPU time for Julia and Python version of Navier-Stokes solver for grid resolution $N_x = 128$ and $N_y = 128$. We note that the codes are written mostly for the pedagogical purpose and the CPU time may not be the optimal time that can be achieved for each of the programming languages.

Programming Language	CPU Time (s)
Julia	5.97
Python (nonvectorized)	1271.55
Python (vectorized)	18.46

Listing 30. Computation of right hand side function in Python without vectorization.

```python
# nx,ny: total number of grids in x and y direction
# dx,dy: grid spacing in x and y direction
# re: Reynolds number
# w: vorticity
# s: streamfunction
def rhs(nx,ny,dx,dy,re,w,s):
aa = 1.0/(re*dx*dx)
bb = 1.0/(re*dy*dy)
gg = 1.0/(4.0*dx*dy)
hh = 1.0/3.0
f = np.empty((nx+3,ny+3))


for i in range(1,nx+2):
for j in range(1,ny+2):
#Arakawa
j1 = gg*( (w[i+1,j]-w[i-1,j])*(s[i,j+1]-s[i,j-1]) \
-(w[i,j+1]-w[i,j-1])*(s[i+1,j]-s[i-1,j]))

j2 = gg*( w[i+1,j]*(s[i+1,j+1]-s[i+1,j-1]) \
- w[i-1,j]*(s[i-1,j+1]-s[i-1,j-1]) \
- w[i,j+1]*(s[i+1,j+1]-s[i-1,j+1]) \
+ w[i,j-1]*(s[i+1,j-1]-s[i-1,j-1]))

j3 = gg*( w[i+1,j+1]*(s[i,j+1]-s[i+1,j]) \
- w[i-1,j-1]*(s[i-1,j]-s[i,j-1]) \
- w[i-1,j+1]*(s[i,j+1]-s[i-1,j]) \
+ w[i+1,j-1]*(s[i+1,j]-s[i,j-1]) )

jac = (j1+j2+j3)*hh

#Central difference for Laplacian
f[i,j] = -jac + aa*(w[i+1,j]-2.0*w[i,j]+w[i-1,j]) \
+ bb*(w[i,j+1]-2.0*w[i,j]+w[i,j-1])

return f
```

Listing 31. Computation of right hand side function in Python with vectorization.

```python
# nx,ny: total number of grids in x and y direction
# dx,dy: grid spacing in x and y direction
# re: Reynolds number
# w: vorticity
# s: streamfunction
```

```python
def rhs(nx,ny,dx,dy,re,w,s):
aa = 1.0/(dx*dx)
bb = 1.0/(dy*dy)
gg = 1.0/(4.0*dx*dy)
hh = 1.0/3.0

f = np.zeros((nx+3,ny+3))

# Arakawa scheme
j1 = gg*((w[2:nx+3,1:ny+2]-w[0:nx+1,1:ny+2]) \
*( s[1:nx+2,2:ny+3]-s[1:nx+2,0:ny+1]) \
-( w[1:nx+2,2:ny+3]-w[1:nx+2,0:ny+1]) \
*( s[2:nx+3,1:ny+2]-s[0:nx+1,1:ny+2]))

j2 = gg*( w[2:nx+3,1:ny+2]*(s[2:nx+3,2:ny+3]-s[2:nx+3,0:ny+1]) \
- w[0:nx+1,1:ny+2]*(s[0:nx+1,2:ny+3]-s[0:nx+1,0:ny+1]) \
- w[1:nx+2,2:ny+3]*(s[2:nx+3,2:ny+3]-s[0:nx+1,2:ny+3]) \
+ w[1:nx+2,0:ny+1]*(s[2:nx+3,0:ny+1]-s[0:nx+1,0:ny+1]))

j3 = gg*( w[2:nx+3,2:ny+3]*(s[1:nx+2,2:ny+3]-s[2:nx+3,1:ny+2]) \
- w[0:nx+1,0:ny+1]*(s[0:nx+1,1:ny+2]-s[1:nx+2,0:ny+1]) \
- w[0:nx+1,2:ny+3]*(s[1:nx+2,2:ny+3]-s[0:nx+1,1:ny+2]) \
+ w[2:nx+3,0:ny+1]*(s[2:nx+3,1:ny+2]-s[1:nx+2,0:ny+1]) )

jac = (j1+j2+j3)*hh

lap = aa*(w[2:nx+3,1:ny+2]-2.0*w[1:nx+2,1:ny+2]+w[0:nx+1,1:ny+2]) \
+ bb*(w[1:nx+2,2:ny+3]-2.0*w[1:nx+2,1:ny+2]+w[1:nx+2,0:ny+1])

f[1:nx+2,1:ny+2] = -jac + lap/re

return f
```

8. Hybrid Arakawa-Spectral Solver

In this section, instead of using a fully explicit method in time for solving 2D incompressible Navier-Stokes equation, we show how to design hybrid explicit and implicit scheme. The nonlinear Jacobian term in vorticity Equation (133) is treated explicitly using third-order Runge-Kutta scheme, and we treat the viscous term implicitly using the Crank-Nicolson scheme. This type of hybrid approach is useful when we design solvers for wall-bounded flows, where we cluster the grid in the boundary layer. For wall-bounded flows, we use dense mesh with a smaller grid size in the boundary layer region to capture the boundary layer flow correctly. First, we can re-write Equation (133) with nonlinear Jacobian term on the right hand side

$$\frac{\partial \omega}{\partial t} = -J(\omega,\psi) + \nu\nabla^2\omega. \tag{141}$$

The hybrid third-order Runge-Kutta/Crank-Nicolson scheme can be written as [51]

$$\omega^{(1)} = \omega^{(n)} + \gamma_1\Delta t(-J^{(n)}) + \rho_1\Delta t(-J^{(n-1)(2)}) + \frac{\alpha_1\Delta t\nu}{2}\nabla^2(\omega^{(1)} + \omega^{(n)}), \tag{142}$$

$$\omega^{(2)} = \omega^{(1)} + \gamma_2\Delta t(-J^{(1)}) + \rho_2\Delta t(-J^{(n)}) + \frac{\alpha_2\Delta t\nu}{2}\nabla^2(\omega^{(1)} + \omega^{(2)}), \tag{143}$$

$$\omega^{(n+1)} = \omega^{(2)} + \gamma_3\Delta t(-J^{(2)}) + \rho_3\Delta t(-J^{(1)}) + \frac{\alpha_3\Delta t\nu}{2}\nabla^2(\omega^{(n+1)} + \omega^{(2)}),\tag{144}$$

where the term $-J^{(n-1)(2)}$ is the nonlinear Jacobian term at the second-step of previous time step at $(n-1)$. We can choose coefficients in Equations (142)–(144) in such a way that ρ_1 vanishes and we do not have to store the solution at second-step of the previous time step. These coefficients are

$$\alpha_1 = \frac{8}{15}, \alpha_2 = \frac{2}{15}, \alpha_3 = \frac{1}{3},\tag{145}$$

$$\gamma_1 = \frac{8}{15}, \gamma_2 = \frac{5}{12}, \gamma_3 = \frac{3}{4},\tag{146}$$

$$\rho_1 = 0, \rho_2 = -\frac{17}{60}, \rho_2 = -\frac{5}{12}.\tag{147}$$

Besides, to use a hybrid scheme for time, we also design a hybrid finite difference and spectral scheme. We use Arakawa finite difference scheme for the nonlinear Jacobian term and spectral scheme for linear viscous terms. In Fourier space, Equations (142)–(144) can be written as

$$\tilde{\omega}^{(1)} = \tilde{\omega}^{(n)} + \gamma_1\Delta t(-\tilde{J}^{(n)}) + \rho_1\Delta t(-\tilde{J}^{(n-1)(2)}) + \frac{\alpha_1\Delta t\nu}{2}(-k^2)(\tilde{\omega}^{(1)} + \tilde{\omega}^{(n)}),\tag{148}$$

$$\tilde{\omega}^{(2)} = \tilde{\omega}^{(1)} + \gamma_2\Delta t(-\tilde{J}^{(1)}) + \rho_2\Delta t(-\tilde{J}^{(n)}) + \frac{\alpha_2\Delta t\nu}{2}(-k^2)(\tilde{\omega}^{(1)} + \tilde{\omega}^{(2)}),\tag{149}$$

$$\tilde{\omega}^{(n+1)} = \tilde{\omega}^{(2)} + \gamma_3\Delta t(-\tilde{J}^{(2)}) + \rho_3\Delta t(-\tilde{J}^{(1)}) + \frac{\alpha_3\Delta t\nu}{2}(-k^2)(\tilde{\omega}^{(n+1)} + \tilde{\omega}^{(2)}),\tag{150}$$

where $k^2 = m^2 + n^2$. Here, m and n are wavenumber in x and y directions, respectively. Rearranging above equations, we get

$$\tilde{\omega}^{(1)} = \frac{1 - \frac{\alpha_1\Delta t\nu k^2}{2}}{1 + \frac{\alpha_1\Delta t\nu k^2}{2}}\tilde{\omega}^{(n)} + \frac{\gamma_1\Delta t}{1 + \frac{\alpha_1\Delta t\nu k^2}{2}}(-\tilde{J}^{(n)}),\tag{151}$$

$$\tilde{\omega}^{(2)} = \frac{1 - \frac{\alpha_2\Delta t\nu k^2}{2}}{1 + \frac{\alpha_2\Delta t\nu k^2}{2}}\tilde{\omega}^{(1)} + \frac{\gamma_2\Delta t(-\tilde{J}^{(1)}) + \rho_2\Delta t(-\tilde{J}^{(n)})}{1 + \frac{\alpha_2\Delta t\nu k^2}{2}},\tag{152}$$

$$\tilde{\omega}^{(n+1)} = \frac{1 - \frac{\alpha_3\Delta t\nu k^2}{2}}{1 + \frac{\alpha_3\Delta t\nu k^2}{2}}\tilde{\omega}^{(2)} + \frac{\gamma_3\Delta t(-\tilde{J}^{(2)}) + \rho_3\Delta t(-\tilde{J}^{(1)})}{1 + \frac{\alpha_3\Delta t\nu k^2}{2}}.\tag{153}$$

We use Arakawa finite difference scheme described in Section 7 to compute the Jacobian term in physical space. The streamfunction is computed from the vorticity field using a spectral method. Once the Jacobian term is available in physical space we convert it to Fourier space using FFT. The relationship between vorticity and streamfunction is Fourier space can be written as

$$\tilde{\psi} = \frac{\tilde{\omega}}{k^2}.\tag{154}$$

The time integration using hybrid third-order Runge-Kutta/ Crank-Nicolson scheme in Julia is listed in Listing 32. The Julia implementation to compute the nonlinear Jacobian term for the Arakawa-spectral solver is given in Listing 33.

Listing 32. Julia implementation of hybrid third-order Runge-Kutta/ Crank-Nicolson scheme for 2D incompressible Navier-Stokes equation with Arakawa-spectral finite difference scheme.

```julia
# nx, ny: number of grid points in x and y directions
# nt: number of time intervals between initial and final time
# re: Reynolds number for vortex-merger problem
# dx, dy: grid spacing in x and y directions
# wn: initial condition for vorticity in physical space
# ns: number of snapshots to store the solution
function numerical(nx,ny,nt,dx,dy,dt,re,wn,ns)
w1f = Array{Complex{Float64}}(undef,nx,ny)
w2f = Array{Complex{Float64}}(undef,nx,ny)
wnf = Array{Complex{Float64}}(undef,nx,ny)
j1f = Array{Complex{Float64}}(undef,nx,ny)
j2f = Array{Complex{Float64}}(undef,nx,ny)
jnf = Array{Complex{Float64}}(undef,nx,ny)
ut = Array{Float64}(undef, nx+1, ny+1)
data = Array{Complex{Float64}}(undef,nx,ny)
d1 = Array{Float64}(undef, nx, ny)
d2 = Array{Float64}(undef, nx, ny)
d3 = Array{Float64}(undef, nx, ny)
k2 = Array{Float64}(undef, nx, ny)

for i = 1:nx for j = 1:ny
data[i,j] = complex(wn[i+1,j+1],0.0)
end~end

wnf = fft(data)
k2 = wavespace(nx,ny,dx,dy) #  wavespace over 2D domain
wnf[1,1] = 0.0

alpha1, alpha2, alpha3 = 8.0/15.0, 2.0/15.0, 1.0/3.0
gamma1, gamma2, gamma3 = 8.0/15.0, 5.0/12.0, 3.0/4.0
rho2, rho3 = -17.0/60.0, -5.0/12.0

for i = 1:nx for j = 1:ny
z = 0.5*dt*k2[i,j]/re
d1[i,j] = alpha1*z
d2[i,j] = alpha2*z
d3[i,j] = alpha3*z
end~end

for k = 1:nt
jnf = jacobian(nx,ny,dx,dy,wnf,k2)

for i = 1:nx for j = 1:ny
w1f[i,j] = ((1.0 - d1[i,j])/(1.0 + d1[i,j]))*wnf[i,j] +
(gamma1*dt*jnf[i,j])/(1.0 + d1[i,j])
end~end

w1f[1,1] = 0.0
j1f = jacobian(nx,ny,dx,dy,w1f,k2)

for i = 1:nx for j = 1:ny
w2f[i,j] = ((1.0 - d2[i,j])/(1.0 + d2[i,j]))*w1f[i,j] +
```

```julia
(rho2*dt*jnf[i,j] + gamma2*dt*j1f[i,j])/(1.0 + d2[i,j])
end~end

w2f[1,1] = 0.0
j2f = jacobian(nx,ny,dx,dy,w2f,k2)

for i = 1:nx for j = 1:ny
wnf[i,j] = ((1.0 - d3[i,j])/(1.0 + d3[i,j]))*w2f[i,j] +
(rho3*dt*j1f[i,j] + gamma3*dt*j2f[i,j])/(1.0 + d3[i,j])
end end
end
ut[1:nx,1:ny] = real(ifft(wnf)) # final solution
ut[nx+1,:] = ut[1,:]
ut[:,ny+1] = ut[:,1]
return ut
end


# function to compute wavespace over 2D domain
function wavespace(nx,ny,dx,dy)
eps = 1.0e-6
kx = Array{Float64}(undef,nx)
ky = Array{Float64}(undef,ny)
k2 = Array{Float64}(undef,nx,ny)

hx = 2.0*pi/(nx*dx)

for i = 1:Int64(nx/2)
kx[i] = hx*(i-1.0)
kx[i+Int64(nx/2)] = hx*(i-Int64(nx/2)-1)
end
kx[1] = eps
ky = kx

for i = 1:nx for j = 1:ny
k2[i,j] = kx[i]^2 + ky[j]^2
end end
return k2
end
```

Listing 33. Julia implementation of computing the nonlinear Jacobian term in physical space and converting it to Fourier space

```julia
# nx, ny: number of grid points in x and y directions
# dx, dy: grid spacing in x and y directions
# wf: vorticity in Fourier space
# k2: wavespace over 2D domain
function jacobian(nx,ny,dx,dy,wf,k2)
s = Array{Float64}(undef, nx+2, ny+2)
sf = Array{Complex{Float64}}(undef,nx,ny)
w = Array{Float64}(undef, nx+2, ny+2)
data = Array{Complex{Float64}}(undef,nx,ny)
jf = Array{Complex{Float64}}(undef,nx,ny)

w[2:nx+1, 2:ny+1] = real(ifft(wf)) # convert vorticity to physical space
# periodic BC
```

```
w[nx+2,:] = w[2,:]
w[:,ny+2] = w[:,2]
# ghost points
w[1,:] = w[nx+1,:]
w[:,1] = w[:,ny+1]

for i = 1:nx for j = 1:ny
sf[i,j] = wf[i,j]/k2[i,j] # streamfunction in Fourier space
end~end

s[2:nx+1, 2:ny+1] = real(ifft(sf)) # convert streamfunction to physical space
# periodic BC
s[nx+2,:] = s[2,:]
s[:,ny+2] = s[:,2]
# ghost points
s[1,:] = s[nx+1,:]
s[:,1] = s[:,ny+1]

# Arakawa finite difference scheme
gg = 1.0/(4.0*dx*dy)
hh = 1.0/3.0
for i = 2:nx+1 for j = 2:ny+1
j1 = gg*((w[i+1,j]-w[i-1,j])*(s[i,j+1]-s[i,j-1]) -
(w[i,j+1]-w[i,j-1])*(s[i+1,j]-s[i-1,j]))

j2 = gg*(w[i+1,j]*(s[i+1,j+1]-s[i+1,j-1]) -
w[i-1,j]*(s[i-1,j+1]-s[i-1,j-1]) -
w[i,j+1]*(s[i+1,j+1]-s[i-1,j+1]) +
w[i,j-1]*(s[i+1,j-1]-s[i-1,j-1]))

j3 = gg*(w[i+1,j+1]*(s[i,j+1]-s[i+1,j]) -
w[i-1,j-1]*(s[i-1,j]-s[i,j-1]) -
w[i-1,j+1]*(s[i,j+1]-s[i-1,j]) +
w[i+1,j-1]*(s[i+1,j]-s[i,j-1]))

data[i-1,j-1] = complex(-(j1+j2+j3)*hh, 0.0)
end~end

jf = fft(data) # convert Jacobian from physical to fourier space
return jf
end
```

We validate our hybrid Arakawa-spectral solver for vortex-merger problem. Figure 27 shows the evolution of two point vortices merging into a single vortex computed using hybridized finite difference and spectral methodology. In the next section, we will eliminate finite difference and present a pseudo-spectral methodology.

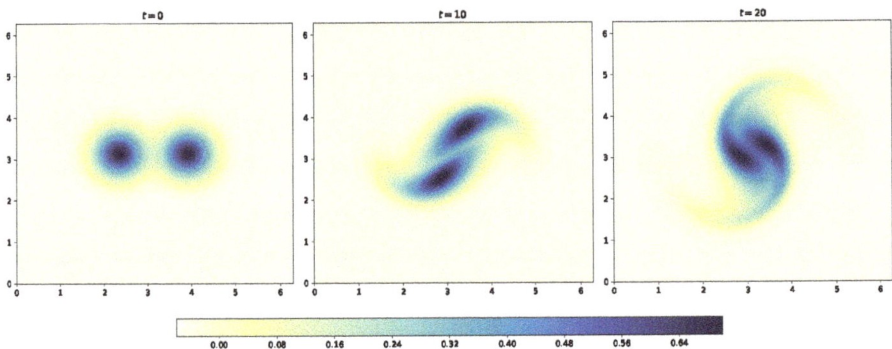

Figure 27. Evolution of the vorticity field at three time steps during the co-rotating vortex interaction. The solution is computed using hybrid Arakawa-spectral solver.

9. Pseudo-Spectral Solver

In Section 8, we computed the Jacobian term using finite difference Arakawa scheme. We can also compute this term using a spectral method to develop the pseudo-spectral solver. The main motivation for the spectral solver is a unique advantage of both accuracy and efficiency considerations in simple geometries. It reduces a PDE problem to a set of ODEs without introducing a numerical discretization error. Furthermore, the code implementation will be much simpler due to the availability of FFT libraries. We refer to a paper by Mortensen and Langtangen [52] for further discussion of high performance pseudo-spectral solver implementation of 3D Navier-Stokes equations. We use the hybrid explicit and implicit Crank-Nicolson scheme for time integration similar to the previous section. The only difference is the computation of the Jacobian term. The Jacobian term involves the first derivative term and can be calculated easily using a spectral method. The first derivative of Equation (88) in x and y direction is calculated as

$$\frac{\partial}{\partial x} u_{i,j} = \frac{1}{N_x} \frac{1}{N_y} \sum_{m=-N_x/2}^{N_x/2-1} \sum_{n=-N_y/2}^{N_y/2-1} (\imath m) \tilde{u}_{m,n} e^{\imath 2\pi \left(\frac{mi}{N_x} + \frac{nj}{N_y} \right)}, \tag{155}$$

$$\frac{\partial}{\partial y} u_{i,j} = \frac{1}{N_x} \frac{1}{N_y} \sum_{m=-N_x/2}^{N_x/2-1} \sum_{n=-N_y/2}^{N_y/2-1} (\imath n) \tilde{u}_{m,n} e^{\imath 2\pi \left(\frac{mi}{N_x} + \frac{nj}{N_y} \right)}. \tag{156}$$

The Jacobian term in the Fourier space can be computed as

$$\tilde{J}_{m,n} = (\imath n \tilde{\psi}) \circledast (\imath m \tilde{\omega}) - (\imath m \tilde{\psi}) \circledast (\imath n \tilde{\omega}), \tag{157}$$

where $\circledast$ is the convolution operation. Using Equation (154), we get

$$\tilde{J}_{m,n} = \left(\imath n \frac{\tilde{\omega}}{k^2} \right) \circledast (\imath m \tilde{\omega}) - \left(\imath m \frac{\tilde{\omega}}{k^2} \right) \circledast (\imath n \tilde{\omega}). \tag{158}$$

The convolution operation is performed in physical space and then converted to physical space. The pseudo-code for calculating the Jacobian term is given as

$$\tilde{J}_{m,n} = \text{fft} \left[\underbrace{\text{ifft} \left(\imath n \frac{\tilde{\omega}}{k^2} \right)}_{\frac{\partial \psi}{\partial y}} \circledast \underbrace{\text{ifft} \left(\imath m \tilde{\omega} \right)}_{\frac{\partial \omega}{\partial x}} - \underbrace{\text{ifft} \left(\imath m \frac{\tilde{\omega}}{k^2} \right)}_{\frac{\partial \psi}{\partial x}} \circledast \underbrace{\text{ifft} \left(\imath n \tilde{\omega} \right)}_{\frac{\partial \omega}{\partial y}} \right]. \tag{159}$$

We can observe that the Jacobian term involves the product of two functions. When we evaluate the product of two functions using Fourier transform, aliasing errors appear in the computation [12]. Aliasing errors refers to the misrepresentation of wavenumbers when the function is mapped onto a grid with not enough grid points. One way to reduce aliasing errors is to use 3/2-rule. In this method, the original grid is refined by a factor $M = 3/2N$ in each direction. The additional grid points are assigned zero value and this is also referred to as zero-padding. For example, if we are using spatial resolution of $N_x = 128$ and $N_y = 128$ in x and y direction, the Jacobian term will be evaluated on grid with $N_x = 192$ and $N_y = 192$ (i.e., a data set consisted of 192^2 points is used in the inverse and forward FFT calls in Equation (159)). The main code for the pseudo-spectral solver is the same as the hybrid solver discussed in Section 8 except for the Jacobian function. The implementation of Jacobian calculation with 3/2 padding is given in Listing 34.

Listing 34. Julia implementation of computing the nonlinear Jacobian term in Fourier space with 3/2 padding rule.

```julia
# nx, ny: number of grid points in x and y directions
# dx, dy: grid spacing in x and y directions
# wf: vorticity in Fourier space
# k2: wavespace over 2D domain
function jacobian(nx,ny,dx,dy,wf,k2)
eps = 1.0e-6
kx = Array{Float64}(undef,nx)
ky = Array{Float64}(undef,ny)

#wave number indexing
hx = 2.0*pi/(nx*dx)

for i = 1:Int64(nx/2)
kx[i] = hx*(i-1.0)
kx[i+Int64(nx/2)] = hx*(i-Int64(nx/2)-1)
end
kx[1] = eps
ky = transpose(kx)

j1f = zeros(ComplexF64,nx,ny)
j2f = zeros(ComplexF64,nx,ny)
j3f = zeros(ComplexF64,nx,ny)
j4f = zeros(ComplexF64,nx,ny)

# x-derivative
for i = 1:nx for j = 1:ny
j1f[i,j] = 1.0im*wf[i,j]*kx[i]/k2[i,j]
j4f[i,j] = 1.0im*wf[i,j]*kx[i]
end~end

# y-derivative
for i = 1:nx for j = 1:ny
j2f[i,j] = 1.0im*wf[i,j]*ky[j]
j3f[i,j] = 1.0im*wf[i,j]*ky[j]/k2[i,j]
end~end

nxe = Int64(nx*1.5)
nye = Int64(ny*1.5)

j1f_p = zeros(ComplexF64,nxe,nye)
```

```
j2f_p = zeros(ComplexF64,nxe,nye)
j3f_p = zeros(ComplexF64,nxe,nye)
j4f_p = zeros(ComplexF64,nxe,nye)

qx = nxe-nx/2+1
qy = nye-ny/2+1

# zero padding with 3/2 rule
j1f_p[1:Int64(nx/2),1:Int64(ny/2)] = j1f[1:Int64(nx/2),1:Int64(ny/2)]
j1f_p[Int64(qx):nxe,1:Int64(ny/2)] = j1f[Int64(nx/2+1):nx,1:Int64(ny/2)]
j1f_p[1:Int64(nx/2),Int64(qy):nye] = j1f[1:Int64(nx/2),Int64(ny/2+1):ny]
j1f_p[Int64(qx):nxe,Int64(qy):nye] = j1f[Int64(nx/2+1):nx,Int64(ny/2+1):ny]

j2f_p[1:Int64(nx/2),1:Int64(ny/2)] = j2f[1:Int64(nx/2),1:Int64(ny/2)]
j2f_p[Int64(qx):nxe,1:Int64(ny/2)] = j2f[Int64(nx/2+1):nx,1:Int64(ny/2)]
j2f_p[1:Int64(nx/2),Int64(qy):nye] = j2f[1:Int64(nx/2),Int64(ny/2+1):ny]
j2f_p[Int64(qx):nxe,Int64(qy):nye] = j2f[Int64(nx/2+1):nx,Int64(ny/2+1):ny]

j3f_p[1:Int64(nx/2),1:Int64(ny/2)] = j3f[1:Int64(nx/2),1:Int64(ny/2)]
j3f_p[Int64(qx):nxe,1:Int64(ny/2)] = j3f[Int64(nx/2+1):nx,1:Int64(ny/2)]
j3f_p[1:Int64(nx/2),Int64(qy):nye] = j3f[1:Int64(nx/2),Int64(ny/2+1):ny]
j3f_p[Int64(qx):nxe,Int64(qy):nye] = j3f[Int64(nx/2+1):nx,Int64(ny/2+1):ny]

j4f_p[1:Int64(nx/2),1:Int64(ny/2)] = j4f[1:Int64(nx/2),1:Int64(ny/2)]
j4f_p[Int64(qx):nxe,1:Int64(ny/2)] = j4f[Int64(nx/2+1):nx,1:Int64(ny/2)]
j4f_p[1:Int64(nx/2),Int64(qy):nye] = j4f[1:Int64(nx/2),Int64(ny/2+1):ny]
j4f_p[Int64(qx):nxe,Int64(qy):nye] = j4f[Int64(nx/2+1):nx,Int64(ny/2+1):ny]

j1f_p = j1f_p*(nxe*nye)/(nx*ny)
j2f_p = j2f_p*(nxe*nye)/(nx*ny)
j3f_p = j3f_p*(nxe*nye)/(nx*ny)
j4f_p = j4f_p*(nxe*nye)/(nx*ny)

j1 = real(ifft(j1f_p))
j2 = real(ifft(j2f_p))
j3 = real(ifft(j3f_p))
j4 = real(ifft(j4f_p))
jacp = zeros(Float64,nxe,nye)

for i = 1:nxe for j = 1:nye
jacp[i,j] = j1[i,j]*j2[i,j] - j3[i,j]*j4[i,j]
end~end

jacpf = fft(jacp)

jf = zeros(ComplexF64,nx,ny)

jf[1:Int64(nx/2),1:Int64(ny/2)] = jacpf[1:Int64(nx/2),1:Int64(ny/2)]
jf[Int64(nx/2+1):nx,1:Int64(ny/2)] = jacpf[Int64(qx):nxe,1:Int64(ny/2)]
jf[1:Int64(nx/2),Int64(ny/2+1):ny] = jacpf[1:Int64(nx/2),Int64(qy):nye]
jf[Int64(nx/2+1):nx,Int64(ny/2+1):ny] =  jacpf[Int64(qx):nxe,Int64(qy):nye]

jf = jf*(nx*ny)/(nxe*nye)

return jf
```

```
end
```

We compare the performance of pseudo-spectral solver for two-dimensional incompressible flow problems written in Julia with the vectorized version of Python code. The FFT operations in Julia are performed using the FFTW package. We use two different packages for FFT operations in Python: Numpy library and pyFFTW package. We measure the CPU time for two grid resolutions: $N_x = N_y = 128$ and $N_x = N_y = 256$. We test two codes at $Re = 1000$ with $\Delta t = 0.01$ from time $t = 0$ to $t = 20$. The CPU time for two grid resolutions for two versions of code is provided in Table 4. It can be seen that the computational time for Julia is slightly better than Python code with the Numpy library for performing FFT operations. The pyFFTW package uses FFTW which is a very fast C library. The computational time required by vectorized Python code with pyFFTW package is better than the Julia code. This might be due to the effective interface between the FFTW library and pyFFTW package. Therefore, each programming language has some benefits over others. However, the goal of this study is not to optimize the code or find the best scientific programming language but is to present how can one implement CFD codes.

Table 4. Comparison of CPU time for Julia and Python version of pseudo-spectral solver at two grid resolutions. We highlight that the codes are not written for optimization purpose and the CPU time may not be the optimal time that can be achieved for each of the programming languages.

Programming Language	CPU Time (s)	
	$N_x, N_y = 128$	$N_x, N_y = 256$
Julia	31.25	163.63
Python (Numpy)	37.05	181.07
Python (pyFFTW)	21.76	105.81

Another common technique to reduce aliasing error which is widely used in practice is to use 2/3-rule. In this technique we retain the data on 2/3 of the grid is retained, while the remaining data is assigned with zero value. For example, if we are using a spatial resolution of $N_x = 128$ and $N_y = 128$ (i.e., $-64 \leq k_x \leq 64$ and $-64 \leq k_y \leq 64$), we will retain the Fourier coefficients corresponding to $(-42$ to $42)$ and remaining Fourier coefficients are set zero. One of the advantages of using this method is that it is computationally efficient since the Fourier transform of the matrix having size as the power of 2 is more efficient than the matrix which has dimension non-powers of 2. If we use 2/3-rule it is equivalent to performing calculation on $N_x = 85$ and $N_y = 85$ grid. However, a data set consisted of 128^2 points is used in the inverse and forward FFT calls in Equation (159). The implementation of Jacobian calculation with 2/3 padding is given in Listing 35. Before we close this section, we would like to note that there is no significant aliasing error in the example presented in this section. Therefore, one might get similar results without performing any padding (1/1 rule).

Listing 35. Julia implementation of computing the nonlinear Jacobian term in Fourier space with 2/3 padding rule.

```julia
# nx, ny: number of grid points in x and y directions
# dx, dy: grid spacing in x and y directions
# wf: vorticity in Fourier space
# k2: wavespace over 2D domain
function jacobian(nx,ny,dx,dy,wf,k2)
eps = 1.0e-6
kx = Array{Float64}(undef,nx)
ky = Array{Float64}(undef,ny)

#wave number indexing
hx = 2.0*pi/(nx*dx)
```

```julia
for i = 1:Int64(nx/2)
kx[i] = hx*(i-1.0)
kx[i+Int64(nx/2)] = hx*(i-Int64(nx/2)-1)
end
kx[1] = eps
ky = transpose(kx)

j1f = zeros(ComplexF64,nx,ny)
j2f = zeros(ComplexF64,nx,ny)
j3f = zeros(ComplexF64,nx,ny)
j4f = zeros(ComplexF64,nx,ny)

# x-derivative
for i = 1:nx for j = 1:ny
j1f[i,j] = 1.0im*wf[i,j]*kx[i]/k2[i,j]
j4f[i,j] = 1.0im*wf[i,j]*kx[i]
end~end

# y-derivative
for i = 1:nx for j = 1:ny
j2f[i,j] = 1.0im*wf[i,j]*ky[j]
j3f[i,j] = 1.0im*wf[i,j]*ky[j]/k2[i,j]
end~end

nxe = Int64(floor(nx*2/3))
nye = Int64(floor(ny*2/3))

# 2/3 padding
for i = Int64(floor(nxe/2)+1):Int64(nx-floor(nxe/2)) for j = 1:ny
j1f[i,j] = 0.0
j2f[i,j] = 0.0
j3f[i,j] = 0.0
j4f[i,j] = 0.0
end~end

for i = 1:nx for j = Int64(floor(nye/2)+1):Int64(ny-floor(nye/2))
j1f[i,j] = 0.0
j2f[i,j] = 0.0
j3f[i,j] = 0.0
j4f[i,j] = 0.0
end~end

j1 = real(ifft(j1f))
j2 = real(ifft(j2f))
j3 = real(ifft(j3f))
j4 = real(ifft(j4f))
jacp = zeros(Float64,nx,ny)

for i = 1:nx for j = 1:ny
jacp[i,j] = j1[i,j]*j2[i,j] - j3[i,j]*j4[i,j]
end~end

jf = fft(jacp)
```

```julia
    return jf
end
```

10. Concluding Remarks

The easy syntax and fast performance of Julia programming language make it one of the best candidates for engineering students, especially from a non-computer science background to develop codes to solve problems they are working on. We use Julia language as a tool to solve basic fluid flow problems which can be included in graduate-level CFD coursework. We follow a similar pattern as teaching a class, starting from basics and then building upon it to solve more advanced problems. We provide small pieces of code developed for simple fundamental problems in CFD. These pieces of codes can be used as a starting point and one can go about adding more features to solve complex problems. We make all codes, plotting scripts available online and open to everyone. All the codes are made available online to everyone on Github (https://github.com/surajp92/CFD_Julia).

We explain fundamental concepts of finite difference discretization, explicit methods, implicit methods using one-dimensional heat equation in Section 2. We also outline the procedure to develop a compact finite difference scheme which gives higher-order accuracy with smaller stencil size. We present a multi-step Runge-Kutta method which has higher temporal accuracy than standard single step finite difference method. We illustrate numerical methods for hyperbolic conservation laws using the inviscid Burgers equation as the prototype example in Section 3. We demonstrate two shock-capturing methods: WENO-5 and CRWENO-5 methods for Dirichlet and periodic boundary conditions. Students can learn about WENO reconstruction and boundary points treatment which is also applicable to other problems.

We show an implementation of the finite difference method for the inviscid Burgers equation in its conservative form in Section 4. Students will get insight into a finite volume method implementation through this example. We present two approaches for computing fluxes at the interface using the flux splitting method and using Riemann solver. We use Riemann solver based on simple Rusanov scheme. Students can easily implement other methods by changing the function for Rusanov scheme. We also develop one-dimensional Euler solver and validate it for the Sod shock tube problem in Section 5. We borrow functions from heat equation codes, such as Runge-Kutta code for time integration. We use WENO-5 code developed for the inviscid Burgers equation for solving Euler equations. This will give students an understanding of how coding blocks developed for simple problems can be integrated to solve more challenging problems.

We illustrate different methods to solve elliptic equations using Poisson equation as an example in Section 6. We describe both direct methods and iterative methods to solve the Poisson equation. The direct solvers explained in this study are based on the Fourier transform. We demonstrate an implementation of FFT and FST for developing a Poisson solver for periodic and Dirichlet boundary condition respectively. We provide an overview of two iterative methods: Gauss-Seidel method and conjugate gradient method for the Poisson equation. Furthermore, we demonstrate the implementation of the V-cycle multigrid framework for the Poisson equation. A multigrid framework is a powerful tool for CFD simulations as it scales linearly, i.e., it has $O(N)$ computational complexity.

We describe two-dimensional incompressible Navier-Stokes solver in Section 7 and validate it for two examples: lid-driven cavity problem and vortex-merger problem. We use streamfunction-vorticity formulation to develop the solver. This eliminates the pressure term in the momentum equation and we can use the collocated grid arrangement instead of the staggered grid arrangement. We use Arakawa numerical scheme for the Jacobian term. The lid-driven cavity problem has Dirichlet boundary condition and hence we use FST-based Poisson solver. We use FFT-based Poisson solver the vortex-merger problem since this problem is periodic in the domain. Most of the functions needed for developing the two-dimensional incompressible Navier-Stokes solver are taken from the heat equation and Poisson equation. Students will learn derivation of streamfunction-vorticity formulation, and its implementation through these examples and it can be easily extended to other problems like Taylor-Green vortex, decaying turbulence, etc. Furthermore, we develop hybrid incompressible Navier-Stokes solver for two-dimensional flow problems and is presented in Section 8. The solver is hybridized using explicit Runge-Kutta scheme and implicit Crank-Nicolson scheme for time integration. The solver is developed by solving the Navier-Stokes equation in Fourier space except for the nonlinear term. The nonlinear Jacobian term is computed in physical space using Arakawa finite difference scheme, converted to Fourier space and then used in Navier-Stokes equations. In addition, Section 9 provides the pseudo-spectral solver with 3/2 padding and 2/3 padding rule for two-dimensional incompressible Navier-Stokes equations. We also compare the computational performance of codes written in Julia and Python for the Navier-Stokes solvers. We should highlight that the codes are written in considering mostly pedagogical aspects (i.e., without performing any additional efforts in optimal coding and implementation practices) and cannot be viewed as a legitimate performance comparison of different languages, which is beyond the scope of this work.

Author Contributions: Data curation, S.P.; Visualization, S.P.; Supervision, O.S.; Writing—original draft, S.P.; and Writing—review and editing, S.P. and O.S.

Funding: This work received no external funding.

Conflicts of Interest: The authors declare no conflict of interest.

Appendix A. Installation Instruction

There are several ways to run Julia on your personal computer. One way is to install Julia on the computer and use it in the terminal using the built-in Julia command line. If you have a file in Julia, you can simly run the code from terminal using the command julia "filename.jl" . Another way is to run Julia code in the browser on JuliaBox.com with Jupyter notebooks. For this method, there is no installation required on the personal computer. One more way is to JuliaPro which includes Julia and the Juno IDE (https://juliacomputing.com/products/juliapro.html). This software contains a set of packages for plotting, optimization, machine learning, database, and much more.

Julia setup files can be downloaded from their website (https://julialang.org/downloads/). The website also includes instructions on how to install Julia on Windows, Linux, and mac operating systems. Some of the useful resources for learning Julia are listed below

- https://docs.julialang.org/en/v1/
- https://www.coursera.org/learn/julia-programming
- https://www.youtube.com/user/JuliaLanguage/featured
- https://discourse.julialang.org/

Oftentimes, one will have to install additional packages to use some of the features in Julia. Pkg is Julia's built in package manager, and it handles operations such as installing, updating and removing packages. We will explain with the help of FTTW library as an example on how to add and use new package. Below are the three commands needed to open Julia using terminal, add new package, and then use that package.

```julia
julia # open julia
Pkg.add("FFTW") # add FFTW package
using FFTW # to use FFTW package
```

Appendix B. Plotting Scripts

We use PyPlot module in Julia for plotting all results. This module provides a Julia interface to the Matploltlib plotting library from Python. The Python Matplotlib has to be installed in order to use PyPlot package. Once the Matplotlib installed, you can just use Pkg.add("PyPlot") in Julia to install

PyPlot and its dependencies. The detailed instruction for installing PyPlot package can be found on https://github.com/JuliaPy/PyPlot.jl.

We mention plotting scripts for two types of plots: XY plot and contour plot. These two plotting scripts are used in this paper to plot all the results. Also, once the student gets familiar with basic plotting scripts, one can easily extend it to more complex plots with examples available on the Internet. Some of the Internet resources that can be helpful for plotting are given below

- https://buildmedia.readthedocs.org/media/pdf/pyplotjl/latest/pyplotjl.pdf
- https://gist.github.com/gizmaa/7214002
- http://faculty.uml.edu/hung_phan/others/ntjuliapyplot.pdf

The scripts for XY plotting and contour plotting are given in Listings A1 and A2, respectively.

Listing A1. Plotting script for XY plot.

```julia
# x: location of grid points
# u_e: exact solution
# u_n: numerical solution
# u_error: discretization error
using PyPlot
rc("font", family="Arial", size=16.0)

fig = figure("FTCS", figsize=(14,6));
ax1 = fig[:add_subplot](1,2,1);
ax2 = fig[:add_subplot](1,2,2);

ax1.plot(x, u_e, lw=4, ls = "-", color="b", label="Exact solution")
ax1.plot(x, u_n, lw=4, ls = "--", color="r", label="FTCS solution")
ax1.set_xlabel("\$x\$")
ax1.set_ylabel("\$v\$")
ax1.set_title("Solution field")
ax1.set_xlim(-1,1)
ax1.legend(fontsize=14, loc=0)

ax2.plot(x, u_error, marker = "o", markeredgecolor="k",
markersize=8, color="g", lw=4)
ax2.set_xlabel("\$x\$")
ax2.set_ylabel("\$\epsilon\$")
ax2.set_title("Discretization error")
ax2.set_xlim(-1,1)
#ax2.legend(fontsize=14, loc=0)

plt[:subplot](ax1);
plt[:subplot](ax2);

fig.tight_layout()
fig.savefig("ftcs.pdf")
```

Listing A2. Plotting script for contour plot.

```julia
# x: location of grid points
# u_e: exact solution
# u_n: numerical solution
# u_error: discretization error
using PyPlot
rc("font", family="Arial", size=16.0)

fig = figure("An example", figsize=(14,6));
ax1 = fig[:add_subplot](1,2,1);
ax2 = fig[:add_subplot](1,2,2);

cs1 = ax1.contourf(xx, yy, transpose(u_e),levels=20, cmap="jet", vmin=-1, vmax=1)
ax1.set_title("Exact solution")
plt[:subplot](ax1); cs1

cs2 = ax2.contourf(xx, yy, transpose(u_n),levels=20, cmap="jet", vmin=-1, vmax=1)
ax2.set_title("Numerical solution")
plt[:subplot](ax2); cs2

fig.colorbar(cs, ax = ax1)
fig.colorbar(cs, ax = ax2)

fig.tight_layout()
fig.savefig("fst_contour.pdf")
```

References

1. Barba, L.A.; Forsyth, G.F. CFD Python: The 12 steps to Navier-Stokes equations. *J. Open Source Educ.* **2018**, *9*, 21. [CrossRef]
2. Oliphant, T.E. *A Guide to NumPy*; Trelgol Publishing: New York, NY, USA, 2006; Volume 1.
3. Ketcheson, D.I. Teaching numerical methods with IPython notebooks and inquiry-based learning. In Proceedings of the 13th Python in Science Conference (SciPy. org), Austin, TX, USA, 6–13 July 2014.
4. Ketcheson, D.I. HyperPython: An Introduction to Hyperbolic PDEs in Python. 2014. Available online: http://github.com/ketch/HyperPython/ (accessed on 25 June 2019).
5. Ketcheson, D.I.; Mandli, K.T.; Ahmadia, A.J.; Alghamdi, A.; Quezada de Luna, M.; Parsani, M.; Knepley, M.G.; Emmett, M. PyClaw: Accessible, Extensible, Scalable Tools for Wave Propagation Problems. *SIAM J. Sci. Comput.* **2012**, *34*, C210–C231. [CrossRef]
6. Bezanson, J.; Edelman, A.; Karpinski, S.; Shah, V.B. Julia: A fresh approach to numerical computing. *SIAM Rev.* **2017**, *59*, 65–98. [CrossRef]
7. Numrich, R.W.; Reid, J. *Co-Array Fortran for Parallel Programming*; ACM Sigplan Fortran Forum; ACM: New York, NY, USA, 1998; Volume 17, pp. 1–31.
8. Sanner, M.F. Python: A programming language for software integration and development. *J. Mol. Graph. Model.* **1999**, *17*, 57–61. [PubMed]
9. Stroustrup, B. *The C++ Programming Language*; Pearson Education: London, UK, 2000.
10. MATLAB. *Version 7.10.0 (R2010a)*; The MathWorks Inc.: Natick, MA, USA, 2010.
11. Versteeg, H.K.; Malalasekera, W. *An Introduction to Computational Fluid Dynamics: The Finite Volume Method*; Pearson Education: New York, NY, USA, 2007.
12. Moin, P. *Fundamentals of Engineering Numerical Analysis*; Cambridge University Press: New York, NY, USA, 2010.
13. Strang, G.; Fix, G.J. *An analysis of the Finite Element Method*; Prentice-Hall: Englewood Cliffs, NJ, USA, 1973; Volume 212.
14. Canuto, C.; Hussaini, M.Y.; Quarteroni, A.; Thomas, A., Jr. *Spectral Methods in Fluid Dynamics*; Springer Science & Business Media: Berlin, Germany, 2012.

15. Strikwerda, J.C. *Finite Difference Schemes and Partial Differential Equations*; Society for Industrial & Applied Mathmatics (SIAM): Philadelphia, PA, USA, 2004; Volume 88.

16. Gottlieb, S.; Shu, C.W. Total variation diminishing Runge-Kutta schemes. *Math. Comput. Am. Math. Soc.* **1998**, *67*, 73–85. [CrossRef]

17. Press, W.H.; Teukolsky, S.A.; Vetterling, W.T.; Flannery, B.P. *Numerical Recipes in Fortran*; Cambridge University Press: New York, NY, USA, 1992; Volume 2.

18. Lele, S.K. Compact finite difference schemes with spectral-like resolution. *J. Comput. Phys.* **1992**, *103*, 16–42. [CrossRef]

19. LeVeque, R.J. *Finite Volume Methods for Hyperbolic Problems*; Cambridge University Press: Cambridge, UK, 2002; Volume 31.

20. Knight, D. *Elements of Numerical Methods for Compressible Flows*; Cambridge University Press: New York, NY, USA, 2006; Volume 19.

21. Hirsch, C. *Numerical Computation of Internal and External Flows: The Fundamentals of Computational Fluid Dynamics*; Butterworth-Heinemann: Burlington, MA, USA, 2007.

22. Pulliam, T.H. Artificial dissipation models for the Euler equations. *AIAA J.* **1986**, *24*, 1931–1940. [CrossRef]

23. Rahman, S.; San, O. A relaxation filtering approach for two-dimensional Rayleigh–Taylor instability-induced flows. *Fluids* **2019**, *4*, 78. [CrossRef]

24. Jiang, G.S.; Shu, C.W. Efficient implementation of weighted ENO schemes. *J. Comput. Phys.* **1996**, *126*, 202–228. [CrossRef]

25. Ghosh, D.; Baeder, J.D. Compact reconstruction schemes with weighted ENO limiting for hyperbolic conservation laws. *SIAM J. Sci. Comput.* **2012**, *34*, A1678–A1706. [CrossRef]

26. Shu, C.W. High order weighted essentially nonoscillatory schemes for convection dominated problems. *SIAM Rev.* **2009**, *51*, 82–126. [CrossRef]

27. San, O.; Kara, K. Evaluation of Riemann flux solvers for WENO reconstruction schemes: Kelvin–Helmholtz instability. *Comput. Fluids* **2015**, *117*, 24–41. [CrossRef]

28. Rahman, S.M.; San, O. A localised dynamic closure model for Euler turbulence. *Int. J. Comput. Fluid Dyn.* **2018**, *32*, 326–378. [CrossRef]

29. Ruasnov, V. Calculation of intersection of non-steady shock waves with obstacles. *USSR Comput. Math. Math. Phys.* **1961**, *1*, 267–279.

30. Toro, E.F. *Riemann Solvers and Numerical Methods for Fluid Dynamics: A Practical Introduction*; Springer Science & Business Media: Berlin, Germany, 2013.

31. Van Der Burg, J.; Kuerten, J.G.; Zandbergen, P. Improved shock-capturing of Jameson's scheme for the Euler equations. *Int. J. Numer. Methods Fluids* **1992**, *15*, 649–671. [CrossRef]

32. Godunov, S.K. A difference method for numerical calculation of discontinuous solutions of the equations of hydrodynamics. *Mat. Sb.* **1959**, *89*, 271–306.

33. Roe, P.L. Approximate Riemann solvers, parameter vectors, and difference schemes. *J. Comput. Phys.* **1981**, *43*, 357–372. [CrossRef]

34. Sod, G.A. A survey of several finite difference methods for systems of nonlinear hyperbolic conservation laws. *J. Comput. Phys.* **1978**, *27*, 1–31. [CrossRef]

35. Peng, J.; Zhai, C.; Ni, G.; Yong, H.; Shen, Y. An adaptive characteristic-wise reconstruction WENO-Z scheme for gas dynamic Euler equations. *Comput. Fluids* **2019**, *179*, 34–51. [CrossRef]

36. Harten, A.; Lax, P.D.; van Leer, B. On upstream differencing and Godunov-type schemes for hyperbolic conservation laws. *SIAM Rev.* **1983**, *25*, 35–61. [CrossRef]

37. Gupta, M.M.; Kouatchou, J.; Zhang, J. Comparison of second-and fourth-order discretizations for multigrid Poisson solvers. *J. Comput. Phys.* **1997**, *132*, 226–232. [CrossRef]

38. McAdams, A.; Sifakis, E.; Teran, J. A parallel multigrid Poisson solver for fluids simulation on large grids. In Proceedings of the 2010 ACM SIGGRAPH/Eurographics Symposium on Computer Animation, Madrid, Spain, 2–4 July 2010; pp. 65–74.

39. Trefethen, L.N.; Bau III, D. *Numerical Linear Algebra*; Society for Industrial & Applied Mathmatics (SIAM): Philadelphia, PA, USA, 1997; Volume 50.

40. Barrett, R.; Berry, M.W.; Chan, T.F.; Demmel, J.; Donato, J.; Dongarra, J.; Eijkhout, V.; Pozo, R.; Romine, C.; Van der Vorst, H. *Templates for the Solution of Linear Systems: Building Blocks for Iterative Methods*; Society for Industrial & Applied Mathmatics (SIAM): Philadelphia, PA, USA, 1994; Volume 43.

41. Hadjidimos, A. Successive overrelaxation (SOR) and related methods. *J. Comput. Appl. Math.* **2000**, *123*, 177–199. [CrossRef]
42. San, O.; Vedula, P. Generalized deconvolution procedure for structural modeling of turbulence. *J. Sci. Comput.* **2018**, *75*, 1187–1206. [CrossRef]
43. Saad, Y. *Iterative Methods for Sparse Linear Systems*; Society for Industrial & Applied Mathmatics (SIAM): Philadelphia, PA, USA, 2003; Volume 82.
44. Arakawa, A. Computational design for long-term numerical integration of the equations of fluid motion: Two-dimensional incompressible flow. Part I. *J. Comput. Phys.* **1966**, *1*, 119–143. [CrossRef]
45. Ghia, U.; Ghia, K.N.; Shin, C. High-Re solutions for incompressible flow using the Navier-Stokes equations and a multigrid method. *J. Comput. Phys.* **1982**, *48*, 387–411. [CrossRef]
46. Hoffmann, K.A.; Chiang, S.T. *Computational Fluid Dynamics*; Engineering Education System: Wichita, KS, USA, 2000; Volume I.
47. Meunier, P.; Le Dizes, S.; Leweke, T. Physics of vortex merging. *Comptes Rendus Phys.* **2005**, *6*, 431–450. [CrossRef]
48. San, O.; Staples, A.E. High-order methods for decaying two-dimensional homogeneous isotropic turbulence. *Comput. Fluids* **2012**, *63*, 105–127. [CrossRef]
49. San, O.; Staples, A.E. A coarse-grid projection method for accelerating incompressible flow computations. *J. Comput. Phys.* **2013**, *233*, 480–508. [CrossRef]
50. Reinaud, J.N.; Dritschel, D.G. The critical merger distance between two co-rotating quasi-geostrophic vortices. *J. Fluid Mech.* **2005**, *522*, 357–381. [CrossRef]
51. Orlandi, P. *Fluid Flow Phenomena: A Numerical Toolkit*; Springer Science & Business Media: Berlin, Germany, 2012; Volume 55.
52. Mortensen, M.; Langtangen, H.P. High performance Python for direct numerical simulations of turbulent flows. *Comput. Phys. Commun.* **2016**, *203*, 53–65. [CrossRef]

 *fluids*

Article

An Efficient Strategy to Deliver Understanding of Both Numerical and Practical Aspects When Using Navier-Stokes Equations to Solve Fluid Mechanics Problems

Desmond Adair [1,*] and **Martin Jaeger** [2]

[1] School of Engineering and Digital Sciences, Nazarbayev University, Nur-Sultan 010000, Kazakhstan

[2] College of Sciences and Engineering, University of Tasmania, TAS 7001, Australia; m.jaeger@ack.edu.kw

[*] Correspondence: desmond.adair@utas.edu.au; Tel.: +7-7172-70-6531

Received: 7 July 2019; Accepted: 27 September 2019; Published: 1 October 2019

Abstract: An efficient and thorough strategy to introduce undergraduate students to a numerical approach of calculating flow is outlined. First, the basic steps, especially discretization, involved when solving Navier-Stokes equations using a finite-volume method for incompressible steady-state flow are developed with the main aim being for the students to follow through from the mathematical description of a given problem to the final solution of the governing equations in a transparent way. The well-known 'driven-cavity' problem is used as the problem for testing coding written by the students, and the Navier-Stokes equations are initially cast in the vorticity-streamfunction form. This is followed by moving on to a solution method using the primitive variables and discussion of details such as, closure of the Navier-Stokes equations using turbulence modelling, appropriate meshing within the computation domain, various boundary conditions, properties of fluids, and the important methods for determining that a convergence solution has been reached. Such a course is found to be an efficient and transparent approach for introducing students to computational fluid dynamics.

Keywords: computational fluid dynamics; finite-volume method; discretization; turbulence modelling; meshing

---

1. Introduction

Computational Fluid Dynamics (CFD) is the simulation of transport phenomena, reacting systems, heat transfer, etc., using modelling, i.e., mathematical physical problem formulation and numerical solution, which include discretization methods, solvers, numerical parameters and mesh generation. The efficient and thorough development, implementation and evaluation of a suitable curriculum, which incorporates both the theoretical and practical aspects of CFD, for undergraduate students as an extension of their knowledge in the thermo-fluids area is warranted and necessary in modern engineering programmes. Numerical methods have an advantage over analytical methods of solution in that analytical methods tend to be restrictive by commonly requiring simple geometries, unrealistic assumptions and until lately assumed linearity, whereas real engineering problems usually have quite complex geometries and highly non-linear phenomena. However, despite the advantages of numerical methods, the latter may miss some finer aspects of fluid dynamics which, from the equations of analytical methods, may be more transparent, for example the question of uniqueness of weak solutions to Navier-Stokes equations in three dimensions. Analytical methods also often task a student to consider the treatment of any nonlinear terms found in partial differential equations, hence encouraging a student's interest in this important topic.

Many areas of engineering education already use mathematical modelling as a teaching tool, for example, for diesel engine studies [1], in fluid mechanics and heat transfer [2–4] and in chemical reactions [5]. It is important when introducing a complex topic such as CFD to have educational user-friendly interfaces, usually in the form of dialogue graphical user interfaces (GUIs), and some work has already been done on this [6,7], where the general aspects of the three main processes of CFD, i.e., building the problem using a pre-processor, solving the problem, and displaying the results using a post-processor were considered. However, although the education benefits associated with integrating computer-assisted learning and simulation technology into undergraduate engineering course are great [8], and, computational fluid dynamics has revolutionized approaches to research and design, its incorporation into the teaching of undergraduate transport phenomena has still difficulties. One of these difficulties, and the one addressed in this paper, is to efficiently and thoroughly combine a sufficient understanding of the theoretical aspects of CFD with its practical use. Another is to engender a critical understanding of obtained results.

To help with an initial understanding of the process of solving Navier-Stokes and related transport equations the symbolic software, *Mathematica* [9] was employed to demonstrate the mathematical aspects of some of the basic steps involved, for example meshing, discretization, application of boundary conditions and the development of a simple solution algorithm. Two-dimensional incompressible flows were calculated for simplicity and the Navier-Stokes, at this stage of the course, were recast in an alternative form, i.e., in terms of streamfunction and vorticity. In many applications, the vorticity-streamfunction form of the Navier-Stokes equations provides a better insight into the physical mechanisms driving the flow than the "primitive variable" formulation which is in terms of the mean velocities u, v and pressure, p. The streamfunction and vorticity formulation is also useful for numerical work, since it avoids some problems resulting from discretization of the continuity equation.

The second part of this work outlines how to best introduce the practical aspects and emphasizes concerns which must be remembered when using CFD. Issues arise when introducing simulation into a curriculum, including learning vs. research objectives, usability vs. predetermined objectives and student demographics. A proper balance must be sought for between these competing objectives; for example, it is just as important that students are taught the systematic and practical ways of using a CFD package in a general sense, as well as achieving a specific result. There is evidence from previous studies that using simulation enhances the curriculum [4]; there is increased learning efficiency and understanding [10,11]; there is effectiveness of hands-on learning methods [12]; and, it is effective to use a combination of physical and simulation laboratories [7].

Other concerns when introducing CFD to students are, for example, when to introduce during the programme, how much to introduce and how much mathematical and thermo-fluid backgrounds are necessary [13]? When a student first commences studying CFD a lot of new knowledge and required skills descend on them from several different curriculum areas, which must be married together hence rendering a steep learning curve. If careful planning is not carried out, this curve may become overwhelming for many of the students. It is also important to ensure that the CFD GUIs are not treated as 'black boxes', without knowledge of what is going on inside the computer software. This is especially true as many commercial codes now provide the opportunity of attaching in-house coding to the main core of the software hence providing a gateway for enhanced research and code development [14]. Without knowledge of CFD fundamentals and the core coding, code development would not be possible. It is important that students maintain a 'feeling' for physical phenomena, remember the necessary assumptions made for the mathematical modelling, and, very importantly, know of the need to verify and validate the calculated results [15]. It has been observed that successful implementation of CFD often requires a re-focus of course objectives and skills taught, and a re-structuring of the course curriculum [16]. An important practical consideration is to engender into the students a healthy skepticism concerning the calculated results and a willingness to be critical of their results until validated [17].

In this paper the mathematical aspects of CFD are first introduced, with a particular emphasize on how to discretize equations. This is followed by a discussion of practical details such as closure of the Navier-Stokes equations using turbulence modelling, appropriate meshing within the computation domain, various boundary condition, properties of fluids, and the important methods for determining that a convergence solution has been reached. An overall description of the course is then given. The paper finishes with conclusions drawn and possible work needed for future development.

2. Numerical Elements of Computational Fluid Dynamics (CFD) Teaching

2.1. Introduction

In this part of the course, which lasted for four weeks, the purpose was to develop the basic steps involved in solving Navier-Stokes equations using a finite-volume approach for incompressible steady-state flow. The students can actually see the development of the mathematical description linked to a programming environment solution process, so often hidden in commercial code used for training CFD students. The students start with discretization of the governing equations, write their own coding to generate a mesh, apply boundary conditions, provide a suitable solution algorithm and to generate contour and x-y plots.

The students had already successfully completed courses on matrix algebra, vector calculus, ordinary and partial differential equations, with the latter solved using a variety of numerical methods, as well as preliminary courses on fluid mechanics and heat transfer. The students also have reasonable experience of using *Mathematica*, which is a necessary prerequisite [18]. Required reading for the course included LeVeque [19] and Ferziger and Peric [20].

In the following sections the theory taught to the students in preparation for them writing appropriate coding is summarized.

2.2. Navier-Stokes Equations

The governing equations for modelling fluid flow are the Navier-Stokes equations and the continuity equation. These laws of motion are valid for all phases including liquids, gases and solids, but there is a fundamental difference between fluids and solids in that fluids have no limit to their distortion. The analysis of fluid flow needs to take into account such distortions. The Navier-Stokes equations can be derived by the consideration of the dynamic equilibrium of a small fluid element and they state that surface and body forces are in equilibrium with inertial forces acting on the fluid element. On consideration of incompressible flow, where the density is constant, the Navier-Stokes equations for three-dimensional flow are as shown in Equations (1)–(4). Here body forces have not been included.

$$\frac{\partial u}{\partial x} + \frac{\partial v}{\partial y} + \frac{\partial w}{\partial z} = 0 \tag{1}$$

$$\frac{\partial u}{\partial t} + u\frac{\partial u}{\partial x} + v\frac{\partial u}{\partial y} + w\frac{\partial u}{\partial z} = -\frac{1}{\rho}\frac{\partial p}{\partial x} + v\left(\frac{\partial^2 u}{\partial x^2} + \frac{\partial^2 u}{\partial y^2} + \frac{\partial^2 u}{\partial z^2}\right) \tag{2}$$

$$\frac{\partial v}{\partial t} + u\frac{\partial v}{\partial x} + v\frac{\partial v}{\partial y} + w\frac{\partial v}{\partial z} = -\frac{1}{\rho}\frac{\partial p}{\partial y} + v\left(\frac{\partial^2 v}{\partial x^2} + \frac{\partial^2 v}{\partial y^2} + \frac{\partial^2 v}{\partial z^2}\right) \tag{3}$$

$$\frac{\partial w}{\partial t} + u\frac{\partial w}{\partial x} + v\frac{\partial w}{\partial y} + w\frac{\partial w}{\partial z} = -\frac{1}{\rho}\frac{\partial p}{\partial z} + v\left(\frac{\partial^2 w}{\partial x^2} + \frac{\partial^2 w}{\partial y^2} + \frac{\partial^2 w}{\partial z^2}\right) \tag{4}$$

where u, v, w are the velocity components in the x, y, z directions, ρ is the fluid density, p is the pressure and v is the kinematic viscosity.

Most flows encountered in engineering problems are turbulent because it is mostly impossible to stop transition from laminar to turbulent flow, or, by intention, as turbulence may aid the engineering application. For turbulent flows however, the variation of the quantities to be calculated with time is often so rapid and with a high degree of randomness that this variation is of little engineering relevance. Hence the quantities are averaged and most often the Reynolds-averaged Navier-Stokes equations as shown in Equations (5)–(8) are used.

$$\frac{\partial \overline{u}}{\partial x} + \frac{\partial \overline{v}}{\partial y} + \frac{\partial \overline{w}}{\partial z} = 0 \tag{5}$$

$$\frac{\partial \overline{u}}{\partial t} + \overline{u}\frac{\partial \overline{u}}{\partial x} + \overline{v}\frac{\partial \overline{u}}{\partial y} + \overline{w}\frac{\partial \overline{u}}{\partial z} = -\frac{1}{\rho}\frac{\partial \overline{p}}{\partial x} + \frac{\partial}{\partial x}\left(v\frac{\partial \overline{u}}{\partial x} - \overline{u'u'}\right) + \frac{\partial}{\partial y}\left(v\frac{\partial \overline{u}}{\partial y} - \overline{u'v'}\right) + \frac{\partial}{\partial z}\left(v\frac{\partial \overline{u}}{\partial z} - \overline{u'w'}\right) \tag{6}$$

$$\frac{\partial \overline{v}}{\partial t} + \overline{u}\frac{\partial \overline{v}}{\partial x} + \overline{v}\frac{\partial \overline{v}}{\partial y} + \overline{w}\frac{\partial \overline{v}}{\partial z} = -\frac{1}{\rho}\frac{\partial \overline{p}}{\partial y} + \frac{\partial}{\partial x}\left(v\frac{\partial \overline{v}}{\partial x} - \overline{u'v'}\right) + \frac{\partial}{\partial y}\left(v\frac{\partial \overline{v}}{\partial y} - \overline{v'v'}\right) + \frac{\partial}{\partial z}\left(v\frac{\partial \overline{v}}{\partial z} - \overline{v'w'}\right) \tag{7}$$

$$\frac{\partial \overline{w}}{\partial t} + \overline{u}\frac{\partial \overline{w}}{\partial x} + \overline{v}\frac{\partial \overline{w}}{\partial y} + \overline{w}\frac{\partial \overline{w}}{\partial z} = -\frac{1}{\rho}\frac{\partial \overline{p}}{\partial y} + \frac{\partial}{\partial x}\left(v\frac{\partial \overline{w}}{\partial x} - \overline{u'w'}\right) + \frac{\partial}{\partial y}\left(v\frac{\partial \overline{w}}{\partial y} - \overline{v'w'}\right) + \frac{\partial}{\partial z}\left(v\frac{\partial \overline{w}}{\partial z} - \overline{w'w'}\right) \tag{8}$$

As can be observed, the instantaneous quantities are replaced by their corresponding time-averaged quantities. Extra terms appear due to the averaging process, e.g., $-\overline{u'v'}$ and these behave like stress terms, often called Reynolds stresses, which need further equations if the system of equations is to be closed. There have, over this last 60–70 years, been many suggestions as to the best way for equation closure, none of which as yet are completely satisfactory.

2.3. Vorticity-Streamfunction Governing Equations

For this work, the Navier-Stokes equations are caste into the vorticity-streamfunction form, where the streamfunction is defined as

$$\psi_A(P) = \int_A^P \mathbf{u} \cdot \mathbf{n} \, ds \tag{9}$$

and where the integral has to be evaluated along a curve C from the arbitrary but fixed point A to point P, $\mathbf{u}$ is the velocity vector, and $\mathbf{n}$ is the unit normal on the curve from A to P. The streamfunction is regarded as a function of the spatial coordinates only. Streamlines are lines that are everywhere tangential to the velocity field, and hence ψ is constant along the streamlines. Invoking the integral incompressibility constraint for an infinitesimally small triangle shows that ψ is related to the two Cartesian velocity components u and v via

$$u = \frac{\partial \psi}{\partial y}; \quad v = -\frac{\partial \psi}{\partial x} \tag{10}$$

Flows that are specified by a streamfunction automatically satisfy the continuity equation since

$$\frac{\partial u}{\partial x} + \frac{\partial v}{\partial y} = \frac{\partial}{\partial x}\left(\frac{\partial \psi}{\partial y}\right) - \frac{\partial}{\partial y}\left(\frac{\partial \psi}{\partial x}\right) = 0 \tag{11}$$

For two-dimensional flows, the vorticity vector $\omega = \nabla \times \mathbf{u}$ only has one non-zero component (in the z direction; i.e., $\omega = \omega e_z$ where

$$\omega = \frac{\partial v}{\partial x} - \frac{\partial u}{\partial y} \tag{12}$$

Using the definition of the velocities in terms of the streamfunction shows that

$$\omega = \frac{\partial}{\partial x}\left(-\frac{\partial \psi}{\partial x}\right) - \frac{\partial}{\partial y}\left(\frac{\partial \psi}{\partial y}\right) = -\nabla^2\psi \tag{13}$$

The Navier-Stokes equations for incompressible steady-state flow in vorticity-streamfunction form are

$$\frac{\partial \psi}{\partial x}\frac{\partial \omega}{\partial y} - \frac{\partial \psi}{\partial y}\frac{\partial \omega}{\partial x} = \frac{1}{Re}\left(\frac{\partial^2 \omega}{\partial x^2} + \frac{\partial^2 \omega}{\partial y^2}\right) \tag{14}$$

$$\frac{\partial^2 \psi}{\partial x^2} + \frac{\partial^2 \psi}{\partial y^2} = -\omega \tag{15}$$

where Re is the Reynolds number.

It should be mentioned that the above formulation is only valid in two-dimensions and some alternation is needed when three-dimensional flow is considered.

2.4. Boundary Conditions

The boundary conditions needed to solve Equations (14) and (15) for the classical computational fluid dynamics case, namely the lid-cavity flow [21], are shown on Figure 1.

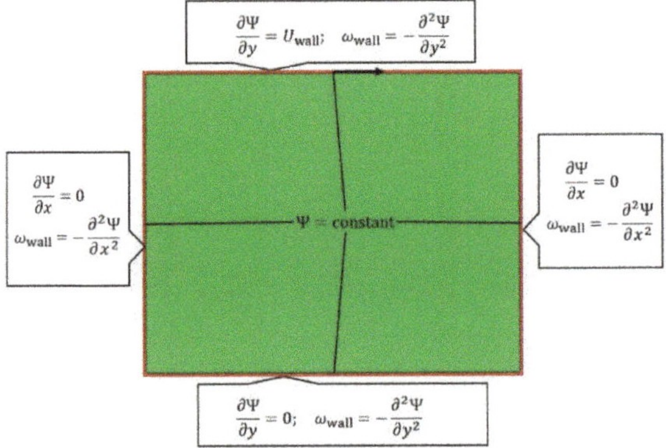

Figure 1. Summary of boundary conditions for cavity with moving lid. (Reproduced with permission from [22]).

A rectangular box, where the lid is allowed to move in the horizontal plane from left to right, is simulated with the moving lid driving the fluid within the box. In Figure 1 the boundary conditions needed for solution are summarized in terms of vorticity and streamfunction. The streamfunction is constant at the walls, as its gradient is velocity, which is zero relative to a given wall (the no-slip condition). The vorticity boundary condition for each wall is also shown in each of the outer boxes and are derived from the streamfunction.

2.5. Discretization of Transport Equations

Before describing the discretization of the governing equations, it is important to note that for convenience Equations (14) and (15) together with the boundary conditions were non-dimensionalized using

$$\widetilde{x} = \frac{x}{b}; \ \widetilde{y} = \frac{y}{a}; \ \widetilde{\psi} = \frac{\psi}{Vb}; \ \widetilde{\omega} = \frac{\omega b}{V}; \ \gamma = \frac{b}{a}; \ Re = \frac{Vb}{\nu}$$

where a and b are the height and width of the cavity, respectively, and V is a reference velocity. The resultant governing equations in vorticity-streamfunction format become

$$\frac{\partial^2 \widetilde{\omega}}{\partial \widetilde{x}^2} + \gamma^2 \frac{\partial^2 \widetilde{\omega}}{\partial \widetilde{y}^2} = Re\gamma\left(\frac{\partial \widetilde{\psi}}{\partial \widetilde{y}}\frac{\partial \widetilde{\omega}}{\partial \widetilde{x}} - \frac{\partial \widetilde{\psi}}{\partial \widetilde{x}}\frac{\partial \widetilde{\omega}}{\partial \widetilde{y}}\right) \tag{16}$$

$$\frac{\partial^2 \widetilde{\psi}}{\partial \widetilde{x}^2} + \gamma^2 \frac{\partial^2 \widetilde{\psi}}{\partial \widetilde{y}^2} = -\widetilde{\omega} \tag{17}$$

defined in the domain $0 \leq \tilde{x} \leq 1$, $0 \leq \tilde{y} \leq 1$.

So as to include the boundary nodes of the cavity, central differencing is used to discretize Equations (16) and (17) with the number of nodes in the x and y directions set at n_x and n_y, respectively. The mesh is a regular Castesian grid with the nodes equally spaced at a distance h, as illustrated on Figure 2.

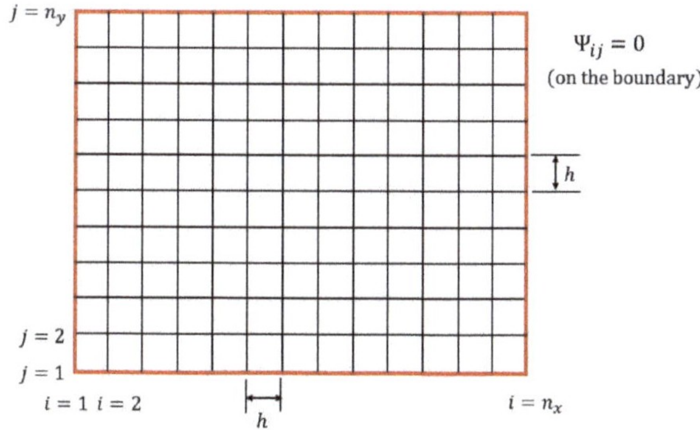

Figure 2. Typical Cartesian mesh used for the lid-cavity flow. (Reproduced with permission from [21]).

To discretize Equations (16) and (17) the following central-difference approximations were introduced for spatial dimensions, obtained by adding or subtracting one Taylor series from another and invoking the intermediate value theorem. As can be seen from Equation (18), the order of accuracy for both the first and second derivatives is $O(h^2)$.

$$\frac{\partial f(x)}{\partial x} = \frac{f(x+h) - f(x-h)}{2h} + O(h^2)$$
$$\frac{\partial f(x)}{\partial x} = \frac{f(x+h) - f(x-h)}{2h} + O(h^2) \tag{18}$$

Using the approximations from Equation (18), the discretized advection-diffusion equation (Equation (16)) at the (i,j) node is

$$\frac{\omega_{i+1,j} - 2\omega_{i,j} + \omega_{i-1,j}}{h^2} + \gamma^2 \frac{\omega_{i,j+1} - 2\omega_{i,j} + \omega_{i,j-1}}{h^2}$$
$$= \text{Re}\gamma \left(\frac{\psi_{i,j+1} - \psi_{i,j-1}}{2h} \frac{\omega_{i+1,j} - \omega_{i-1,j}}{2h} - \frac{\psi_{i+1,j} - \psi_{i-1,j}}{2h} \frac{\omega_{i,j+1} - \omega_{i,j-1}}{2h} \right) \tag{19}$$

The elliptic equation (Equation (17)) at the (i,j) node is

$$\frac{\psi_{i+1,j} - 2\psi_{i,j} + \psi_{i-1,j}}{h^2} + \gamma^2 \frac{\psi_{i,j+1} - 2\psi_{i,j} + \psi_{i,j-1}}{h^2} = -\omega(x_i, y_j) \tag{20}$$

The equations are applied at the internal nodes of the Cartesian mesh; i.e., $2 \leq i \leq n_{x-1}$ and $2 \leq i \leq n_{y-1}$.

2.6. Solution Algorithm

The iterative solution used here is based on Burggraf's proposal [23]. The solution method uses residual functions; i.e., if the values of $\psi_{i,j}$ and $\omega_{i,j}$ are exact on the nodes spanned by the residual functions $\mathcal{R}_{ij}$ and $\mathcal{L}_{ij}$, then

$$\mathcal{R}_{ij}^{k+1} = \mathcal{L}_{ij}^{k+1} = 0 \tag{21}$$

where

$$\mathcal{R}_{ij}^{k+1} = \frac{1}{2(1+\gamma^2)}\left(\psi_{i+1,j}^{k} + \psi_{i-1,j}^{k+1} + \gamma^2\left(\psi_{i,j+1}^{k+1} + \psi_{i,j-1}^{k+1}\right)\right) + h^2\omega_{i,j}^{k} - \psi_{i,j}^{k} \tag{22}$$

$$\begin{aligned}
\mathcal{L}_{ij}^{k+1} &= \frac{1}{2(1+\gamma^2)}\left(\left(\omega_{i+1,j}^{k} + \omega_{i-1,j}^{k} + \gamma^2\left(\omega_{i,j+1}^{k} + \omega_{i,j-1}^{k}\right)\right)\right. \\
&\quad \left. -\frac{Re}{4}\gamma\left(\left(\psi_{i,j+1}^{k} - \psi_{i,j-1}^{k+1}\right)\left(\omega_{i+1,j}^{k} - \omega_{i-1,j}^{k+1}\right) - \left(\psi_{i+1,j}^{k} - \psi_{i-1,j}^{k+1}\right)\left(\omega_{i,j+1}^{k} - \omega_{i,j-1}^{k+1}\right)\right)\right) - \omega_{i,j}^{k}
\end{aligned} \tag{23}$$

The following fixed-point iterative procedure, based on the Gauss-Seidel scheme (see Appendix A for a short general explanation), is then constructed

$$\begin{aligned}
\psi_{i,j}^{k+1} &= \mathcal{F}^{k}\left(\psi^{(k)}, \psi^{(k+1)}, \omega^{(k)}, \omega^{(k+1)}\right) \equiv \psi_{i,j}^{k} + p\mathcal{R}_{ij}^{k} \\
\omega_{i,j}^{k+1} &= \mathcal{G}^{k}\left(\psi^{(k)}, \psi^{(k+1)}, \omega^{(k)}, \omega^{(k+1)}\right) \equiv \omega_{i,j}^{k} + p\mathcal{L}_{ij}^{k}
\end{aligned} \tag{24}$$

where p is a relaxation parameter lying in the range $0 < p \leq 1$, and $k+1$ and k refer to the respective iterations. The use of this relaxation parameter is to improve the stability of a particular computation, particularly when solving steady-state problems.

The boundary conditions now need to be determined. For the streamfunction

$$\psi_{i,1} = 0 \text{ for } i = 1,\dots,n_x,$$

$$\psi_{n_x,j} = 0 \text{ for } j = 1,\dots,n_y,$$

$$\psi_{i,n_y} = 0 \text{ for } i = 1,\dots,n_x,$$

$$\psi_{1,j} = 0 \text{ for } j = 1,\dots,n_y.$$

For vorticity, the following needs to be considered for, say, the left wall of the cavity with one node outside the computational domain.

$$\omega_{i,j} = \frac{1}{h^2}\left(\psi_{2,j} - 2\psi_{1,j} + \psi_{0,j} + \gamma^2\left(\psi_{1,j+1} - 2\psi_{1,j} + \psi_{1,j-1}\right)\right) \tag{25}$$

The value of $\psi_{0,j}$ can be accounted for by using the no-slip boundary condition on the cavity wall, i.e.,

$$v = -\frac{\partial\psi}{\partial x} = 0, \text{ at } x = 0 \tag{26}$$

Using central differencing, this condition can be written as

$$-\left(\frac{\partial\psi}{\partial x}\right)_{1,j} = 0 \Rightarrow -\left(\frac{\psi_{2,j} - \psi_{0,j}}{2h}\right) = 0 \Rightarrow \psi_{0,j} = \psi_{2,j} \tag{27}$$

where again the order of accuracy is $O(h^2)$.

This finding, together with $\psi_{1,j} = \psi_{1,j+1} = \psi_{1,j-1} = 0$ gives

$$\omega_{1,j} = -\frac{2}{h^2}\left(\psi_{2,j}\right) \text{ for } j = 2, n_{y-1} \tag{28}$$

Similarly, for the other walls of the cavity the boundary conditions for vorticity can be deduced from

$$\omega(x_i, y_i) = -\left(\frac{\psi_{i+1,j} - 2\psi_{i,j} + \psi_{i-1,j}}{h^2} + \gamma^2\frac{\psi_{i,j+1} - 2\psi_{i,j} + \psi_{i,j-1}}{h^2}\right) \tag{29}$$

to give

$$\omega_{i,1} = -\frac{2}{h^2}\gamma^2(\psi_{i,2}) \text{ for } i = 1,\ldots, n_x$$
$$\omega_{n_x,j} = -\frac{2}{h^2}\gamma^2\left(\psi_{n_x-1,j}\right) \text{ for } j = 2,\ldots, n_{y-1} \tag{30}$$
$$\omega_{i,n_y} = -\frac{2}{h^2}\gamma^2\left(\psi_{i,n_{y-1}} + \frac{h}{\gamma}\right) \text{ for } i = 1,\ldots, n_x$$

It should be noted that the final boundary condition shown in Equation (30) is for the moving lid, and the extra term is due to the tangential velocity being non-zero.

3. Practical Elements of CFD Teaching

3.1. Introduction

For Computational Fluid Dynamics laboratories, each student had a desktop computer on which was loaded software which provides the interface as a 'three-dimensional' fully interactive environment. The interface uses the concept of 'virtual reality' and allows a student to simulate a flow, including geometry, properties, boundary conditions, closure of equations, numerics, etc, from beginning to end without having to resort to specialized codes, or code writing. The code used can be used for professional engineering consultancy, so giving students useful skills and experience which contribute to their preparation for the workplace. The virtual reality (VR) environment is designed as a general purpose CFD interface, which can handle such problems as fluid flow, heat transfer, reacting flows, or a combination of all, and consists of the VR-Editor (pre-processor), the VR-Viewer (post-processor) and the solver module which performs the actual calculations. The VR-Editor allows a student to set the overall computational domain size, define the position, size and properties of object introduced into the domain, specify the material properties which occupy the domain, specify inlet and outlet boundary conditions, specify initial conditions, select turbulence models (if necessary), specify the fineness and type of computational mesh, and set the numerics by choosing a suitable solution algorithm and which influence the speed of calculation to obtain a convergence solution. Once the student is satisfied with problem set-up in the pre-processor, calculations are made using the solver module and the progress of the calculations is clearly monitored on-screen until convergence is reached. The results of the flow simulation are then displayed on the VR-Viewer graphically or they can be analysed from the raw numerical data. The post-processing graphical capabilities are vector plots, contour plots, iso-surfaces, streamlines and x-y plots.

3.2. Interface Design Features

The Computational Fluid Dynamics laboratory is designed in a user-friendly way, the computational features are easy to access and implement. It is important to keep parameter settings at the top level, and not to use multi-levels, as much as possible, as the setting of vital parameters necessary to the success of a particular simulation may well get hidden or even forgotten about, thus frustrating the student with strange results. Also it is important to show students that CFD methodology needs to be systematic and rigorous, as the laboratory must reflect what is found in engineering practice.

The CFD interface was designed, as shown on Figure 3, so that the features systematically inform, it is vocationally sound and is easy to use.

Each simulation process reflected the situation in engineering practice with students systematically setting up the problem, solving and finally analysing the results. The students interact with the software using mouse and keyboard input with no requirement for the distraction of computer language coding so enabling them to concentrate on the CFD methodology and procedures and how close to reality their results are. An important feature of the CFD interface is that it is stand-alone, by which is meant that the problem-building, solving and post-processing are all combined in the same virtual reality environment. Also important is that all the data representing the results can be moved to the familiar Microsoft Office environment for further processing and reporting. The CFD software package operates on the Windows OS platform using PCs with relatively low computer power, and so the solvers have to be reasonably fast because the students have limited guided time in the laboratories, and also to keep their level of interest heightened, calculated results should come back reasonably quickly. In addition to producing contour plots, vector plots, etc., the post-processor could also produce animation for transient flows.

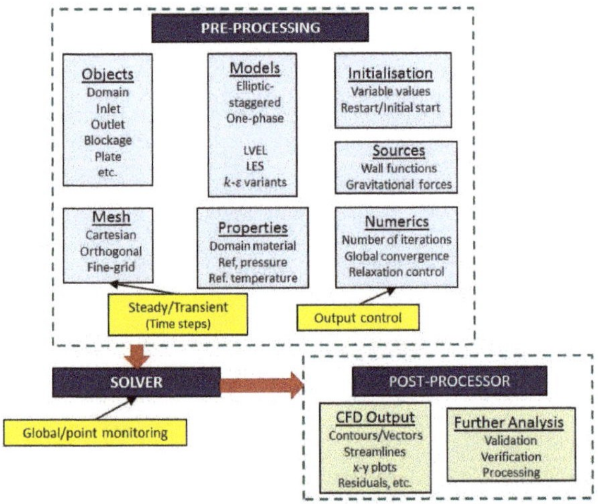

Figure 3. Summary of Computational Fluid Dynamics (CFD) interface. (Reproduced with permission from [24]).

3.3. Example to illustrate use of the Interface

An example of flow through a labyrinth with heat exchange is given to illustrate the simplicity of the interface. The geometry is 2D with dimensions and the content of the computation domain is summarized on Figure 4.

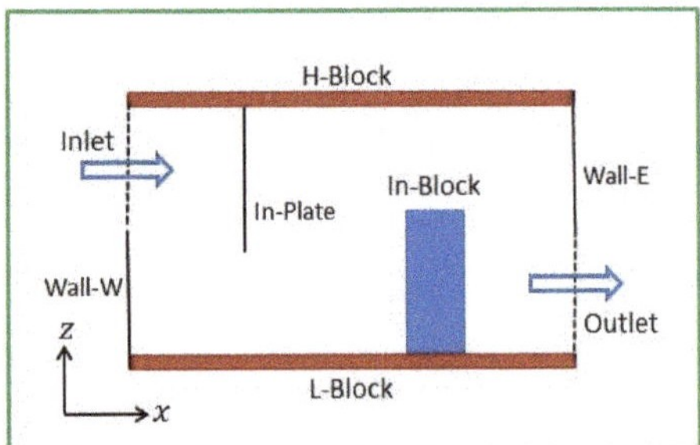

Figure 4. Geometry and contents of a 2D computational domain with steady turbulent flow of air and heat exchange.

Initially the pre-processing virtual reality GUI is opened and the case title immediately entered. With reference to Figure 3, under 'Objects' the x-domain size is changed to 2.0 m and under 'Mesh' a Cartesian grid is chosen. Under 'Properties', the user now checks that the default properties for this case are chosen, i.e., 'Air at 20 °C, 1 atm', the ambient air is set at 20 °C, and the ambient pressure is set at 0 Pa relative to 1 atm.

Under 'Models' the 'solution for velocities and pressure' is left ON, the 'energy equation' is clicked and 'Temperature' is selected. The turbulence model selection is left as the KE variant 'KECHEN'.

Under 'Objects' each of the objects within the domain are selected, sized and positioned and where necessary given the relevant attributes. The objects introduced in order are, the H-BLOCK whose attribute is 'ALUMINIUM at 27 °C. the L-BLOCK whose attribute is 'ALUMINIUM and which is a HEAT SOURCE of 100 W', the WALL-E which is a plate, the WALL-W which is a plate, the IN-PLATE, the IN-BLOCK whose attribute is 'COPPER', the INLET whose attribute is velocity in the x-direction at 0.5 m/s and whose temperature is 'AMBIENT', the OUTLET which has 'PRESSURE' as its attribute.

Once the basic geometry is in place, a computational grid is automatically suggested and can be inspected by the 'MESH' toggle, which for the present case, generates a picture as shown on Figure 5. The grid can be customized to the user's wishes simply by clicking on 'Grid Mesh Setting' and adjusting the default numbers.

Figure 5. Grid automatically generated for labyrinth case.

In this example, 'Initialisation' and 'Sources' are taken as the defaults, and so only 'Numerics' is left to inspect and adjust. Here the number of iterations is set and the convergence tolerances are left at default. Importantly the probe used as one measure of convergence is appropriately positioned, i.e., not too close to the inlet or not placed in an anticipated region of recirculation.

The solver is now activated to perform the simulation. The progress of the calculations is displayed in relation to convergence, number of sweeps of the computation domain completed and the estimated time left to complete the simulation.

Once the solver has closed, the GUI post-processor can now be opened to view the results. Toggles are available to view 'Vectors', 'Contours', 'Streamlines', Iso-surfaces', and further toggles are available to select which calculated variable is of interest. The results can be viewed on any plane or at any angle by the use of direction and rotation buttons respectively. A typical result is shown on Figure 6 for a vector plot.

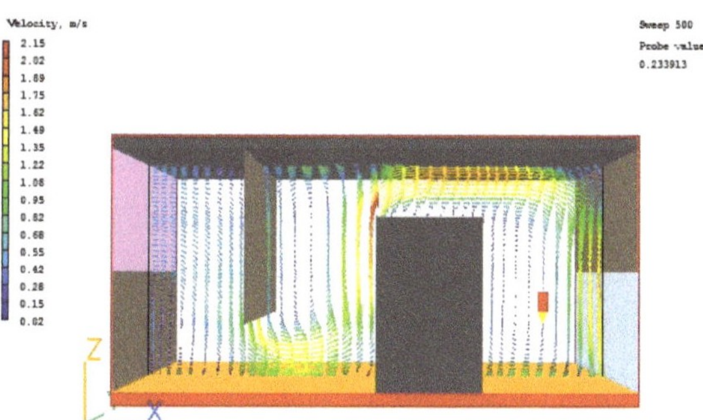

Figure 6. A typical vector plot.

The students are shown how to save their work, and how to transfer results suitable for reports and further analysis.

3.4. Implementation of the CFD Laboratories

The Computational Fluid Dynamics laboratories were introduced using a combination of short-lectures on back-ground theory, demonstrations, student group work and discussion, and assisted hands-on computer usage over an intensive four week period. Gradually, as time passed, the students became more confident and were able to navigate their way around the software efficiently and with assurance. The format of the demonstrations was that the instructor built simple simulations while students, watching a large overhead screen, did the same operations on their own computer. Group work was also used where perhaps a group of four students were given an exercise and asked to build a simulation. This proved a very effective and efficient way of learning. Gradually the method of delivering instruction changed over the four weeks, from demonstrations, to the use of detailed instruction sheets fully supported by faculty and teaching assistant intervention, to the distribution of descriptions of problems without instructions or hints and only the occasional help from support staff. There was never any restriction in seeking help from other students (again something which happens in engineering practice).

The main learning outcomes during this part of the course were to become familiar with the equations that govern steady and transient fluid flow and heat transport (conservation of mass, momentum, heat, species and energy) and be able to apply these equations to a range of practical problems in the areas of fluid flow, heat transfer and species transport. Seminars were given to inform students of what was expected during the CFD laboratory. It was stressed that the students should not compartmentalize theory, experiments and computational fluid dynamics but rather view them as mutually complimentary. The main bulk of these seminars was the introduction of the students to standard CFD methodology and procedures and the splitting of the CFD process into three distinct procedures, i.e., pre-processing, solving and post-processing.

During this four week section of the course, theoretical fundamental topics covered included:

- Getting started;
- CFD notation;
- CFD equations-continuity, momentum, energy, concentration of species;
- Finite differencing;
- The finite-volume method;
- Boundary conditions;
- Accounting for the pressure term;
- Closure of averaged equations;
- Time-stepping techniques; and,
- Properties of numerical methods.

This section of the course included discussions on CFD good practices and importantly the students were taught to be critical of their results and how to examine if what they got as results is what they might have expected. During the hands-on sessions, the following areas listed in Table 1 were emphasized.

Table 1. Areas for systematic consideration.

Geometry	Solid and Other Fluid Boundaries, Special Shapes of Objects.
Physics	Incompressible/compressible fluid, which quantities to be solved for, closure of equations, materials within the computational domain, initial and boundary conditions
Mesh	The choice here is Cartesian meshing or orthogonal meshing. The Cartesian mesh can be automatically generated or built by hand, and can be refined close to solid objects or in areas of high velocity gradient. Fine-grid embedding can also be used when appropriate.
Numerics	Convergence monitoring, selection of numerical scheme, maximum number of iterations, convergence criteria for each variable.
Post-processing	Flow visualization, analysis, validation using published experimental data or analytical calculation data.

4. Discussion

4.1. Overall Course Description

The main learning outcomes of this course are to understand how to apply the equations of transport governing conservation of mass, momentum and energy in an open and transparent way, and to be able to apply them correctly to a range of practical projects, including

- predicting drag forces on bluff, streamline bodies and flat plates;
- analyzing the flow in pipe systems;
- analyzing performance of radial flow pumps and turbines; and,
- matching pumps and turbines for particular applications.

The overall course last twelve weeks and was conducted during the fifth semester for students of mechanical and aerospace engineering, with 38 students taking the course. The first four weeks were used to teach the numerical aspects of computational fluid dynamics using close integration between theoretical lectures and hand-on computer laboratories, using as already described, mainly *Mathematica*, although some exercises using the versatile Python computer language were also introduced. The course during the second four weeks still had theoretical aspects within it, mainly in the form of turbulence modelling, but tended to concentrate on giving the students skills so as to solve real, though simple practical problems. The last four weeks of the course involved the students carrying out the series of projects outlined above, either individually or in small groups, which at first were supplemented with guidance and hints from the instructors on how to solve, but gradually this moved to problem description only, and it was left entirely to the student or students on how to enact a solution. In addition to set class time during these four weeks, the laboratory was open to students at any time, and discussion between the students was actively encouraged.

Assessment of students was by a series of bi-weekly quizzes and by assignment. No final examination was used.

4.2. Use of Mathematica

Based on the governing equations and boundary conditions established in Section 2, the students had to individually write coding using the symbolic software, *Mathematica* [9], to solve the problem of a lid-cavity flow [21]. The students had to write coding which set-up the required geometry, the mesh parameters and initial and boundary conditions. Next came coding for Eqns. [14] and [15], i.e., for $\mathcal{R}_{ij}^{k+1}$ and $\mathcal{L}_{ij}^{k+1}$ followed by coding for the iterative algorithm.

The mesh size for this problem and grid spacing had to be defined, and initially ω and ψ were set to zero for all nodes, except the top wall, where the vorticity is non-zero. The Reynolds number was set to 100, the relaxation parameter, p to 1 and the aspect ratio γ to 1. The students also had to set an appropriate residual value e and carefully nest the loops used to execute the iterative algorithm. At every iteration step k, if the maximum value of the absolute value of $\mathcal{L}[i, j]$ is less than e, then the calculations are halted. The iterative loops were wrapped with the Mathematica function Timing to give an estimate of the time taken to do the calculations.

Following completion of the calculations, students had to then write coding to display contour and vector plots of their results. An example of a contour plot obtained for the streamfunction is given on Figure 7.

Extraction of the centerline velocities was also instructive for the students in that first, students were given more experience of viewing velocity profiles, and second there are ample calculations and measurements of the lid-cavity flow for comparison with their results [25–27]. The centerline velocities u (in the x direction) and v (in the y direction) were derived from the streamfunction values using the equations

$$u_{i,j} = \frac{\partial \psi}{\partial y} = \frac{\psi_{i,j+1} - \psi_{i,j-1}}{2h} \quad j = 2, \ldots, n_y \tag{31}$$

$$v_{i,j} = -\frac{\partial \psi}{\partial x} = -\frac{\psi_{i+1,j} - \psi_{i-1,j}}{2h} \quad i = 2, \ldots, n_x \tag{32}$$

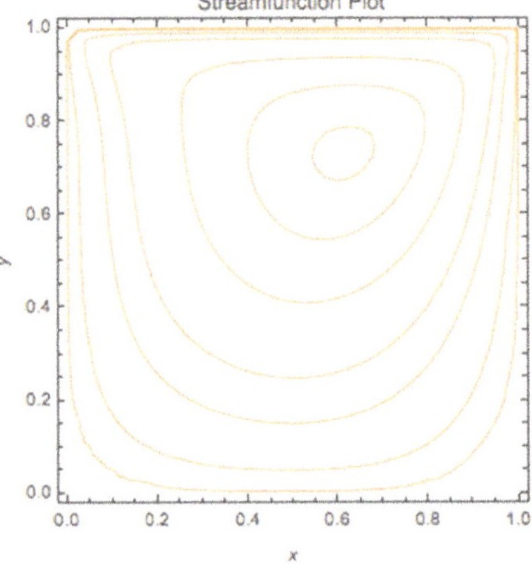

Figure 7. Contour plot of streamfunction for lid-cavity problem. (Reproduced with permission from [22]).

The velocity distributions for u and v in Figure 8 are drawn along the vertical and horizontal centerlines, respectively.

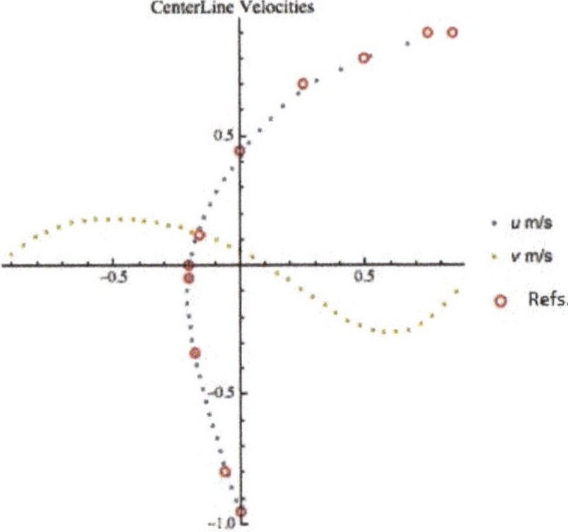

Figure 8. Calculated centerline velocity results compared with those of Refs. [25] and [27]. (Reproduced with permission from [21]).

Importantly, the students were made aware that the profiles they calculated must be compared, preferably with experimental measurements, to test the validity of their calculations. As can be seen from Figure 8, the calculations for the horizontal centerline velocities compare very favourably with those reported by [25,27] for Re = 100. Finally, the students were asked to produce coding which displayed the formation of the streamfunction contour plot with time.

4.3. Computational Fluid Dynamics methodology

In the second third of the course, the students moved on to the skills involved when using CFD in a professional consultancy capacity, while still learning necessary underpinning theory. The idea of fluid mechanics theory, experimental work and the current CFD skills and knowledge being acquired as complementary was reinforced. CFD now has well established methodology coupled with good practice and procedures, and it was emphasized that following these carefully would produce results much more efficiently and with a better degree of confidence. The students were taught how to use CFD judicially and appropriately, and to check their work against experimentation where possible. The idea of breaking the CFD process into three distinct tasks was again reinforced, i.e., using the pre-processor to build the problem, using the solver to obtain a converged solution and the use of the post-processor to obtain informative results. The numerical process of discretization was especially reinforced, as was how to activate the acquired conservation equations, geometrical aspects, fluid and solid properties, activation of sources, especially the gravity or body force, numerical details, which included initialization, the iterative process and how to use under-relaxation when necessary. Setting of the boundary conditions correctly was taught as was how to close the conservation equations by using an appropriate turbulence model [28,29]. When results were achieved it was extremely important that students could interpret what they had got in light of, say, geometric considerations, and also to be open-minded and critical.

During the demonstrations on how to use the GUI interface it was felt important that students could have 'hand-on' experience rather than just watching. Several simple three-dimensional flows were used as exemplars to give an over-arching view of the CFD process and the virtual reality environment which facilitates the VR-Editor (pre-processor), the VR-Viewer (post-processor) and the solver module which performs the calculations was introduced.

It was demonstrated how the VR-Editor is used to set the size of the computational domain, define the position, size and properties of objects introduced into the domain, specify the materials which occupy the domain, specify inlet(s) and outlet(s) and other boundary conditions, specify chosen initial conditions, select a turbulence model (when appropriate), specify the fineness of the computational grid and set parameters which define convergence and also influence the speed of convergence of the solution procedure. Viewing movement controls were also discussed which zoom, rotate, and allow for vertical and horizontal translations. It was emphasized how to systematically create a new environment for a particular simulation and finally set an appropriate number for the maximum number of iterations, set the position of the monitor for convergence. Finally, the students were encouraged to review the computational domain, especially in regard to position and attributes of the objects and the refinement of the grid close to solid surfaces, which at this stage of the demonstration would not be appropriate.

The convergence to a solution was monitored in two ways. First a point in the computation domain was chosen and values of the variables updated after each successive sweep. Once the values became constant it was assumed a converged solution was obtained. This was backed up by calculating the 'global errors' or changes between successive sweeps through the computation domain for each variable and when these errors were deemed sufficiently small, convergence was declared.

Once a converged solution was obtained, the simulation results were viewed by way of the post-processor facilities which displayed vector plots, contour plots, iso-surfaces, streamlines and x-y plots. The students were encouraged to find the optimum way of presenting their results and were also instructed on how to print the results. Very importantly, they were taught to always save and catalogue all input and output files.

4.4. Individual and Group Projects

For the last four weeks of the course the students were given a series of assignments which got progressively more difficult. The projects were wide ranging and covered important topics, such as the use of orthogonal grids, fine-grid embedding, cut-cell techniques, further exploration of turbulence models and how appropriate they are, the volume of fluid technique and two-phase flows (solid/gas and liquid/gas). Several projects covered how to import complicated geometries, built using a computer-aided drawing (CAD) package. In addition to problems involving purely fluid flow, projects were included which expanded to flow combined with heat transfer, the modelling of species transport and gaseous combustion. Use was made of the non-premixed combustion model where a natural gas combustion problem was set and solved using the non-premixed combustion model for the reaction chemistry. A project was also included which modelled an evaporating liquid spray where several discrete-phase models, including wall surface reactions were included.

During this part of the course the teaching model used emphasized active and collaborative learning in addition to students and instructors discovering and constructing knowledge and techniques together. Outside of assignments set for grading, the students were encouraged to collaborate and cooperate as much as they though appropriate.

4.5. Online Questionnaire

An anonymous online survey was conducted before students who had taken the course were given their final grades for the course to help with its formative evaluation. The fifteen questions/statements listed in Table 2 were used during the survey, with each student requested to mark as they thought appropriate, on a five-point Likert scale, whether they: strongly agree, agree, neutral, disagree, strongly disagree with the question/statement. There was also a column to mark if they had no opinion, and a space to make short comments for qualitative feedback.

Table 2. A list of questions/statements used in the survey for students' feedback.

No.	Question/Statement
1	I have used CFD modelling before.
2	I found the use of vorticity and streamfunction to be instructive.
3	I could relate vorticity and streamfunction to the primitive variables easily.
4	I found the use of collaboration with other students useful.
5	I preferred working on my own to working in groups.
6	Do you think the balance between CFD theory and practice was correct?
7	This CFD course enhances my understanding of fluid mechanics theory.
8	This CFD course is a useful addition to the fluid mechanics laboratories.
9	The 'hands-on' aspects of this CFD course has taught me extra skills.
10	Do you feel you can continue to model basic flows without much help?
11	Does this course give you confidence that you can model more difficult flows?
12	By using CFD I have learned things that could not be taught through traditional theory or experiments.
13	I have now a sound knowledge of CFD good practice.
14	I have enjoyed this course.
15	I would recommend this CFD course to others.

Generally, student feedback surveys have a very low response rate [30,31]. However the response rate here was relatively high (77%) and overall the results of the survey showed a positive attitude of the students towards the course. The numerical responses to the survey are shown on Figure 9 and indicate that students felt they had benefited from their exposure to this CFD course. In additional comments made by the students the view was expressed that the amount of material introduce was about right, although some felt that some of the exercises were unnecessarily long. Students did not seem to worry about the concept of streamfunction being used but some did express concern about their understanding of vorticity in a physical sense. The students were generally appreciative that they could easily visualize flow using contour and vector plots and generally agreed that the combination of basic theory combined with computer coding gave them more confidence and understanding. Student showed some enthusiasm about learning more about CFD and in particular how they could incorporate CFD into projects from other subjects and capstone projects.

It was noted that the students liked the hands-on and self-discover approach, although at times some frustration was also noted, and they were fairly neutral on working in groups as opposed to working on their own. In fact it was very noticeable that students took pride in performing the simulations by themselves with a real sense of achievement ensuing.

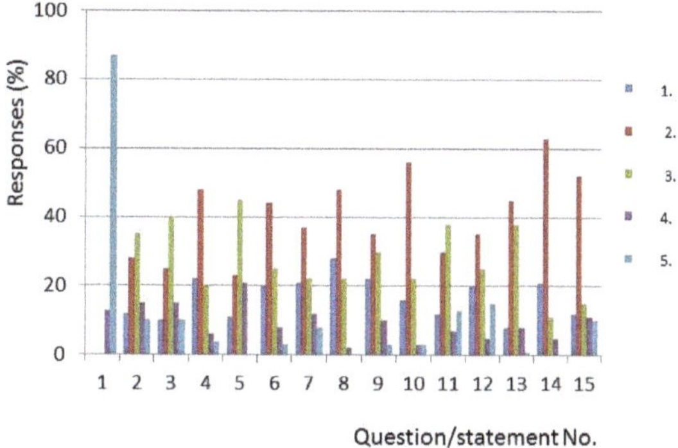

Figure 9. Chart showing survey results (N = 29).

Several points should also be made: the traditional view of CFD is that it has a steep learning gradient, but it has been shown here that with proper interfaces and a limited depth of flow difficulty that this gradient can be brought to an acceptable level. Also, it should be pointed out that for the practical software used no mention of code development was made to the students, as the learning outcome was to become proficient in the use of the existing coding. This would need to be remedied by a later course which involves ways of interacting with the existing, quite complicated core coding.

Evaluation of the course will continue in the future and hence be developed according to results. Only student views have been collected up to now but it is intended to take into account both student and instructor views in the future. It is felt more attention to how interest is aroused in general concerning the use of CFD is needed in future surveys as is the promotion of the use of CFD as a research tool during say capstone projects and post-graduate work. There will also be an attempt to elicit feedback from students who are at the 'beyond-completion' stage of the course.

5. Conclusions

In this paper the approach to delivering understanding of the numerical and practical aspects of the use of Navier-Stokes equations to solve fluid mechanics problems is outlined. For the numerical understanding, a well-defined sequence of steps needed to solve the Navier-Stokes equation cast in vorticity-streamfunction form was outlined. The sequence was integrated with the software package *Mathematica* at appropriate stages to take care of tedious computations, and hence allowed the students to concentrate on the overall details of the solution process. In addition, it is now important generally to integrate computer technology so as to complete and compliment lectures and theory. This has the advantages of helping with the computations, aiding presentation for reports and the ability to continue with further analysis as well as motivation for the students. Incorporating Mathematica also takes away the 'black-box' approach so often encountered by students with full CFD commercial codes, which give no real understanding of the numerical mathematics involved.

In addition to teaching the students some of the numerical aspects of CFD, this course also allowed for an introduction to the type of CFD carried out in professional offices, and taught additional topics like turbulence modelling, the idea of obtaining a grid independent solution and how critical it is to validate results obtained by the CFD approach. One of the main reasons for the inclusion of the second part of this course was to give the students practical skills and it was found that the simplified interface design does provide students with hands-on experience, gained through an interactive and user-friendly environment, and encourages student self-learning. It was noted from a survey that students liked getting hands-on experience and the self-discovery approach, although at times some frustration was noted when setting a problem up.

Author Contributions: Conceptual, D.A. and M.J.; methodology and software D.A.; validation and analysis, D.A. and M.J.; original draft preparation, D.A.; review and editing, M.J.

Funding: This research received no external funding.

Conflicts of Interest: The authors declare no conflict of interest.

Appendix A

In general, when solving for the nodal quantities, this is in effect solving a system of n linear equations, where n is the total number of nodes in the system. The system of equations can easily be described by

$$A\mathbf{x} = \mathbf{b} \tag{A1}$$

where A is a $n \times n$ matrix of coefficients, $\mathbf{x}$ is a vector of n unknown values, and $\mathbf{b}$ is a vector of n constants. Rather than inverting the matrix A directly, the Gauss-Seidel method can be used to solve the equations iteratively, which is especially desirable for large systems. In general the Gauss-Seidel method employs forward substitution and is given by

$$x_i^{k+1} = \frac{1}{a_{ii}}\left(b_i - \sum_{j=1}^{i-1} a_{ij}x_j^{k+1} - \sum_{j=i+1}^{n} a_{ij}x_j^{k} \right) \tag{A2}$$

for $i = 1, 2, \ldots, n$. The iterations are performed until the difference of the quantity value, often called the residual error (E), between iterations reaches a specified tolerance.

References

1. Assanis, D.N.; Heywood, J.B. Development and use of a computer simulation of turbo-compounded Diesel system for engine performance and component heat transfer studies. *SAE Trans.* **1986**, *2*, 451–476.
2. Devenport, W.J.; Schetz, J.A. Boundary layer codes for students in Java. In Proceedings of the ASME Fluids Engineering Division Summer Meeting, Washington, DC, USA, 21–25 June 1998.
3. Zheng, H.; Keith, J.M. JAVA-based heat transfer visualization tools. *Chem. Eng. Educ.* **2004**, *38*, 282–290.
4. Rozza, G.; Huynh, D.B.P.; Nguyen, N.C.; Patera, A.T. Real-time reliable simulation of heat transfer phenomena. In Proceedings of the ASME-American Society of Mechanical Engineers-Heat Transfer Summer Conference, San Franscisco, CA, USA, 19–23 July 2009.
5. Qian, X.; Tinker, R. Molecular dynamics simulations of chemical reactions for use in education. *J. Chem. Educ.* **2006**, *81*, 77–90.
6. Pieritz, R.A.; Mendes, R.; Da Silva, R.F.A.F.; Maliska, C.R. CFD: An educational software package for CFD analysis and design. *Comput. Appl. Eng. Educ.* **2004**, *12*, 20–30. [CrossRef]
7. Stern, F.; Xing, T.; Yarbrough, D.B.; Rothmayer, A.; Rajagopalan, G.; Otta, S.P.; Caughey, D.; Bhaskaran, R.; Smith, S.; Hutchings, B.; et al. Hands-on CFD educational interface for engineering courses and laboratories. *J. Eng. Educ.* **2006**, *95*, 63–83. [CrossRef]
8. Curtis, J.C. Integration of CFD into undergraduate chemical engineering curriculum. In Proceedings of the Annual AIChE Meeting, Philadelphia, PA, USA, 16–21 November 2008.
9. Wolfram Research, Inc. *Mathematica, Version 12.0*; Wolfram Research, Inc.: Champaign, IL, USA, 2019.
10. Jaeger, M.; Adair, D. Human factors simulation for construction management. *Eur. J. Eng. Educ.* **2010**, *35*, 299–310. [CrossRef]
11. Jaeger, M.; Adair, D. Communication simulation in construction management simulation: Evaluating the learning effectiveness. *Australas. J. Eng. Educ.* **2012**, *18*, 1–14. [CrossRef]
12. Patil, A.; Mann, L.; Howard, P.; Martin, F. Assessment of hands-on activities to enhance students' learning for the first year engineering skills course. In Proceedings of the 20th Australasian Association for Engineering Education Conference, Adelaide, Australia, 6–9 December 2009; pp. 286–292.
13. Edgar, T.F. Enhancing the undergraduate computing experience. *Chem. Eng. Educ.* **2006**, *40*, 231–238.
14. Adair, D.; Jaeger, M. Incorporating computational fluid dynamics code development into an undergraduate engineering course. *Int. J. Mech. Eng. Educ.* **2015**, *43*, 153–167. [CrossRef]

15. Parulekar, S.J. Numerical problem solving using Mathcad in undergraduate reaction engineering. *Chem. Educ. Eng.* **2006**, *40*, 14–23.

16. Rockstraw, D.A. Aspen Plus in the curriculum-suitable course content and teaching methodology. *Chem. Eng. Educ.* **2005**, *39*, 68–75.

17. Finlayson, B.A.; Rosendall, B.M. Reactor/transport models for design: How to teach students and practitioners to use the computer wisely. In *AIChE Symposium Series*; American Institute of Chemical Engineers: New York, NY, USA, 2000; pp. 176–191.

18. Pomeranz, S. Using a computer algebra system to teach the finite element method. *Int. J. Eng. Educ.* **2000**, *16*, 362–368.

19. LeVeque, R.J. *Finite Difference Methods for Ordinary and Partical Differential Equations: Steady-State and Time-Dependent Problems*; Siam: Philadelphia, PA, USA, 2007.

20. Ferziger, J.H.; Peric, M. *Computational Methods for Fluid Dynamics*; Springer: New York, NY, USA, 1996.

21. Bozeman, J.D.; Dalton, C. Numerical study of viscous flow in a cavity. *J. Comput. Phys.* **1973**, *12*, 348–363. [CrossRef]

22. Adair, D.; Jaeger, M. Developing an understanding of the steps involved in solving Navier-Stokes equations. *Math. J.* **2015**, *17*, 1–19. [CrossRef]

23. Burggraf, O.R. Analytical and numerical studies of the structure of steady separated flows. *J. Fluid Mech.* **1966**, *24*, 113–151. [CrossRef]

24. Adair, D. Incorporation of Computational Fluid Dynamics into a Fluid Mechanics Curriculum. In *Advances in Modeling of Fluid Dynamics*; InTech: Berlin, Germany, 2012; ISBN 980-953-307-311.

25. Gia, U.; Ghia, K.N.; Shin, C.T. High-Re solutions for incompressible flow using the Navier-Stokes equations and a multigrid method. *J. Comput. Phys.* **1982**, *48*, 387–411. [CrossRef]

26. Spotz, W.F. Accuracy and performance of numerical wall boundary conditions for steady, 2D, incompressible streamfunction vorticity. *Int. J. Numer. Methods Fluids* **1998**, *28*, 737–757. [CrossRef]

27. Wan, D.C.; Zhou, Y.C.; Wei, G.W. Numerical solution of incompressible flows by discrete singular convolution. *Int. J. Numer. Methods Fluids* **2002**, *38*, 789–810. [CrossRef]

28. Launder, B.E.; Spalding, B. *Mathematical Models of Turbulence*; Academic Press: London, UK, 1972.

29. Agonafer, D.; Liao, D.G.; Spalding, B. *The LVEL Turbulence Model for Conjugate Heat Transfer at Low Reynolds Number, Concentration*; Heat and Momentum Ltd.: London, UK, 2008.

30. Gamliel, E.; Davidovitz, L. Online versus traditional teaching evaluation: Mode can mater. *Assess. Eval. High. Educ.* **2005**, *30*, 581–592. [CrossRef]

31. Nulty, D.D. The adequacy of response rates to outline and paper surveys: What can be done? *Assess. Eval. High. Educ.* **2008**, *33*, 301–314. [CrossRef]

Article

Fluids in Music: The Mathematics of Pan's Flutes

Bogdan G. Nita [1,*] and Sajan Ramanathan [2]

[1]   Department of Mathematical Sciences, Montclair State University, Montclair, NJ 07043, USA
[2]   Department of Mathematics, Yale University, New Haven, CT 06520, USA; sajan.ramanathan@yale.edu
*   Correspondence: nitab@montclair.edu; Tel.: +1-973-655-7261

Received: 15 July 2019; Accepted: 30 September 2019; Published: 10 October 2019

Abstract: We discuss the mathematics behind the Pan's flute. We analyze how the sound is created, the relationship between the notes that the pipes produce, their frequencies and the length of the pipes. We find an equation which models the curve that appears at the bottom of any Pan's flute due to the different pipe lengths.

Keywords: Pan's flute; music; sound frequencies; wind instruments

1. Introduction

Fluids impact our everyday lives at any moment, from the air we breathe, to the blood circulating in our bodies, to the rivers and oceans that we swim in and drink from. Most of the examples concerning moving fluids are extremely complex with behavior governed by complicated equations involving many parameters. Applications of fluid mechanics are extremely varied and this translates in a plethora of methods and approaches to teach the subject. In this paper, we focus on the relation between fluids and musical instruments and present a somewhat atypical way to introduce fluids to undergraduate students.

Fluids, and in particular flowing air, are indispensable in the creation of sound in wind instruments and human voicing. The interaction between flowing air and different structures, e.g., inside of a tube, a reed or human vocal folds, produces air vibrations which are perceived as sounds by the human ear. The study of these interactions are useful to understanding and perfecting musical instruments or to identifying and curing vocal disorders. In this paper, we look at the geometry of Pan's flutes and the relation between fluid dynamics and equal temperament scales brought together by this musical instrument.

Pan flutes are wind instruments consisting of multiple pipes which are gradually decreasing in length (see Figure 1). The pipes are typically made from bamboo, maple or other wood varieties. Other materials include plastic, metal or ivory. The tubes are stopped at one end, and are tuned to correct pitch by placing small pieces of rubber inside. Very often, the pipes are positioned to form a curved surface to allow quick and easy access to all pipes. The instrument featured in Figure 1 is a maple tenor Pan's flute tuned in the key of G-major. The longest pipe measures about 29 cm and produces the lowest sound of the instrument, D4. Table 1 shows the notes produced by each pipe and corresponding frequencies measured by any physical tuner or a tuner app.

Table 1. The notes and their corresponding frequencies (rounded to the nearest integer) produced by the Pan's flute.

Notes	D4	E4	F#4	G4	A4	B4	C5	D5	E5	F#5	G5	A5	B5	C6	D6	E6	F#6	G6
Frequencies	294	330	370	392	440	494	523	587	659	740	784	880	988	1047	1175	1319	1480	1568

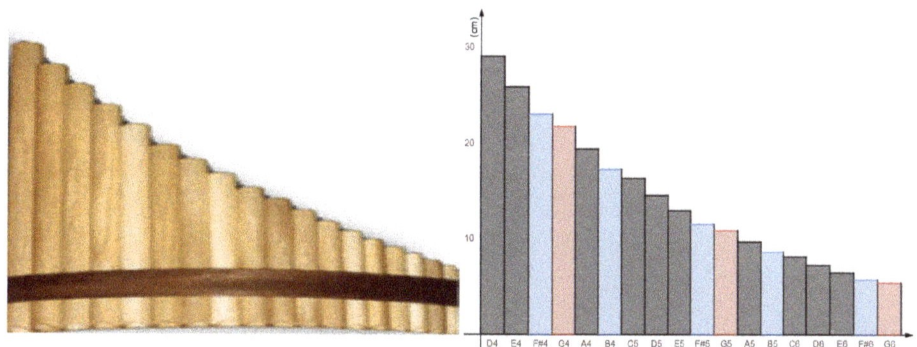

Figure 1. An 18-pipe Pan's flute tuned in G-major and a bar graph representing the lengths of individual pipes in cm. The red bars show the root notes (in this case G's), the gray bars show the location of whole tone intervals, and the blue bars show the location of semitone intervals.

The goal of this research is to determine the relation between the length of the pipes and the notes played by each pipe, therefore providing a functional formula for the curve created at the top of the instrument in Figure 1. We will do this in two steps: first, determine a relation between the frequency and the length (function f in Figure 2) and then a relation between the notes and corresponding frequencies (function L in Figure 2). The composition between the two functions should provide the desired relationship.

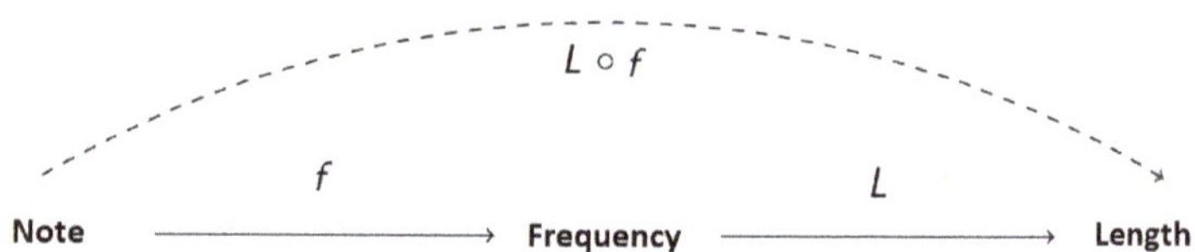

Figure 2. The relation between the notes played by a Pan's flute, their corresponding frequency, and the length of pipes as a composition of functions.

The benefits of this apparent simple problem are twofold. First, one can learn about several math-physics areas and methods starting with fluid dynamics, wave equation, partial differential equations and continuing with elements of algebra in particular frequency ratios, and logarithmic and exponential functions. Second, the application of mathematics to a seemingly unrelated field like music (and art in general) is extremely exciting, novel and attractive to students who otherwise would not consider working on a mathematical research problem.

In this paper, we assume all pipes are ideal, i.e., the wavelength of the sound produced is proportional to the length of the tube. This is equivalent to assuming a zero diameter for the pipes of the Pan's flute, which is an acceptable approximation for the current research. In reality, the diameter of the pipes plays a role in the pitch produced by the musical instrument: a pipe with a non-zero diameter appears to be acoustically longer than its physical length. When designing musical instruments consisting of large diameter pipes (organ, boom-whacker, etc.), an end-correction of the pipes needs to be applied. For an in-depth discussion of this effect, see [1].

2. The Relationship between the Frequency and the Length of a Pipe

Sound in pipes is produced by vibrating air pressure due to the air interaction with the walls of the pipe. The partial differential equation that models this physics is the wave equation for the acoustic pressure. The equation can be developed from the linearized one-dimensional continuity equation, the linearized one-dimensional force equation, and the equation of state (see for example [2]), and takes the form

$$\frac{\partial^2 p}{\partial x^2} = \frac{1}{c^2}\frac{\partial^2 p}{\partial t^2},\tag{1}$$

where $p(x,t)$ is the air pressure due to air moving inside of the pipe, x is the coordinate along the pipe, t is time, and $c \approx 340$ m/s is the speed of sound in air. The boundary conditions associated with this differential equation are based on the physics of the problem: at the open end, the moving air has nothing to push against and therefore the pressure is zero; at the closed end, the air piles up and increases the pressure to a maximum value. Therefore, for a Pan's flute pipe of length L, open at $x = L$ and closed at the $x = 0$, the boundary conditions are

$$p(L,t) = 0,\tag{2}$$
$$p_x(0,t) = 0.$$

It is worth mentioning that for other wind instruments, such as the flute, clarinet, etc., the partial differential equation given in Equation (1) is the same while the boundary conditions in Equation (2) change to reflect different constructions. The solutions of the boundary value problem described by Equations (1) and (2) are linear combinations of the normal modes [3]

$$p_n(x,t) = \cos\left(\frac{2n-1}{2}\pi\frac{x}{L}\right)\left(b_n\cos\left(\frac{2n-1}{2L}\pi ct\right) + c_n\sin\left(\frac{2n-1}{2L}\pi ct\right)\right),\tag{3}$$

where $n = 1,2,3\ldots$ and b_n and c_n are constants. It is easy to see now that the frequency of the nth mode is

$$f_n = \frac{(2n-1)c}{4L}.\tag{4}$$

These frequencies have a simple explanation related to musical instruments. Notes played on conventional instruments (wind, string, etc.) produce one main frequency, the fundamental, and a series of lower amplitude frequencies, partials, which are integer multiples of the fundamental frequency. For example, when blowing air into the largest pipe of the Pan's flute in Figure 1, you will only hear the fundamental frequency of f_1 (the note D4) even though an infinite series of partials, having frequencies $3f_1$, $5f_1$, etc., are produced.

The 1st mode, the fundamental, is described by the frequency

$$f_1 = \frac{340}{4L},\tag{5}$$

which is our first desired equation relating the frequency of the musical note produced by a pipe and the length of that pipe. For example, the longest pipe in our instrument produces a frequency of approximately 294 Hz, which means that its length is approximately

$$L = \frac{340}{4 \times 294} = 0.29\ m,\tag{6}$$

which agrees with the measurements in Figure 1.

3. The Relationship between the Musical Notes and the Frequencies

It was long postulated that different notes whose frequency ratios are expressible as ratios of small integers are consonant (for a more in depth discussion see [4]). For example, f_1 and $2f_1$ attain the ratio of the smallest possible integers, namely 2:1, the interval (the distance between them) being designated as an octave. Another example of a consonant interval is the perfect fifth, which is represented by two notes whose frequencies achieve a 3:2 ratio. Based on this condition of consonance, it was natural to build scales using the most consonant intervals possible, namely the octave and the perfect fifth. This produced the Pythagorean scale (see Table 2).

Table 2. One octave of the Pythagorean scale of G major.

Note	G4	A4	B4	C5	D5	E5	F#5	G5
Ratio	1:1	9:8	81:64	4:3	3:2	27:16	243:128	2:1

The scale in Table 2 is obtained by choosing an octave (say G4–G5), then adding to the fist note (G4) a perfect fifth interval, thus obtaining the second note (D5). Adding another perfect fifth interval to D5, we obtain a frequency ratio of 9:4 (with respect to G4) that is outside of our octave. Reducing this interval by an octave (by dividing this frequency ratio by 2), we obtain a ratio of 9:8, which is our third note in the scale (A4). We continue the process until we discover all the notes in the scale (for more details, refer to [4]). In this system, adding/subtracting intervals amounts to multiplying/dividing frequency ratios, which is why it is commonly referred to as a logarithmic scale. One important observation is that the Pythagorean scale is based on the assumption that going up 12 fifths and coming down 7 octaves will return you exactly to the same starting note, or, in other words, that

$$2^{19} \approx 3^{12}. \tag{7}$$

While these numbers are close, they are not exactly equal, which gives rise to certain problems in the Pythagorean scale. Among the issues are other musical intervals, like the third, which are perceived as dissonant in this scale. Other scales were later developed to alleviate some of these problems—for example, just intonation, meantone and the equal temperament scale.

The equal temperament scale makes all the semitones equally distant (see Table 3). It is designed to 'fix' all the problems associated with other scales by distributing the dissonance occurring in different intervals throughout the entire scale. This implies that, while classical intervals like the perfect fifth or the third are no longer going to have frequency ratios represented by small integer ratios, they are going to sound the same (relatively consonant) in any key and any pitch.

Table 3. One octave of the equal temperaments scale of G major.

Note	G4	A4	B4	C5	D5	E5	F#5	G5
Ratios	1:1	$2^{\frac{1}{6}} : 1$	$2^{\frac{1}{3}} : 1$	$2^{\frac{5}{12}} : 1$	$2^{\frac{7}{12}} : 1$	$2^{\frac{3}{4}} : 1$	$2^{\frac{11}{12}} : 1$	2:1

All the modern musical instruments, our Pan's flute included, are tuned in the equal temperament scale. In this scale, the frequency f of any note can be obtained from the frequency f_1 of any reference note (we chose D4 as our reference note as it corresponds to the note produced by the first pipe of the Pan's flute) from the formula

$$f = 2^{\frac{x}{12}} f_1, \tag{8}$$

where x represents the number of semitones that separates the note with the frequency f from the first note: $x = 0$ represents the first note, D4, $x = 2$ represents the second note, E4, $x = 4$ represents the third note, F#4, $x = 5$ represents the fourth note, G4, and so on. Note that x does not cover all the positive integer values since the scale contains only the main notes in the G-major scale and does not include the sharps. This is consistent with the construction of the Pan's flute that we use in this paper. Because of this, when plotting the lengths of the pipes as a function of x, we will need to adjust for these apparent irregular values taken by the variable x.

4. Putting It All Together

Combining Equations (5) and (8), we find

$$L = \frac{340}{4 \times 2^{\frac{x}{12}} \times f_1},$$

(9)

which is the desired relation between the length of a pipe and the note played by that pipe. Plotting this curve on the same coordinate system with the model of the musical instrument, we obtain Figure 3, which shows a perfect match with our initial model of the Pan's flute.

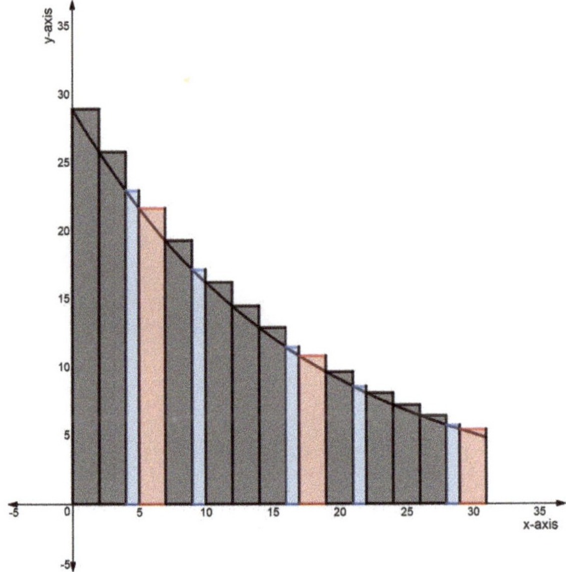

Figure 3. The curve in Equation (9) superimposed on the model of the Pan's flute. The semitones, represented in blue, were adjusted to indicate the correct musical intervals.

5. Conclusions

In this paper, we discuss a mathematical model of Pan's flutes. The interaction of mathematics and music makes it an interesting application for undergraduate students as they find refuge from the mathematical rigor into the flexibility and creativity of arts. The study of wind instruments provides a rich environment for teaching fluid mechanics and other areas of applied mathematics in a fun and relaxed way.

Author Contributions: This work was performed while S.R. was an undergraduate student at Montclair State University. B.G.N. and S.R. read and approved the final manuscript.

Funding: This research received no external funding.

Conflicts of Interest: The authors declare no conflict of interest.

References

1. Fletcher, N.H.; Rossing, T.D. *The Physics of Musical Instruments*; Springer: Berlin/Heidelberg, Germany, 2010.
2. Feynman, R. Chapter 47: Sound. The wave equation. In *Lectures in Physics*; Caltech: Pasadena, CA, USA, 1963; Volume 1.

Fluids **2019**, *4*, 181

3. Boyce, W.E.; Di Prima, R.C.; Meade, D.B. *Elementary Differential Equations and Boundary Value Problems*, 11th ed.; Wiley: Hoboken, NJ, USA, 2017.
4. Benson, D. *Music: A Mathematical Offering*; Cambridge University Press: Cambridge, UK, 2006.

Article

Bottle Emptying: A Fluid Mechanics and Measurements Exercise for Engineering Undergraduate Students

Hans C. Mayer

Department of Mechanical Engineering, California Polytechnic State University, San Luis Obispo, CA 93407, USA; hmayer@calpoly.edu

Received: 12 September 2019; Accepted: 30 September 2019; Published: 11 October 2019

Abstract: A comprehensive exercise, suitable for an undergraduate engineering audience studying fluid mechanics, is presented, in which participants were tasked with emptying a bottle. That simple request yielded data collected by students and the author for $N = 454$ commercially available bottles, spanning nearly four orders of magnitude for volume V, and representing the largest experimental dataset available in the literature. Fundamental statistics are used to describe the emptying time, $\overline{T}_e$, for any single bottle. Dimensional analysis is used to transform the raw data to yield a predictive trend, and a method of least-squares regression analysis is performed to find an empirical correlation relating dimensionless time $\overline{T}_e\sqrt{g/d}$ and dimensionless volume V/d^3. We find that volume, V, and neck diameter, d, can be used to estimate the emptying time for any bottle, although the data suggests that the shape of the neck plays a role. Furthermore, two basic analytical models found in the literature compare favorably to our data and empirical correlation when recast using our dimensionless groups. The documented exercise provides students with the opportunity to use basic engineering statistics and to see the utility of dimensional analysis.

Keywords: multiphase flow; slugging; flooding; bottle; bubbles

---

1. Introduction

1.1. A Familiar Scene

The scene depicted in Figure 1a is a familiar one. Here, we see a sequence of still images, captured during the emptying of an inverted bottle, that show the complex behavior of water leaving and air entering through the single opening at the neck. When we observe such an emptying event in its entirety, we cannot help but marvel at what unfolds—in particular, the violent motion within the bottle as rising bubbles reach the surface and the oscillatory flow at the outlet, all of which is accompanied by a series of "glugs" until at last the bottle has been evacuated. We do our best to illustrate this behavior in the superposition of sequenced images in Figure 1b, which from start to finish takes ∼10 s. The images of Figure 1c present a strikingly different scene. Here, we have removed the bottom of the bottle, and we now find the flow to be orderly and predictable, without the competition between fluid phases at the neck. This emptying process takes only ∼2 s, Figure 1d showing the superposition of images and progression of the surface. Thus, if we ask, "what is the emptying time for a bottle?" The answer is readily accessible for the vessel of Figure 1c,d, but can be a daunting task for that of Figure 1a,b, in particular for an undergraduate engineering student at the start of their study of fluid mechanics.

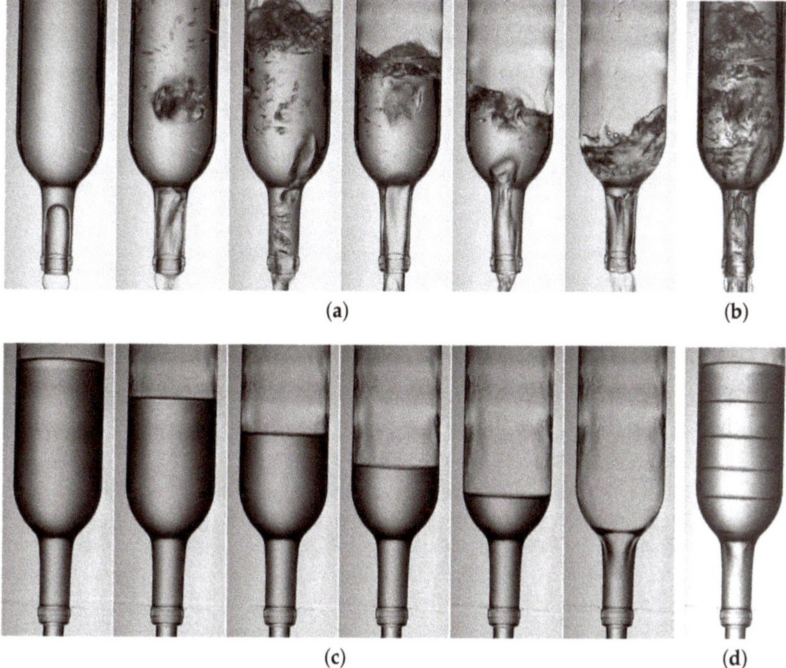

Figure 1. Photographs of water emptying from a bottle (wine bottle: 750 mL volume and 20 mm diameter opening at neck). (**a**) A sequence of images from the emptying event (1.67 s apart). A slug of air can be seen to exist in the neck and the bulk of the liquid is populated with bubbles both large and small. Superposition of sequential images during a typical emptying event is shown in panel (**b**). When the bottom of the bottle is removed, making it an open-top container when inverted, the scene in panel (**c**) is strikingly different (0.23 s between images). The air–water interface within the bottle is flat, its position more readily predictable (**d**), and has a uniform jet of liquid issues from the opening in the neck rather than a behavior. All images were recorded at 60 fps using a Nikon D3400.

1.2. Literature Review

The complexity of bottle emptying, as exemplified in Figure 1, must have casually intrigued scientists and laypersons for a very long time, but a review of the literature only shows a handful of papers dedicated to the scientific study of the subject, and these were published primarily over the last three decades. Although Davies & Taylor [1] and Zukoski [2] investigated the drainage of tubes via the passage of long bubbles, Whalley [3,4] was the first to investigate the emptying times of real bottles, i.e., commercially available bottles characterized by a neck with a diameter smaller than that of the body. His studies were concerned with the overall emptying time of these vessels, which ranged in size and shape, and he recognized that the total time was governed by the nature of the flow at the outlet—in particular, he identified the closely related phenomena of flooding and slugging and their relation to this flow. The occurrence of flooding, a type of concurrent flow of gas and liquid phases within a tube, can be predicted using a semiempirical relation attributed to Wallis [5] that incorporates a Wallis flooding "constant" C (see Section 2.5). For an air–water system, this results in values of $C \sim \mathcal{O}(1)$. The values of C calculated from the experiments of Whalley, using a small number of bottles, yielded consistent results strengthening the idea that bottle emptying is a flooding process. Furthermore, Whalley recognized that emptying is also similar to slugging: another regime of two-phase flow in which large slugs of gas form in tubes. If thought of in this context, the emptying process can be modeled as liquid flowing out of the bottle and around a stationary slug of air (cf. Figure 1b, looking closely at the bottleneck). Whalley's experiments also demonstrated that emptying time, and thus C, can be influenced by water temperature, neck length (artificially created using tube inserts), and angle of inclination. He further considered differences between bottle emptying and filling, the later process was accomplished by submerging an upright empty bottle below the free surface of water in a large tank. Tehrani [6] also measured the emptying time for a very limited number of bottles and found, similar to the results of Whalley, that increases in water temperature significantly influence emptying time.

More rigorous experiments concerning the emptying of bottles were undertaken by Schmidt and Kubie [7], but to carefully investigate the details of emptying, they performed controlled experiments with "ideal" bottles, i.e., various sizes of plastic tubes of constant cross section with centrally located outlets that provided short necks (with either sharp or filleted/profiled entrances). These authors recognized the cyclical nature of the emptying process, but were concerned primarily with gross behavior of the air–water interface within the body of the bottles and not the physics of the process on the single-cycle scale. In particular, their key findings showed that during emptying the velocity of the air–water interface (within the bottle) is nearly constant with respect to the level in the bottle, implying that height does not play a role in the total emptying time beyond increasing the volume of the container; that the velocity changes with the shape of the outlet (increasing velocity with more rounded entrances); and that the average liquid discharge velocity increase weakly with both bottle diameter and neck diameter. These key results were echoed in the findings of Kordestani and Kubie [8] and Tang and Kubie [9], who also used "ideal" bottles. Unlike Schmidt and Kubie, Kordestani and Kubie presented many of the results of their experiments in the form of the Wallis flooding constant C and found favorable comparison to the results of Whalley (cf. Section 2.5), noting that C increases with a smoother outlet profile (increasing C indicating a faster emptying time). Their experiments with liquids emptying from bottles into a liquid environment demonstrated that the nature of the emptying process is the same but slower than the gas–liquid case. Changes in C for air–water systems due to the inclination of the outlet was investigated by Tang & Kubie [9].

Clanet & Searby [10], without reference to the multiple works of Whalley [3,4] or Kubie [7–9], thoroughly investigated the emptying of ideal bottles both theoretically and experimentally, considering both the long timescale associated with complete emptying and the short timescale responsible for the familiar "glug-glug". The basic construction of their bottles was the same as that of Kubie et al., i.e., long tubes with centrally-located outlets; however, Clanet and Searby used thin plates with sharp edges for their outlets rather than the thicker plates of Kubie. Not surprisingly, owing to the similarity in apparatus, their experiments revealed the same quasi-constant velocity of the air–water interface within the tube and noted that this velocity decreased with reduction in outlet diameter. These authors also developed a simple analytical model to predict the emptying time for these bottles which we will make use of later. Considerable detail is provided by Clanet & Searby regarding the physical origin of the oscillations (in mass flow rate and pressure) characterizing the short timescale glug-glug, but those results are not directly relevant to the new work presented here. Most recently, computational fluid dynamics (CFD) has been shown by Geiger et al. [11] to accurately simulate the bottle emptying process, including the transient behavior of the complex two-phase flow and the overall emptying time. A CFD analysis of bottle emptying has also been published by Mer et al. [12], along with experiments to quantify transient and oscillatory behavior associated with the variety of two-phase flow phenomenon encountered during emptying [13]. These experimental results are expected to help validate future CFD studies.

1.3. Motivation—Establishing a Curricular Thread

Inspiration for the elements of this paper originated from a homework problem buried within S. Middleman's "Introduction to Fluid Dynamics" [14], which begins "a classroom demonstration on the time to drain water from a set of initially filled inverted bottles yielded the following results..." and is concluded with a set of six data points (bottle volume and emptying time) extracted from the work of P.B. Whalley [4]. Expanding on this example, we have developed a complete exercise that provides the opportunity for a class of students to consider a complex, but commonplace, fluid flow phenomenon from the standpoint of dimensional analysis; collect and analyze data using basic engineering statistics; combine data to form a large set; plot the data; and extract a predictive trend using basic regression techniques. Such an exercise, presented as a complete unit, forms what might be considered a curricular thread, stitching together various aspects of a variety of science and engineering courses, in particular an undergraduate fluid mechanics course. However, as the emphasis of the exercise is not a detailed investigation of the fluid dynamics of a multiphase flow, the exercise described here is accessible to a broad audience and would be of value beyond a fluid mechanics course. Furthermore, because of the nature of the experiments, there is no need for expensive laboratory equipment or an elaborate demonstration apparatus—a plethora of bottles in a wide variety of shapes and sizes are readily available. In addition, the instructions for data collection are simple, and as such give the project a feel of a public study [15]. This particular fluid mechanics example has the added advantage that the data can be compared directly to a select number of sparse datasets. If an instructor is so inclined, there also exists the opportunity to encourage students to research the open literature for the two models provided for comparison. In doing so, they will find evidence of the rich and complex phenomena that accompany an event that might previously have been considered mundane. In what follows, we present this complete exercise and results from data collected by students and the author, over many academic quarters and years, who were asked a simple but intriguing question: "What is the emptying time for a bottle?"

1.4. Outline

The work presented here is divided conveniently into several main parts. In Section 2.1, we introduce the protocol for data collection. In Section 2.2, an analysis of the basic statistics associated with the data obtained from a few single bottles is presented. A complete picture of the emptying time data for the entire collection of bottles is contained in Section 2.3 as well as a summary of the measurement uncertainties from experiments. A review of the dimensional analysis used to obtain relevant dimensionless groups is presented ending with a graphical summary of that analysis. In Section 2.4, we focus on regression analysis of the dimensionless data to obtain a predictive trend for bottle emptying. Finally, in Section 2.5, we compare our data and the results from the regression analysis to the limited bottle emptying data available in the literature. We also provide comparisons of simple models published in the literature to our extensive dataset and find reasonable agreement.

The primary audience for this paper are instructors and students in an undergraduate engineering fluid mechanics course; however, the data by itself would be useful in an engineering statistics/measurements course. The paper is structured so that different portions can be used to direct exercises in those topics. For example, Sections 2.2 and 2.4 can be considered as standalone projects for an engineering statistics class, whereas Sections 2.3 and 2.5 are suited for a dimensional analysis example or homework problem in a fluid mechanics lecture. Alternatively, all aspects (Sections 2.2–2.5) together form a complete investigation and represent a project that when applied at the appropriate level leverages content from a variety of engineering courses.

2. Results and Discussion

What follows is a suggested presentation of the data and analysis for use as a student exercise and/or guided discussion. It is intended for an undergraduate engineering audience. Familiarity with basic statistics and fluid mechanics are helpful; however, sufficient details, guidance, and references have been provided for the reader. All quantitative analysis can be completed using a basic computer spreadsheet program, e.g., Excel®, or programmed in MATLAB®, and do not require any statistics-specific software. All experiments can be completed using simple tools and instruments (e.g., a stopwatch, ruler, graduated cylinder, scale, etc.). Components of the complete exercise presented here were implemented in various forms during sophomore or junior level classes at California Polytechnic State University San Luis Obispo, the University of California at Santa Barbara, and Rutgers University.

2.1. Data Collection Protocol

Regardless of the format of the exercise (e.g., an extended statistics-only exercise, a fluid mechanics dimensional analysis problem, or both), students were provided with a basic data sheet and a short handout explaining the overall goal—to gain insight into emptying of single outlet vessels via collective experiments using commercially available bottles. Each student was tasked with finding a bottle or single-outlet container for their experiment (from here on the term "bottle" will be used exclusively to denote any single-outlet container used for experiments). Certain limitations applied: ordinary soda/beer bottles were discouraged to promote a wide variety of shapes and sizes (no restriction was placed on the shape), and it was suggested that the bottles be rigid (e.g., glass or hard plastic) to reduce, or avoid altogether, flexing of the sidewalls during emptying. Because the interest is in single-outlet containers, students were encouraged to use any single-outlet vessel. Thus, many of the "bottles" collected for use were not intended for drinking or even holding liquids. Example include bottles for cooking sauces and oils, jars, soap/shampoo bottles, vases, etc.

Once a bottle was found, students were asked to measure the internal neck diameter, d, and report the nominal value and the measurement uncertainty (further details are provided in Section 2.3 where relevant). Measurements of internal bottle volume V were performed gravimetrically or volumetrically as permitted by availability of equipment. Students were instructed to fill the bottle completely for these internal volume measurements, as all emptying experiments would use filled bottles. Gravimetric measurements were performed using digital balances, and volumetric measurements were obtained using graduated cylinders, measuring cups, or syringes. In all cases, a nominal value and measurement uncertainty for the bottle volume was obtained. Later versions of student instructions asked for measurements of overall bottle height, H, and maximum diameter, D, along with a reporting of the measurement uncertainties. The later inclusions of bottle height and maximum diameter mean that there are more measurements of emptying time that correspond to values of d and V than those that correspond to d, V, H, and D. In these later iterations, students were also asked to submit photographs of bottles to confirm shapes and reported dimensions.

A simple data collection protocol was followed. To start, the bottle was completely filled (up to the outlet of the neck) with room temperature tap water, assuring approximate consistency in liquid viscosity and surface tension. Next, the bottle opening was blocked with a hand or card to prevent outflow and then the bottle was inverted and oriented vertically. Inversion was followed by a short delay to minimize swirl during emptying. It is known that emptying time for bottles will vary with the orientation angle as well as swirl [4]. The protocols students were asked to follow attempt to reduce the effect of both variables from the experimental results. Finally, the students let the bottle empty while recording the overall emptying time T_e. This process was repeated for $n = 3, 5, 10,$ or 25 trials depending on the theme of the exercise and/or difficulty in handling filled bottles (this is especially true for very large bottles, i.e., of several gallons, where $n = 3$–5 was considered reasonable).

Not surprisingly, the students participating in these exercises were able to find a diverse range of bottle shapes and sizes. Several hundred students, distributed over multiple academic quarters, have contributed to the bottle emptying data presented here, along with many bottles and containers tested by the author. The result is that for the present study we have data from $N = 454$ bottles (The entire $N = 454$ collection of bottle information, including the dimensions (d, D, H, and V), shape designation, and emptying times, are available in Table S1 in the Supplementary Materials section). Efforts have been made to reduce to an insignificant number duplicate bottles from this dataset. This was accomplished by either a direct visual examination of the bottle (to assess shape, brand, etc.) or a review of a photograph provided by the student. This dataset is larger than any other reported in the literature [3,4,6] and represents a vast range of shapes and sizes. In particular, bottle volume spans $10 \lesssim V \lesssim 27,000$ mL, with emptying times ranging from less than a second to nearly four minutes.

2.2. Exercise 1—Single Bottle Experiments (Statistics Focus)

This section provides an overview of statistics generated from multiple measurements of emptying time using a single variable, e.g., descriptive statistics and basic inferential statistics based on the emptying time considered as a normally distributed variable. Several graphical means of describing the data are also presented. A complete listing of equations used to generate the reported statistics is found in Appendices A and B. We start here, before proceeding to the results, using dimensional analysis, as this could be treated as a standalone exercise in basic statistics.

2.2.1. Single Bottle Statistics—Descriptive

The data and analysis from experiments performed by the author is now provided, in which a single bottle was emptied 100 times. When students were asked to perform similar experiments for the purpose of single bottle statistics, the number of trials was reduced to 10–25 in the interest of limiting experimentation time outside of class. However, larger number of trials are preferable from the standpoint of obtaining statistically significant results. Here, the measurand of interest is the emptying time, T_e. As demonstrated in what follows, data collected from a single bottle can be used to introduce and/or reinforce the key elements of descriptive statistics, graphical presentation of data, and basic inferential statistics for a single variable. Later, we show how these findings are used to estimate overall emptying time uncertainty.

Before considering in detail the data collected during the course of experiments, it is worthwhile to generate a sequential plot to ensure that the emptying time does not itself change with time/order, i.e., to ensure that time is not a lurking variable [16] in the analysis that follows (although it is not anticipated, based on the physics of bottle emptying, that the emptying time will vary with the order of collection). Pedagogically, this is an instance in which, prior to generating the plot, students can reflect on what they expect to see from a sequential plot of the data and explain their expectations. Figure 2a depicts the sequential plot of bottle emptying time for 100 trials performed with the same bottle (cf. Figure 7a for a diagram of the bottle). The data appears as expected, possessing a time-invariant central tendency and spread, quantified later, with the scatter in the data likely caused by subtle differences in the initial conditions of the experiment, for example, differences in filled volume, initial fluid motion within the bottle, and geometry of the air–water interface at the bottle opening. Reaction time likely plays a role as well. We can therefore proceed with further analysis and turn our attention to descriptive statistics and graphical presentations of those results.

At this point, we make no assumption regarding the distribution of the data, nor do we make any inference from the data. Rather, we are interested in providing concise summary measures of the data at hand [16]. The summary measures of interest will describe the central tendency of the data, the spread—or variability—about the central tendency, and the shape of the data. More specifically, these measures will be the sample mean , median, sample standard deviation, five-number summary, and the skewness and kurtosis. We can also graphically present the data to highlight these metrics.

The mean (A1) is easily calculated using a spreadsheet, and for the data shown in Figure 2a, the mean value is $\overline{T}_e = 6.61\,\text{s}$. The median—the middle value in the ordered set of data—is $\tilde{T}_e = 6.63\,\text{s}$. A difference of only 0.3% exists between the two measures of the central tendency. There are many ways to characterize the spread, or variability, of the data about the measure of the central tendency. We will begin by generating a "five-number summary" after the data has been ordered from minimum to maximum (j will be the index for the ordered dataset). Once ordered, we can easily obtain the range $T_{e,\max} - T_{e,\min}$, and the interquartile range, i.e., $IQR = Q_3 - Q_1$, where Q is used to denote a quartile. The complete five-number summary is presented in Table 1.

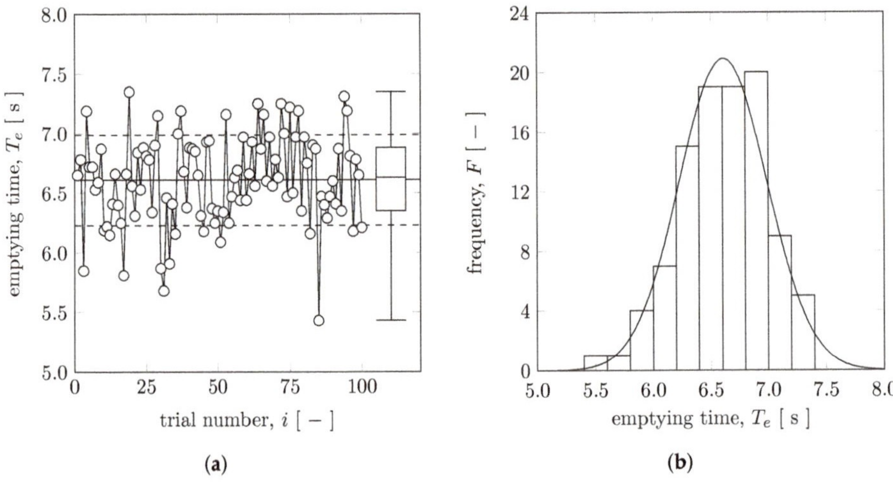

Figure 2. (a) Sequential plot of emptying time recorded from 100 trials using a single bottle (Brer Rabbit Molasses, 375 mL, and d = 18.3 mm; cf. Figure 7a). Emptying time was measured using a digital stopwatch with 0.01 s resolution. No apparent trend over time (i.e., trial number i) is suggested by the plot. The sample mean ($\overline{T}_e$ = 6.61 s) is shown as the solid horizontal line, and the sample standard deviation (s = 0.38 s) is indicated symmetrically about the mean by the dashed lines. Counting of the data points reveals that 71% are located within $\overline{T}_e \pm s$. A box-and-whisker plot is located at the right edge. The extent of the box is reflective of the interquartile range and the line within the box indicates the value of the median ($\tilde{T}_e$ = 6.63 s). (b) Frequency distribution created from the single bottle data (n = 100). For the distribution shown, the bin number is k = 10 and ΔT_e = 0.20 s. Superimposed on the histogram is a normal distribution generated using $\overline{x}$ and s, in place of μ and σ within Equation (A8), and it is evident that there is agreement.

Table 1. The five-number summary for the single bottle data collected by the author ($n = 100$).

Summary Item	j [-]	Value [s]	IQR [s]	Range [s]
$T_{e,min}$ (minimum)	$j = 25$	5.43		
Q_1 (1st quartile)	$j = 25$	6.35		
$\tilde{T}_e$ (median)	$j = 50/51$	6.63	0.53	1.92
Q_3 (3rd quartile)	$j = 76$	6.88		
$T_{e,max}$ (maximum)	$j = 100$	7.35		

The five-number summary can be depicted graphically in the form of the box-and-whisker plot [16]. This type of plot is provided on the right edge of Figure 2a. The shape of the box-and-whisker plot reinforces the notion that the distribution of emptying times measured for a single bottle is nearly symmetric. Aside from the five-number summary, we can calculate a single value to characterize the spread, namely, the sample standard deviation s (A2), which for the data has a value of $s = 0.38$ s. The standard deviation can be considered as a kind of "average" distance the data sits, as a whole, from the central tendency. The extent of the standard deviation about the sample mean, i.e., $\overline{T}_e + s$ and $\overline{T}_e - s$, is also shown in Figure 2a, and we find that 71 out of 100 data points, i.e., 71%, fall within this region.

The symmetry and peakedness of the measurand distribution, both measures of the shape, can each be expressed by a single value—the skewness Sk (A3) and kurtosis Ku (A4), respectively. We find, for the data in Figure 2a, the value of the skewness to be $Sk = -0.34$ and the kurtosis $Ku = 3.03$. Skewness can take on both positive and negative values. Positive skewness implies a distribution shape that is shifted to the right, or, more specifically, the mean value exceeds the mode. Negative skewness indicates the reverse–the value of the mode exceeds the mean. For a symmetric distribution, $Sk = 0$, i.e., the mean value is equal to the mode value [17]. To interpret the value of the skewness obtained from our data, let us consider the following rules of thumb [18]; highly skewed, $|Sk| > 1$; moderately skewed, $1 > |Sk| > 0.5$; approximately symmetric, $0.5 > |Sk|$. Using this interpretation with a value of $Sk = -0.34$, we can conclude that the distribution is approximately symmetric. The kurtosis calculated from a set of data is often compared against the value for a known distribution, the normal distribution being a frequent comparator. For a normal distribution, $Ku = 3$, with values of $Ku < 3$ representative of a flatter distribution, and values of $Ku > 3$ for more peaked distributions [17]. Therefore, the calculated value for the kurtosis ($Ku = 3.03$) is reasonably close to that of a normal distribution. It should not be surprising then that we found 71% of the data to fall within $\overline{T}_e \pm s$, as 68% of data will fall within $\mu \pm \sigma$ for a normal distribution (and for $n = 100$ we expect that $\overline{x} \to \mu$ and $s \to \sigma$; in other words, the sample is approaching a population).

The quantitative measures of the central tendency, spread, and shape can also be graphically presented, as we have seen with the sequential plot and box-and-whisker plot. However, a frequency distribution (i.e., histogram) more clearly illustrates the mode and shape, and it is worth developing one here. Later, we will use this histogram from our experimentally determined data to support the selection of a standard probability density function, so as to further interpret the data.

For the student, it is important to recognize the bin number (or number of classes), k, should be selected to adequately convey the nature of the distribution of data. Little can be learned from a histogram with too many or too few bins [16]. Although the development of a histogram can be highly subjective—the selection of bin number and boundaries is at the discretion of the student—there are a few rules of thumb to consider: (1) in general [19], the bin number should be some value between 5–20; (2) for $n > 40$, consider using $k = 1.87(n - 1)^{0.4} + 1$ as a starting estimate [20]; and (3) for large enough n, try $k \sim \sqrt{n}$ as an initial estimate [21]. Regardless of method, one should take care to consider the resolution with which the measurand was obtained, as this can influence the frequency of values contained in bins as the value of k gets large with increasing n, i.e., bin sizes smaller than the resolution of the measuring instrument can lead to artificially empty bins.

For $n = 100$, the rules of thumb suggest that $k = 10$ or 13 will be reasonable. Using the value of the range (cf. Table 1), this yields bins with widths of 0.20 and 0.15 s. The data from Figure 2a was binned using $k = 10$, and the frequency F of occurrences in each bin was tabulated. The result is shown in Figure 2b, and the distribution demonstrates that the data is unimodal and approximately symmetric (consistent with the calculated value of Sk and the box-and-whisker plot shape).

2.2.2. Single Bottle Statistics—Inference

In the previous section, we did not utilize a standard distribution to interpret the data; we simply described the data quantitatively and graphically. However, the purpose of obtaining a sample set of emptying times is to ultimately develop an estimate of the "true" emptying time for any particular bottle in the complete set of $N = 454$, i.e., an estimate of the population mean including a statistical uncertainty (which will be combined with a measurement uncertainty). This will be used in later portions of this work when the results of different bottles are used collectively.

The evidence presented thus far suggests that the emptying times are, or can be modeled as being, normally distributed. Specifically, the skewness $Sk = -0.34$ (near zero) implies the distribution is approximately symmetric, and the kurtosis $Ku = 3.03$ suggests the distribution is near-normal; the number of data points contained within $\overline{T}_e \pm s$ and the frequency distribution of Figure 2b hint at the characteristic Gaussian shape. If the assumption that emptying times are normally distributed is valid, then with the data from a sample, we can estimate the population mean using $T_{e,\mu} = \overline{T}_e \pm \left[(t_\nu s) / \sqrt{n} \right]$, where the μ subscript implies a population parameter and the degrees of freedom is expressed as $\nu = n - 1$. The Student-t variable t_ν is tabulated and is a function of the degrees of freedom and percent confidence specified [21]. The sample mean and sample standard deviation have already been calculated; to utilize this estimate we must clearly demonstrate normalcy in the data.

To supplement the evidence, we can strengthen the visual assessment of normalcy in two ways. The first involves superimposing atop the data from Figure 2b a continuous curve representing a normal frequency distribution. We can generate such a curve by starting with the probability density function for a normally distributed variable (see Appendix B). Recalling that the probability density p can be related to the relative frequency f and therefore the frequency F for a finite sample, we can create the normal distribution $F(T_e)$. Added to the frequency distribution in Figure 2b is the continuous curve representing a normal distribution. A visual inspection of Figure 2b shows that the frequency distribution for the data is consistent with the shape of a normal distribution.

A second graphical method to evaluate the normalcy of data is the normal probability plot. This is a plot of the standard normal variable, z, versus the measurand, i.e., T_e. Each value of z is obtained by assigning every ordered data point an equal cumulative probability and then calculating the corresponding value of z based on the cumulative probability, assuming that the data is normally distributed. Specific details for developing the plot can be found in standard statistics textbooks [16,21]. The purpose of the normal probability plot is that if the data is approximately normal, the plot of z vs. T_e will be linear according to the definition of the standard normal variable, i.e., $z = (T_e - \mu)/\sigma$. Therefore, rather than relying on a comparison of curved shapes as in Figure 2b, we can visually inspect how the data falls against a straight line. A normal probability plot for the bottle emptying data is given in Figure 3a along with a straight line to guide eye. This graphical assessment of normalcy is reasonable if there at least 20 data points [16], as is the case here ($n = 100$), and under conditions when students are asked to perform 25 measurements (e.g., if the instructor is intending to emphasize the statistics aspect of this exercise). The majority of the points in Figure 3a fall along the straight line and would pass a "fat pencil test", with only a few points at the upper and lower tails showing deviation. Little weight is placed on the extreme points if they fail to fall along the straight line [21].

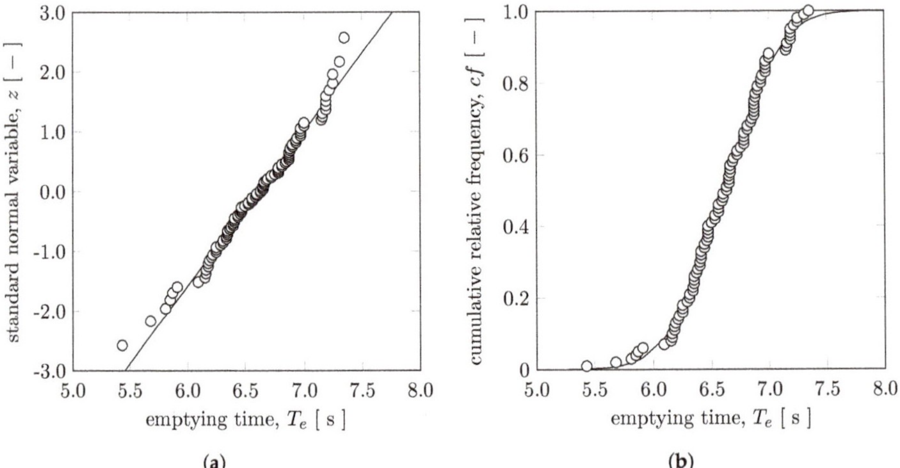

Figure 3. (a) A normal probability plot and (b) a cumulative relative frequency plot of the bottle emptying time data. The normal probability plot, z vs. T_e, shows a linear relationship between the two variables, with the exception of a few data points contained in each of the tails. Typically, little emphasis is placed on the agreement of data in the tails [21]. This linear trend is indicative of data that can be considered normally distributed for the purposes of inference.

Lastly, a final quantitative justification for treating the bottle emptying time as being normally distributed can be had by determining whether or not the empirical distribution when compared to a normal distribution passes the Kolmogorov–Smirnov test, cf. Appendix B and Equation (A9). This involves comparing the maximum deviation between the empirical cumulative frequency and the cumulative frequency for a continuous normal distribution. For the $n = 100$ points, the maximum deviation was found to be $d_{KS} = 0.043$, which is below the critical value of $d_\alpha(n) = 0.136$ for a level of confidence of $\alpha = 0.05$. This indicates that the normal distribution passes the Kolmogorov–Smirnov test. A visual comparison of the empirical cumulative frequency and the theoretical normal cumulative frequency can be found in Figure 3b.

In summary, the collective evidence presented in this section suggests that bottle emptying times can be reasonably modeled as normally distributed, for the sake of estimating the population mean for emptying time. Thus, using Equation (A5), we can determine that the emptying time for this particular bottle, accounting for the statistical uncertainty is $T_{e,\mu} = 6.61 \pm 0.08$ s (with 95 % confidence).

2.2.3. Single Bottle Statistics—Additional Examples

Given the many academic quarters that the author has used this exercise, there have been numerous instances in which the same type of bottle has been used by students for data collection. Although these redundant sets of data are not included in the sections that follow (where we attempt to use unique bottles for each data point), here they present the opportunity to show that a similar presentation of single bottle statistics (descriptive and inferential) can be had, without subjecting one experimenter to the tedious task of emptying a bottle 100 times. Examples of data collected by multiple students using the same type of bottles is shown in Figure 4. Within Figure 4a we see sequential plots and box-and-whisker plots for three different bottles (milk, tea, and soda). Figure 4b shows the corresponding histograms and normal distributions. Again, the data suggests that bottle emptying time can be reasonably modeled as normally distributed. Perhaps the only distinct feature of Figure 4a as compared to Figure 2a are instances of localized increases in the magnitude of the data spread and abrupt changes in the value of the central tendency (e.g., $i = 60$–70 "tea" data and $i = 70$–80 "soda" data). It is thought that these instances are due to differences in student interpretation of when a bottle has fully emptied.

Another permutation of the single-bottle experiment is to provide students with the same type of bottle and to group the emptying time data that is collected. In this case, given the size of a class, a significant number of data points for a single bottle can be obtained in a short period of time. An example of the data that results is shown in Figure 5, where 75 students were provided with wine bottles of the same style and size. Each student was asked to contribute $n = 10$ values of bottle emptying time, yielding a total of $n = 750$. The data arranged in sequential order is provided in Figure 5a. Here, again we see the instances of abrupt changes in the magnitude of the central tendency for each student data cluster (e.g., $i = 200$–210), which can be attributed to differences in student interpretation of the end of an emptying experiment. Given a large enough set of emptying times, the tendency is for the data to appear as though all values were collected from a single experimenter; this is especially true if the data is randomized. To see this, compare the sequential plots for the sequential and randomized data in Figure 5a. The histogram for this data is shown in Figure 5b where a subtle skewness is observed, when compared to the normal distribution, favoring a longer emptying time by a small fraction of students.

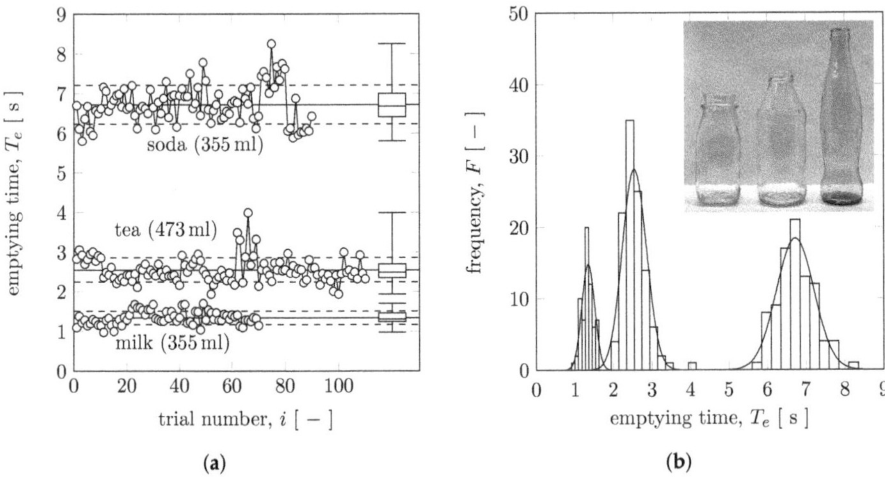

Figure 4. Emptying time data for several common bottles used by students. (**a**) Sequential plots and box-and-whisker plots suggest that emptying time is normally distributed, even when data is collected by many students instead of a single experimenter. Each student collected $n = 10$ values of emptying time. (**b**) Histograms and normal distributions also support the normalcy of the data. The inset shows the shape of the milk, tea, and soda bottles (arranged from left-to-right).

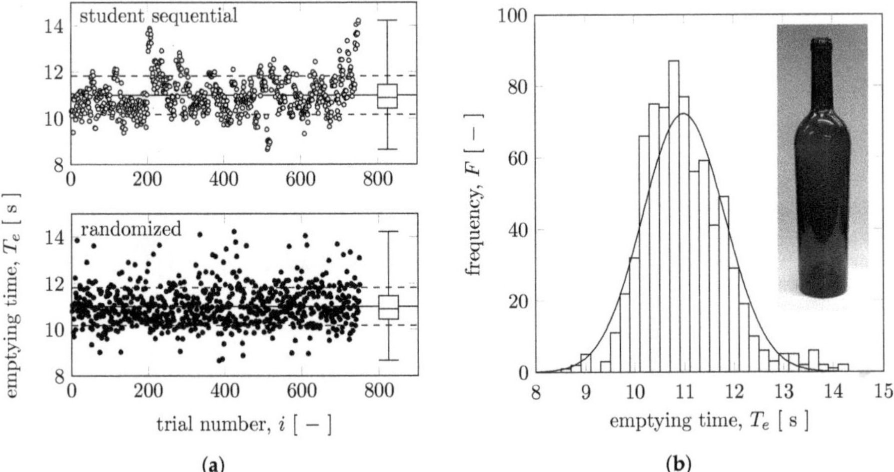

Figure 5. Emptying time data for wine bottles issued to 75 students. The unused wine bottles were procured in cases to ensure consistency in size and shape ($V = 750\,\text{mL}$ and $d = 18.4\,\text{mm}$). (**a**) Sequential plots with box-and-whisker diagrams. The student sequential data (top, open circles) is ordered so that each group of 10 values belongs to a single student. When the sequential data is randomized (bottom, closed circles), it appears as though it were collected from a single individual. (**b**) The histogram suggests some skewness which is likely due to several students interpreting a longer emptying time.

2.3. Exercise 2—Interpreting the Body of Results (Multiple Bottles and Dimensional Analysis Focus)

2.3.1. Quantifying Experimental Uncertainty

Up to this point, we have considered the results from a few experiments performed using single bottles, and considered only statistical variations associated with the emptying time. However, there are other uncertainties in the measured quantities (e.g., emptying time, bottle volume, neck diameter, etc.)—we must at least account for the measurement uncertainties or resolution uncertainties connected to the bottle emptying experiments. Thus, before presenting the results from all student experiments, let us discuss the uncertainties associated with the individual measurements and combine those, when appropriate, with the statistical uncertainties developed in the previous section.

With regard to emptying time T_e, we have already considered the statistical uncertainty associated with multiple measurements; this is simply the confidence interval, $\pm t_\nu s / \sqrt{n}$. Recall that students were asked to perform $n = 3, 5, 10$, or 25 trials and could use the results from these experiments to calculate the statistical uncertainty associated with the emptying time (using 95% confidence). There still exists the resolution uncertainty of the timers employed by the students. As these were typically cell phone app stopwatches, with stated resolutions of 0.01 or even 0.001 s, the resolution uncertainty, treated as one-half the instrument resolution [20], is negligible compared to the statistical uncertainty. However, due to manual timing of each trial, it would be more appropriate to consider errors due to human reaction time, as the dominant factor in the emptying time measurement error, not simply the resolution uncertainty. Although this can be the subject of further experimental study by students engaged in this exercise, a value of 0.25 s can be used to account for the total uncertainty in the reaction to the beginning and end of emptying (and, inherently, the perception of the emptying event, which is subject to interpretation). This value is typical of the range of human reaction time errors associated with stopwatch timing [22]. The reaction time was summed in quadrature with the timer resolution uncertainty and the statistical uncertainty to obtain an overall uncertainty in emptying time $u_{\overline{T}_e}$. The average value for this uncertainty for nearly all bottles tested was found to be 9.3%, although there are a number of instances (42 of 454 bottles) for which the uncertainty represents a significant fraction of the time, i.e., greater than 25%, due to fast emptying times (e.g., emptying times on the order of 1 s).

For bottle volume, students were provided access to either a graduated cylinder or a digital scale to make volumetric or gravimetric measurements. In general, using these techniques, the resolution uncertainty can be quite small. However, this can under predict the actual variation in filled volume that can occur during testing with the repeated inversion and waiting required by the procedure (which can lead to leaks). Because of this, it was decided to assign a reasonable uncertainty for bottle volume. Those bottles with $V \leq 50\,\text{mL}$ were assigned an uncertainty of 1 mL, whereas those with $V > 50\,\text{mL}$ were assigned an uncertainty of 5 mL. Many students reported using household measuring cups to determine volume, and the author even used bathroom and mechanical scales for certain large-volume bottles. Typically, these instances yielded values of uncertainty larger than those assigned. In all instances where the reported uncertainty exceeded the conservative values assigned based on volume, the larger uncertainty was used. For all of the bottles tested, the average uncertainty in bottle volume is 2.1%.

Similarly, the uncertainties in the measurements of neck diameter, u_d; maximum diameter, u_D; and height, u_h, were low (averages of 1.4%, 0.6%, and 0.4%, respectively), owing to the tools used by the students (e.g., ruler or caliper) in comparison to the size of the nominal values. All of the uncertainties discussed are summarized in Table 2 and are represented by error bars in the dimensional plots that follow. Having sorted out the uncertainties associated with the student measurements, we are now at a point where we can present the collection of bottle emptying data in dimensional form from all bottles, of various shapes and sizes, collected by students and the author.

Table 2. Measurement and statistical uncertainties for dimensional bottle emptying data.

Uncertainty	Average Value	Comment
u_{V}	2.1%	gravimetric/volumetric estimate
u_d	1.4%	measuring tool resolution uncertainty
u_D	0.6%	measuring tool resolution uncertainty
u_H	0.4%	measuring tool resolution uncertainty
$u_{\overline{T}_e}$	9.3%	overall uncertainty in $\overline{T}_e$

2.3.2. Bottle Emptying—Dimensional Results

From our most basic understanding of fluid mechanics, we expect that the emptying time for a bottle will increase with volume. This is essentially what is suggested by the data presented in Figure 6a, in which the average emptying time, $\overline{T}_e$, is plotted versus bottle volume, V. The error bars include measurement and statistical uncertainties as previously discussed. Despite our expectation regarding the relationship between emptying time and volume it is apparent that, without taking into account other geometric properties of the bottle, no useful predictive trend yielding a reasonably accurate estimate of emptying time can be extracted from such a dataset. There is simply too much scatter in the data as plotted. In fact there can exist more than an order of magnitude variation in emptying time for any fixed volume (e.g., see variation in $\overline{T}_e$ for $V \sim 10^3$ mL). The lack of availability of very large commercially available bottles limits the data to the region $V \lesssim 3 \times 10^4$ mL.

Similarly, the lack of predictive power is also apparent in the plot of emptying time, $\overline{T}_e$, versus neck diameter, d, cf. Figure 6b. Again, our intuition leads us to believe that emptying time should decrease with increase in neck diameter; however, the overall picture in Figure 6b is not satisfactory for our purposes. Perhaps the only piece of information we can glean is that the majority of bottles have neck diameters in the neighborhood of $d \sim 20$ mm, which is not altogether surprising if we recognize that most of the bottles collected by students for these experiments are made for direct human consumption or convenient pouring. Finally, the remaining geometric measurements of maximum diameter, D, and height, H, are also not useful, by themselves, at establishing a predictive trend for emptying time, as shown in Figure 6c,d.

Recalling the motivation for this study, Figure 6 rules out the usefulness of the dimensional data to obtain a predictive relation for bottle emptying time (if one does exist). The figures clearly demonstrate that an assumption that the emptying time correlates with only one geometric variable is of no particular use. More than one dimension may be necessary. This conclusion leads us naturally to a discussion of dimensional analysis.

2.3.3. Dimensionless Groups Using Buckingham-Pi

Many undergraduate engineering students are first exposed to dimensional analysis in a fluid mechanics course, where the subdiscipline is considered in the context of experimental design and the arrangement of data to yield useful phenomenological relationships [23]. However, during this introduction, there is often a limited chance to see first-hand the power of dimensional analysis through application, i.e., the design of an experiment and the reduction of data. More often students complete typical textbook problems, the end result of which is the determination of dimensionless groups only. Without an applied experience, the main concepts and steps can be confusing, e.g., selection of repeating parameters in the use of the Buckingham-Pi theorem—a step described as requiring the "application of engineering judgment" and "experience" for a successful outcome [24], qualities that students are only beginning to develop.

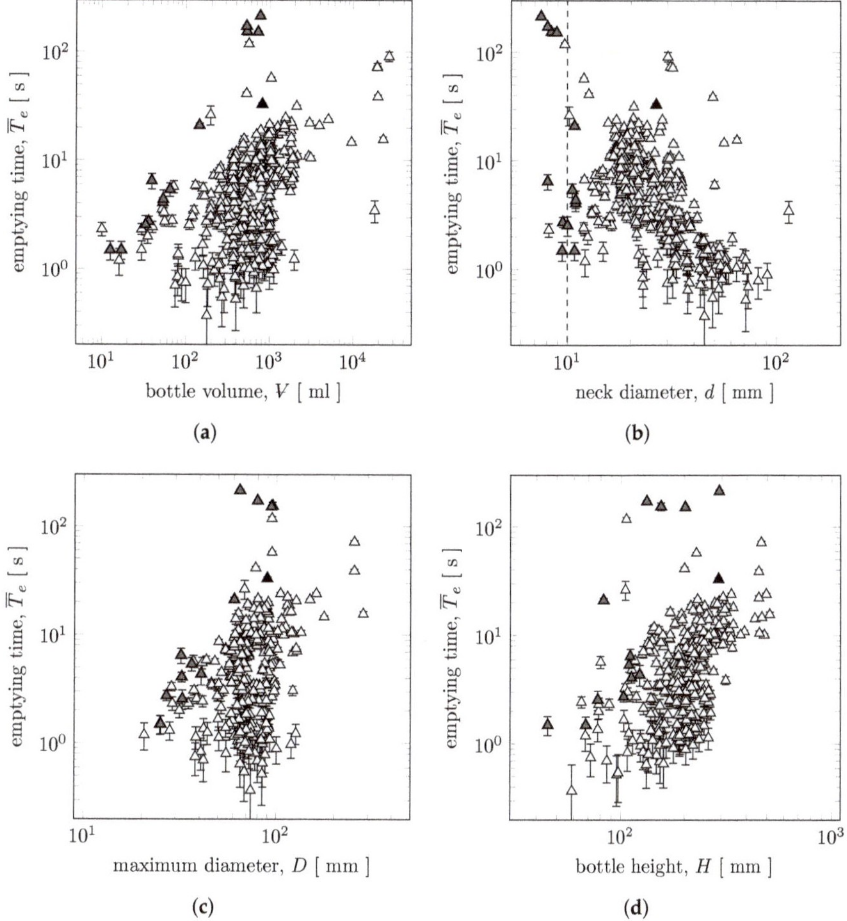

Figure 6. Emptying time versus the measured geometric dimensions for all bottles tested ($N = 454$). (**a**) Bottle emptying time generally increases with volume, but no predictive trend is apparent. The thirteen data points marked with gray triangles are the results from emptying small bottles with 70% isopropyl alcohol instead of water. The single black triangle denotes a data point for a bottle that has an internal decorative feature that can disrupt the flow. This was the only bottle tested with such a feature. (**b**) Neck diameter, d, alone is also not a good predictor of bottle emptying time; We observe that most bottles/containers collected by students tend to cluster around $d{\sim}20$ mm, most likely a result of their intended use (liquids for direct consumption). It is also apparent that neck diameters are well in excess of the capillary length, $\sqrt{\gamma/(\rho g)}{\sim}2.7$ mm. A number of small bottles tested by the author (gray triangles) have neck diameters less than the stable cut-off length associated with the Rayleigh–Taylor instability for a circular interface [25,26], i.e., $d_{RT}{\sim}3.7\sqrt{\gamma/(\rho g)}{\sim}10$ mm (dashed vertical line). For these cases, the emptying using water is essentially infinite, and so measurements have been made using 70% isopropyl alcohol to create the condition $d > d_{RT}$ by reducing the interfacial tension γ. This can be used to prompt added discussion among students during a presentation of the data. Note that three bottles with $d < d_{RT}$ were reported by students using water. It is suspected by the author that this was possible given nonideal neck conditions (e.g., a chip in one instance and a ground glass surface in another) or subtle off-vertical orientations that can readily occur during experiments; (**c**) maximum diameter D and (**d**) bottle height H do not, alone, provide a useful correlating variable for emptying time.

Given the lack of predictive trend apparent in the dimensional plots of Figure 6, the present goal is now to find relevant dimensionless groups that characterize the bottle emptying process–specifically the long timescale that characterizes complete emptying and not the short timescale associated with the familiar "glug-glug" [10], of which the latter phenomenon tends to dominate our casual observations. To be relevant for a discussion in an undergraduate fluid mechanics setting, the procedure for the Buckingham-Pi theorem as outlined in a typical fluid mechanics textbook is followed [24].

We can anticipate that for a liquid-filled bottle emptying into the atmosphere at room temperature and pressure, the emptying time T_e will be a function of many variables. For example, we might consider,

$$T_e = \phi_1(\rho, \mu, \gamma, g, H, D, d, S),\tag{1}$$

where ρ, μ, and γ are the relevant liquid properties (water density, viscosity, and liquid–air surface tension, respectively) and g is the gravitational acceleration. The geometry of the bottle is represented by an overall height and major diameter, H and D; the neck diameter, d; and a shape factor, which we designate S (cf. Figure 7). This term accounts for the complexity of the true bottle shape, including in particular the shape, length, and taper of the neck, and we will consider S as being dimensionless. In past studies on bottle emptying not involving the dimensional analysis approach taken here, the various shapes of the bottle used were presented diagrammatically with no effort made to quantify the details of the shape [3,4]. Or, conversely, specific experimental apparatuses were fabricated to provide well characterized containers for study. In the latter case, the exit condition (essentially an orifice plate, not a bottleneck) was described by a discharge coefficient [7]. Here, the sheer number of bottles ($N = 454$) prevents us from presenting a comprehensive diagram detailing each individual bottle shape (which can vary dramatically), and lack of control over shape by the use of commercially available bottles prohibits the specification of a known discharge coefficient and, therefore, the lumping of these qualities into S. Furthermore, given the myriad variations of real bottles and containers, we can expect that isolating separate trends based on characteristic bottle length and diameter will be too difficult for the present study. So, although we recognize it is a gross simplification, we will replace H and D with the total bottle volume V. This reduces the total number of variables in (1) from 9 to 8. The variables considered in (1) do not constitute an exhaustive list, and in the classroom setting it would be useful for an instructor to ask students to consider all other possible variables (including, but not limited to angle of inclination, amount of initial rotation/swirl, and surface roughness, all of which could be tested as additions to the work published in this study).

Continuing with the Buckingham-Pi methodology, after selecting three repeating parameters—ρ, g, and d—that characterize the fluid, flow field, and geometry for the emptying phenomenon (as well as capturing all primary dimensions in the set of variables), respectively, we see that there will be five dimensionless groups. Thus,

$$T_e\sqrt{\frac{g}{d}} = \phi_2\left(\frac{\mu}{\rho\sqrt{gd^3}}, \frac{d}{\sqrt{\gamma/(\rho g)}}, \frac{V}{d^3}, S\right),\tag{2}$$

which do not present themselves as the typical dimensionless numbers encountered by undergraduates (e.g., Reynolds number, Re). Here, it should be mentioned that $\sqrt{\gamma/(\rho g)}$ is also referred to as the capillary length, l_c, and represents a length scale below which capillary forces (i.e., interfacial tension forces) become significant. For the present study, we are only interested in emptying bottles containing water into air at room temperature, and we will utilize bottles with neck diameters sufficient to rule out capillary effects (cf. Figure 6 caption). Therefore, the dimensionless group d/l_c will play no role. The inverse of the group $\mu/(\rho\sqrt{gd^3})$ is analogous to a Reynolds number (where $\sqrt{gd}$ is a characteristic velocity). For all of the bottle emptying experiments, the value of $(\rho\sqrt{gd^3})/\mu$ exceeds 2000, and most (423 of 454) have values in excess of 5000. Thus, we can anticipate insignificant variations in experiments caused by this dimensionless parameter as nearly all flows will be turbulent. As a result, we neglect the independent dimensionless groups listed in (2) containing the fluid properties. Further, neglecting for now the shape factor S (which we will return to later), the dimensional analysis suggests that we should present the data in the following form,

$$T_e\sqrt{\frac{g}{d}} = \phi_3\left(\frac{V}{d^3}\right),\tag{3}$$

to look for a relationship between emptying time and bottle geometry. Whether or not we have picked the most relevant repeating parameters, or if we have over simplified our analysis, remains to be seen when we try to plot the data in dimensionless form.

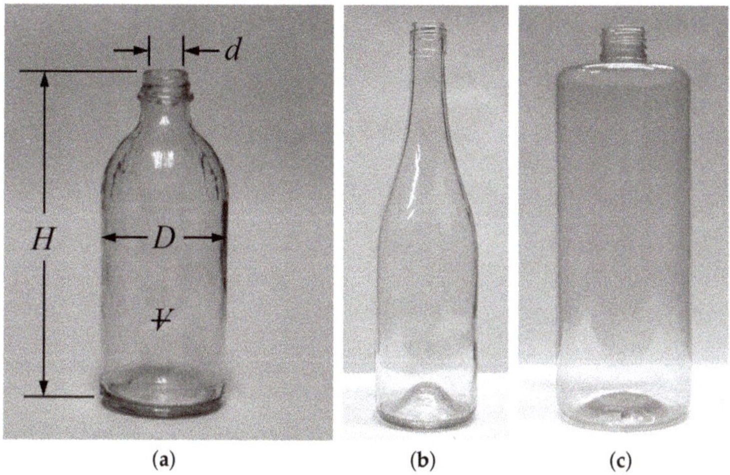

(a)   (b)   (c)

Figure 7. A visual description of bottle geometry relevant to the analysis of bottle emptying. (**a**) Dimensions associated with bottles used for emptying experiments: d, neck diameter (internal); D, major diameter; H, overall bottle height; and V, internal volume. It was used by the author for 100 measurements of emptying time (cf. Figure 2). (**b**) A bottleneck qualitatively classified as "smooth" due to the gradual evolution from D to d. This in sharp contrast to (**c**), which shows a bottleneck classified as "square", as a result of the abrupt change from D to d. The bottle shown in (**a**) would be considered as something "other" than smooth or square.

2.3.4. Data Presented in Dimensionless Form

We can now plot the data in dimensionless form, and we do so according to the dimensionless groups presented in Equation (3). The result is shown in Figure 8. Error bars for dimensionless time and volume have been calculated by propagation of the uncertainties from each of the variables forming the dimensionless groups. Inspection of the figure shows it is clear that we arrived at relevant dimensionless groups in Section 2.3.3, based on the collapse of nearly all of the data points about, what appears to be, a single dominant trend—far different from the dimensional results in Figure 6. However, if we look at the data more closely, we can also see what looks like a difference in bottle emptying time based on the shape S. Earlier we had ignored the influence of the shape factor S, and the wide variety of shapes makes detailed classification based on such a factor difficult. However, for now, we can broadly classify the bottles as having a "smooth" neck (e.g., a wine bottle with a gradual taper from the average diameter to the neck), a "square" neck (i.e., minimal taper), or "other" if the bottle fails to be easily classified into the former groups. Examples of these classifications are shown in Figure 7, and we make no effort here to rigorously classify the shapes as it is beyond the scope of this work. The inset of Figure 8 indicates that many of the bottles with a more smooth neck appear to empty faster than square neck bottles of similar dimensionless volume. Further analysis of the data based on a more rigorous classification of shape may provide a better prediction of bottle emptying time. Yet, for the purposes of this exercise, when the data is viewed as a whole it does suggest a general trend that is sufficient for engineering estimates of emptying time.

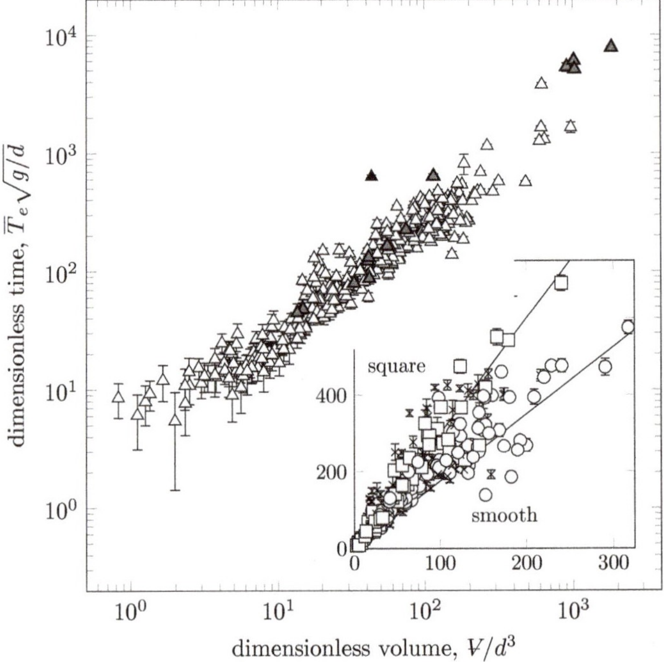

Figure 8. Data from bottle emptying experiments presented in dimensionless form according to the groups given in Equation (3). It is apparent that the data as a whole, despite variations in bottle shape and size, tend to collapse onto a single trend. However, greater scrutiny of the data, apparent in the inset (note the linear axes and reduced range and the symbol change—□ for square and ○ for smooth), shows what appears to be a difference in dimensionless emptying times based on bottle shape S. Smooth neck bottles empty faster than square necks. The solid lines in the inset are meant to guide the eye.

From the dimensionless data of Figure 8, it appears that a trend has emerged that allows us to make quantitative predictions for bottle emptying time as a function of volume and neck diameter. Our focus now shifts to determining an appropriate correlation using standard curve fitting techniques.

2.4. Exercise 3—Developing an Empirical Correlation (Regression Analysis)

The goal of the analysis in this section is to obtain a curve fit for the data presented in Figure 8, i.e., to elucidate the relationship between $\overline{T}_e\sqrt{g/d}$ and V/d^3, so that we can determine with sufficient accuracy the emptying time for any future bottle or single outlet container. A curve fit for the data will be determined using a method of least-squares analysis and we have provided details and equations in Appendix C.

2.4.1. Regression Analysis

In general, we can utilize the known physics of a phenomenon to guide the appropriate choice of a model for curve fitting of data. However, in the case of bottle emptying, where the fluid dynamics appear complex, we are left to wonder what is the appropriate choice for a curve fit. The method of least-squares, which is what we will eventually use, can only be applied after the order of the polynomial curve fit has been decided, and it is unacceptable to arbitrarily choose the order. What then are we left with? In this instance, let us consider the span of the data for both coordinates.

For both $\overline{T}_e\sqrt{g/d}$ and $V\!\!\!/\,/d^3$, the data extends over several orders of magnitude, so a log–log format is the best choice for plotting the data. Inspection of the data, as shown in Figure 8, plotted on the log–log axis and appearing linear suggests that indeed a power–law curve fit between the dimensionless groups may be appropriate [20]. Thus, regression analysis will be used to find the best-fit coefficient A and the exponent B for a power–law function of the form $\overline{T}_e\sqrt{g/d} = A\left(V\!\!\!/\,/d^3\right)^B$. Only the data for small values of dimensionless volume, i.e., $V\!\!\!/\,/d^3 \lesssim 3$ hint at a deviation from linearity. This change in behavior is likely based on a change in the fundamental shape of the bottles at these scales. For these values of $V\!\!\!/\,/d^3$ the bottles are really jars, mugs, or cups, instead of a bottle with what we think of as a neck, e.g., with $d < D$. These vessels, when emptying, do not "glug". Rather, drainage proceeds as a single bubble of air rises into vessel along one side and liquid drops from the other. This is also in contrast to the emptying of a long narrow tube [1,2].

The power–law curve fit can be obtained using a method of least-squares approach for a linear regression, e.g., $y = a_0 + a_1x$, after a standard linear transformation has been performed. In other words, to make use of the equations for the regression coefficients a_0 and a_1 in the standard first-order curve fit (familiar to most students in a first course on statistics), we must first transform our variables such that $\overline{T}_e\sqrt{g/d} \rightarrow y$ and $V\!\!\!/\,/d^3 \rightarrow x$. For a power law, this is accomplished using natural logarithms. Thus,

$$\overline{T}_e\sqrt{g/d} = A\left(V\!\!\!/\,/d^3\right)^B \rightarrow y = a_0 + a_1x \tag{4}$$

with $y = \ln[\overline{T}_e\sqrt{g/d}]$, $x = \ln[V\!\!\!/\,/d^3]$, and $a_0 = \ln A$ and $a_1 = B$. What remains is to solve for the regression coefficients [16] a_0 (Equation (A10)) and a_1 (Equation (A11)), and then transform back to find A and B. This analysis yields the following power–law curve fit to all of the data,

$$\overline{T}_e\sqrt{\frac{g}{d}} = (3.8 \pm 0.4)\left(\frac{V\!\!\!/}{d^3}\right)^{(0.90 \pm 0.02)}, \tag{5}$$

which is plotted in Figure 9a, along with uncertainty bands (based on 95% confidence) qualitatively demonstrating how well the regression equation (solid line) predicts the mean of the data (inner set of dashed lines) and the uncertainty that the regression equation can predict any single value (outer dashed line). Not surprisingly, given the appearance of the dimensionless data in Figure 8, the power law correlation fits the data over the nearly four orders of magnitude spanned by $V\!\!\!/\,/d^3$.

The variation of individual data points about the curve fit line is presented in Figure 9b. Here, the variation is expressed as a normalized residual r, defined as the difference between the experimental and curve fit values of $\overline{T}_e\sqrt{g/d}$, normalized by the curve fit value. The purpose of normalizing the residuals is to account for the difference between experimental values, and curve fit values inherently increasing in magnitude with increase in dimensionless volume. What we find from a review of these figures is that the normalized residuals are scattered somewhat symmetrically about zero, without any apparent trend that would indicate a more appropriate curve fit model exists. More specifically, the values of r appear to be distributed normally about zero when a histogram is generated over the entire range of $V\!\!\!/\,/d^3$. The average value is calculated to be $r = 5\%$ with an average absolute value of $|r| = 25\%$.

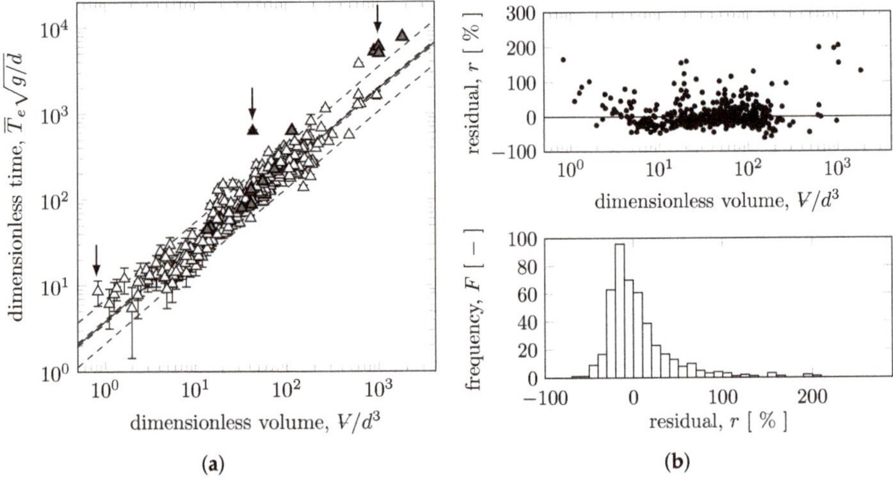

Figure 9. (**a**) Data from bottle emptying experiments presented in dimensionless form according to the groups given in Equation (3) along with the power–law curve fit of Equation (5) (solid line). Added to these plots are confidence intervals for the curve fit (95% confidence). The inner set of dashed lines, almost indistinguishable from the solid line, indicates the confidence that the curve fit can estimate the mean value of the population, whereas the outer set of dashed lines denotes the confidence that the curve fit can predict any individual value of the ordinate. Arrows highlight bottles that deviate significantly from the power–law fit. The bottle with the internal feature (black triangle) was not used in the curve fit analysis but is shown to indicate the effect of the internal feature on the emptying time. (**b**) Discrepancy between experimental data and power–law curve fit (Equation (5)) expressed as a normalized residual, r (% units). (Top) The scatter of r about zero for the entire range of the dimensionless volume shows no apparent trend, but a few data points show considerable deviation from the curve fit (i.e., r exceeds 100%). (Bottom) Values of r appear to be symmetrically, almost normally, distributed about a value near zero.

2.4.2. Trend Deviations

The data points with the most significant values of r, e.g., $\sim 200\,\%$, are at the extreme ends of the V/d^3 range. These are highlighted by the arrows in Figure 9a at $V/d^3 \sim 1$ and $/d^3 \sim 10^3$. Note that these single-outlet vessels are the furthest, in shape/style, from what we might think of as a typical "bottle". As shown in Figure 10a, the data point at $V/d^3 \sim 1$ corresponds to a pint glass, where $d \approx D$ without a true neck. The data points clustered near $V/d^3 \sim 10^3$ are for reusable tumblers with a thin lid and rounded orifice intended for a straw (cf. Figure 10b). Although for this container $d << D$, the shape is still far from a traditional bottle since, like a pint glass, there is also no true neck.

Every bottle in the dataset ($N = 454$) used to generate Equation (5) shared the characteristic that the internal space of the vessel was empty and that there were no obstructions within the neck. A single bottle provided by a student, highlighted by the arrow in Figure 9a at $V/d^3 \sim 40$, contained an internal decorative feature as well as a bridge of glass material that spanned the neck at its base. Both of these features can be seen in Figure 10c. Because of these features, the data collected from experiments in this bottle was not used in the analysis to obtain a curve fit (and does not constitute one of the $N = 454$ data points). However, it has been included in nearly all of the dimensional and dimensionless plots, with the exception of Figure 9b, to show the effects of the features on the emptying time. A comparison of the dimensionless emptying time for this bottle to that predicted by Equation (5) yields a value of $r = 450\%$. Looking back at the sequence of images in Figure 1, it is easy to imagine how disruptive the internal decorative feature would be on the inflow of air and outflow of water near the bottleneck. In addition, it is estimated that the bridge of glass within the neck reduces the flow passage area by nearly 1/3–1/2 the area calculated using the neck diameter. Both explanations are likely responsible for causing the large value of r for this particular bottle. The neck diameter for this bottle was measured as $d = 26.55$ mm, along with a volume of $V = 814$ mL, giving the value of $V/d^3 \sim 40$ reported. If we estimate that the flow passage area of the neck is restricted by 1/3, we arrive at an equivalent neck diameter of $d = 21.7$ mm, a value of $V/d^3 \sim 80$, and $r = 218\%$. With a flow passage area restricted by 1/2, an equivalent neck diameter of $d = 18.8$ mm results, yielding $V/d^3 \sim 120$ and $r = 118\%$. Although these are estimates, we can anticipate that with $r \gtrsim 120\%$, the discrepancy is due not only to the obstruction of the neck but the internal decorative feature as well.

(a)   (b)   (c)

Figure 10. Images of three "bottles" that yield significant values of r when compared to the power–law curve fit of the dimensionless data (indicated with arrows in Figure 9a). (**a**) Typical pint glass with $V/d^3 \sim 1$. (**b**) Reusable plastic tumbler ($V/d^3 \sim 10^3$) with a small hole in the lid intended for a straw. (**c**) A bottle with an large and obstructive internal decorative feature (left) along with a glass bridge across the flow passage at the base of the neck (right—imaged from side of bottle looking up at the base of the neck).

2.5. Exercise 4—Comparison of Empirical Correlation to Available Results

2.5.1. A Substantial Data Set

We now depart from the statistics and dimensional analysis content for a review of the sparse datasets and simple analytical models available in the literature, which we will compare to our experimental findings presented in Figures 8 and 9. This section, although not necessary for an exercise in either engineering statistics or dimensional analysis, does provide some physical insight into the emptying of a single-outlet vessel and reinforces the selection of a power–law curve fit to best describe the data. The choice to present this content to the student group or to let them research on their own would depend on the academic level of the fluid mechanics course. Note, here, the dramatic difference between datasets in the literature and the substantial set of new data in the present work. Figure 11a shows this difference. Here, we can see the range of both V and d spanned by the $N = 454$ bottles from the present work, as compared to the data extracted from the works of Whalley [3,4] (16 bottles), Tehrani et al. [6] (four bottles), and Geiger et al. [11] (two bottles simulated using CFD).

2.5.2. Published Results and Models

Whalley [3,4] investigated the emptying time of bottles via simple experiments and related the outflow to the phenomenon of "flooding" in tubes—a well-researched subject in the field of multiphase flows. He presented his experimental results in dimensional form (i.e., emptying time) and also calculated a Wallis flooding "constant", C, using

$$C = \frac{\left(\rho_G^{1/4} + \rho_L^{1/4}\right)}{\left[(\rho_L - \rho_G)\, gd\right]^{1/4}} \left(\frac{4V}{\pi d^2 T_e}\right)^{\frac{1}{2}}, \tag{6}$$

where ρ_G and ρ_L are the densities of air and water, respectively. This semiempirical relationship has its basis in a correlation for flooding that utilizes dimensionless superficial gas and liquid velocities, and is consistent with the simple picture that during the entire course of emptying, a stationary air slug exists in the bottleneck with a continuous outflow of water around the slug [3,4] (i.e., slugging). The velocity of liquid outflow is equivalent, by continuity, to the rising velocity of the gas slug (as given by Davies and Taylor [1]). To visualize this, consider the fourth image in Figure 1a with the slug in the bottleneck and envision that this picture of the flow persists over the entire course of emptying as hinted at by the overlay of images shown in Figure 1b. Both sets of data from Whalley's experiments, specifically the 11-bottle set from his earlier work [3] and the 5-bottle set from his later work [4], show a range of calculated values for C from 0.76 to 1.02 with an average value of $\overline{C} = 0.91$. Kordestani and Kubie [8], who also expressed their results for ideal bottles in terms of C, found a similar range of values spanning approximately 0.8 to 1.15. Using the values for ρ_G and ρ_L of air and water at room temperature, the average value of C from Whalley's works, and rearranging Equation (6) to reflect the dimensionless groups obtained from our analysis in Section 2.3.3, reveals the following,

$$T_{e,\mathrm{W}}\sqrt{\frac{g}{d}} = 2.2\left(\frac{V}{d^3}\right)^1. \tag{7}$$

The W subscript in $T_{e,\mathrm{W}}$ is intended to distinguish this Whalley result from others. Given the variation in values of C, the coefficient in Equation (7) has a range of 1.7 to 3.1 with the exponent of V/d^3 unaffected by changes in C. Although the coefficient differs by a little less than a factor of 2 from that of Equation (5), the overall trend is similar to that exhibited by our data as demonstrated by the dashed line in Figure 11b (Equation (7) plotted). Quantitative comparison of our data with the result of Whalley, using our previously defined normalized variable, yields $r = 30\%$ and $\overline{|r|} = 36\%$. We can also see in Figure 11b how well our data overlaps and extends the limited set of data from Whalley [3,4].

Clanet & Searby [10] analyzed the emptying of "ideal" vertically-oriented bottles (i.e., bottles with constant diameter and very short sharp-edged openings manufactured specifically for laboratory experiments). Their model of the long timescale, the overall emptying time T_e–not the timescale associated with passage of individual slugs of air through the opening, is based on similar physics to that of Whalley, but not described from the standpoint of accepted two-phase flow correlations. Rather, these authors build their simple model from the required balance between the incoming air at the opening and the downward motion of the air–water interface in the bottle. They use the rise speed of a long bubble in an infinite medium (again, from Davies and Taylor [1]) to characterize the incoming airflow, and make the assumption that the length of the incoming bubble is equivalent to the opening diameter (recall that their "ideal" bottles have no true neck length—therefore the diameter of the opening serves as a relevant scale for the length of the incoming bubbles). Clanet & Searby present their model for the long timescale as $T_e/T_{e,0} = (D/d)^{5/2}$. In their model, $T_{e,0} \simeq 3.0L/\sqrt{gD}$ is the emptying time of an unrestricted cylinder of length L and diameter D. This model yields the following estimate for the emptying time (using $V = (\pi/4)D^2 L$ and after rearrangement to be consistent with present form),

$$T_{e,\mathrm{CS}}\sqrt{\frac{g}{d}} = 3.8\left(\frac{V}{d^3}\right)^1. \tag{8}$$

We also plotted this trend in Figure 11b for visual comparison, and, quantitatively, we find that $r = -26\%$ and $\overline{|r|} = 34\%$. Equation (8) has the same scaling for V/d^3 as the result of Whalley, and is similar to the results of the present study, although there is incomplete agreement with the power–law exponent. The exponent associated with $\mathcal{V}/d^3$ from both Whalley and Clanet & Searby does not fall within the error of the experimentally determined exponent in the present work and differs by 10%.

The remaining data found in the literature that pertain to emptying of real bottles is found in the CFD analysis of Geiger et al. [11] and the experiments of Tehrani et al. [6], and this too is found to compare favorably with the new data presented here. A final inspection of Figure 11b shows that nearly all of the plotted data points tend to fall between the predictions of Whalley and Clanet & Searby, and that the power–law fit to our experimental data spans the two models over the range investigated.

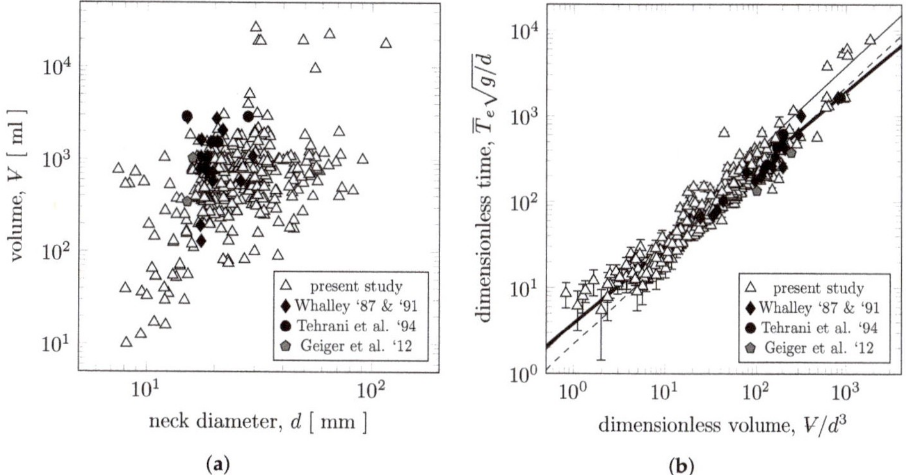

Figure 11. (**a**) Dimensions of the bottles from experiments of students (present work) plotted with limited datasets found in the literature, specifically those of Whalley [3,4], Tehrani et al. [6], and Geiger et al. [11]. The dataset collected by students is nearly twenty times larger than the collective data of the literature and contains bottles that span approximately 0.25–10× the volumes previously tested. The range of bottle diameters tested are also much larger, and smaller, than those previously explored. (**b**) Data from bottle emptying experiments of students plotted with limited datasets found in the literature, specifically those of Whalley [3,4], Tehrani et al. [6], and Geiger et al. [11]. The trend line from our data, Equation (5), is shown (thick solid line), as is the model of Clanet and Searby, Equation (8) (thin solid line) and that of Whalley, Equation (7) (dashed line).

2.5.3. Revisiting Regression Analysis

As a final note, given the exponent associated with $\mathcal{V}/d^3$ in the models of Whalley and Clanet and Searby, perhaps it is worth revisiting our regression analysis, now that we have some knowledge of the physics at hand (recall from Section 2.4.1 that we typically rely on the physics to guide our choice of curve fit function). What the models of Whalley and Clanet and Searby suggest is that we could also seek a curve fit of the form $\overline{T}_e\sqrt{g/d} = A\left(\mathcal{V}/d^3\right)^1$, where the exponent has been set to 1. This is analogous to finding a linear curve fit with zero intercept, i.e., $y = a_{1*}x$ (note that this is not the same a_1 as defined by Equation (A11); therefore, we must use Equation (A12)). If we perform such an analysis, we arrive at the following fit to our data,

$$\overline{T}_e\sqrt{\frac{g}{d}} = (3.7 \pm 0.2)\left(\frac{\mathcal{V}}{d^3}\right)^1.$$

(9)

This is nearly identical and, given the uncertainty range, it is equivalent to the model of Clanet and Searby (Equation (8)). The nominal value of the coefficient differs by nearly a factor of two compared to the equation of Whalley (Equation (7)). A summary of all of the empirical models and analytical models from the literature is provided in Table 3.

Table 3. Summary of the regression models developed in this work and the simple analytical models obtained from the literature. All models have the form of a power–law function, with A as the coefficient and B the exponent.

Source	Equation	Coefficient A	Exponent B
present work	(5)	3.8 ± 0.4	0.90 ± 0.02
present work	(9)	3.7 ± 0.2	1
Whalley [3,4]	(7)	2.2 $(1.7 - 3.1)$	1
Clanet & Searby [10]	(8)	3.8	1

2.6. Very Large Bottles—Does the Trend End?

A goal of the present study was to experiment with only commercially and readily available bottles so that no special equipment or experimental apparatus is necessary (an example of this equipment would be the "ideal" bottles used in the literature). This goal was achieved, and any class that is asked to perform this exercise will likely find a similar range of bottle shapes and sizes yielding consistent results. Ignoring the data corresponding to very small values of V/d^3, the trend in the dimensionless data of Figure 8 appears to hold for the entire range of V/d^3, and it has always piqued the interest of the author to ask if there is an upper limit to the applicability of this trend. However, addressing this question requires going beyond commercially available bottles, and building bottles to suit the needs of the experiment. There are only two ways to achieve ever larger values of V/d^3: (1) reduce the neck diameter d of a bottle and (2) increase the volume V of a bottle. As we have seen, there is a limit to the reduction in the neck diameter governed by capillary effects and the Rayleigh–Taylor instability, which will prevent emptying. Therefore, the only practical way of achieving very large values of V/d^3 is to use large volumes.

To explore bottle emptying for large values of V/d^3, i.e., $V/d^3 \gtrsim 5 \times 10^3$ (which is approximately the limit for the commercially available bottles found by students and the author), the author resorted to the fabrication of bottles using both a 15-gallon and a 55 gallon plastic drum (these are nominal volumes—the measured volumes were found to be 64,500 mL and 225,000 mL representing an increase in volume of 2.4×-8.3× compared to the largest bottle in the $N = 454$ set). These drums, shown in Figure 12a, were outfitted with interchangeable 3D printed necks with measurements of $d = 12.5$, 25, and 37.5 mm. The necks were fitted to the drums using machine screws and sealed with gaskets. These inverted bottles were secured in a vertical orientation on an elevated stand. Garbage cans directly below the bottles were used to collect and recycle the water during the experiments. Each bottle was filled by means of a centrifugal pump using the bung hole (which was sealed with the threaded and gasketed bung).

The emptying times for these bottles ranged from approximately 110–1500 s (15 gallons) to 400–6100 s (55 gallons), with the emptying time decreasing with increasing d. The combination of bottle volume and neck diameters result in values of V/d^3 spanning from nearly 10^3 to 10^5—almost two orders of magnitude larger than the commercially available bottle set. Despite the dramatic increase in dimensionless bottle volume, and to the delight of the student audience that this is presented to, the trend persists and the agreement between Equation (5) is qualitatively on display in Figure 12b. Quantitatively, for the 15-gallon bottles, $\overline{|r|} = 17\%$, and for the 55-gallon bottles, $\overline{|r|} = 21\%$. Both values of $\overline{|r|}$ are within the value of 25 % reported earlier for the $N = 454$ dataset. It is left to another group of experimentalists, perhaps a motivated student group, to continue to push the limits of V/d^3.

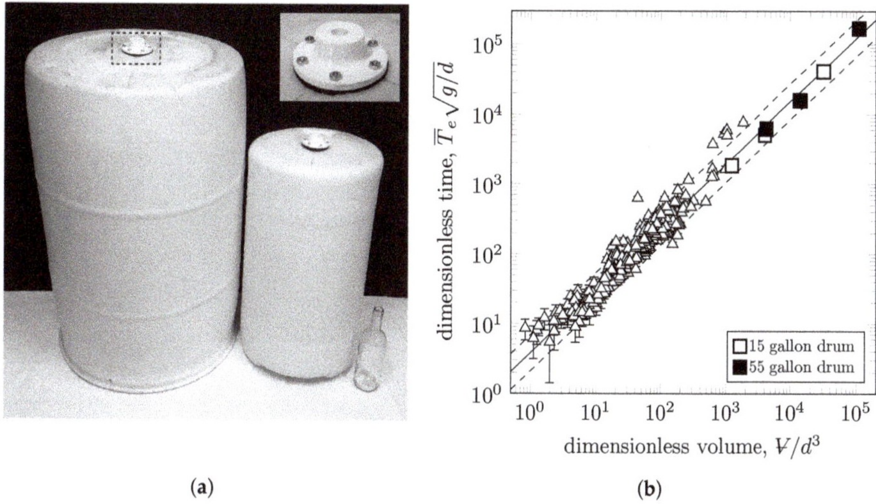

(a)   (b)

Figure 12. Very large bottles that can significantly extend the values of V/d^3 from the $N = 454$ dataset are not readily available. (**a**) The author resorted to manufacturing bottles using commercially available plastic drums (15-gallon and 55-gallon) and 3D printed necks (see inset image) to achieve over an order of magnitude increase in V/d^3. A wine bottle is shown in the image for scale. (**b**) The results of these large bottle emptying experiments are consistent with the data and curve fit presented.

3. Conclusions

We have presented an exercise suitable for an undergraduate engineering statistics class, as well as a course in which the basics of dimensional analysis are presented—the ideal venue being an introductory fluid mechanics course. The exercise involves a large dataset that can be gathered safely and easily by students, without resorting to expensive laboratory equipment or complex procedures. Basic descriptive and inferential statistics are suitable to describe the data. The every day phenomenon under investigation is rather complex and, to an undergraduate student first encountering fluid mechanics, it may not appear to be accessible by normal analytic means. Results presented here clearly demonstrate the power of dimensional analysis in reducing the data to yield a predictive trend and the results agree well with the limited set of data available and simple models developed in the literature. The variations of this particular exercise are myriad, and many of the experiments could still be performed with basic equipment. For example, an instructor could task the students with confirming the variation in emptying time with water temperature [3,4,6], or variations associated with inclination [3,4,9]. Students could be directed to establish the relevance of another dimensionless group by experimenting with differences in liquid viscosity (or ratios of density or viscosity). Filling time, rather than emptying time, could also be explored. Greater detail as to the role shape plays can be elucidated, and, given the appearance of g within the dimensionless emptying time, perhaps bottles could be emptied with the aid of centrifugal forces. There is no doubt that students could be tasked with, and would be eager about, finding a modification to this exercise that is worthy of study. These and other variations can be accomplished in the spirit of the work presented here and we encourage others to explore similar activities for the benefit of engineering students at their respective universities.

Supplementary Materials: The following are available online at www.mdpi.com/xxx/s1.

Author Contributions: H.C.M. is responsible for all aspects of this article except for portions of the raw data contributed by students.

Funding: This research received no external funding.

Acknowledgments: The author would like to thank the numerous students at Cal Poly SLO, UCSB, and Rutgers University from 2007 to 2019 who helped gather data during their undergraduate statistics/measurements and fluid mechanics courses. Without their help over the last decade, the dataset would not have been as large or as interesting. The author would also like to acknowledge Professors G. Thorncroft, P. Lemieux, J. Maddren (Cal Poly SLO), S. Pennathur (UCSB), R. Krechetnikov (U. Alberta), and S. Shojaei-Zadeh (Rutgers) for providing useful feedback on earlier drafts of the manuscripts.

Appendix A. Descriptive Statistics

Equations for the basic statistics used throughout the paper are presented here for the interested reader. In what follows in this section, emphasizing descriptive statistics, we will use the typical variable notation x to denote our measured quantity under investigation (whereas in the body of the paper we used T_e).

The most commonly used measure of the central tendency is the arithmetic mean (i.e., the mean) and is defined as

$$\bar{x} = \frac{1}{n} \sum_{i=1}^{n} x_i, \tag{A1}$$

where n is the size of the sample and i denotes the number in the sequential dataset.

The sample standard deviation s is a single measure characterizing the spread of the data about the central tendency. For a sample of size n, it can be calculated using

$$s = \sqrt{\frac{1}{n-1} \sum_{i=1}^{n} (x_i - \bar{x})^2}. \tag{A2}$$

The standard deviation can be considered as a kind of "average" distance the data sits, as a whole, from the central tendency.

Although we can visualize the shape of the data by developing a frequency distribution, we can also quantify the symmetry and peakedness of the distribution by calculating the skewness Sk and kurtosis Ku, respectively. We use the following definitions for calculations of each measure,

$$Sk = \frac{1}{(n-1)s^3} \sum_{i=1}^{n} (x_i - \bar{x})^3, \tag{A3}$$

and

$$Ku = \frac{1}{(n-1)s^4} \sum_{i=1}^{n} (x_i - \bar{x})^4. \tag{A4}$$

Appendix B. The Student-T and Gaussian (Normal) Distributions

Here, we present equations related to inferential statistics and standard distributions used throughout the paper, in particular to estimate the statistical uncertainty associated with multiple measurements of bottle emptying time. Again, we use x in lieu of T_e.

For a sample (implying finite size) of x_i of size n whose parent population is normally distributed, we can estimate the population mean μ using the Student's t variable. This estimate is given as

$$\mu = \bar{x} \pm \frac{t_\nu s}{\sqrt{n}}, \tag{A5}$$

where t_ν is tabulated for various levels of confidence. Often the term $\pm(t_\nu s)\sqrt{n}$ is referred to as the "confidence interval".

The probability density function for a normally distributed variable,

$$p(x) = \frac{1}{\sigma\sqrt{2\pi}} e^{-(x-\mu)^2/2\sigma^2}. \tag{A6}$$

is familiar to many students in science and engineering disciplines. As a refresher, the probability density $p(x)$ is related to the probability that x will lie between values a and b, i.e., $P(a < x < b)$ via

$$P(a < x < b) = \int_a^b p(x)dx. \tag{A7}$$

Additionally, for small values of Δx, we can write that $P(x < X < x + \Delta x) = p(x)\Delta x$. We are interested in relating this back to the finite data that we have to work with, specifically so as to visually compare to the frequency distribution in Figure 2b. Thus, we must use $P = F/n$, which then allows us to write that $F(x) = np(x)\Delta x$. Using this finding and Equation (A6) we can create a continuous curve $F(x)$ that can be superimposed onto a frequency distribution, i.e.,

$$F(x) = n\left(\frac{1}{\sigma\sqrt{2\pi}}e^{-(x-\mu)^2/2\sigma^2}\right)\Delta x. \tag{A8}$$

Alternatively we can transform the frequency data for x into the corresponding probability density. The probability density for each bin, l, in the frequency distribution (where $l = 1...k$) can be calculated using our definition of the probability density $p_l = f_l/\Delta x$, where $f_l = F_l/n$ is the relative frequency. One advantage to this approach is that bar charts of probability density, rather than frequency F, demonstrate that probability density values are nearly independent of bin width and will approach limiting values i.e., a limiting distribution, as $n \to \infty$ and $\Delta x \to 0$.

To quantitatively assess whether a normal distribution adequately fits the data, we used the Kolmogorov–Smirnov test [27]. In this test, we first compute the maximum of the deviation, d_{KS}, between the empirical cumulative relative frequency, cf, to theoretical value from a standard distribution. For our case this is the normal distribution, but this test is not restricted to this distribution only. The maximum of the deviation just described can be expressed as

$$d_{KS} = \max_{1 \le j \le n} \left|cf(x_j) - cf_n(x_j)\right|, \tag{A9}$$

where $cf(x_j)$ is the cumulative relative frequency (i.e., bounded by 0 and 1) based on the theoretical distribution for a population, and $cf_n(x_j) = j/n$ is the empirical cumulative relative frequency. Here, the index j denotes the ordered dataset. We then compare the value of the maximum deviation d_{KS} to the critical value $d_\alpha(n)$ based on the sample size, n, and the level of significance α (critical values are tabulated [27]). If $d_{KS} < d_\alpha(n)$, the theoretical distribution passes the test. If $d_{KS} > d_\alpha(n)$, the discrepancy between the empirical cumulative frequency distribution and the theoretical distribution is considered significant and the theoretical distribution fails to represent the data.

Appendix C. Regression Analysis

For the necessary material related to regression analysis, which is needed to find the empirical correlation describing the entire set of our bottle emptying data, we transition from the statistics of a single variable x to the relationship between two variables $y(x)$. In what follows, the notion that we use here now reflects $x \to \Psi/d^3$ and $y \to \overline{T}_e\sqrt{g/d}$.

The method of least squares can be used to find the best-fit polynomial function of the form $y = a_0 + a_1 x + a_2 x^2 + ... + a_m x^m$ through a set of n data points (x_i, y_i) once the order of the polynomial, m, has been selected. For a linear regression of the form $y = a_0 + a_1 x$, the method is straightforward and reduces to solving for the regression coefficients [16] a_0 and a_1, which minimizes the deviation between the curve fit equation and the data points. Here, we provide the definitions for a_0 and a_1:

$$a_0 = \frac{\sum x_i \sum x_i y_i - \sum x_i^2 \sum y_i}{\left(\sum x_i\right)^2 - n\sum x_i^2} \quad \text{and} \tag{A10}$$

$$a_1 = \frac{\sum x_i \sum y_i - n \sum x_i y_i}{\left(\sum x_i\right)^2 - n \sum x_i^2}. \tag{A11}$$

The summation symbols have been simplified but all apply from $i = 1 \ldots n$. Although a program (e.g., Excel$^{\circledR}$ or MATLAB$^{\circledR}$) can be used to automatically solve for these coefficients (and this may be the appropriate route to consider if the emphasis of the student project is dimensional analysis and fluid mechanics); a basic spreadsheet can be used to compute the necessary summations.

For the case where the linear regression we seek has the form $y = a_{1*}x$, where $a_{1*} \neq a_1$, the single regression coefficient can be solved for using

$$a_1 = \frac{\sum x_i y_i}{\sum x_i^2}. \tag{A12}$$

References

1. Davies, R.; Taylor, G. The mechanism of large bubbles rising through liquids in tubes. *Proc. R. Soc. Lond.* **1950**, *200A*, 375–390.
2. Zukoski, E. Influence of viscosity, surface tension, and inclination angle on motion of long bubbles in closed tubes. *J. Fluid Mech.* **1966**, *25*, 821–837. [CrossRef]
3. Whalley, P.B. Flooding, slugging, and bottle emptying. *Int. J. Multiph. Flow* **1987**, *13*, 723–728. [CrossRef]
4. Whalley, P.B. Two-phase flow during filling and emptying of bottles. *Int. J. Multiph. Flow* **1991**, *17*, 145–152. [CrossRef]
5. McQuillan, K.; Whalley, P.B. Flow patterns in vertical two-phase flow. *Int. J. Multiph. Flow* **1984**, *11*, 161–175. [CrossRef]
6. Tehrani, A.; Wragg, A.; Patrick, M. How fast will your bottle empty? *Proc. ICHEME* **1994**, 1069–1071.
7. Schmidt, O.; Kubie, J. An experimental investigation of outflow of liquids from single-outlet vessels. *Int. J. Multiph. Flow* **1995**, *21*, 1163–1168. [CrossRef]
8. Kordestani, S.; Kubie, J. Outflow of liquids from single-outlet vessels. *Int. J. Multiph. Flow* **1996**, *22*, 1023–1029. [CrossRef]
9. Tang, S.; Kubie, J. Further investigations of flow in single inlet/outlet vessels. *Int. J. Multiph. Flow* **1997**, *23*, 809–814. [CrossRef]
10. Clanet, C.; Searby, G. On the glug-glug of ideal bottles. *J. Fluid Mech.* **2004**, *510*, 145–168. [CrossRef]
11. Geiger, F.; Velten, K.; Methner, F.J. 3D CFD simulation of bottle emptying process. *J. Food Eng.* **2012**, *109*, 609–618. [CrossRef]
12. Mer, S.; Praud, O.; Neau, H.; Merigoux, N.; Magnaudet, J.; Roig, V. The emptying of a bottle as a test case for assessing interfacial momentum exchange models for Euler-Euler simulations of multi-scale gas–liquid flows. *Int. J. Multiph. Flow* **2018**, *106*, 109–124. [CrossRef]
13. Mer, S.; Praud, O.; Magnaudet, J.; Roig, V. Emptying of a bottle: How a robust pressure-driven oscillator coexists with complex two-phase flow dynamics. *Int. J. Multiph. Flow* **2019**, *118*, 23–36. [CrossRef]
14. Middleman, S. *An Introduction to Fluid Dynamics: Principles of Analysis and Design*; John Wiley & Sons: Hoboken, NJ, USA, 1998.
15. Tobin, S.; Meagher, A.; Bulfin, B.; Möbius, M.; Hutzler, S. A public study of the lifetime distribution of soap films. *Am. J. Phys.* **2011**, *79*, 819–824. [CrossRef]
16. Levine, D.; Ramsey, P.; Schmidt, R. *Applied Statistics for Engineers and Scientists*; Prentice Hall: Englewood Cliffs, NJ, USA, 2001.
17. Dunn, P. *Measurement and Data Analysis for Engineering and Science*; CRC Press: Boca Raton, FL, USA, 2010.
18. Bulmer, M. *Principles of Statistics*; Dover Publications: New York, NY, USA, 1979.
19. Walpole, R.; Myers, R. *Probability and Statistics for Engineers and Scientists*; The Macmillan Company: New York, NY, USA, 1972.
20. Figliola, R.; Beasley, D. *Theory and Design for Mechanical Measurements*; John Wiley and Sons: Hoboken, NJ, USA, 2000.
21. Montgomery, D.; Runger, G. *Applied Statistics and Probability for Engineers*; John Wiley and Sons: Hoboken, NJ, USA, 1999.

22. Naatanen, R. An explanation for differences between stop-watch and mechanical timing of races. *Percept. Mot. Skills* **1969**, *28*, 79–89. [CrossRef] [PubMed]
23. Fay, J.A. *Introduction to Fluid Mechanics*; MIT Press: Cambridge, UK, 1994.
24. Pritchard, P. *Fox and McDonald's Introduction to Fluid Mechanics*, 8th ed.; John Wiley & Sons: Hoboken, NJ, USA, 2011.
25. Faber, T.E. *Fluid Dynamics for Physicists*; Cambridge University Press: Cambridge, UK, 2004.
26. Yih, C. *Stratified Flows*; Academic Press: Cambridge, MA, USA, 1980.
27. Massey, F. The Kolmogorov–Smirnov test for goodness of fit. *J. Am. Stat. Assoc.* **1951**, *46*, 68–78. [CrossRef]

Essay

Teaching and Learning Pressure and Fluids

Petros Kariotoglou [1],* and Dimitris Psillos [2]

[1] School of Education, University of Western Macedonia, 53100 Florina, Greece
[2] School of Education, Aristotle University of Thessaloniki, 54124 Thessaloniki, Greece; psillos@auth.gr
* Correspondence: kariotog@gmail.com; Tel.: +30-23850-5080

Received: 8 October 2019; Accepted: 22 November 2019; Published: 25 November 2019

Abstract: This essay is a synthesis of more than twenty years of research, already published, on teaching and learning fluids and pressure. We examine teaching fluids globally, i.e., the content to be taught and its transformations, students' alternative conceptions and their remediation, the sequence of educational activities, being right for students' understanding, as well as tasks for evaluating their conceptual evolution. Our samples are junior high school students and primary school student-teachers. This long-term study combines research and development concerning teaching and learning fluids and has evolved through iteratively based design application and reflective feedback related to empirical data. The results of our research include several publications.

Keywords: teaching and learning fluids; teaching and learning pressure; alternative conceptions; teaching learning sequences; conceptual change; constructivism

1. Introduction

In the present paper, we describe and discuss a synthesis of several investigations we conducted concerning the teaching and learning of the conceptually demanding topic of fluids. Our aim was to enhance the conceptual evolution of junior high school students and primary school student teachers in the field of science through a proposed Teaching–Learning Sequence (TLS). Our incentive for organizing such a long-term research study was due to a number of factors, namely, the inclusion of fluids in the Greek curriculum as well as in that of several other countries, the variety of everyday phenomena related to fluids, and students' alternative conceptions identified in studies prior to our own [1]. The theoretical framework which our study is based on regarding the teaching and learning of science is individual and emphasizes the importance of learners' prior conceptions and reasoning in facilitating or preventing their understanding of new scientific concepts and models, as well as their active participation in the construction of new knowledge [2–4]. Moreover, it is argued that models that have been accepted by the scientific community may potentially undergo transformations based on research evidence, in order to become knowledge to be taught comprehensibly to target populations [5,6].

In this context, the present synthesis focuses on: the research results on students' domain-specific conceptions and reasoning in the field of fluids; content analysis of the relevant scientific models; the reconstructed research-based scientific model to be taught. In addition, the interactions between the model to be taught and students' conceptions and reasoning led to the development of a pioneering Teaching Learning Sequence (TLS) in fluids, with selected effective tasks. Moreover, we also present selected research results concerning students' and primary school teachers' conceptual evolutions in their various applications of the TLS. At the outset of this paper, we would like to clarify that, despite the presentations essentially having a linear/sequential format, during the numerous studies, there were cycles of research and development concerning students' understanding and iterative development of instructional interventions [7].

2. Students' Domain-Specific Conceptions and Reasoning in the Field of Fluids

Since fluids are included in several curricula worldwide, a number of researchers, as early as the 1980s, have examined students' conceptions related to liquids or gases prior to or after instruction, mainly in compulsory education. Briefly, a considerable number of students, before and after instruction, consider water 'pressure' to be a 'force' or a 'weight' [8], they think that 'pressure' has a preferred downwards direction, increases with depth [1,9] and that the value of 'pressure' increases with the total volume of liquid. Furthermore, other students ignore the incompressibility of liquids or consider them not to have a constant volume.

An essential aspect of these difficulties is students' confusion about pressure and force, an issue that has only slightly, but not explicitly, been approached by researchers. In the present study, we believe that students' conceptions in this domain should be examined systematically in terms of how they relate and/or differentiate the concepts of force and pressure [10]. We, thus, set out to further investigate the vector and scalar features of students' conceptions with regard to liquids in equilibrium using paper and pencil tasks, as well as semi-structured interviews. In order to corroborate the findings, the data were triangulated. For instance, we asked students (13–14 year-olds) before and after instruction, to compare the values of pressure at the same depth in two vessels containing water, for several different situations, e.g., at two different points of the water, at the backs of two divers, in small vessels, in a well, and in the sea, etc. In addition to the descriptive results, we articulated a number of mental models that are possibly used by the students in this domain, named "packed crowd model", "pressing force model" and "liquidness model", as has extensively been presented elsewhere [11].

Briefly, students who adopted the "packed crowd model" consider the density of the water to be variable. This means that the density in the narrow vessel is greater than in the wide one, thus, making the pressure greater in the former than in the latter. These students regard pressure as having no direction, it is considered or calculated on a surface. Students, who adopted the "pressing force" model, consider pressure in the wider vessel to be greater than in the narrower one, as a result of the larger amount of water. They also regard pressure as having a direction, depending on the amount of liquid, is considered or calculated on a surface, divided or shared, and it is expressed as "accepts/exerts pressure".

Students, who adopted the "liquidness model", consider the value of pressure at the same depth of a wide and a narrow vessel to be equal, and the pressure to be the property of the liquid, they calculate it on a point and express it as "has/exists pressure".

The "liquidness model" is relatively close to the scientific one, whereas the "packed crowd" one is primitive. The frequency of evidence in each model varies according to the period of teaching before the test. Usually, after the relevant instruction, approximately 40% of students understand the "pressing force model", 30% the "liquidness model", and 15% the "packed crowd model" [11]. The same models were identified among primary school student-teachers [12], though their wording was richer than that of the junior high school students.

The above-mentioned results show the major difficulties that students face when it comes to understanding this important concept, being a prerequisite for describing and interpreting fluids-related phenomena.

3. Scientific Content: Representations and Transformation

In the context of constructivism, clarification of the gap between students' prior knowledge and the representations of scientific knowledge is of great importance, for elaborating teaching, adapted to students' ideas and reasoning [13]. Taking such a theoretical perspective, we carried out a content analysis of six well-known textbooks on introductory physics, looking at how the concept of pressure is introduced and treated (for more details: [13]). The six textbooks refer to general physics in different countries at three levels, i.e., junior and senior high school and tertiary education. Briefly, the findings of this investigation show that all of the textbooks introduce pressure via the known formula:

$$P = F/S. \tag{1}$$

(**P** is Pressure, **F** is pressing Force, and **S** is Surface)-half of them also present it through liquids, while the other half through solids. All of the books, with the exception of one, attribute characteristics of a vector to pressure, which is a scalar quantity, and use the expressions "exerted or accepted pressure". Finally, the calculation of pressure on a surface, accompanied with arrows reinforcing the characteristics of a vector to pressure, appears in all of the six books.

Of course, the textbooks also use elements that attribute features of a scalar quantity to pressure. Such an approach is explicitly followed in four of the six books. It is mainly linked with the equation:

$$P = d \times g \times h, \tag{2}$$

where **P** is Pressure, **d** is density, **g** is gravitational acceleration, and **h** is the depth of the liquid. However, in only two of the six books do the authors use the expressions "...it has pressure..." or "...there is pressure...", which imply that pressure is a scalar quantity. The content analysis showed that the concept of pressure is articulated and utilized in the textbooks in two distinct ways: as a vector and as a scalar quantity. This leads us to conclude that textbooks often include implicit transformations of the accepted scientific model. The main transformation is the attribution of vector characteristics to pressure, which is close to students' dominant conception of the "pressing force model". It would be interesting to investigate the reasons that guided the authors of the textbooks to adopt such transformations for the concept of pressure.

Students who study introductory physics as part of their secondary compulsory education are required to understand the concept of pressure as it is a prerequisite for their conceptual development in the field of fluid mechanics. From the previous description of pressure models in liquids, it is evident that the most significant conceptual obstacle is the non-differentiation of pressure from the resultant pressing force. Therefore, the students who adopted this model could not conceive the meaning of pressure as a scalar magnitude, in comparison to the resultant pressing force, which is a vector. Considered at a point of liquid, the former is expressed as "has or exist pressure", while the latter as "accepted or exerted force".

In order to help students comprehend the differentiation between pressure and force in liquids, we have proposed a didactic transformation or educational reconstruction [5,14,15]. Specifically, we propose the introduction of pressure as a primary concept, measured directly, qualitatively and experimentally, rather than introducing it as a derivative magnitude from force with the Equation (1) [13,16,17]. Through this approach, we expect that the concept of pressure will be more comprehensible to students and will subsequently be differentiated from force, which appears to be the dominant concept in students' minds.

For an integrative approach concerning the teaching of fluids, further didactical transformations, although perhaps not as radical as the introduction of pressure as a primary concept, need to be made. One of these is the unification of liquids and gases in the category of fluids. This is because pressure has a unifying meaning in fluids, i.e., a scalar quantity, which differs from that of solids, where it is considered as stress, i.e., a vector quantity. In this way, we avoid using examples such as the pin of a pin on the board, or the destruction of a wooden floor with a stiletto heel, which are commonly presented in textbooks.

Another conceptual obstacle is the understanding of liquids as non-compressible, meaning that the density of a liquid does not change when it is contained in vessels of a different size, which is essential scientific knowledge. This obstacle may be overcome much more easily than the ones above, with the following experiment: a syringe filled with water is presented to the students, when force is exerted on the plunger, no compression of the liquid is observed This, along with the appropriate discussion, can make students understand that liquids are a practical, incompressible substance.

4. An Innovative Teaching–Learning Sequence for the Teaching of Fluid Related Phenomena

A contemporary, widespread strategy for developing innovative teaching approaches based on actively engaging students to change their domain specific ideas using several methods and techniques strategies, such as Predict–Observe–Explain physical phenomena, cognitive conflict, and the use of analogies are Teaching–Learning Sequences (TLS) [18,19]. TLSs are medium-scale (5–15 teaching hours) curriculum packages. According to Psillos and Kariotoglou, "A TLS is often both a research process and a product, which includes research-based structured teaching-learning activities. Frequently, a TLS develops iteratively out of several implementations, in accordance with a cyclical evolutionary process based on research data, which results in its being improved, with empirically validated expected students' outcomes from the planned activities" [18,20]. Several researchers agree with this outline of a TLS [20,21]. In addition, researchers have argued that grand pedagogical theories are too general in their proposals and that models or frameworks are necessary to design, apply, and evaluate a TLS [21]—one which is most used is "The Model of Educational Reconstruction" [12]. Following this model, we have identified the learning objectives, clarified students' domain-specific alternative conceptions, carried out a content analysis of textbooks, conceived an appropriate didactic transformation mentioned in the previous sections, as well as developed teaching strategies and activities, which are selectively presented further below. While the proposed TLS for junior high and primary school teachers is in accordance with accepted scientific models in that it focuses on the force–pressure relationship, its structure is based on a sound educational reconstruction of the treatment of these two concepts. Due to time restrictions, the TLS does not include pressure transmission, or the sinking and floating of solids immersed in fluids, although it does set the conceptual basis for studying such principles and phenomena later.

4.1. Basic Teaching–Learning Objectives

4.1.1. Differentiating Pressure from Pressing Force

The main innovative conceptual objective of the TLS is the differentiation of two overlapping concepts, i.e., pressure with the resultant force which is a hard task, demanding thorough instructional design. For this reason, we propose a didactic transposition/transformation for the introduction of pressure as a primary concept to students. Our instructional design is based on the constructivist approach of teaching and learning, and, in particular, creating cognitive conflicts between students' predictions and the observation of experimental evidence [4]. More specifically, the suggested experiments are: (1) the comparison/measurement of pressure at two points, at the same depth, in a narrow and in a wide vessel, and (2) the comparison of the forces required to detach two different surface sucks (Figure 1).

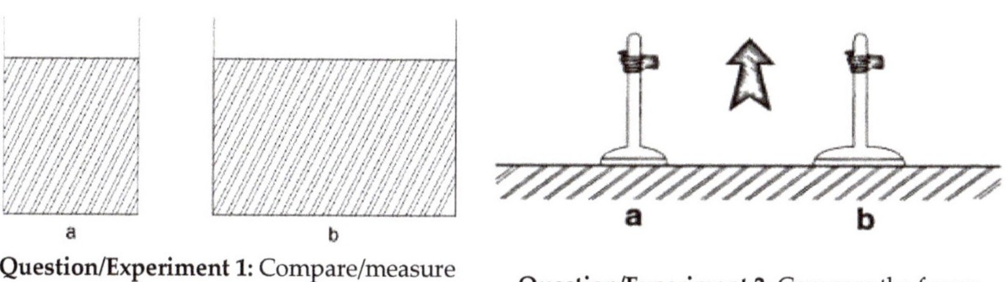

Figure 1. The two tasks/experiments that stimulate cognitive conflict in students [22].

We wish to note that this simple experiment of the comparison/measurement of pressure on the surface of two different vessels is usually not presented in textbooks. We consider this experiment as a significant intervention because it not only elicits students' alternative conceptions (see pressure models), but also helps them to differentiate force from pressure.

4.1.2. Comprehending the Compressibility of Liquids

Another innovative objective in the proposed TLS is to assist students to go beyond their conception of the compressibility of liquids, as some seem to think that, when water is transferred from a wider vessel into a narrower one, it becomes denser, i.e., its density increases. This conceptual obstacle may be overcome much more easily than the others mentioned above with the following experiment: the teacher shows students a syringe filled with water, then when force is exerted on the plunger, students observe that this does not result in compression of the water. Such an experimental result followed by the appropriate discussion can lead students to deduce that liquids are a practical, incompressible substance.

4.1.3. Classifying Both Liquids and Gases as Fluids

A third innovative objective is that of unifying liquids and gases into the category of fluids. Here, we aim to associate pressure (as a scalar quantity) with fluids, in contrast to solids, which are better described through stress, i.e., the force distributed on a surface (a vector quantity). As in several approaches, fluids are used explicitly and not implicitly as a unifying category of liquids and gases, the use of which may foster cognitive economy in students. This objective is attained by excluding all examples/phenomena or experiments where pressure is introduced via solids, for example, an object's impression in the sand or the pin on a wall.

4.1.4. Establishing the Relationship between Pressure and Pressing Force

The TLS also includes more common learning objectives, such as the dependence of hydrostatic pressure on depth: Equation (2). This rule is explained with the relevant experiment and measurements, at the same time, highlighting the experimental methodology, such as the distinction and control of variables.

4.1.5. Enhancing the Dependence of Hydrostatic Pressure on Depth

By the end of the sequence, pressure is related to the resultant pressing force with the following equation, where $\mathbf{F}$ is the Force, $\mathbf{P}$ is the Pressure, and $\mathbf{S}$ is the Surface:

$$\mathbf{F} = \mathbf{P} \times \mathbf{S}. \tag{3}$$

We state that this formula relates the two magnitudes as a linear relation, whereas, in the classical curricula, the formula used is that of Equation (1) in order to introduce pressure as a derivative magnitude via force, which is considered to be a primary concept.

4.2. Selected Features of the Structure and Teaching of the TLS

Using all the above-mentioned elements, we have developed the TLS, which consists of three units each lasting about two teaching hours. A major feature of the TLS is the active involvement of students in the experiments, which were intentionally selected to comprise tasks consisting of liquids or gases but not solids, which are commonly included in standard as well as in some constructivist curricula. Teaching is based on experimentation, group work, and teacher-led discussions. Qualitative and semi-quantitative experiments are conducted, which allow students to focus on different aspects of the conceptual model, providing them with concrete references in order to grasp the meaning, as well as discover the connection between their observations and their deductions. Quantitative experiments are used for students' conceptual growth and familiarization with the design of scientific investigations, while the effective constructivist teaching strategy Predict, Observe, Explain (POE) is applied in the activities [23]. The experimental activity begins by students predicting what they believe will happen in the two tasks described above. Should they follow the pressure–force model, it means that they identify pressure with force, and thus they would predict that the pressure and the force in the wider vessel is greater than that of the narrower one. A detailed discussion between the teacher and students, as well as among the students themselves, will reveal the inconsistency of their reasoning. The actual case is that the pressure in the two vessels is equal, while the force required to detach the two sucks is not, leading to the existence of two different sizes: pressure (P) and pressing force (F).

In the first unit of the TLS, the aim is to familiarize students with the phenomena being studied and for them to understand that liquids and gases are fluids. To this end, the teacher performs relevant experiments and takes measurements, while the students do some experiments in small groups. The findings are discussed firstly among the group members and then, with the teacher's coordination, before the whole class. The aim of the second unit is the following: students reinforce the concept of pressure and become familiar with the experimental methodology, focusing on the distinction and control of the variables in the Equation (2), (basic hydrostatic law) [24]. Lastly, in the third unit, students are expected to differentiate pressure from the pressing force by engaging in the process described in Section 4.1.1. In brief, the same procedure is followed as that described above, while the next step is to facilitate students to relate pressure with force and introduce the Equation (3). The content and activities of the TLS are summarized in Table 1.

Table 1. Objectives, structure content and activities of TLS, for teaching fluids and pressure, in junior high school (retrieved with permission from [12], www.tandfonline.com).

Unit/Time Duration	Objectives	Content	Activities
1st unit	Familiarization with phenomena Unification of fields	Experiments in the field of gases and liquids in terms of pressure Measurement of Phydr. and Patm Pressure difference causes the movement of the fluid Pressure has no direction	Discussion/performance in groups and classroom discussion Student experiments
2nd unit	Reinforcement of weak concept (pressure) Familiarization with the experimental methodology	Variables affecting hydrostatic and atmospheric pressure Distinction of variables Experimental verification of the empirical law for hydrostatic pressure	Discussion and performance of experiments before the whole class Demonstration of Experiments
3rd unit	Distinction between pressure and force Theory of pressing forces	Comparison between pressures/forces in a narrow/wide vessel and a small/big sucker Introduction of the relation between pressure and force	Predictions in groups and classroom discussion Demonstration and group experiments

5. Enhancing Students' Understanding of Fluids: Selected Results

The TLS was developed iteratively and tested on two target samples with some modifications. The first sample consisted of 58 students (13–14 year-olds) from the second year of Junior High School (a.k.a. Gymnasium), which in Greece constitutes one of the compulsory education tiers. The second sample comprises primary school student-teachers studying in the Department of Primary Education, at the Aristotle University of Thessaloniki, Greece. As in numerous other countries, we think that 13–14 year old students, as well as primary school student teachers in Greece have difficulty understanding science. The main presumption for this in terms of the student teachers is that they enter their university studies through studying the Theoretical Orientation in the last two years of senior High school (a.k.a. Lyceum), which comprises arts and humanities subjects, rather than the Applied Orientation, comprising math and science subjects. Furthermore, their tertiary studies to become "one-teacher-to-teach-all subjects" offer a limited number of science courses [25,26]. The data analysis of applying the TLS on the two sample groups gave satisfactory results [12,27]. In the present paper, due to space constraints, we present only the findings of the TLS intervention on second year junior high school students.

5.1. Qualitative Results

An effective method to facilitate the way pressure is perceived, which is a weak concept, is to have students experimentally validate pressure dependence on depth in a fluid, as well as introducing to them the fundamental law of hydrostatics. By participating in the taking of measurements, students are encouraged to apply their ideas, which is essential for their conceptual evolution. The data analysis of the recorded discussion of students and their teacher during instruction strongly suggests that such activities helped students to consolidate their initial ideas as regards the change of the magnitude of pressure in water with depth. Additionally, taking measurements of pressure enhanced students' notion of the existence of pressure. We believe that taking measurements greatly enhances students' assurance in the validity of the relationship between pressure and depth in different types of fluid. Obviously, the procedure requires further elaboration, because students realize that they need to have two measurements to be able to confirm a relationship; however, they do not seem to realize the need to have more than two in order to enable mathematical treatment. Thus, at this stage, the teacher coordinates the relevant discussion about the experimental measurements and subsequently shows the relationship between pressure and density via Equation (2).

Distinction between Pressure and Pressing Force

Considering questions 1 and 2 after teaching, most of the students answered them correctly (see Figure 1, [22]). As regards question 1, they responded that pressure is the same because the water heights are the same, as explained in the instruction. They also agreed that it is more difficult to detach the larger sucks than the smaller ones because of their difference in size (question 2). There were some students, however, who insisted on their initial conceptions in accordance with the "pressure–force model". Following their teacher's prompts in order to find a way out of such an impasse, the students proposed to measure the pressure in the two vessels. This shows that they deem experimental measuring as a suitable procedure as a way of checking their own ideas.

It is important to note that students who used the pressure–force model came up against an important discrepancy. By claiming that the pressures in question 1 are equal, and noting that the forces required for question 2 are unequal, in order for these students do be consistent, they would have to have responded that the pressures and forces in both questions are either equal or smaller regarding the smaller area. This discrepancy is not directly perceptible to the students, which is why it is crucial at this stage that the teacher makes it easy for students to understand the discrepancy in their responses. When this contradiction becomes obvious to the students, it leads to disputing their initial ideas. The discussion between students and teacher gradually and slowly leads them to dispel the confusion regarding the two conceptions.

5.2. Quantitative Results

The results of the posttest (just after the instruction) and the post-posttest (eight months later) for both the experimental and control groups are presented in Table 2. Table 2 also includes the percentage of correct answers, as well as the corresponding confidence intervals, to the questions that are required the prediction and interpretation of phenomena involving the concept of pressure. Below are the results for questions 3, 4, 5, 6, and 7 (see Figure 2). These questions are related to pressure comparisons in various situations and are part of a larger questionnaire used in our fluid research study.

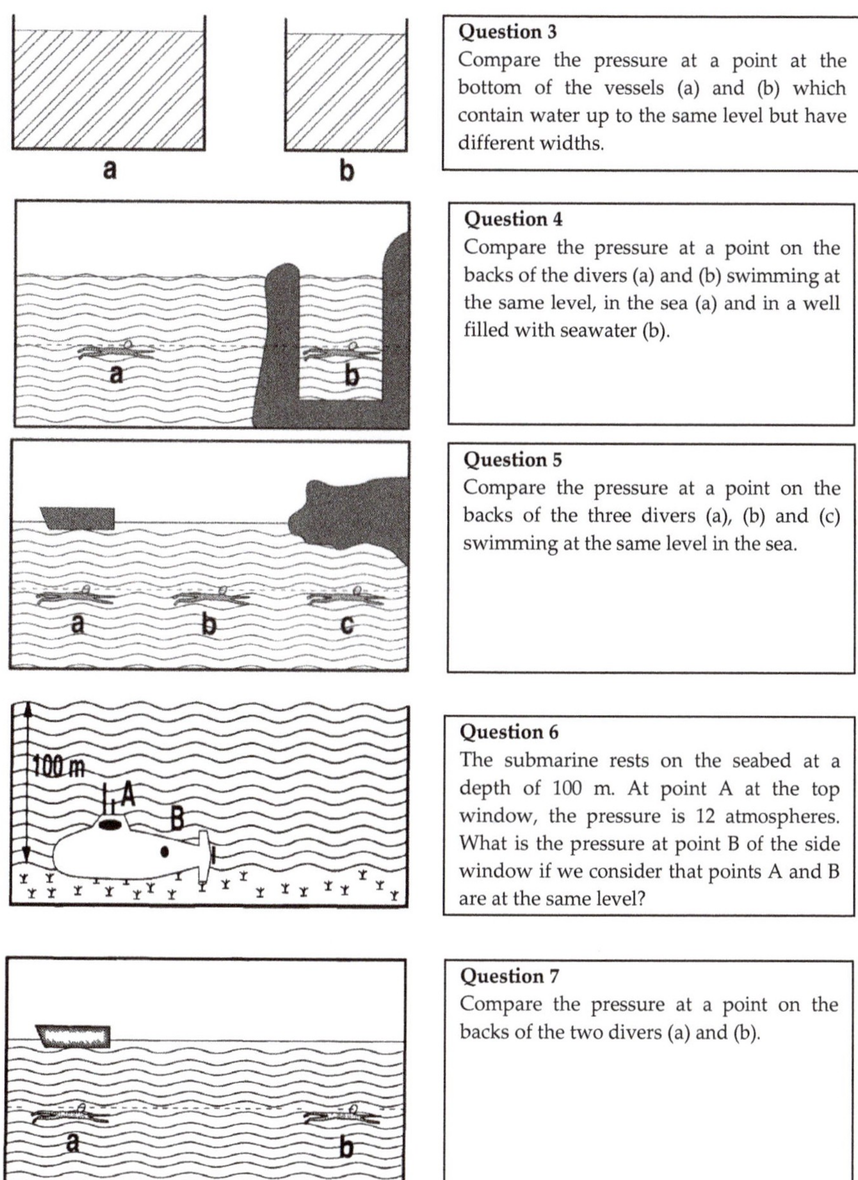

Figure 2. The five tasks of the quantitative questionnaire [22].

Table 2 shows the percentages of students, who answered the five questions correctly, as well as the confidence intervals corresponding to the significance level a = 0.05. The 58 students in the experimental group answered all five questions correctly after the teaching instruction, in a range of 60.5% to 83%, whereas the corresponding percentages for the control group were 20% to 54%. For all of the questions, the difference between the experimental and control groups is statistically significant at the level a = 0.05. The findings for the post-posttest show that the percentages of correct answers for the experimental group remained at a similar high level, with only a few regressions eight months after teaching, whereas, for the control group, these percentages decreased significantly regarding four of the five questions (with the exception of Q 4).

Table 2. Quantitative results [22].

	POST TEST		POST-POST TEST	
Question	**Experimental Group** ($n = 58$)	**Control Group** ($n = 214$)	**Experimental Group** ($n = 58$)	**Control Group** ($n = 214$)
3	78.0 ± 10.5	40.0 ± 6.8	65.5 ± 12.0	25.0 ± 5.8
4	80.5 ± 10.5	53.0 ± 6.7	93.0 ± 6.5	43.5 ± 6.9
5	60.5 ± 12.5	20.0 ± 5.4	62.0 ± 12.5	0
6	83.0 ± 9.5	54.0 ± 6.7	84.5 ± 9.5	40.0 ± 6.8
7	71.0 ± 12.5	38.0 ± 6.5	68.0 ± 12.5	15.0 ± 5.0

6. Conclusions

In this paper, we present and analyze a research and development strategy for the teaching and learning of fluids, with an emphasis on the concept of pressure. Our strategy, a result of our research, development, and teaching, was iteratively evolved over a period of twenty years and was applied iteratively and successfully on two target groups: second year junior high school students and primary school student-teachers. The combination of qualitative and quantitative results strongly indicates that our Teaching and Learning Sequence had a significant level of success. The students recognized that pressure is a non-additive magnitude in various contexts and situations, simple or complex. They also learned that pressure has no direction, which is characteristic of pressing force but not of pressure. Moreover, students were able to discriminate between the concepts of pressure and pressing force.

We consider that our strategy is successful on account of the following: firstly, the systematic study, analysis, and understanding of students' reasoning, secondly, the transformation of the content to be taught in order for students to overcome difficulties, and, thirdly, the adoption of a Teaching–Learning Sequence as a research and development framework, which leads to a comprehensive approach of teaching and learning scientific knowledge [18–21]. Moreover, we suggest that the presented Teaching Learning Sequence, as an example of effective good teaching practice, can help physics teachers to elaborate on their teaching and inspire innovative treatment of the topic of fluids and pressure in science curricula for junior high school physics. Finally, the proposed research strategy and educational materials are experimentally validated resources appropriate for teaching in methodology courses addressed to under- and post-graduate primary education teachers.

Author Contributions: The authors contributed equally to the preparation and writing of present paper.

Funding: This research received no external funding.

Conflicts of Interest: The authors declare no conflicts of interest.

References

1. Engel-Clough, E.; Driver, R. What do children understand about pressure in fluids? *Res. Sci. Technol. Educ.* **1985**, *3*, 133–144.
2. Chi, M.T.H.; Slotta, J.D.; De Leeuw, N.D. From things to processes: A theory of conceptual change for learning science concepts. *Learn. Instr.* **1994**, *4*, 27–43. [CrossRef]
3. Chi, M.T.H. Three types of conceptual change: Belief revision, Mental model transformation and Categorical shift. In *International Handbook of Research on Conceptual Change*; Vosniadou, S., Ed.; Routledge: New York, NY, USA, 2008; pp. 61–82.
4. Vosniadou, S. Conceptual Change and Education. *Hum. Dev.* **2007**, *50*, 47–54. [CrossRef]
5. Chevallard, Y. *La Transposition Didactique*; La Pensée Sauvage: Grenoble, France, 1985.
6. Chevallard, Y.; Bosch, M. *Encyclopedia of Mathematics Education*; Lerman, S., Ed.; Springer: Dordrecht, The Netherlands, 2014.

7. Kariotoglou, P.; Psillos, D.; Tselfes, V. Modelling the evolution of a teaching—Learning sequence: From discovery to constructivist approaches. In *Science Education in the Knowledge Based Society*; Psillos, D., Kariotoglou, P., Tselfes, V., Fassoulopoulos, G., Hatzikraniotis, E., Kallery, M., Eds.; Kluwer: Dordrecht, The Netherlands, 2003; pp. 259–268.

8. Mayer, M. Common sense knowledge versus scientific knowledge: The case of pressure, weight and gravity. In Proceedings of the Second International Seminar on Misconceptions and Educational Strategies in Science and Mathematics, Ithaca, NY, USA, 26–29 July 1987; Novak, J., Ed.; Cornell University: Ithaca, NY, USA, 1987; Volume 1, pp. 299–310.

9. Giese, P.A. Misconceptions about water pressure. In Proceedings of the Second International Seminar on Misconceptions and Educational Strategies in Science and Mathematics, Ithaca, NY, USA, 26–29 July 1987; Novak, J., Ed.; Cornell University: Ithaca, NY, USA, 1987; Volume 2, pp. 143–148.

10. Kariotogloy, P.; Koumaras, P.; Psillos, D. A constructivist approach for teaching fluid phenomena. *Phys. Educ.* **1993**, *28*, 164–169. [CrossRef]

11. Kariotoglou, P.; Psillos, D. Pupils' Pressure Models and their implications for instructions. *Res. Sci. Technol. Educ.* **1993**, *11*, 95–108. [CrossRef]

12. Psillos, D.; Kariotoglou, P. Teaching Fluids: Intended knowledge and students' actual conceptual evolution. *Int. J. Sci. Educ. Spec. Issue Concept. Dev.* **1999**, *21*, 17–38. [CrossRef]

13. Kariotogloy, P.; Psillos, D.; Vallassiades, O. Understanding pressure: Didactical transposition and pupils' conceptions. *Phys. Educ.* **1990**, *25*, 92–96. [CrossRef]

14. Duit, R. Science education research internationally: Conceptions, research methods, domains of research. *Eurasia J. Math. Sci. Technol. Educ.* **2007**, *3*, 3–15. [CrossRef]

15. Duit, R.; Gropengießer, H.; Kattmann, U.; Komorek, M.; Parchmann, I. The Model of Educational Reconstruction—A Framework for improving teaching and learning science. In *Science Education Research and Practice in Europe: Retrospective and Prospective*; Jorde, D., Dillon, J., Eds.; Sense Publishers: Rotterdam, The Netherlands, 2012; pp. 13–37.

16. Hewitt, P.G. *The Concepts of Physics*, 3rd ed.; PEK: Herakleion, Greece, 2004; (Greek Translation).

17. Haliday, D.; Resnick, R. *Physics*; Wiley: New York, NY, USA, 1978.

18. Psillos, D.; Kariotoglou, P. (Eds.) *Iterative Design of Teaching-Learning Sequences: Introducing the Science of Materials in European Schools*; Springer: Berlin/Heidelberg, Germany, 2016.

19. Meheut, M.; Psillos, D. Teaching-Learning Sequences: Aims and tools for science education research. *Int. J. Sci. Educ.* **2004**, *26*, 515–535. [CrossRef]

20. Guisasola, J.; Almudi, J.M.; Ceberio, M.; Zubimendi, J.L. Designing and evaluation research-based instructional sequences for introducing magnetic fields. *Int. J. Sci. Math. Educ.* **2009**, *7*, 699–722. [CrossRef]

21. Tiberghien, A.; Vince, J.; Gaidioz, P. Design-based Research: Case of a teaching sequence on mechanics. *Int. J. Sci. Educ.* **2009**, *31*, 2275–2314. [CrossRef]

22. Kariotoglou, P.; Koumaras, P.; Psillos, D. Differenciation conceptuelle: Un enseignement d' hydrostatique, fonde sur le development et la contradiction des conceptions des eleves. *Didaskalia* **1995**, *7*, 63–90. [CrossRef]

23. White, R.; Gunstone, R. Probing understanding. In *Chapter 3. Prediction—Observation—Explanation*; Falmer Press: London, UK, 1992; pp. 44–64.

24. Boudreaux, A.; Shaffer, P.; Heron, P.; McDermott, L. Student understanding of control of variables: Deciding whether or not a variable influences the behavior of a system. *Am. J. Phys.* **2008**, *76*, 163–170. [CrossRef]

25. Harlen, W.; Elstgeest, J. *UNESCO Sourcebook for Science in the Primary School: A Workshop Approach to Teacher Education*; UNESCO: Paris, France, 1992.

26. Vosniadou, S.; Pnevmatikos, D.; Makris, N. The role of executive function in the construction and employment of Scientific and Mathematical concepts that require conceptual change learning. *Neuroeducation J. Spec. Issue Exec. Funct. Acad. Learn.* **2018**, *5*, 62–72. [CrossRef]

27. Kariotoglou, P. Issues of Teaching and Learning of Fluids Mechanics in High School. Ph.D. Thesis, Department of Physics, AUTh, Thessaloniki, Greece, 1991. Unpublished.

 *fluids*

Article

Forty Years' Experience in Teaching Fluid Mechanics at Strasbourg University

Daniel G. F. Huilier

ICUBE laboratory-CNRS, Faculty of Physics & Engineering, Strasbourg University, 67084 Strasbourg, France; huilier@unistra.fr; Tel.: +33-637597063

Received: 8 November 2019; Accepted: 27 November 2019; Published: 29 November 2019

Abstract: A summary of the personal investment in teaching fluid mechanics over 40 years in a French university is presented. Learning and Teaching Science and Engineering has never been easy, and in recent years it has become a crucial challenge for curriculum developers and teaching staff to offer attractive courses and optimized assessments. One objective is to ensure that students acquire competitive skills in higher science education that enable them to compete in the employment market, as the mechanical field is a privileged sector in industry. During the last decade, classical learning and teaching methods have been coupled with hands-on practice for future schoolteachers in a specific course on subjects including fluid mechanics. The hands-on/minds-on/hearts-on approach has demonstrated its effectiveness in training primary school teachers, and fluids are certainly a nice source of motivation for pupils in science learning. In mechanical engineering, for undergraduate and graduate students, the development of teaching material and the learning and teaching experience covers up to 40 years, mostly on fluid dynamics and related topics. Two periods are identified, those prior to and after the Bologna Process. Most recently, teaching instruction has focused on the Fluid Mechanics Concept Inventory (FMCI). This inventory has been recently introduced in France, with some modifications, and remedial tools have been developed and are proposed to students to remove misconceptions and misunderstandings of key concepts in fluid mechanics. The FMCI has yet to be tested in French higher education institutions, as are the innovative teaching methods that are emerging in fluid mechanics.

Keywords: fluid mechanics; science teaching; hands-on/minds-on/hearts-on; inquiry-based problem; science learning; educational activities; FMCI; Strasbourg University

---

1. Introduction

The present paper summarizes a 40 year learning and teaching experience in a French university, with the occasional assistance of foreign professors on sabbatical and on-the-job learning that took place in the mid-1970s. Teaching fluid mechanics—connected to mathematics, computer science, and numerical methods—is summarized for undergraduate and graduate students in mechanical engineering. However, over the last 15 years, learning how fluids behave based on hands-on methods was also shown to be a challenge for future school professors [1]. Details on this specific teaching activity will be developed in Section 2. Critical aspects related to drastic changes in the curricula, often a consequence of the Bologna Process [2,3], are also highlighted in fluid mechanics engineering. The development of computer science, the internet, and innovative pedagogical methods in learning and teaching science (SoTL, [4]) will bring to light that novel methods are needed to persuade students to pursue the subject of fluids in their studies in higher science education, as well as to improve young people's interest in science studies and attract them to fluid mechanical engineering and related professions. A recent proposal, presently tested on a remediation program, to fill in the gaps and deficits in fluid mechanics will also be developed; this is inspired by the Fluid Mechanics Concept

Inquiry (FMCI) test initiated by Martin, Mitchell, and Newell [5,6]. The first part of this paper is devoted to the teacher training that started in 2007, a rather unusual investment for an academic scientist specialized in physics and fluid mechanics. The experience of teaching fluid mechanics in an engineering faculty, at undergraduate and graduate levels, is described in the second part, which is subdivided into Sections 3 and 4, teaching prior to and after the implementation of the Bologna Process. This part concerns the main activity of the author's career. Section 5 focuses on the tools being developed to improve teaching and learning fluid mechanics in regard to overcoming students' misconceptions with help of the FMCI, remediation tools, and the acquired experience in educational sciences. A short review of recent innovative teaching and learning methods applied to fluids completes the paper, with the goal of encouraging collaboration with young colleagues.

2. Teaching and Learning Fluids with Future Schoolteachers

Undergraduate students willing to become primary schoolteachers in France must complete a Master's in Education Science and succeed in a competitive exam. One option suited to students in biology, life science, chemistry, engineering or physics is to follow a course entitled "Experimenting and Understanding Physical Sciences", which was founded 15 years ago. It is based on the hands-on/minds-on method coupled with constructivism, science learning, inquiry-based problems, hypothetico-deductive (HD) reasoning, methods of observation, initial representation, verification by experimentation, and feedback. The group asserts that a change in school science teaching pedagogy from mainly deductive to inquiry-based methods will provide the means to increase interest in science. Indeed, the "learning by doing method", in which the teacher accompanies the pupil and leads him to discover science for him/herself, stimulates the child's observation skills, imagination, and reasoning capacity [7]. The course unit consists of four lectures, mainly via an introduction to Education Science and different scientific investigation approaches (hands-on, inquiry-based, problem-based, inductive, deductive, and HD processes). The course is further organized into five practical sessions in physics and chemistry (among one in Newtonian and fluid mechanics); The required education level of scientific knowledge being the third year of the Batchelor's Degree. The students learn to apply the HD and hands-on during these sessions, last they learn to create lesson cards for elementary classrooms at home. Preference is given to educational science as proposed by Marguerite Altet, Jean-Pierre Astolfi, Philippe Meirieu, Philippe Perrenoud, and Michel Serres.

Hands-on [1] was introduced in France by Georges Charpak (Nobel prize, 1992) after visiting schools in Chicago, where he met Leon Lederman (Nobel prize, 1988) and Karen Worth, the daughter of physicist Victor Weisskopf. Karen Worth is an important figure in the development of inquiry-based and hands-on learning in schools. They also worked with Jerry Pine at Caltech, who successfully developed a K–12 systemic reform of science learning and inquiry-based science teaching with the hands-on experimentation model. Jerry Pine pioneered this model by offering experimental kits to his students [8].

2.1. Hypothetico-Deductive (H-D) Method in Sciences

The hypothetico-deductive (H-D) method, broadly used in science research, is classically a seven-step process (according to Popper [9–11] and further contributions [12,13]):

- Identify a broad problem area;
- Define the problem statement;
- Develop hypotheses;
- Determine measures;
- Data collection;
- Data analysis.

In our case, we slightly modified the different steps of the scientific investigation method, associating theory with experimentation (Figure 1). For a given problem (experiment to be carried out, evocation/display of a phenomenon, of a technical object, scientific question) related to an available theoretical structure (model, concept, knowledge), we present the following items:

- Observation and questioning related to initial representations, problem identification;
- Theoretical framework, background research, preliminary information gathering;
- Elaboration of hypotheses (testable, falsifiable afterwards);
- Definition, design of test experiments;
- Predictions of results associated to the test experiences;
- Experimentation, data recording and collection, and analysis/interpretation;
- Validation or invalidation of hypotheses and predictions, as well as their assessment;
- Elaboration of a theory or law based on robust validations.

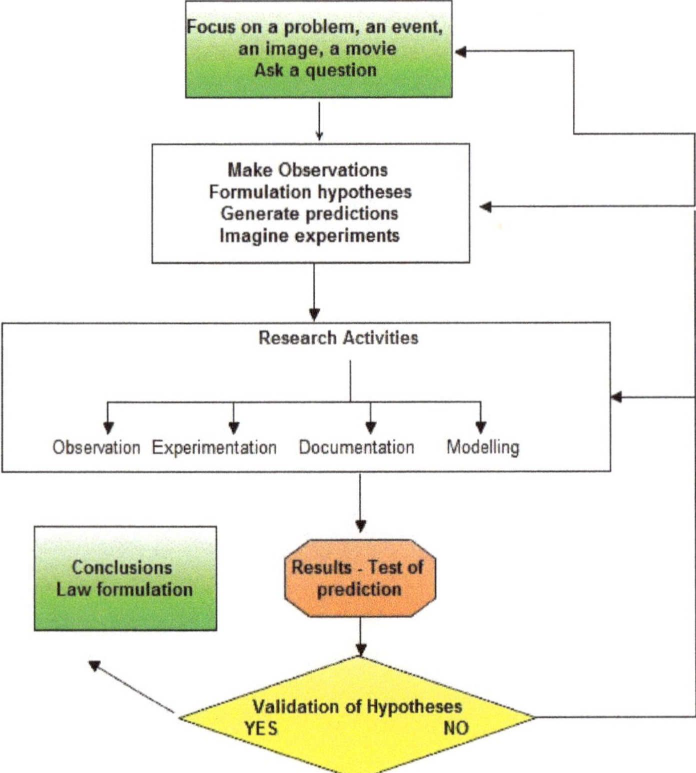

Figure 1. Schematic of the scientific method.

2.2. Practical Work and Projects in Fluid Mechanics

In fluid mechanics, the students, except for the last item on lesson cards, must provide a quality work at a Bachelor's degree, and they:

- ➢ Have two weeks to update their knowledge in fluids (Newton's laws, free fall bodies in a void, fall in air or water/oil, the Archimedes' principle, the Bernoulli law of perfect fluids) and investigate the concepts of viscosity and drag force (Reynolds number, Stokes' law, terminal fall velocity), writing all in a science notebook.

➤ Must imagine simple inexpensive experiments that identify and highlight relevant science notions and concepts, with some available material being listed and some experience items being proposed, such as:

- Galileo's experiments of falling objects, using sport balls in air from different heights up to 20 m (to observe differences in terminal velocities for different spheres (diameters, density));
- Making a classic Cartesian diver (understanding pressure, Archimedean buoyancy);
- Building a parachute with fabric and string/building a wind turbine;
- Observing the settling of sand grains in water;
- Observing bubbles rising in oil or water, or the fall of a golf ball in water (viscosity calculation);
- Galileo's experiment with rolling balls down an inclined plane (the calculation of gravity);
- The levitation of a ping pong ball (Coanda effect, Bernoulli equation);
- The measurement of the density of objects (cork, lead sinkers) based on the Archimedes' principle.

➤ Investigate and execute creative experiments while setting up the hypothetico-deductive teaching strategies.
➤ Elaborate several lesson preparation cards for pupils in primary schools on a topic related to fluid mechanics.

Aside from the available materials, students can use computer and web resources, textbooks and lecture notes, scientific articles, exercises, and educational videos, which are made available on a personal website [14], to prepare their practical work and experiments during their practical session. The course is intended to develop skills in fluid mechanics as well as in science education. The benefits of the course include:

- Improvement in organization of their working hours;
- Efficient use of available information and resources (in books, articles, the internet);
- Investment and teamwork (organizing small students' groups for interactive pedagogy learning);
- Scientific curiosity, creativity, motivation, and the pleasure of learning;
- Acquisition of reflective practice and scientific questioning;
- Implementing the HD method coupled with a project-based approach;
- Problem solving, formulating and testing hypotheses, theories, or laws;
- Analyzing and understanding scientific phenomena;
- Learning to measure and calculate with efficiency;
- Developing various abilities and skills, including the acquisition of transversal competences such as creativity and critical thinking.

2.2.1. Example of an Experiment: Free Fall of Bodies

The present example is dedicated to the laws governing falling bodies in a vacuum, gas, or liquid. Students have to imagine experiments such that highlight the different forces acting on a falling body (gravitational force, buoyancy, drag/air or liquid resistance). Several text extracts such as passages from the Principia, popular and scholarly articles, and some handouts on Newtonian mechanics help them to recall Newton's laws of motion. In a vacuum—as on the moon, in the absence of an atmosphere—all objects fall at the same rate, as astronaut Dave Scott demonstrated by dropping a hammer and a feather in the Apollo mission XV in 1971. This experiment, shown in NASA or Youtube videos [15], can help students to understand Newton's second law of motion in a vacuum.

Without any other force besides gravitation, and air resistance being neglected, in Earth's atmosphere, spheres dropped at zero speed, have a velocity V and impact time T, as given in Table 1. Although simplified, Table 1 yields an order of magnitude of T and highlights the difficulties in making observations with only a chronometer and a measuring tape for small values of T. Furthermore, students would be interested and excited to experience the simultaneous fall of different spheres from the third or fourth floor of a university building, just like the Galileo's hypothetical Leaning Tower of Pisa experiment.

Table 1. Time and velocity reached in free fall from different heights.

Height (in meters)	$T = (2\,h/g)^{1/2}$ (s)	V (m/s) = gT
1	0.45	4.4
2	0.64	6.3
3	0.78	7.6
5	1	9.8
10	1.43	14

A rain droplet has a fall velocity of about 10 m/s [16]. Students have to investigate the reasons behind this value and become familiar with the concept of air resistance inducing a drag force that increases with velocity, as well as the concept of terminal velocity of bodies falling in fluids. Supplementary papers or textbook chapters are made available to students on drag, sport ball ballistics, and the famous Reynolds number as well [17–20]. Many spheres of different sizes and material composition (golf balls, tennis balls, table tennis balls, baseballs, whiffle balls; plastic, rubber, glass, or metal balls; hollow balls) are available to students to experience the dynamics of falling bodies (Figure 2). As potential distractors, other objects are usually added to the panel (a badminton shuttlecock, a crumpled sheet of paper, hollow balls with many holes, etc.).

Figure 2. Set of sport balls and other spheres.

For most spheres and sport balls, released from less than 5 m in air, gravity is the only force acting on them. Some distractors do not really behave differently. However, by releasing spheres from at least 15 m outside a building, students begin to notice a difference. They nevertheless have some doubt about the nature of the other forces acting on the spheres, Archimedes' buoyancy or drag forces due. revealing misconceptions of surface forces, pressure, viscosity, and viscous effects. Complementary experiments in water, 'does it sink or float', the levitation of a ping pong ball with a hair dryer, are essential to deepen the students' understanding of the different observed phenomena. It is the teacher's role as mediator and facilitator to provide explanations and clarifications on the nature of these other forces. Once Newton's second law of motion is well written, calculations can be then made to determine atmospheric or hydrodynamic drag force, the coefficient of drag for the object, the instantaneous or terminal velocity of the object.

The bouncing of balls on different surfaces (ceramics, natural stone, concrete, parquet or laminate, sand, etc.) can also be analyzed, and the effect of various drop heights on the bounce height of a ball can be studied [21–23].

2.2.2. Example of an Experiment: Viscosity Calculation

Small air bubbles released from the bottom of a vertical tube of water or oil reach their terminal velocity in a quite short distance. By measuring the time elapsed from a given height, the terminal velocity can be easily calculated; supposing the viscosity of water or oil is unknown Newton's second law of motion states that the equilibrium of the external forces on a bubble (weight, buoyancy, and drag) enables the estimation of the viscosity. The only cautions are that bubbles remain spherical, which is a good approximation for diameters less than 0.5 mm. Furthermore, Stokes' law (at small Reynolds number (Re)) is justified. For larger bubbles, other drag coefficient laws such as that of Schiller–Naumann [24] can be applied, though trial-and-error or iterative calculations are needed to obtain the viscosity. By solving Newton's law, the terminal velocity U is given by

$$U = \sqrt{\frac{4}{3}g\frac{D}{C_D(\mathrm{Re})}\left(\frac{\rho_p}{\rho_f} - 1\right)} \tag{1}$$

where D is the sphere diameter, ρ_f and ρ_p are the fluid and sphere bulk density (volume weight), and C_D is the drag coefficient as function of the Reynolds number Re. The derived equation for U is nevertheless implicit since the drag coefficient C_D is itself a function of U.

Starting with the equation of the motion of a sphere under gravity, the buoyancy and drag of a fluid at rest can be determined as

$$\rho_p\pi\frac{D^3}{6}\frac{du_P}{dt} = \pi\frac{D^3}{6}\left(\rho_p - \rho_f\right)g - \pi\frac{D^2}{8}\rho_f C_D(\mathrm{Re})\left|u_P\right|u_P \tag{2}$$

At equilibrium, it is

$$\pi\frac{D^3}{6}\left(\rho_p - \rho_f\right)g = \pi\frac{D^2}{8}\rho_f C_D(\mathrm{Re})|U|U \tag{3}$$

In the case of Stokes' drag, $C_D = 24/\mathrm{Re}$, the drag force is reduced to $F = 3\pi\mu DU$, the terminal velocity is given by $U = \frac{gD^2}{18\mu}(\rho_p - \rho_f)$, and the dynamic viscosity is $\mu = \frac{gD^2}{18U}(\rho_p - \rho_f)$. The Stokes' relaxation time is $\tau_P = \frac{\rho_p D^2}{18\mu}$, after which a falling sphere reaches 63% of its maximal velocity. Other possible experiments could be the measurement of the viscosity of milk [25] or water—for this experiment, it is best to choose a small sphere with a bulk density close to that of the fluid. The sedimentation/settling of sand grains in water can be an attractive activity too, wherein the settling velocity measurements and water viscosity allows the calculation of the diameter of the sand grains, for instance.

2.2.3. Some Scientific Questions Treated in a Preparation Sheet

A preparation sheet is school lesson plan, created by the teacher, involving the learning objectives of the lesson, step-by-step experiments and activities for the pupils, and details of the materials to be used and possibly what should be written in child's notebook. It may involve a formative assessment. It must be in concordance with the curricula fixed by the French Ministry of National Education. When well written, it can allow a colleague to step in and lead the planned lesson in the event of absence or disease.

Preparing a sheet on a subject related to fluid mechanics for an elementary classroom is a difficult task. It is very time-consuming, requires imagination and skill, and requires inexpensive materials for experiments to be conducted by several groups of three to four children. These preparation sheets are intended for primary classrooms. The concepts and vocabulary must be popularized and adapted to the age of the children to be taught, which represents a real challenge for novice teachers. The teacher's role as facilitator is important for this task and time consuming for both parties, learner, and mediator. Here the learning cycle, play (heart-on learning), explore (hands-on Learning), and discuss (minds-on) is crucial.

Aside from their well-known textbook in physical hydrodynamics [26], Etienne Guyon et al. published a popularizing science book on fluids (unfortunately only available in French) entitled *What Fluids Tell Us* [27]. This textbook contains a vast selection of real-world, engaging examples of fluid mechanics, such as the falling of snowflakes and bubbles in a champagne glass, which are invaluable for novice teachers when writing their lesson plans. Another French textbook, now available in English, illustrates the wonder of fluids through "drops, bubbles, pearls, and waves" [28].

3. Teaching Fluid Mechanics from 1976 to 2002

From early period (1976) to the Bologna Process, teaching fluid mechanics to undergraduate and graduate students was quite well established. The library resources were reduced to some lecture notes and to a few fundamental books. For undergraduate students, most of the topics were covered in French books or in a few available English books (from Curle-Davies, Shapiro, Anderson, White, etc.). For graduates, classical books (such those of Batchelor, Hinze, Schlichting [29–32] or Lumley, Monin-Yaglom, Chandrasekhar, etc.) and the *Journal of Fluid Mechanics*, as well as the scientific papers of G.I. Taylor and microfilm issues, provided support for the teaching of fluid mechanics and turbulence. A large allocated volume of teaching hours and well-equipped laboratory experiments using pressure–velocity measurement techniques (Pitot, hot-wire, and even laser anemometry) favored extensive learning in fluids.

The advent of microcomputers fostered computational fluid dynamics, and the internet brought about the creation of effective websites. Besides a myriad of new textbooks, very serious scientific websites delivering much material on homework, courses, handouts, and exercises with corrections could be used for the content of a curriculum on fluid mechanics. Furthermore, enough teaching time was still allocated to enable in-depth training in many fields (statics of fluids, kinematics, dynamics, ideal fluids, viscous flows, internal and external flows, incompressible or compressible, turbulence, two-phase flows). The curriculum was coupled with numerical methods, applied mathematics, and computer science. The onset of turbulence, chaos, non-linear phenomena, and hydrodynamic instabilities appeared around 1990 and these courses, although a challenge for university professors, found a broad audience among students. It should be noted that undergraduate students had a solid background and skills in mathematics and physics, and master students were able to easily assess/handle partial differential equations (Navier-Stokes equations). Moreover, neither laminar/turbulent boundary layer theories (Blasius, Falkner-Skan, von Karman equation, Polhausen, or Twaites methods), plane or round jets, self-similarity approaches, nor turbulence models (k-ε, RANS, etc.) posed major problems to most of these students. Learning to apply finite difference or finite volume techniques as well as remeshing and studying stability problems of numerical schemes also found enthusiastic participation.

4. Teaching and Learning Fluids after the Bologna Process

The Bologna Process [2,3] is an intergovernmental higher education reform and cooperation process, launched in 1998 (Sorbonne declaration)–1999 (Bologna declaration). Its principal purpose is to promote the quality and recognition of the higher education systems in 48 European countries and to implement a common degree level system for undergraduates (Bachelor's degree) and graduates (Master's and Doctoral degrees), based on the so-called European Credits Transfer and Accumulation System (ECTS). In addition, it aims to facilitate student and teacher mobility, increase the recognition of academic degrees and qualifications, and increase employability, enhancing the attractiveness of in higher education in Europe. A number of European organizations (UNESCO, Council of Europe, ENQA, EUA, etc.) are also involved in this reform. The European Higher Education Area (EHEA) was launched in 2010 at the ministerial summit in Budapest, Vienna.

Several studies and recent reports based on TIMSS (Trends in Mathematics and Science Study) have shown that in France, the educational achievements of 15-year-olds and the share of high school students that go on to university are now below the EU average. They are lacking mathematics and science abilities. Since the Bologna process started twenty years ago, the French government and the Ministry of National Education modified several times all the curricula in primary and secondary schools, reducing the training hours allocated to science, engineering, and mathematics. Very recently the "Baccalauréat", the French secondary school diploma/ high-school degree (A-levels in UK) curricula again. Furthermore, the best high school pupils often opt for non-scientific disciplines and young people turn their backs on science. The progressive changes in European higher education led all French universities and some engineering schools to reconsider the programs and curricula on a national level and to revisit the contents and objectives of undergraduate and graduate programs, especially in STEM (Science, Technology, Engineering, and Mathematics). As a consequence, Bachelor of Science programs were affected and modified, and changes are still underway, at least in the French universities. These are some real causative factors for the observed changes in student ability in science and technology.

The progressive changes in European higher education led all French universities and engineering schools to reconsider the programs and curricula on a national level and to revisit the contents and objectives of undergraduate and graduate programs. Furthermore, even Bachelor of Science programs were affected and modified, and changes are still underway. Shortcomings in mathematics are known to be an essential hindrance to young people's interest and success in science and technology. In physics and engineering, at several universities, including Strasbourg, mechanical engineering really suffered from the curricula changes. Fluid mechanics felt to an optional course in undergraduate studies, with a reduced volume of teaching hours. Even main courses such as "Mathematical Methods for Physicists" had to be revisited. Almost every year, teachers like me had to scale down our ambitions in terms of the contents of our courses in linear algebra, analysis, computer science, and applied mathematics, as well as gradually adapt the learning and teaching level to meet students' lower competencies in scientific knowledge, especially in basic mathematics. Under these unforeseen circumstances, teaching methods, skill evaluation, and curricula content must be continuously adjusted.

4.1. Case of Undergraduate Students

The curricula contents for undergraduate students have been progressively reduced to statics of fluids, simple viscous flows, flows in pipes and open channels, and flows around immersed bodies. For instance, the solution of Stokes' creeping flow field around a sphere at a very small Reynolds number is no longer calculated mathematically. The first and second Stokes' problems (flow over a suddenly accelerated plate, flow above an oscillating infinite plate) and unsteady Couette flow between parallel plates are no longer addressed analytically, due to insufficient abilities in mathematics. Misconceptions about the ideal gas law, fluid statics, Bernoulli's equation, and Archimedes' principle are frequently observed in undergraduate students. Confusion between relative density or specific gravity (in hydraulics 1 for water) and volume weight or density (1000 kg/m^3 for water), as well as between kinematic and dynamic viscosity, is not uncommon. More worrying is that students are unable to write down the volume of a sphere $V = 4\pi R^3/3$ or the surface of a disk $S = \pi R^2$ without the support of an internet connection or a smartphone. Archimedes' buoyancy force is seldom understood, and students think that the buoyancy force increases with depth in water, as is the case with static pressure.

It is the author's opinion that no correlation directly exists between these observations and the Bologna Process, but that they coincide with the generation of 'digital natives' or the 'millennial generation' and indicate that these students approach and structure their studies and everyday life in a new way. Reducing the volume of teaching hours and setting the fluid mechanics course as optional did not made things any better in several French universities. The latter change is, however, linked with the reform process. Moreover, reduced financial support has affected the maintenance of experimental equipment and computer equipment.

4.2. Case of Graduate Students: A Revealing Illustration

The following short example is one among others that are intended to demonstrate what could be learned by master students only 10 years ago. Many master and PhD students have been working with the author in multiphase flows and particle-laden turbulent flows, on Lagrangian modeling or turbulent dispersion. In Section 2.2.2, the equation of motion of a sphere in a fluid at rest was simplified, including the gravitational, drag, and buoyancy forces. This concept is generally well understood by undergraduate students, even in education sciences:

$$\frac{\pi}{6}D_p^3\rho_p\frac{dV}{dt} = \frac{\pi D_p^3}{6}(\rho_p - \rho_f)g + \frac{1}{2}C_d\frac{\pi D_p^2}{4}\rho_f V^2 \tag{4}$$

However, in most cases, the equation is much more complicated and referred to as the Basset–Boussinesq–Oseen (BBO) equation. Thus, the trajectory $X(t)$ of a spherical particle of density ρ_p, viscosity v_f, and diameter D_p, located at X at time t, and moving with velocity vector $V(t)$ in a fluid of density ρ_f with the fictitious velocity at the center of the particle $U(t)$, is obtained by solving the set of equations

$$\frac{\pi}{6}D_p^3(\rho_p + C_a\rho_f)\frac{dV}{dt} = -\frac{1}{2}C_d\frac{\pi D_p^2}{4}\rho_f|V - U|(V - U) + \frac{\pi D_p^3}{6}\rho_f\frac{DU}{Dt}$$
$$+C_a\frac{\pi D_p^3}{6}\rho_f\frac{dU}{dt} + \frac{\pi D_p^3}{6}(\rho_p - \rho_f)g - C_h\frac{D_p^2}{4}\rho_f(\pi v_f)^{\frac{1}{2}}\int_0^t \frac{d}{d\tau}(V - U)\frac{d\tau}{\sqrt{t-\tau}} \tag{5}$$

$$\frac{dX(t)}{dt} = V(t) \text{ with } \frac{d}{dt} = (\frac{\partial}{\partial t} + V_j\frac{\partial}{\partial x_j}) \text{ and } \frac{D}{Dt} = (\frac{\partial}{\partial t} + U_j\frac{\partial}{\partial x_j})$$

where C_d is the drag coefficient, C_a the added mass coefficient, C_h the Basset history term coefficient, and the particle Reynolds number is based on the particle–fluid slip velocity $|V - U|$. Master's and even PhD students in mechanical engineering now have real difficulty in understanding and handling this type of equation, not to mention solving it numerically. Adding a Magnus force (and torque equation) is even more complicated for them.

Working with Generation Y, another term designating the digital natives, has some positive returns on learning and teaching in general. Teachers have to change their pedagogical methods, switching from self-centered teaching to a student-centered learning environment. They have to question their own teaching techniques and find strategies to stimulate the students' interests. Insofar as the author had the possibility to get involved in education sciences over the past 10 years, it made sense to him to join the teaching and learning committee of the French Mechanics Association in order to foster education in fluid mechanics.

5. Fluid Mechanics Remediation Test

The French Mechanics Association, regrouping many academic university members as well as industrial companies, has a committee (GTT-AUM) working on the subject of learning and teaching mechanics at university. One of their aims is to develop tests for students that enable them to summarize their knowledge in different fields (Newton's laws, solid statics, fluid mechanics, etc.). A further the idea is to propose simple tools of remediation to students and university instructors, consisting of course refreshers to go over basic concepts, exercises, videos, and simple experiments. In the past six years, the commission has investigated concept inventories (CI) that have been well implemented in North America. The lack of interest in science and technology, as well as in mathematics, among young people profoundly affects the mechanical engineering sector. Instructors are aware of the need to revitalize traditional education and rethink pedagogical strategies.

5.1. Conceptual vs. Procedural Knowledge, Misconceptions

Conceptual knowledge, as defined by Rittle-Johnson and Wagner-Alibali [33], is an "explicit and implicit understanding of the principles that govern a domain and of the interrelations between pieces of knowledge in a domain". Other complementary definitions exist [34–36]. In contrast, procedural knowledge is defined as "action sequences for solving problems" [37]. Conceptual knowledge plays a key role in solving engineering problems. Learning science is not a rote memorization of figures, facts, laws, equations, and experimental or computational procedures. Teachers have to develop learning strategies and apply innovative pedagogies to guide students to acquire knowledge, add knowledge of new concepts, fill gaps, and avoid misconceptions [38–40].

5.2. Concept Inventories (CI)

Concept inventories are designed to be efficient tools in physics and engineering education and have been developed to identify key student misconceptions. They "are research-based assessment instruments that probe students' understanding of particular physics concepts" [41]. They use multiple-choice questions (MCQs) on the topic of one or a set of concepts. They have multiple-choice answers including distractor answers (incorrect choices) that ideally match misconceptions [42]. Moreover, they are resources for teaching evolution [43]. They are used as pre-tests to assess students' prior knowledge and as post-tests at the end of courses to estimate the changes in students' conceptual understanding. They originated in physics education research (PER) and became popular as efficient tools in engineering education for capturing conceptual understanding. They are rather carefully designed, multiple-choice instruments that require students to select the correct answer for a particular problem among distractors based on known student misconceptions.

In the 1980s, Halloun and Hestenes [44,45] developed a "multiple-choice mechanics diagnostic test" to assess students' concepts about motion. They later developed a force concept inventory (CI) and a basic mechanical test (BMT) based on Newton's laws. The FCI, a test of conceptual understanding of the three Newton's laws in mechanics, was further improved [46] and is now widely used as a reference, a so-called 'gold standard' conceptual inventory in the physical sciences. It consists of 30 MCQs with five answer choices (including distractors) for each question and is effective in testing students' understanding of the mechanical concepts of force, motion, velocity, and acceleration of Newton's three laws. It has been administered to thousands of physics students worldwide and is a strong indicator of misconceptions. In addition to the famous force concept inventory (FCI), the Foundation Coalition (FC), funded by the National Science Foundation (NSF), has introduced a variety of assessments for specific discipline domains within engineering and created resources and development projects for partner campuses. The development of CIs is one objective [47]. Members of the Foundation Coalition have introduced a variety of assessments for domains within engineering, including electronic circuits, signals and systems, fluid mechanics, acoustics, waves, computer engineering, electromagnetics, dynamics, civil engineering, heat transfer, thermodynamics, and material science.

5.3. The American Fluid Mechanics Concept Inventory (FMCI)

Among the many tests developed for engineering and physics education in the United States [48–55], one is devoted to fluid mechanics [5,6]. Initiated by a cooperative effort between Jay Martin and John Mitchell at the University of Wisconsin-Madison and Ty Newell at the University of Illinois, Champaign-Urbana in 2001, it led to a final version in 2006 with the aim of promoting student understanding of fluid mechanics, as taught in mechanical engineering in the United States. The purpose of these inventories is to check whether fundamental concepts are understood by students, without any calculations. The results of the FMCI test are intended to modify and improve the way fluid mechanics courses are taught.

Like the FCI, it consists of 30 MCQs including graphics with five answer choices for each question, and it tests student understanding of disparate concepts in fluid mechanics, with no numerical calculation being necessary. Watson et al. [56] recently proposed to modify the FMCI test for civil engineers. They started by regrouping the main concepts for each question. Table 2 summarizes the different concepts. Many distracters in the test items embody commonsense beliefs about the nature of force and its effect on motion.

Table 2. Concepts tested on the existing American Fluid Mechanics Concept Inventory (FMCI) [56].

Targeted Concepts	
3 Continuity; compressible	18 Manometry; compressible
4 Bernoulli; incompressible	19 Drag force; compressible
5 Boundary conditions	20 Boundary layer; compressible
6 Momentum; incompressible	21 Boundary layer; incompressible
7 Pressure definition	22 Continuity; incompressible
8 Boundary layers; incompressible	23 Continuity/Bernoulli; incompressible
9 Pascal's Law	24 Boundary layer; compressible
10 Manometry; compressible	25 Impulse-momentum; incompressible
11 Bernoulli; incompressible	26 Boundary layer; compressible
12 Forces on submerged surface	27 Continuity/Bernoulli; incompressible
13 Ideal Gas Law	28 Drag force; compressible
14 Manometry; compressible	29 Drag force; compressible
15 Shear stress; compressible	30 Pressure measurement; compressible
16 Boundary layers	31 Continuity/Temperature variations; compressible
17 Bernoulli; incompressible	32 Fluid properties (viscosity)

To get an idea of the FMCI, a specific example of a multiple-choice question (question 1 of the FMCI) is given by Martin et al. [6] for a compressible flow in a pipe of constant section, with the downstream density being half of the upstream density. Answers concern the relationship between the up- and downstream velocities. Distractor answers state that the velocity would be smaller too, and both factor 2 and 4 are proposed. Wrong answers are typical for mass flux misinterpretations.

Another example developed in [57] refers to the flow of liquids in a contraction; a common misconception claims that the greater the velocity is, the greater the pressure will be. Several studies indicate that, for students, either liquids are compressible or pressure is a force and force is linked to velocity and not to acceleration, a typical misconception in the FCI. Here, Bernoulli's equation was clearly not understood.

Based on the last draft of the FMCI (version 3.4) from John Mitchell in 2016, a French version was written with small modifications and issued in January 2019. The previous concept table was modified (Table 3) and the questions were classified into four groups:

- Statics of fluids, seven questions (Q7, Q9, Q10, Q12, Q14, Q18, Q30);
- Ideal fluids and conservation laws (Bernouilli, Euler), 10 questions (Q3, Q4, Q6, Q13, Q17, Q22, Q23, Q25, Q27, Q31);
- External viscous flows, seven questions (Q5, Q19, Q20, Q24, Q26, Q28, Q29);
- Internal viscous flows, six questions (Q8, Q11, Q15, Q16, Q21, Q32).

The present classification was optimized according to the French curricula in in higher education institutions. Remediation tools (complementary lecture notes and numerous exercises with solutions) were recently implemented on a Moodle platform. The French and English FMCI will be tested in several French university classrooms this autumn and winter. Following this, the results will have to be processed.

Table 3. Concepts revisited for the French FMCI test (version VF1-2019).

Targeted Concepts	
	19 Laminar drag force over a flat-plate
3 Continuity; compressible	20 Boundary layer; control volume approach
4 Bernoulli; incompressible	21 Flow between two moving flate-plates, velocity
5 Boundary conditions for boundary layer	profile
6 Momentum; incompressible, Euler	22 Continuity; incompressible
7 Pressure definition, statics	23 Continuity/Bernoulli; vertical diffuser,
8 Viscous flow between flat-plates; incompressible	incompressible
9 Pascal's Law, statics	24 Boundary layer; velocity profile
10 Manometry; compressible, Bernoulli	25 Impulse-momentum; incompressible
11 generalized Bernoulli; pressure loss in pipes	26 Boundary layer and wall stress profile
12 Forces on submerged surface	27 Continuity/Bernoulli; vertical contraction,
13 Ideal Gas Law	incompressible
14 Manometry; statics	28 Drag force of different profiles in air
15 Shear stress	29 Drag force; viscous drag
16 Flow between two flat-plates, velocity profile	30 Pressure measurement; Prandtl/Pitot tube
17 Bernoulli; flow in a horizontal diffuser	31 Continuity/Temperature variations; compressible
18 Manometry; compressible, statics	flow through a diffuser
	32 Fluid properties (viscosity)

The FMCI will certainly be revisited after expertise and results processing, as well as the collection of instructor and student feedback. Several versions could emerge due to the numerous concepts in fluid mechanics; they will have to fit the level (undergraduate or graduate) and the scientific engineering field (mechanical, civil, chemical engineering, hydrodynamics, aerodynamics). Currently, no question concerns open-channel flows, supersonic flows, instabilities, or turbulence. Only one question refers to pipe flows.

6. Work in Progress in Learning and Teaching Fluid Mechanics

A personal objective of the author, in collaboration with a young colleague, is to split the FMCI and provide several new versions of the FMCI—an ambitious project. For undergraduates, the versions would concern statics and ideal flows, then viscous flows, last compressible flows, and aerodynamics. For graduates, the versions would address turbulence, multiphase flows (according to textbooks like that of Clayton Crowe from Washington State University), and last numerical methods in fluids. In our Faculty of Physics and Engineering, for the 2019/2020 autumn/winter semester, undergraduate students will be FMCI-tested twice, at the beginning and at the end of the fluid mechanics course. Furthermore, a specific multiple-choice question test of 15 questions on statics and ideal fluid flow will be created for formative assessment. At the same time, MCQs will be created for our Master's students concerning numerical methods in fluid flow (CFD—Computation Fluid Dynamics). Last, students will be required to anonymously complete a teaching evaluation form, specific to the course, in addition to the standard assessment proposed by the faculty. These measures will hopefully help our professional development.

As a potential tutor of the young colleague for the coming year, the author has the will to implement a student-centered teaching, also referred to as learner-centered teaching. The idea is to take inspiration from flipped classrooms, as well as active and cooperative learning strategies. To reinforce these beliefs, a summary of more general own opinions and presently explored topics related to the recent literature is given here. It concerns education sciences applied to engineering and fluid mechanics, and more generally to science, technology, engineering, and mathematics (STEM).

In college learning, the replacement of true experiments by virtual experiments and numerical simulations was certainly a wrong choice for many university engineering science programs. True experimental work based on the use of experimental equipment, and measurement techniques are an invaluable step to really understand the physics and mechanisms of causal processes and predict phenomena. Establishing cause and effect relations is a key principle in a scientific approach. In fluid mechanics, experimental investigations have always fostered theoretical studies, modelling and more recently complex numerical simulations. Relevant examples are experiments conducted in the space station science NASA programs in microgravity on microfluidics, Maragoni effects among others. Unfortunately, experimental equipment and maintenance are expensive, much more than a computer room and software dedicated to numerical simulations, and must also be equipped with computers for data processing.

New assessment tools, skill assessments, and concept inventories certainly have to be reviewed and optimized. Thus, the use of student interviews can help the development of so-called FASI-type teaching instruments (FASI—Formative Assessment of Instruction) and to measure student learning, expert-like thinking, and the effectiveness of instruction [58], rather than only employing a summative assessment of student learning. It is also necessary to understand what we are assessing, by using newly developed tools such as concept inventories, three-dimensional learning (3DL), and 3DLAPs (3D learning assessment protocols) [58,59].

Even the physics of ideal fluids has revealed misconceptions, and difficulties encountered by students in classical hydrodynamics (Bernoulli's equation) have been reported in detail [57]. This paper corroborates several misconceptions on pressure and velocity in ideal fluids and naive interpretations of Bernoulli's equation, and includes a valuable (though incomplete) bibliography.

In fluid mechanics, literature on teaching, formative assessment, remedial instruction, and tools are expanding much more than excepted. It has been reported that concepts from pressure and fluid statics [57,60–65], to Archimedes' principle [66], to Bernoulli's law often lead to misunderstandings [67–72]. To combat such misconceptions, various studies have developed approaches such as CFD instrumentation, e-learning, experimental hands-on implementation, interview analyses, student engagement, and interactive learning techniques, among others [73–80]. New instructional methods have been developed in the past 10 years based on active pedagogy [81], problem-based approaches [82], real-world problems, and flipping classes [83–86]. Societal changes and students' needs and demands request both a richer and more efficient learning environment, necessitating adjustments in teaching. In France, serious research on educational sciences, related to learning and teaching in Universities, started late after 2008, in the field of fluid mechanics and more generally in mechanics.

Educational research and physics education research (PER) must offer guidance on how students should think and learn, and on how we can develop true assessments and coherent curricula in physics and fluid mechanics in order to promote the emergence of innovative teaching techniques and to incite interest in science from audiences of all ages (from childhood to higher education). Several handbooks have been published recently on science learning and teaching research [87–90]; on evidence-based e-learning design [91]; on cognition and metacognition in science, technology, engineering, and mathematics (STEM) [89]; and on hands-on methods to improve education at the elementary, secondary, and collegiate levels [91]. There is no doubt that promoting science interest in primary and secondary education will give the necessary impetus to enhancing student contribution and participatory learning. Science laboratory exercises are fundamental. Over the past 20 years, collaborative, cooperative, peer-instruction/learning/teaching, problem-based, project-based learning and teaching have been promoted. These concepts can be seen as extensions of hands-on and inquiry-based activities, aligning well with other teaching methods such as ICT (Information and Communication Technology)-supporting approaches. Moreover, self-, peer-, and group-assessments are being explored to enhance the quality of teaching and learning in higher education. Fluid mechanics, a rather complex science that is necessarily connected to other engineering fields such as thermal science and chemical and computer engineering, is perfectly suited to reinvigorate and foster students' curiosity and knowledge if such methods are further developed.

7. Conclusions

Personal investment in teaching fluid mechanics over 40 years in France has brought satisfaction in different areas. In mechanical engineering, undergraduate and graduate activities have been performed, with enthusiasm, prior to and following the Bologna Process (a reform which sometimes altered the traditional way of teaching and learning) combined with changes in the everyday social life of students. At the same time, more recently, during a period of more than 10 years, a hands-on approach to education was implemented with success for undergraduates wishing to become schoolteachers. The opportunity of learning and teaching this specific method to a younger generation—and the desire to transfer experimental sciences in pupils' classrooms—greatly contributed to the knowledge acquisition of educational sciences and changing teaching methodologies. This also contributed to a desire to develop and implement recently developed tools, derived from the Fluid Mechanics Concept Inventory, which hopefully will help to remove students' misconceptions and misunderstanding of the key concepts in fluid mechanics in some French universities. For the coming years, the challenge will be to apply innovative teaching methods to fluid mechanics, based on a learner-centered teaching approach, in order to foster students' motivation to learn about fluid engineering at Strasbourg University. Furthermore, the proposed teaching evaluations will also hopefully foster the professional development of university teaching staff.

Funding: This research received no external funding.

Acknowledgments: The author expresses his thanks to John Mitchell at the University of Wisconsin-Madison for allowing the use of FMCI in French higher education institutions. The present paper is dedicated to Henri Burnage (1935–2018), from Strasbourg University and to Emeritus Barry Bernstein (1930–2014), from the Illinois Institute of Technology.

Conflicts of Interest: The author declare no conflict of interest.

References

1. Flick, L.B. The Meanings of Hands-On Science. *J. Sci. Teach. Educ.* **1993**, *4*, 1–8. [CrossRef]
2. Sorbonne Declaration. Available online: http://www.ehea.info/cid100203/sorbonne-declaration1998.html (accessed on 15 June 2019).
3. EHEA. The Framework of Qualifications for the European Higher Education Area (PDF). Archived from the Original (PDF) on 23 September 2015. May 2005. Available online: http://www.ehea.info/media.ehea.info/file/BFUG_Meeting/49/6/BFUG_BG_SR_61_4c_AppendixIII_947496.pdf (accessed on 18 March 2016).
4. McKinney, K. Attitudinal and structural factors contributing to challenges in the work of the scholarship of teaching and learning. *New Dir. Inst. Res.* **2006**, *129*, 37–50. [CrossRef]
5. Martin, J.; Mitchell, J.; Newell, T. Development of a Concept Inventory for Fluid Mechanics. In Proceedings of the 33th ASEE/IEE Frontiers in Education Conference, Boulder, CO, USA, 5–8 November 2003; p. 20.
6. Martin, J.K.; Mitchell, J.; Newell, T. Work in Progress: Analysis of Reliability of the Fluid Mechanics Concept Inventory. In Proceedings of the 34th ASEE/IEEE Frontiers in Education Conference, Savannah, GA, USA, 20–23 October 2004.
7. New Approach to Science Teaching Needed in Europe, Say Experts. Available online: http://europa.eu/rapid/press-release_IP-07-797_en.htm?locale=en (accessed on 15 June 2019).
8. Caltech Mourns Professor Emeritus Jerry Pine. Available online: https://www.caltech.edu/about/news/caltech-mourns-professor-emeritus-jerry-pine-80363 (accessed on 28 June 2019).
9. Popper, K. *Conjectures and Refutations: The Growth of Scientific Knowledge*, 5th ed.; Routledge: London, UK, 1989.
10. Popper, K. *Objective Knowledge: An Evolutionary Approached, Revised Edition*; Oxford University Press: Oxford, UK, 1979; p. 30.
11. Popper, K.R. *The Logic of Scientific Discovery*; Hutchinson & CO: London, UK, 1959.
12. Lawson, A.E. The generality of hypothetico-deductive reasoning: Making scientific thinking explicit. *Am. Biol. Teach.* **2000**, *62*, 482–495. [CrossRef]
13. Lewis, R.W. Biology: A hypothetico-deductive science. *Am. Biol. Teach.* **1988**, *50*, 362–366. [CrossRef]

14. Available online: http://daniel-huilier.fr (accessed on 28 November 2019).

15. The Apollo 15 Hammer-Feather Drop. Available online: https://nssdc.gsfc.nasa.gov/planetary/lunar/apollo_15_feather_drop.html (accessed on 3 July 2019).

16. Beard, K.V. Terminal Velocity and Shape of Cloud and Precipitation Drops. *J. Atmos. Sci.* **1976**, *33*, 851–864. [CrossRef]

17. Clanet, C.H. Sport ballistics. *Annu. Rev. Fluid Mech.* **2015**, *45*, 455–478. [CrossRef]

18. Moreau, R.; Sommeria, J. Traînée subie par les corps en mouvement. *Encyclopédie de L'environnement.* 2019. Available online: https://encyclopedie-environnement.org (accessed on 28 November 2019).

19. Munson, B.R.; Okiishi, T.H.; Huebsch, W.W.; Rothmayer, A.P. *Fundamentals of Fluid Mechanics*, 7th ed.; Wiley: Singapore, 2013; Chapter 9.

20. Rott, N. Note on the History of the Reynolds Number 1990. *Annu. Rev. Fluid Mech.* **1990**, *22*, 1–12. [CrossRef]

21. Striking Results on Bouncing Balls. Available online: https://staff.fnwi.uva.nl/a.j.p.heck/Research/art/BouncingBall.pdf (accessed on 25 June 2019).

22. Bouncing Ball Experiment. Available online: https://www.brooklyn.k12.oh.us/userfiles/66/Classes/13814/Lab%20-Bouncing%20Ball.pdf (accessed on 25 June 2019).

23. Hands-on Activity: Reverse Engineering. Available online: https://www.teachengineering.org/activities/view/ball_bounce_experiment (accessed on 25 June 2019).

24. Schiller, L.; Naumann, Z. VDI Zeitung 1935. *Drag Coeff. Correl.* **1935**, *77*, 318–320.

25. Milk Viscosity. Available online: https://www.teachengineering.org/activities/view/nyu_milk_activity1 (accessed on 28 June 2019).

26. Guyon, E.; Hulin, J.P.; Petit, L.; Mitescu, C. *Physical Hydrodynamics*, 2nd ed.; Oxford University Press: Oxford, UK, 2015.

27. Guyon, É.; Hulin, J.P.; Petit, L. *Ce Que Disent Les Fluides*, 2nd ed.; Editions Belin: Paris, France, 2011.

28. De Gennes, P.G.; Brochard-Wyart, F.; Quéré, D. *Gouttes, Bulles, Perles Et Ondes*; Editions Belin: Paris, France, 2002.

29. Batchelor, G. *An Introduction to Fluid Dynamics*; Cambridge Press University: Cambridge, UK, 2000.

30. Schlichting, H. *Boundary Layer Theory*, 4th ed.; McGraw-Hill: New York, NY, USA, 1979.

31. Hinze, J.O. *Turbulence*, 2nd ed.; McGraw-Hill: New York, NY, USA, 1975.

32. Batchelor, G.K. *The Scientific Papers of Sir Geoffrey Ingram Taylor*; Cambridge Press University: Cambridge, UK, 1960; Volume 2.

33. Rittle-Johnson, B.; Wagner-Alibali, M.W. Conceptual and procedural knowledge of mathematics: Does one lead to the other? *J. Educ. Psychol.* **1999**, *91*, 175–189. [CrossRef]

34. Barr, C.; Doyle, M.; Clifford, J.; de Leo, T.; Dubeau, C. *There Is More to Math: A Framework for Learning and Math Instruction*; Waterloo Catholic District School Board: Waterloo, ON, Canada, 2003.

35. Arslan, S. Traditional instruction of differential equations and conceptual learning. *Teach. Math. Appl. Int. J. IMA* **2010**, *29*, 94–107. [CrossRef]

36. Star, J.R. Re-conceptualizing procedural knowledge: The emergence of "intelligent" performances among equation solvers. In Proceedings of the 28th Annual Meeting of the North American Chapter of the International Group for the Psychology of Mathematics Education, Columbus, OH, USA, 26–29 October 2002; Mewborn, D., Sztajn, P., White, D., Wiegel, H., Bryant, R., Nooney, K., Eds.; ERIC Clearinghouse for Science, Mathematics, and Environmental Education: Columbus, OH, USA, 2002; pp. 999–1007.

37. Rittle-Johnson, B.; Siegler, R. The relations between conceptual and procedural knowledge in learning mathematics: A review, chapter 4. In *Studies in Developmental Psychology. The Development of Mathematical Skills*; Donlan, C., Ed.; Psychology Press: Hove, UK; Taylor & Francis: London, UK, 1998; pp. 75–110.

38. Rittle-Johnson, B. Promoting transfer: Effects of self-explanation and direct instruction. *Child Dev.* **2006**, *77*, 1–15. [CrossRef] [PubMed]

39. Piaget, J.; Cook, M.; Norton, W. *The Origins of Intelligence in Children*; International Universities Press: New York, NY, USA, 1952; Volume 8.

40. Chi, M.T.; Slotta, J.D.; de Leeuw, N. From things to processes: A theory of conceptual change for learning science concepts. *Learn. Instr.* **1994**, *4*, 27–43. [CrossRef]

41. Madsen, A.; McKagan, S.; Sayre, E. Best Practices for Administering Concept Inventories. *Phys. Teach.* **2017**, *55*, 530. [CrossRef]

42. Sadler, P.M. Psychometric models of student conceptions in science: Reconciling qualitative studies and distractor-driven assessment instruments. *J. Res. Sci. Teach.* **1998**, *35*, 265–296. [CrossRef]

43. Furrow, R.E.; Hsu, J.L. Concept inventories as a resource for teaching evolution. *Evol. Educ. Outreach* **2019**, *12*, 11. [CrossRef]

44. Halloun, I.A.; Hestenes, D. The initial knowledge state of college physics students. *Am. J. Phys.* **1985**, *53*, 1043–1055. [CrossRef]

45. Halloun, I.A.; Hestenes, D. Common sense concepts about motion. *Am. J. Phys.* **1985**, *53*, 1056–1065. [CrossRef]

46. Hestenes, M.; Wells, G. Swackhamer, Force Concept Inventory. *Phys. Teach.* **1992**, *30*, 141–158. [CrossRef]

47. Coalition Foundation. Available online: http://fc.civil.tamu.edu/ (accessed on 25 June 2019).

48. Lindell, R.S.; Peak, E.; Foster, T.M. Are They All Created Equal? A Comparison of Different Concept Inventory Development Methodologies. In *AIP Conference Proceedings 883*; McCullough, L., Heron, P., Hsu, L., Eds.; AIP Press: Melville, NY, USA, 2007; pp. 14–17. [CrossRef]

49. Adams, W.K.; Wieman, C.E. Development and Validation of Instruments to Measure Learning of Expert-Like Thinking. *Int. J. Sci. Educ.* **2011**, *33*, 1289–1312. [CrossRef]

50. Available online: www.physport.org/assessments (accessed on 7 July 2019).

51. Henderson, C. Common Concerns about the Force Concept Inventory. *Phys. Teach.* **2002**, *40*, 542. [CrossRef]

52. Laverty, J.T.; Caballero, M.D. Analysis of the most common concept inventories in physics: What are we assessing? *Phys. Rev. Phys. Educ. Res.* **2018**, *14*, 010123. [CrossRef]

53. Gray, G.L.; Evans, D.; Cornwell, P.; Costanzo, F.; Self, B. Toward a nationwide dynamics concept inventory assessment test. In Proceedings of the American Society for Engineering Education Annual Conference & Exposition, Nashville, TN, USA, 22–25 June 2003.

54. List of Concept Inventories. Available online: http://fc.civil.tamu.edu/home/keycomponents/concept/index.html (accessed on 28 June 2019).

55. David Sands, D.; Parker, M.; Hedgeland, H.; Jordan, S.; Galloway, R. Using concept inventories to measure understanding. *High. Educ. Pedagog.* **2018**, *3*, 173–182. [CrossRef]

56. Watson, K.; Mills, R.; Bower, K.; Brannan, K.; Woo, M.; Welch, R. Revising a Concept Inventory to Assess Conceptual Understanding in Civil Engineering Fluid Mechanics. In Proceedings of the 122nd ASEE Annual Conference & Exhibition, Seattle, WA, USA, 14–16 June 2015.

57. Suarez, A.; Kahan, S.; Zavala, G.; Marti, A.C. Student Conceptual Difficulties in Hydrodynamics. *Phys. Rev. Phys. Educ. Res.* **2017**, *13*, 020132. [CrossRef]

58. Laverty, J.T.; Underwood, S.M.; Matz, R.L.; Posey, L.A.; Carmel, J.H.; Caballero, M.D.; Fata-Hartley, C.L.; Ebert-May, D.; Jardeleza, S.E.; Cooper, M.M. Characterizing College Science Assessments: The Three-Dimensional Learning Assessment Protocol. *PLoS ONE* **2016**, *11*, e0162333. [CrossRef]

59. National Research Council. *A Framework for K-12 Science Education*; The National Academies Press: Washington, DC, USA, 2012. [CrossRef]

60. Besson, U. Students' conceptions of fluids. *Int. J. Sci. Educ.* **2004**, *26*, 1683–1714. [CrossRef]

61. Loverude, M.E.; Heron, P.R.L.; Kautz, C.H. Identifying and addressing student difficulties with hydrostatic pressure. *Am. J. Phys.* **2010**, *78*, 75–85. [CrossRef]

62. Matewsky, M.; Boyer, A.; Bazan, Z.; Wagner, D.J. Exploring student difficulties with pressure in a fluid. *AIP Conf. Proc.* **2013**, *1513*, 154–157.

63. Nihous, G.C. Notes on hydrostatic pressure. *J. Ocean Eng. Mar. Energy* **2016**, *2*, 105–109. [CrossRef]

64. Young, D.; Meredith, D.C. Using the resources framework to design, assess, and refine interventions on pressure in fluids. *Phys. Rev. Phys. Educ. Res.* **2017**, *13*, 010125. [CrossRef]

65. Koumaras, P.; Pierratos, T.H. How Much Does a Half-Kilogram of Water "Weigh"? *Phys. Teach.* **2015**, *53*, 174–176. [CrossRef]

66. Loverude, M.E.; Kautz, C.H.; Heron, P.R. Helping students develop an understanding of Archimedes' principle. I. Research on student understanding. *Am. J. Phys.* **2003**, *71*, 1178–1187. [CrossRef]

67. Recktenberg, G.W.; Edwards, R.C.; Howe, D.; Faulkner, J. Simple Experiment to Expose Misconceptions about the Bernoulli Equation. In Proceedings of the ASME 2009 International Mechanical Engineering Congress and Exposition, Lake Buena Vista, FL, USA, 13–19 November 2009; Volume 7, pp. 151–160. [CrossRef]
68. Bedford, D.; Lindsay, R. A misinterpretation of Bernoulli's theorem. *Phys. Educ.* **1977**, *12*, 311–312. [CrossRef]
69. Eastwell, P. Bernoulli? Perhaps but What about Viscosity? *Sci. Educ. Rev.* **2007**, *6*, 1–13.
70. Eastwell, P. Thinking Some More about Bernoulli. *Phys. Teach.* **2008**, *46*, 555–556. [CrossRef]
71. Kamela, M. Thinking about Bernoulli. *Phys. Teach.* **2007**, *45*, 379–381. [CrossRef]
72. Martin, D.H. Misunderstanding Bernoulli. *Phys. Teach.* **1983**, *21*, 37. [CrossRef]
73. Kusairi, S.; Alfad, H.; Zulaikah, S. Development of Web-Based intelligent (iTutor) to help students learn fluid statics. *J. Turkish Sci. Educ.* **2017**, *14*, 1–11. [CrossRef]
74. Fraser, D.M.; Pillay, R.; Tjatindi, L.; Case, J.M. Enhancing the Learning of Fluid Mechanics Using Computer Simulations. *J. Eng. Educ.* **2007**, *96*, 279–289. [CrossRef]
75. Stern, F.; Xing, T.; Yarbrough, D.B.; Rothmayer, A.; Rajagopalan, G.; Prakashotta, S.; Caughey, D.; Bhaskaran, R.; Smith SHutchings, B.; Moeyke, S. Hands-On CFD Educational Interface for Engineering Courses and Laboratories. *J. Eng. Educ.* **2006**, *95*, 63–83. [CrossRef]
76. Cranston, G.; Lock, G. Techniques to encourage interactive student learning in a laboratory setting. *Eng. Educ.* **2012**, *7*, 2–10. [CrossRef]
77. Lu, H.; Zhang, X.; Jiang, D.; Zhao, Z.; Wang, J.; Liu, J. Several Proposals to Improve the Teaching Effect of Fluid Mechanics. In *Emerging Computation and Information Chnologies for Education. Advances in Intelligent and Soft Computing*; Mao, E., Xu, L., Tian, W., Eds.; Springer: Berlin/Heidelberg, Germany, 2012; Volume 146. [CrossRef]
78. Burgher, J.; Finkel, D.; van Wie, B.J.; Adescope, O.O. Comparing misconceptions in fluid mechanics using interview analysis pre and post hands-on learning module treatment. In Proceedings of the 121st ASEE Annual Conference & Exposition, Indianapolis, IN, USA, 15–18 June 2014.
79. Brown, A. Enhancing learning of fluid mechanics using automated feedback and by engaging students as partners. In Proceedings of the Higher Education Academy Annual Conference, Birmingham, UK, 2–3 July 2014; p. 2.
80. Albers, L.; Bottomley, L. The Impact of Activity Based Learning, A New Instructional Method, In an Existing Mechanical Engineering Curriculum for Fluid Mechanics. In Proceedings of the 118th ASEE Annual Conference & Exposition, Vancouver, BC, Canada, 26–29 June 2011; Available online: https://www.asee.org/public/conferences/1/papers/781/view (accessed on 25 June 2019).
81. Bondehagen, D.L. Inspiring students to learn fluid mechanics through engagement with real world problems. In Proceedings of the 118th ASEE Annual Conference & Exposition, Vancouver, BC, Canada, 26–29 June 2011; Available online: https://peer.asee.org/18190 (accessed on 24 June 2019).
82. McClelland, C.J. Flipping a large-enrollment fluid mechanics course—Is it effective? In Proceedings of the 120th ASEE Annual Conference & Exposition, Atlanta, GA, USA, 23 June 2013; Available online: https://www.asee.org/public/conferences/20/papers/7911/view (accessed on 24 June 2019).
83. Webster, D.R.; Majerich, D.M.; Madden, A. Flippin'Fluid Mechanics—Comparison Using Two Groups. *Adv. Eng. Educ.* **2016**, *5*, 1–20.
84. Webster, D.R.; Majerich, D.M.; Luo, J. Flippin' Fluid Mechanics—Quasi-experimental Pre-test and Post-test Comparison Using Two Groups. In Proceedings of the DFD14 Meeting of The American Physical Society, San Francisco, CA, USA, 23–25 November 2014.
85. Wachs, F.L.; Fuqua, J.L.; Nissenson, P.M.; Shih, A.C.; Ramirez, M.P.; DaSilva, L.Q.; Nguyen, N.; Romero, C. Successfully Flipping a Fluid Mechanics Course Using Video Tutorials and Active Learning Strategies: Implementation and Assessment? In Proceedings of the 125th ASEE Annual Conference & Exhibition, Salt Lake City, UT, USA, 24–27 June 2018; Available online: https://www.asee.org/public/conferences/106/papers/23046/view (accessed on 8 July 2019).
86. Pietrocola, M. *Upgrading Physics Education to Meet the Needs of Society*; Springer: Berlin/Heidelberg, Germany, 2019.
87. Kalman, C.S. *Successful Science and Engineering Teaching: Theoretical and Learning Perspectives*; Springer International Publishing: Cham, Switzerland, 2018.
88. Dori, Y.J.; Mevarech, Z.R.; Baker, D.R. *Cognition, Metacognition and Culture in STEM Education: Leraning, Teaching and Assessment*; Springer international Publishing: Cham, Switzerland, 2018.

89. Pietrocola, M.; Gurgel, I. *Crossing the Border of the Traditional Science Curriculum: Innovative Teaching in Basic Science*; Sense Publishers: Rotterdam, The Netherlands, 2017.
90. Clark, R.C.; Mayer, R.E. *E-Learning and the Science of Instruction: Proven Guideline for Consumers and Designers of Multimedia Learning*; Wiley: Hoboken, NJ, USA, 2016.
91. De Silva, E. *Cases on Research-Based Teaching Methods in Science Education*; IGI Global: Hershey, PA, USA, 2014.

 *fluids*

Article

Using Legitimation Code Theory to Conceptualize Learning Opportunities in Fluid Mechanics

Robert W. M. Pott [1,*] **and Karin Wolff** [2]

1   Department of Process Engineering, Stellenbosch University, Stellenbosch 7600, South Africa
2   Faculty of Engineering, Stellenbosch University, Stellenbosch 7600, South Africa; wolffk@sun.ac.za
*   Correspondence: rpott@sun.ac.za; Tel.: +27-0218082064

Received: 29 September 2019; Accepted: 3 December 2019; Published: 6 December 2019

Abstract: With widespread industry feedback on engineering graduates' lack of technical skills and research demonstrating that higher education does not effectively facilitate the development of open-ended problem-solving competencies, many educators are attempting to implement measures that address these concerns. In order to properly formulate sensible interventions that result in meaningful improvements in student outcomes, useful educational measurement and analysis approaches are needed. Legitimation Code Theory (LCT) has rapidly emerged as an effective, theoretically informed 'toolkit' offering a suite of dimensions through which to observe, analyze, interpret, and design teaching and learning practices. LCT Semantics has been used to help engineering educators unpack both levels of engineering knowledge abstraction and the complexity of engineering terms, while LCT Specialization focuses on knowledge practices (using the epistemic plane) and enables a visualization and differentiation between kinds of phenomena and the fixed versus open-ended methods with which to approach a particular phenomenon. Drawing on a range of initiatives to enable an improved practical grasp of fluid mechanics concepts, this paper presents a description and graphic LCT analysis of student learning that has been designed to anchor the 'purist' principles underpinning applied fluid mechanics concepts (such as in piping and pump network design) by way of concerted 'doctrinal' practices, and the exposure to more open-ended practical situations involving peer learning/group work, allowing educators to visualize the code clash between the curriculum and the world of work.

Keywords: fluid mechanics; Legitimation Code Theory; undergraduate teaching

---

1. Introduction

Science and engineering education battles with a number of disjunctures in purpose, epistemology, and implementation. Many courses (for instance, fluid mechanics) have a significant focus on theoretical 'content'—teaching the material with an approach often called scientific realism [1]. Others present and examine material in a way that is more applied (either practically or in calculation exercises linked to examples), but which can descend into 'cookbook exercises' of dubious educational benefit [2]. The approaches used are for the most part manifestations of educators attempting to 'face both ways' [3]: toward both the theoretical knowledge base and increasingly complex application contexts, requiring working practical or pragmatic knowledge. Thus, one sees more theory and more practice being introduced into an already full curriculum and occasionally a shift in one direction or the other, which is sometimes to the detriment of the students' epistemic access [4] or value in the workplace. Courses can no longer focus only on tools (for instance, manipulating the mathematics of Navier–Stokes equations or calculating pump efficiency); they also need to focus on problems (system design, application to poverty alleviation or unemployment, development of new technologies or application of old to new systems, and so forth) in socioeconomic contexts, as indicated by the holistic International Engineering

Alliance competency profiles [5]. Then, the tools would be used only to the limit of their relevance for analyzing such problems, and not for their own sake: tools are only as useful as the problems they can solve.

As these debates illustrate, knowledge-building practices are neither homogeneous nor royal roads to cumulative knowledge building [6]: stronger epistemic relations (i.e., a strong focus on disciplinary science knowledge) do not by themselves guarantee intellectual progress. Cumulative knowledge building requires both a diversity of ideas and a shared means of navigating among these—in other words, multiple repertoires and an expanding collective reservoir [7]. To have these tools to hand, as a rephrasing of 'practical reasoning' [8], is to have access to the tool that solves the problem, while also having a deep and conscious understanding of the scientific and epistemic underpinnings: the 'know why' that is essential for 21st century complexity [9].

The educational literature, and in-practice educators themselves, have examined and trialed a number of methodologies and approaches that may address (part(s) of) this concern, including approaches such as problem-based learning e.g., [10], project-based learning e.g., [11], or flipped classrooms e.g., [12]. Many of these show real improvements in student knowledge and skill development. However, one shortcoming in many of these studies is the chronocentricity inherent in their set-up and evaluation. The efficacy of the program is commonly assessed within the same subject/module period. Progressivist pedagogical initiatives may appear successful within the timeframe of the actual course, but perhaps the true measure of transformative epistemic access is its continued usefulness to students and the manifestation of successful epistemic access and integration. Very often this is only evident after graduation, or in later academic years. Few academics have the benefit of a (un)planned longitudinal view, but where it is found, this view can be insightful. A second shortcoming is that active learning approaches—while intuitively appropriate in professional education designed to bridge the theory–practice divide—have not necessarily been adequately problematized from the perspective of forms of knowledge and associated practices. That is to say, our tools for analyzing how and why these interventions work is often limited to a 'show and tell'—analytical tools would enhance this discussion.

The discussion below firstly draws on collaborative work across the authors' department, its undergraduate program, academic staff access to initiatives, and their effects by way of observation, practice-sharing, and discussion, to give insight into gaps that persist in undergraduate skills and knowledge, even post 'educational initiatives', in a longer term view. Further, and most principally, the emerging field of Legitimation Code Theory (LCT) facilitates understanding of the nature and implications of these initiatives. LCT gives insight into what the connection between theory and practice is, as experienced by students and manifested throughout their undergraduate careers. LCT provides the tools for the investigation, analysis, and interpretation of 'knowledge practices' [13].

This paper aims to accomplish two objectives: it firstly draws on the experience and observation of interventions carried out in a second-year fluid mechanics module, as it plays out specifically in the final Capstone research project, to draw insight into fluid mechanics teaching more generally. Secondly, and potentially more usefully, the paper conceptualizes an extended approach to using LCT dimensions to visualize teaching strategies intended to facilitate improved learning outcomes, illustrated using examples from fluid mechanics, with the intent of giving educators an additional tool for analyzing their practices and curricula, and potentially bridging gaps between the curriculum and the world of work.

2. Theoretical Framework in Context

Legitimation Code Theory has emerged in the sociology of education as a powerful and illustrative 'toolkit' to analyze the organising principles of knowledge practices, their associated meaning-making systems, and the nature of knower dispositions [13]. For readers new to LCT, there are several excellent and accessible texts introducing these ideas, such as for instance [13–15]. The key to LCT is the idea of making visible *what* or *who* legitimates practices, and *how* they do so, so as to enable epistemic and social access to knowledge practices. Educators who understand the forms of and relationship between knowledge, knowing, and knowers are better equipped to address teaching, learning, and assessment challenges, particularly in increasingly diverse and complex 21st century educational contexts. In this paper, we suggest that LCT offers a practical set of tools through which engineering educators can interrogate forms of knowledge and knowing.

Over the past decade, as challenges in the recruitment, retention, and successful graduation of Science, Technology, Engineering and Mathematics (STEM) students became a global concern, education researchers began to use the LCT dimensions of Semantics and Specialization to interrogate existing practices. Hence, we find researchers looking at the gaps between school and the first-year biology curriculum [16], unpacking conceptual difficulties in physics teaching [17], interrogating difficult chemistry concepts [18], examining the link between theory and practice in fluid mechanics [19], and analyzing engineering problem solving [20]. Most of this initial research used the dimension of Semantics to differentiate between strengths of context-dependency (semantic gravity) and/or of complexity (semantic density). In the research site in question, the 'semantic wave'—common currency for theory–practice linking strategies that can lead to cumulative knowledge building [13]—has been used to conceptualize the *semantic range* between abstract, context-independent concepts and specific context-dependent examples. By explicitly unpacking the different levels in a particular disciplinary semantic range, a number of engineering educators have been able to structure their teaching across a particular course or sequence of courses in such a manner as to build a *semantic wave* intended to enable cumulative learning. Table 1 illustrates the use of Semantics in a number of engineering case studies:

Table 1. Modification of existing semantic range examples based on [21].

Semantic Range	Levels of Meaning	Civil Engineering	Process Engineering	Mechanical Engineering
Weaker *semantic gravity*	Principle	Structural forces determining bracing	Conservation of mass and energy	Principle of projection
	Formula and Calculations	C_r = ∅Af_y (1 + λ^2n)^(−1/n)	Mathematical expressions of process control	First and third angle projection
	Representation	Technical schematic drawings	Block diagram schematic of process control	Orthographic drawing showing different views of an object
	Model	3D/Simulations of structural behavior	Software simulation system	CAD model of the object (orthographic views derived from the model)
Stronger *semantic gravity*	Real	Physical structure (real building)	Physical process control systems	Physical object

Although the semantic range in the given examples enables a view of particular higher-order concepts and general formulae, as we move down the range, the potential representations, models, and physical contexts actually begin to proliferate. In order to design a richer cumulative learning experience that addresses the need to encourage more open-ended problem solving, engineering educators in the research context have added a second LCT dimension to their repertoire, one that enables the simultaneous view of multiple possible contexts. The LCT Specialization dimension offers insights into forms of knowledge (epistemic relations) and kinds of knowers (social relations). Researchers in the Faculty have used the latter, for example, to analyze how engineering educators 'build' kinds of 'knowers' [22]—in other words, how the field interprets and applies the development of engineering expertise. For the purpose of this paper, we will now focus on the epistemic plane, which enables a view of the relationship between phenomena and their approaches—the so-called *what* and *how* of knowledge practices.

The epistemic plane (shown in Figure 1) differentiates between strengths of ontic relations (the identity of a phenomenon) and discursive relations (procedures for approaching a phenomenon). Represented as a Cartesian plane, the epistemic plane reveals four quadrants with which to differentiate between practices requiring different *insights* or knowledge-practice codes:

- *Purist insight* is required where a phenomenon has strong ontic relations (universally accepted identity) and strong discursive relations (a standardized approach);
- *Doctrinal insight* describes practices where the approach or procedure takes precedence, such as the rules of the scientific method or differentiation in Calculus;
- *Situational insight* denotes practices dictated by phenomena with strong ontic relations, but where the discursive relations are weaker (there are more possibilities and open-ended approaches);
- *No/Knower insight* is evident when a practice is either 'anything goes' or not determined by knowledge, but rather *knowers* (or social relations) [13].

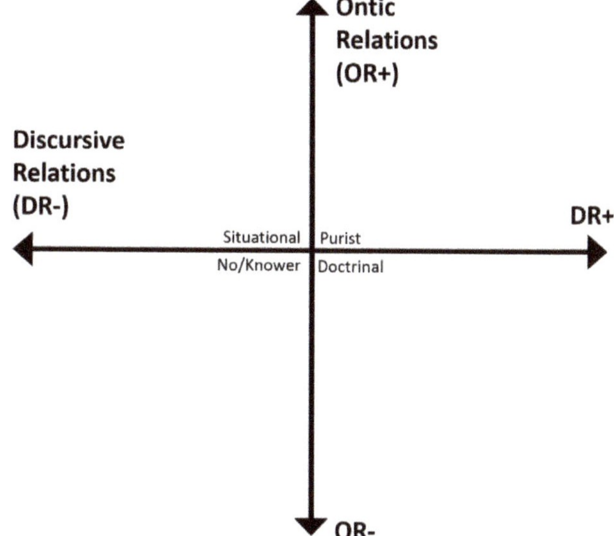

Figure 1. The epistemic plane, from [13], showing the four quadrants determined by relative strength of ontic and discursive relations. Reproduced with permission [13].

There is no right or wrong insight or quadrant. Each insight—as with problem-solving tools—is merely a means to an end. Recent industrial research using the epistemic plane has indicated that successful engineering practitioners navigate the plane in a nonlinear, iterative fashion drawing on different insights as the different contexts demand [23]. In other words, successful engineers practice 'code shifting' (the movement between the quadrants) [13], and the question is whether or not higher education is in fact facilitating adequate code-shifting practices during coursework, simulating what is needed in engineering practice.

3. Research Context

This discussion is situated within the experiences of a four-year chemical engineering degree program at a research-intensive traditional university in South Africa. The program is aligned to the International Engineering Alliance standards and accredited by the Engineering Council of South Africa, which is a signatory of the Washington accord. The fluid mechanics course is in the second year, second semester, and aims to instruct students in the fundamentals of fluid mechanics, for application in chemical engineering specifically. The course deals both with conceptual, abstract topics such as the Navier–Stokes equations, and how to manipulate and use them, as well as more practical calculations and topics, for example, pressure drop calculations, design and calculations around piping networks and pump sizing. While there are context and societally-specific aspects within this program, research conducted with these cohorts is likely to be broadly applicable to other global institutions and engineering programs.

The department in which this course is situated runs a number of educator development programs and initiatives, which are facilitated through the Faculty of Engineering. Several of the lecturers are active contributors to the educational literature, and a common sociological tool (in LCT) is shared as a language of analysis by these researchers. Further, as in most departments, a healthy vibrant discussion of educational practices, experiences, shortcomings, and effects is both apparent and encouraged. This continuing space for co-discussion has allowed for a more longitudinally oriented focus to educational initiatives—initiatives are put in place (for instance, in second-year fluid mechanics), but are not only evaluated in that course; these cohorts of students are followed until (and after) final year, and insights into their continued development can be garnered from these discussions.

Feedback from engineering-related industries suggests that graduates consistently have difficulties 'applying theory' [24] and adapting to technical workplace requirements (manpower.com, 2015). This supports the well-reported science–engineering disjuncture [25], which has led to such poor retention rates on engineering programs. Despite the holistic intention of the IEA engineering graduate competency profiles [5], the engineering curricula at research-intensive institutions are predominantly framed by the now 150-year-old science-based tradition [26]. In other words, the core curriculum is interpreted as requiring *purist insight*: well-known and established phenomena with standardized approaches. Large class sizes, a lack of adequate human and technical resources, and undergraduates not adequately prepared for the transition into tertiary levels of study have meant that traditional, assessment-driven pedagogies requiring doctrinal insight remain the status quo, with a great deal of work focusing on computational application.

Higher education practices limited to the right-hand side of the epistemic plane are in stark contrast to the insights that the industry demands of graduates, such as curiosity, agility, and collaboration [27]. These attributes are seldom explicitly taught or encouraged, particularly in the research context in question. The global endeavor to address Sustainable Development Goals requires engineers who are flexible and adaptable, and who are able to engage with stakeholders in multiple, complex sociotechnical contexts. Problem-solving or critical-thinking engineering professionals are those who are able to code shift: to draw on fundamental principles (*purist insight*) and apply standardized methodologies (*doctrinal insight*) in multiple contexts that require situational and knower insights.

In response to student feedback (including observations of student frustrations at being restricted to purist and doctrinal insights), a number of educators in the department have begun to include code-shifting strategies in their classrooms. A collaborative initiative to develop unit conversion and estimation competencies among first-year Chemical Engineering students has seen the establishment of an online question bank explicitly designed using the epistemic plane to encourage students to code-shift between what a unit represents (*purist insight*), how it is converted (*doctrinal insight*) and locally, contextualized estimation challenges (*situational* and *knower insights*) [28]. Another study with final-year students sought to introduce students to a range of contextual examples of core principles and procedures as applied in various mining and metallurgy industries [21]. The same authors went on to examine the third and fourth-year curriculum using the epistemic plane (Figure 2) and found the majority of courses foreground *purist* and *doctrinal insights* [29]. Here, we have amended their figure to include a plotting of fluid mechanics in comparison to other, later-year courses.

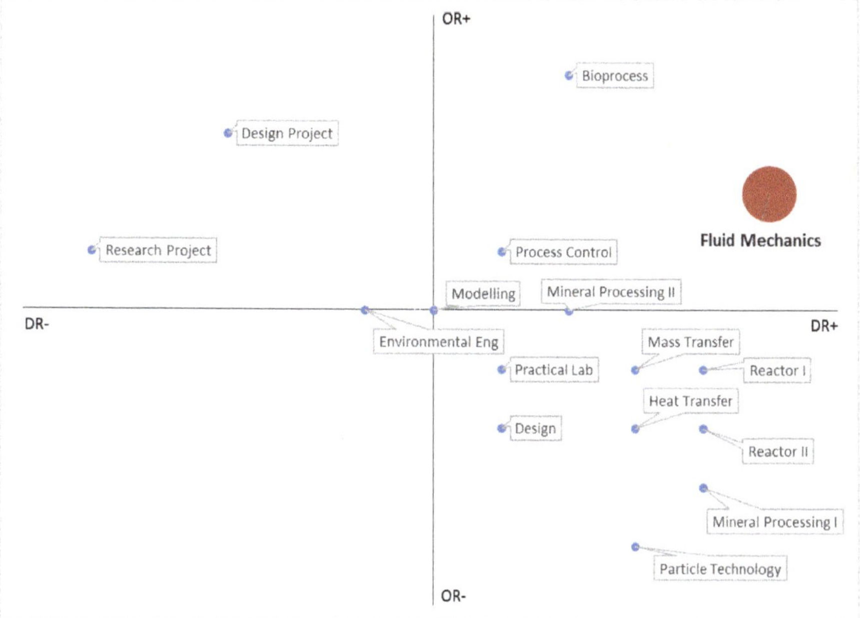

Figure 2. Modified from [29], plot of course content and approach on the epistemic plane for several third and fourth year chemical engineering courses, as well as second-year fluid mechanics. Reproduced with permission [29].

4. Analysis of Fluids Mechanics Using the Epistemic Plane

The major focus of this article is to introduce fluids educators, and engineering educators more generally, to the sociological tool of Legitimation Code Theory, through a demonstration of its usefulness in analysing course activities in fluid mechanics. Let us use an analogy here, for context: say you were to perform a catalysis experiment. In order to measure the extent of reaction, you would take samples and have these analyzed, perhaps on a gas chromatograph or using some other analysis method. What we propose here is that LCT is that analysis method—the tool by which we measure whether our experimentis working.

To achieve a demonstration of LCT's use as an analysis method, the fluid mechanics course will here be broken down into activities—lectures, tutorials, assessments, practicals, assignments, and discussion—and each of these will be analyzed using the lens of LCT, specifically the epistemic plane. A comparison against the transversal of the epistemic plane in industrial application will be presented as a foil to illustrate the gap between the course and its intended site of application.

4.1. Lectures

For the most part, lectures are dialectic discussions of high-level concepts that are occasionally brought down to the level of example or calculation. As such, they sit firmly in the *purist insight quadrant* (Figure 3)—the things being discussed are well defined and the methodology employed to solve the same are well defined and constrained. There may be occasional threads pulled from the situational insight quadrant (for instance, the lecturer may ask, "How could we decide on the pump requirements for a particular reactor configuration?"; the ontic relations are stronger, as in 'well defined' (the system), but the approach may be many-fold). There is generally little discussion that could be considered in the *Knower insight quadrant*—the type of person performing the calculations is not foregrounded.

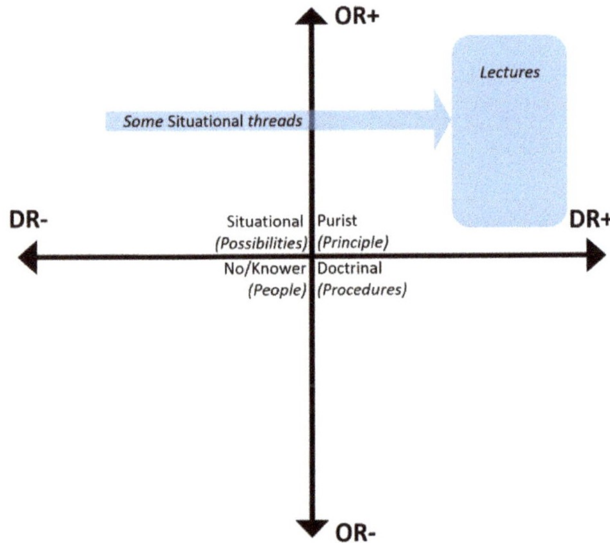

Figure 3. Epistemic plane detailing ontic and discursive movement within fluid mechanics lectures.

Educators will recognize that attending lectures is not sufficient for excellent learning—perhaps stemming from the relatively stationary situation within the *purist insight quadrant* (although there is clearly more to it than presentation without context). Dale [30] espoused the use of multiple media and then solid application to lead to a solid knowledge base, and perhaps a similar argument could be made here. For this reason, it is common to follow lectures with a weekly tutorial.

4.2. Tutorials

Tutorials are a space in which students are given a number of problems and questions to work through, while the lecturer and student assistants (usually post-graduate students) roam the room to answer questions. From the students' perspective, this for the most part sits within the *doctrinal insight quadrant*—there is much "recognize and reproduce", with students seeking cues from previously done examples. However, with appropriate question setting, some motion toward the *purist insight quadrant* can be achieved. This occurs when the students do not simply recognize the 'shape' of the question and impose an algorithmic approach to problem solving, but rather recognize the underlying principles and their relevance to each question.

A recurring issue during undergraduate engineering education is students' adopting a superficial learning route to problem solving through 'pattern recognition' [31,32]. Students can hide behind recognizing the pattern of the algorithm, rather than grappling with (and therefore undergoing 'cumulative learning' [6]) fundamentals. Superficial learning methods can give rise to 'correct answers' but shallow understanding. Within the epistemic plane (Figure 4), this represents a narrow transversal between the *doctrinal* and *purist insight quadrants*. Then, what is perhaps needed is some motion toward the situational insights—where students are required to pull on the range of tools they have learned, select one, and apply it to a defined problem.

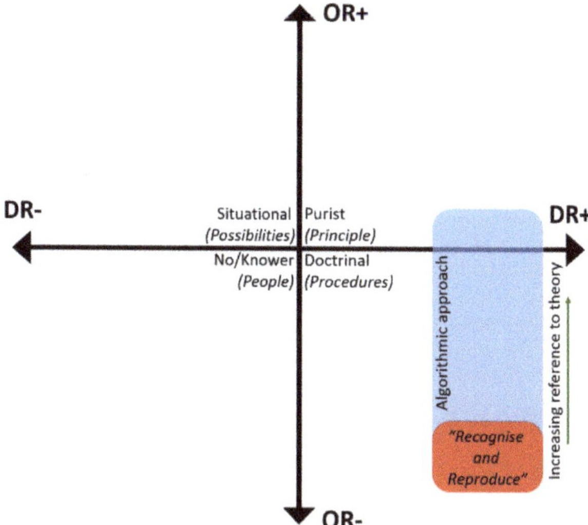

Figure 4. Epistemic plane detailing ontic and discursive movement within fluid mechanics tutorials.

4.3. Practicals

One method to pull students both toward a more situational insight as well as strengthen the semantic gravity of concepts (see [19]) is through the use of practicals. In an ideal situation, students would be presented with an open-ended task or experiment (*situational insight*), in which they would then need to employ the theory (*purist insight*) and calculation ability (*doctrinal insight*) to solve. Thus, they would be traversing a significant portion of the epistemic plane (depicted in Figure 5) and potentially improving their knowledge through the exercise. Further, if there is group work involved, then there may be some development of 'soft skills', which require *knower insight*.

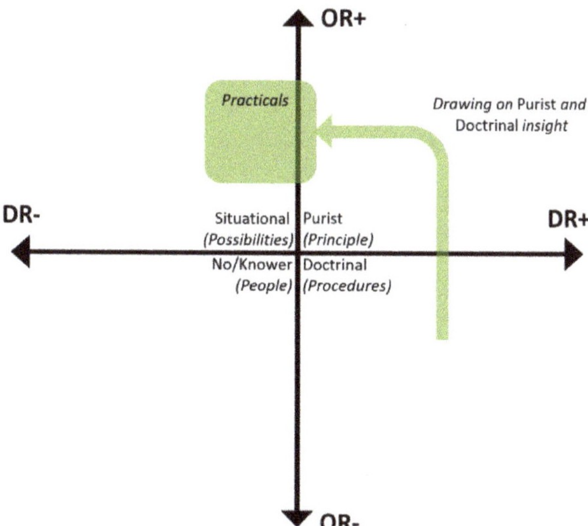

Figure 5. Epistemic plane detailing ontic and discursive movement within fluid mechanics practicals under good conditions.

However, as is discussed in [19,33,34], practicals can often become much less active than envisioned in the above figure. Students frequently do little preparation for practicals (i.e., they do not perform the tour through the *purist insight* quadrant). They will frequently arrive and follow the practical experimental procedure with little or no understanding or insight. This is possible in many practicals due to large class numbers using limited equipment under significant time constraints. In this case, the experience of the students is either a *doctrinal insight* one (where calculations are performed almost by rote), or even devolving into *no insight*, where little learning is taking place, as illustrated in Figure 6.

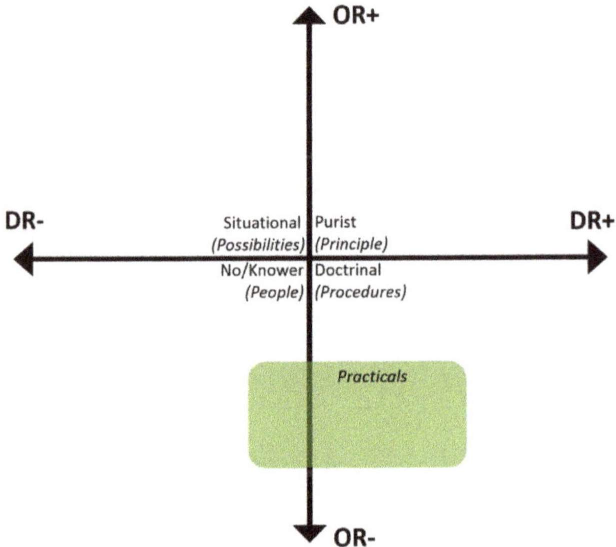

Figure 6. Epistemic plane detailing ontic and discursive movement within fluid mechanics practicals under poor conditions.

An understanding of this shift from a useful exercise to a wasted effort could allow educators to put in place mechanisms stimulating the sorts of desirable activities that lead to real cumulative learning while discouraging the wasted effort of poor learning. For instance, putting in place a pre-practical test to determine whether a student has engaged with the theoretical content that would allow them to understand the practical they are about to do may prevent a weakly epistemically underpinned afternoon in the laboratory.

4.4. Assignment

An assignment is another area where an open-ended, significantly situational problem can be posed, allowing the students to spend time exploring the various possible routes to solutions (a *situational insight*), and once a methodology is arrived at, solving the problem using *purist* and *doctrinal insights*. An example of such an assignment could be to design a piping network and select an appropriate pump for a molasses plant (see the supplementary information for the full assignment). In this case, the problem is an open one, although it is partially bounded by the requirement for piping and pump selection. There are a number of approaches that students could take to solve this (fairly well-defined) problem (*situational insight*). In order to solve it well, they will need to pull on explicitly covered material (calculate frictional pressure losses, use pump characteristic curves to choose an appropriate pump), thus pulling them through *purist* and *doctrinal insight quadrants*, as they conceive of the correct approach and perform the appropriate calculations through a route such as that illustrated in Figure 7.

However, there are other concepts that they need to consider in the assignment that are not explicitly part of the fluids curriculum; these include questions such as, "Will the fluid corrode the piping material? Is there a seasonal variation in flow rates? Is the effect of temperature significant, and if it is, how do we mitigate against this? Is the pump I have chosen very expensive or not available?" These questions pull the student deeper into the *situational insight quadrant,* and as shall be discussed below, deeper into the realm of where real-life engineering problem solving often lies.

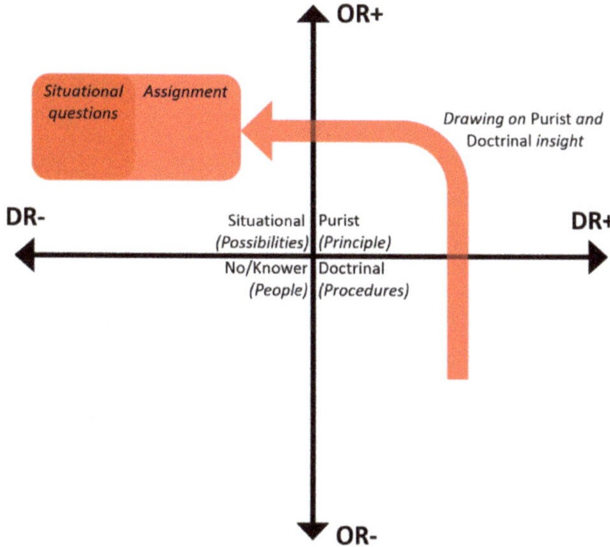

Figure 7. Epistemic plane detailing ontic and discursive movement within a fluid mechanics assignment.

At this point, the reader might be tempted to say that since a full tour of the epistemic plane has taken place through the sum of the course, then we might assume that cumulative learning has taken place. Of course, the answer is more complex than this, since students can successfully complete all required tasks and pass the course with comparatively little long-term learning having taken place (as evidenced in later years where students are unable to make reasonable judgments about pump and piping networks, for instance, in their final-year design module). However, it appears that through a more intentional pursuit of epistemic touring through course structure, student knowledge retention, interest, and focus is improved [19,28,29].

4.5. Application in the World of Work

Having examined some of the significant contact portions of a fluid mechanics course, how does this compare with the real-world application of fluid mechanics in, for instance, a design house? A significant portion of what employers consider important requires *knower insight*: teamwork, communication, ethics, and other 'soft skills'. The problems presented to working engineers are firmly within the situational insight quadrant; questions such as, "We have wastewater that needs treating; what should we do?" Clearly, there are multiple routes to a solution here, even though the problem in question is a specific one. In order to solve it, design engineers pull on detailed knowledge of specific methodologies, and test their applicability and suitability (*purist insight*). These separate portions of the job are bridged in technical communication, and product design specifically focused on the customers' requirements.

It is clear from Figure 8 that a significant portion of what is needed in industry is focused on the left-hand side of this plane, while teaching and learning in lectures and tutorials (the primary modes of learning) are almost entirely on the right. This code clash [13] may be one of the reasons for industry's (and recent graduates') complaints that graduates are not sufficiently prepared for the world of work [35,36]. However, it cannot be the case that educators address all of these concepts in every course. It is the role of the overall curriculum to ensure that sufficient opportunities are created across a program to allow for scaffolded code-shifting. Visualizing the existing dominant codes and clashes or mismatches is a necessary first step. It provides an opportunity for designing progressive pedagogies and interventions that bridge these gaps and improve the potential for substantive cumulative learning.

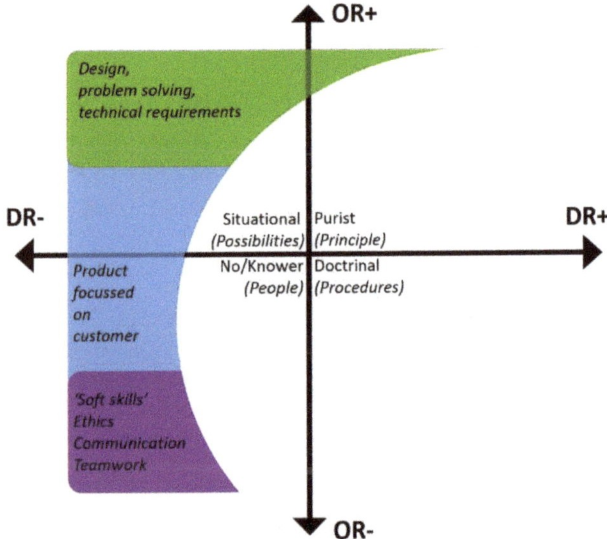

Figure 8. Epistemic plane detailing ontic and discursive movement within the world of work.

5. Conclusions

This paper has set out to illustrate the application of a theoretically informed approach to analyzing an engineering curriculum and pedagogic practices. Focusing on the student learning experience in fluid mechanics at a traditional research-intensive university, the Legitimation Code Theory dimension of Specialization (Epistemic Relations) has been used to demonstrate possible code-shifting strategies that can facilitate the cumulative knowledge building necessary for engineering graduates in complex 21st-century contexts. The analysis was driven by experiences in teaching fluid mechanics and follow-up discussions with colleagues and students regarding knowledge and skill retention beyond the course in question through the degree and post-graduation. The benefit of LCT beyond the course context is its illustrative power in allowing the conceptualization and graphing of educationally lived experience versus intended or needed experience. This may allow educators to consider a priori the potential effects and impact of proposed interventions in progressive pedagogy.

Author Contributions: The authors conceived of and wrote this article together. Conceptualization, R.W.M.P. and K.W.; Investigation, R.W.M.P. and K.W.; Methodology, K.W.; Writing—original draft, R.W.M.P. and K.W.; Writing—review and editing, R.W.M.P.

Funding: This research was funded by the Stellenbosch University Fund for Innovation and Research into Learning and Teaching.

Acknowledgments: The authors would like to thank Goosen for his significant contribution to fluids education and conceptualization at Stellenbosch.

Conflicts of Interest: The authors declare no conflicts of interest.

References

1. Vamvakeros, X.; Pavlatou, E.A.; Spyrellis, N. Survey Exploring Views of Scientists on Current Trends in Chemistry Education. *Sci. Educ.* **2010**, *19*, 119–145. [CrossRef]
2. Hodson, D. Assessment of practical work. *Sci. Educ.* **1992**, *1*, 115–144. [CrossRef]
3. Barnett, M. *Knowledge, Curriculum and Qualifications for South African Further Education*; Young, M., Gamble, J., Eds.; HSRC Press: Cape Town, South Africa, 2006.
4. Morrow, W.E. *Bounds of Democracy: Epistemological Access in Higher Education*; HSRC Press: Cape Town, South Africa, 2009.
5. Alliance, I.E. Graduate Attributes and Professional Competencies. Available online: https://www.ieagreements.org/assets/Uploads/Documents/Policy/Graduate-Attributes-and-Professional-Competencies.pdf (accessed on 28 September 2019).
6. Maton, K. Cumulative and segmented learning: Exploring the role of curriculum structures in knowledge-building. *Br. J. Sociol. Educ.* **2009**, *30*, 43–57. [CrossRef]
7. Bernstein, B. *Pedagogy, Symbolic Control and Identity: Theory, Research, Critique*; Taylor & Francis: London, UK, 2000.
8. Brickhouse, N.W.; Stanley, W.B.; Whitson, J.A. Practical reasoning and science education: Implications for theory and practice. *Sci. Educ.* **1993**, *2*, 363–375. [CrossRef]
9. Muller, J. Forms of knowledge and curriculum coherence. *J. Educ. Work* **2009**, *22*, 205–226. [CrossRef]
10. Perrenet, J.C.; Bouhuijs, P.A.J.; Smits, J.G.M.M. The Suitability of Problem-based Learning for Engineering Education: Theory and practice. *Teach. High. Educ.* **2000**, *5*, 345–358. [CrossRef]
11. Kokotsaki, D.; Menzies, V.; Wiggins, A. Project-based learning: A review of the literature. *Improv. Sch.* **2016**, *19*, 267–277. [CrossRef]
12. Karabulut-Ilgu, A.; Cherrez, N.J.; Jahren, C.T. A systematic review of research on the flipped learning method in engineering education. *Br. J. Educ. Technol.* **2018**, *49*, 398–411. [CrossRef]
13. Maton, K. *Karl Maton: Knowledge and Knowers: Towards a Realist Sociology of Education*; Routledge: London, UK; New York, NY, USA, 2014.
14. Maton, K. Making semantic waves: A key to cumulative knowledge-building. *Linguist. Educ.* **2013**, *24*, 8–22. [CrossRef]
15. Maton, K.; Hood, S.; Shay, S. *Knowledge-Building: Educational Studies in Legitimation Code Theory*; Routledge: London, UK; New York, NY, USA, 2016.
16. Kelly-Laubscher, R.F.; Luckett, K. Differences in Curriculum Structure between High School and University Biology: The Implications for Epistemological Access. *J. Biol. Educ.* **2016**, *50*, 425–441. [CrossRef]
17. Conana, C.H. Using Semantic Profiling to Characterize Pedagogical Practices and Student Learning: A Case Study in Two Introductory Physics Courses. Available online: http://etd.uwc.ac.za/handle/11394/4962 (accessed on 28 September 2019).
18. Blackie, M.A.L. Creating semantic waves: Using Legitimation Code Theory as a tool to aid the teaching of chemistry. *Chem. Educ. Res. Pract.* **2014**, *15*, 462–469. [CrossRef]
19. Pott, R.W.; Wolff, K.E.; Goosen, N.J. Using an informal competitive practical to stimulate links between the theoretical and practical in fluid mechanics: A case study in non-assessment driven learning approaches. *Educ. Chem. Eng.* **2017**, *21*, 1–10. [CrossRef]
20. Wolff, K. Engineering problem-solving knowledge: The impact of context. *J. Educ. Work.* **2017**, *30*, 840–853. [CrossRef]
21. Dorfling, C.; Wolff, K.; Akdogan, G. Expanding the semantic range to enable meaningful real-world application in chemical engineering. *S. Afr. J. High. Educ.* **2019**, *33*, 42–58. [CrossRef]
22. Wolff, K.; Basson, A.H.; Blaine, D.; Tucker, M. Building a national engineering educator community of practice. In Proceedings of the 8th Research in Engineering Education Symposium, Cape Town, South Africa, 10–12 July 2019; pp. 781–789.
23. Wolff, K. Negotiating Disciplinary Boundaries in Engineering Problem-Solving Practice. Ph.D. Thesis, University of Cape Town, Cape Town, South Africa, 2015.
24. Griesel, H.; Parker, B. *Graduate Attributes—A baseline Study on South African Graduates from the Perspective of Employers*; Higher Education South Africa & the South African Qualifications Authority: Pretoria, South Africa, 2009.

25. Bernold, L.E.; Spurlin, J.E.; Anson, C.M. Understanding Our Students: A Longitudinal-Study of Success and Failure in Engineering With Implications for Increased Retention. *J. Eng. Educ.* **2007**, *96*, 263–274. [CrossRef]
26. UNESCO. Engineering: Issues, Challenges and Opportunities for Development. Available online: http://www.unesco.org/new/en/natural-sciences/science-technology/engineering/engineering-education/unesco-engineering-report/ (accessed on 28 September 2019).
27. Beaton, C. Top Employers Say Millennials Need These 4 Skills in 2017. Available online: https://www.forbes.com/sites/carolinebeaton/2017/01/06/top-employers-say-millennials-need-these-4-skills-in-2017/#596e2e257fe4 (accessed on 28 September 2019).
28. Tadie, M.; Pott, R.; Goosen, N.; van Wyk, P.; Wolff, K.E. Expanding 1st year problem-solving skills through unit conversions and estimations. In Proceedings of the IEEE Global Engineering Education Conference, EDUCON (2018), Canary, Spain, 17–20 April 2018; pp. 1035–1043.
29. Wolff, K.; Dorfling, C.; Akdogan, G. Shifting disciplinary perspectives and perceptions of chemical engineering work in the 21st century. *Educ. Chem. Eng.* **2018**, *24*. [CrossRef]
30. Dale, E. *Audiovisual Methods in Teaching*; Dryden Press: New York, NY, USA, 1969.
31. Felder, R.M.; Woods, D.R.; Stice, J.E.; Rugarcia, A. The future of engineering education: Part 2. Teaching methods that work. *Chem. Eng. Educ.* **2000**, *34*, 26–39.
32. Case, J.; Marshall, D. Between deep and surface: Procedural approaches to learning in engineering education contexts. *Stud. High. Educ.* **2004**, *29*, 605–615. [CrossRef]
33. Abdel-Salam, T.; Kauffman, P.J.; Crossman, G. Does the lack of hands-on experience in a remotely delivered laboratory course affect student learning? *Eur. J. Eng. Educ.* **2006**, *31*, 747–756. [CrossRef]
34. Chen, W.; Shah, U.; Brechtelsbauer, C. The discovery laboratory—A student-centred experiential learning practical: Part I—Overview. *Educ. Chem. Eng.* **2016**, *17*, 44–53. [CrossRef]
35. Martin, R.; Maytham, B.; Case, J.; Fraser, D. Engineering graduates' perceptions of how well they were prepared for work in industry. *Eur. J. Eng. Educ.* **2005**, *30*, 167–180. [CrossRef]
36. Manpower Group. 10th Annual Talent Shortage Survey. Available online: https://www.manpowergroup.com/wps/wcm/connect/db23c560-08b6-485f-9bf6-f5f38a43c76a/2015_Talent_Shortage_Survey_US-lo_res.pdf?MOD=AJPERES (accessed on 28 September 2019).

 *fluids*

Article

Computing Effective Permeability of Porous Media with FEM and Micro-CT: An Educational Approach

Rafael S. Vianna [1,2,*], **Alexsander M. Cunha** [1], **Rodrigo B. V. Azeredo** [3], **Ricardo Leiderman** [2] and **Andre Pereira** [2,*]

[1]   School of Engineering, Fluminense Federal University, Rua Passo da Patria 156, Niterói 24210-240, Brazil; amarciano@id.uff.br

[2]   Institute of Computing, Fluminense Federal University, Rua Passo da Patria 156, Niterói 24210-240, Brazil; leider@ic.uff.br

[3]   Institute of Chemistry, Fluminense Federal University, R. São João Batista, s/n, Niterói 24020-150, Brazil; rbagueira@id.uff.br

*   Correspondence: rafaelvianna@id.uff.br (R.S.V.); andre@ic.uff.br (A.P.)

Received: 21 December 2019; Accepted: 20 January 2020; Published: 24 January 2020

Abstract: Permeability is a parameter that measures the resistance that fluid faces when flowing through a porous medium. Usually, this parameter is determined in routine laboratory tests by applying Darcy's law. Those tests can be complex and time-demanding, and they do not offer a deep understanding of the material internal microstructure. Currently, with the development of new computational technologies, it is possible to simulate fluid flow experiments in computational labs. Determining permeability with this strategy implies solving a homogenization problem, where the determination of the macro parameter relies on the simulation of a fluid flowing through channels created by connected pores present in the material's internal microstructure. This is a powerful example of the application of fluid mechanics to solve important industrial problems (e.g., material characterization), in which the students can learn basic concepts of fluid flow while practicing the implementation of computer simulations. In addition, it gives the students a concrete opportunity to work with a problem that associates two different scales. In this work, we present an educational code to compute absolute permeability of heterogeneous materials. The program simulates a Stokes flow in the porous media modeled with periodic boundary conditions using finite elements. Lastly, the permeability of a real sample of sandstone, modeled by microcomputed tomography (micro-CT), is obtained.

Keywords: fluid mechanics; permeability; finite element method; homogenization; microstructure; micro-CT

---

1. Introduction

Permeability is an important macro parameter usually used to characterize porous media. This parameter is a measure of the resistance that fluid faces when flowing through a porous medium. Therefore, this parameter plays an essential role in a great variety of fields, such as soil mechanics, oil prospection, petrophysics, material characterization, etc. The permeability is usually determined in routine laboratory tests by simply applying Darcy's law. However, laboratory tests can be complex, expensive, time-demanding, and sometimes destructive, and they do not offer a deep understanding of the internal microstructure of the material.

Computer simulations are increasingly applied as an alternative to laboratory tests due to their capability to overcome many difficulties presented in physical testing. Some of the advantages of numerical simulations are their non-destructive characteristic, low cost, ease of performing repeatable analyses, and possibility to perform analyses that cannot (or are very difficult to) be performed in experimental tests. In addition, it is also possible to run simulations on both synthetic and realistic models. Although synthetic models are simpler to be generated, they may introduce inaccuracies in the simulation due to the fact that they are an approximation of reality. Therefore, more accurate results are expected to be obtained with realistic models, which may be generated by means of imaging technologies on several scales (e.g., microcomputed tomography (micro-CT)). Considering those advantages, computational simulations based on numerical methods are an attractive alternative to assess the permeability of porous materials, since this methodology can consider the material internal structure through realistic virtual models.

The work of Andreassen and Andreasen [1], Aarnes et al. [2], Sheng and Zhi [3], Akanji and Mattha [4], and Yang et al. [5] are some examples that illustrate the use of the finite element method (FEM), finite volume method, and lattice Boltzmann method to determine permeability through computational homogenization.

The development of computational codes to determine permeability of porous materials has several peculiarities. Finding documents in the literature that address the subtlety in programming this problem is not an easy task. In addition, most of the codes found in the literature have in common the fact that they usually focus on computational efficiency. This often leads to a lack of important steps for the understanding of the reader, who possibly does not have extensive experience with all the subjects involved in the process of permeability determination. Although the work of Andreassen and Andreasen [1] may be partially included in this context, it presents one of the most relevant educational descriptions of the problem. They described a compact educational program, which adopts a set of interesting strategies, developed in MatLab. For these reasons, the work of Andreassen and Andreasen [1] was used as a starting point and basic reference for the developments described in our paper.

The main objective of this paper is to present the step-by-step development and implementation of an educational program to determine the effective absolute permeability of porous materials based on the concepts of numerical homogenization. The program aims to determine the mean velocity field value in a single-phase fluid flowing through the connected pores of a given material. This flow is assumed to be governed by the Stokes equation, which is solved with FEM. The program also covers the implementation of periodic boundary conditions, in which the basic assumption is that the numerical model is a representative volume of a non-contoured (statistically homogeneous) medium. By didactic choice, the program was developed for two-dimensional (2D) cases, facilitating the understanding of students, but the extension for three-dimensional (3D) cases is almost straightforward. After program validation, the permeability of a real sandstone sample, modeled by microcomputed tomography, is performed. This paper is suggested as a use case in lectures and trainings for senior undergraduate and graduate students. Since it covers the application of fluid mechanics, numerical methods, and multiscale problems from both theoretical and practical perspectives, it can be a powerful tool when teaching subjects such as fluid dynamics, finite elements, petrophysics, and other classes.

2. Effective Permeability through Numerical Homogenization

The determination of permeability through material microstructure analysis is a multiscale problem. The determination of this macro parameter depends on the determination of the velocity field, which is a function of the material's microstructure. This means that, to solve the permeability problem of a medium, it is necessary to solve the governing equations of the microscale. The connection between micro and macroscales must be done through homogenization techniques.

Computational homogenization techniques seek to determine relationships between the micro and the macrostructure of the material, which consist of the determination of the effective properties of representative elementary volumes (REV) by considering characteristics and properties of microscopic elements. For determining the effective permeability of a porous medium, a known pressure gradient or body force is applied to the material REV. Under this condition, a velocity field is induced in the domain. The homogenized effective property of the material $\mathbf{K}$ is defined as the relationship between the applied pressure gradient ∇p_Ω or body force vector $\mathbf{b}_\Omega$, the fluid viscosity μ, and the average velocities $\mathbf{u}_\Omega$ at the macroscale, known as Darcy's law (Ω denotes representative volume).

$$\langle \mathbf{u} \rangle_\Omega = \frac{1}{\mu} \overline{\mathbf{K}} \left(\langle \nabla p \rangle_\Omega - \langle \mathbf{b} \rangle_\Omega \right), \tag{1}$$

where $\mathbf{u}_\Omega = \{u_1, u_2\}^T_\Omega$ is the average velocity vector resulted from the microscopic scale, i.e.,

$$\langle \mathbf{u} \rangle_\Omega = \frac{1}{\Omega} \int_\Omega \mathbf{u}\, d\Omega. \tag{2}$$

Darcy's law, given by Equation (1), states that the total velocity (units of displacement by time, e.g., m/s) is equal to the product of the medium permeability (m^2), the pressure gradient (unit of pressure per length, e.g., Pa/m), and the inverse of the fluid viscosity (Pa·s).

Considering a two-dimensional problem, the matrix $\overline{\mathbf{K}}$ is obtained by two analyses, in which either a unit pressure gradient or a unit body force per direction is applied. The mean velocity vector determined in each analysis is equivalent to that column of matrix

$\overline{\mathbf{K}}$ related to the respective direction of the pressure gradient (or body force) applied in each case. To determine the velocity field in the problem domain, the governing equations of the problem must be solved through numerical methods, as described in the next section.

3. Computational Fluid Dynamics with Finite Element Method (FEM)

The continuity equation and the Navier-Stokes equation are the equations that govern the fluid flow problem for most cases of transport in porous medias. As the scale of the problem to be modeled reduces (e.g., nano scales), it is important to be aware that some effects may emerge in the system (e.g., capillarity, chemical interaction). However, this work focuses on viscous fluids at a scale of micrometers; therefore, it can be considered that there is no other effect than those that the Navier-Stokes equation is capable of contemplating. As this work aims to evaluate low-speed incompressible viscous flows only, the convective effects may be neglected (due to low Reynold's number) and, therefore, the Navier-Stokes equation is reduced to the Stokes equation. Although this simplifies the problem, for the great majority of practical applications, these differential equations that govern the problem still have no analytical solution; thus, numerical methods such as the finite element method (FEM) are employed as an essential tool to obtain reliable and approximate solutions.

The FEM proposes to approximate the problem solution by a finite set of elements connected by a finite number of points (commonly called nodes). The collection of finite elements that represents the continuous space is commonly called the discretized domain. Unlike the continuous domain, in the discretized domain, the variables are calculated only in the nodes of the finite elements. The continuous domain of the problem is represented by interpolation functions, called shape functions, which interpolate the calculated variables at the element nodes.

One of the FEM basic requirements is to transform the differential form (strong form) of the governing equations of the problem into integral equations (weak form). This process can be done by applying the weighted residual method, which allows the exact solution of the problem to be approximated by one with null residuals if integrated in the domain of interest. Therefore, the boundary value problem of the continuous medium can be solved by the mean values of a defined range.

The governing equations of the incompressible Stokes flow problem arise from conservation of linear momentum and conservation of mass, which are expressed, in a closed domain Ω, by Equations (3) and (4), respectively.

$$-\mu \nabla^2 \mathbf{u} + \nabla p = \mathbf{b}, \tag{3}$$

$$\nabla \cdot \mathbf{u} = 0, \tag{4}$$

where μ is the viscosity, $\mathbf{u} = \{u_1, u_2\}^T$ is the velocity vector, p represents the pressure, and $\mathbf{b} = \{b_1, b_2\}^T$ represents the body force. Note that Equation (4) represents the mass continuity equation simplified to incompressible flows, i.e., the density is assumed to be constant (independent of space and time) and the conservation of mass is simplified to the equation of conservation of volume. That assumption is possible because the permeability of porous media is mostly determined in steady-state flow, in which the compressibility effect can be ignored (Akanji and Mattha [4]).

The solution of a Stokes flow problem requires the specification of boundary conditions in addition to Equations (3) and (4). As mentioned before, there are basically two ways to compute the average velocity field in order to determine the permeability. It is either by applying a known pressure gradient as boundary conditions or by simply applying known body forces in the domain. The latter approach was chosen because applying domain quantities (such as body forces) instead of applying boundary conditions is the simplest and the most natural choice when using a domain-based discretization procedure such as the FEM to compute the velocity fields.

By applying the weighted residual method, the governing equations are multiplied by continuous and differentiable weight functions ψ. The sum of residues generated by the approximation is considered null within a defined range, resulting in

$$\int_\Omega \boldsymbol{\psi}^T \left(-\mu \nabla^2 \mathbf{u} + \nabla p - \mathbf{b} \right) d\Omega = 0, \tag{5}$$

$$\int_\Omega \psi (\nabla \cdot \mathbf{u}) \, d\Omega = 0. \tag{6}$$

Using the divergence theorem in the first term of the integral in Equation (5) in combination with the compact support of ψ, i.e., the vanishing boundary integrals, yields

$$\int_\Omega \mu (\nabla \boldsymbol{\psi} : \nabla \mathbf{u}) d\Omega - \int_\Omega \nabla \boldsymbol{\psi}^T p \, d\Omega - \int_\Omega \boldsymbol{\psi}^T \mathbf{b} \, d\Omega = 0. \tag{7}$$

Different weight functions can be used in the weighted residual method. One of the most common strategies is to assign to the weight functions the interpolation functions of the nodal values, i.e., the shape functions. This method is known as the Galerkin method. The matrix of shape functions is called $\mathbf{N}$. In classical formulations of the FEM, there arises a matrix whose elements contain derivatives of the shape functions, known as the $\mathbf{B}$ matrix. These matrices establish the following relationships:

$$\mathbf{u} = \mathbf{N}^u \bar{\mathbf{u}}, \tag{8}$$

$$p = \mathbf{N}^p \bar{\mathbf{p}}, \tag{9}$$

$$\mathbf{B} = \mathbf{L}\mathbf{N}, \tag{10}$$

where the values $\bar{\mathbf{u}}$ and $\bar{\mathbf{p}}$ represent the nodal values of $\mathbf{u}$ and p, respectively, and $\mathbf{L}$ is the matrix of the differential operators.

According to Reddy [4], in order to make an analogy with the virtual powers principle, the weight function vector ψ presented in Equation (5) can be understood as virtual velocities taken by $\mathbf{u}$; therefore, $\psi = \mathbf{N}^u \bar{\psi}^u$. The weight function ψ presented in Equation (6) is associated with the volume change of the fluid element; thus, it should be understood as a virtual hydrostatic pressure that causes this volume variation. That way, we define $\psi = \mathbf{N}^p \bar{\psi}^p$.

Applying Equations (8)–(10) and the chosen weight functions to Equations (6) and (7), the weak form of the governing equations can be represented as functions of matrices $\mathbf{B}$ and $\mathbf{N}$. Equations (7) and (6) should be respectively rewritten, in the element domain, as follows:

$$(\bar{\boldsymbol{\psi}}^{u})^{T}\left(\int_{\Omega}(\mathbf{B}^{u})^{T}\mu\mathbf{B}^{u}d\Omega\,\bar{\mathbf{u}} - \int_{\Omega}(\mathbf{B}^{u})^{T}\mathbf{N}^{p}d\Omega\,\bar{\mathbf{p}} - \int_{\Omega}(\mathbf{N}^{u})^{T}\mathbf{b}\,d\Omega\right) = 0, \tag{11}$$

$$(\bar{\boldsymbol{\psi}}^{p})^{T}\int_{\Omega}(\mathbf{N}^{p})^{T}\mathbf{B}^{u}\,d\Omega\,\bar{\mathbf{u}} = 0. \tag{12}$$

As the nodal weight values are arbitrary, the terms in the parentheses of Equations (11) and (12) should vanish, resulting in

$$\int_{\Omega}(\mathbf{B}^{u})^{T}\mu\mathbf{B}^{u}d\Omega\bar{\mathbf{u}} - \int_{\Omega}(\mathbf{B}^{u})^{T}\mathbf{N}^{p}d\Omega\,\bar{\mathbf{p}} = \int_{\Omega}(\mathbf{N}^{u})^{T}\mathbf{b}\,d\Omega, \tag{13}$$

$$\int_{\Omega}(\mathbf{N}^{p})^{T}\mathbf{B}^{u}\,d\Omega\,\bar{\mathbf{u}} = 0. \tag{14}$$

The terms on the right-hand side of the conservation of momentum equations are the result of the effect of the body forces in each element, called **f**.

Finally, solving the partial differential Equations (3) and (4) with the application of FEM is simplified to solve the following system of linear algebraic equations:

$$\begin{bmatrix} \mathbf{K} & \mathbf{G} \\ \mathbf{D} & 0 \end{bmatrix}\begin{Bmatrix} \bar{\mathbf{u}} \\ \bar{\mathbf{p}} \end{Bmatrix} = \begin{Bmatrix} \mathbf{f} \\ 0 \end{Bmatrix}, \tag{15}$$

where **K** is the viscosity matrix,

$$\mathbf{K} = \int_{\Omega}(\mathbf{B}^{u})^{T}\mu\mathbf{B}^{u}\,d\Omega, \tag{16}$$

G is the gradient matrix,

$$\mathbf{G} = -\int_{\Omega}(\mathbf{B}^{u})^{T}\mathbf{N}^{p}\,d\Omega, \tag{17}$$

and **D** is the velocity divergent matrix and is equal to the transpose of **G** matrix,

$$\mathbf{D} = -\int_{\Omega}(\mathbf{N}^{p})^{T}\mathbf{B}^{u}\,d\Omega. \tag{18}$$

The above model is known as mixed formulation. By mixed formulation, it explains that both pressure and velocity are unknown variables. Considering that, usually, fluid dynamic problems have boundary conditions in terms of both velocity and pressure, it becomes the most intuitive formulation to be used. However, the coupling of velocity and pressure variables brings new difficulties to the resolution of the system of algebraic equations.

According to Reddy [6], one way to interpret the coupling between velocity and pressure is to understand the continuity equation as a constraint of incompressibility to momentum conservation equations. Due to the fact that the incompressibility constraint does not depend on pressure variables, constructing a finite element model becomes complicated. The direct solution of the system cannot be performed since the coefficient matrix that multiplies the variable vector has a sub-matrix with a diagonal of null coefficients, which makes the coefficient matrix singular and non-inversible. Thus, it is necessary to use other strategies to solve the equation system.

Another common difficulty of mixed formulation is the choice of shape functions. The relationship between the second derivative of the velocity field and the first derivative of the pressure field stated in the Stokes equation shows that the velocity field must have a greater degree than the pressure field. The use of equal-order approximation functions for both pressure and velocity fields generates instability. This instability can be explained by the solvability analysis of the equation system. The solvability of a linear system is explained by several authors as the guarantee of the inf-sup condition, also known as the LBB condition, thanks to the works of Ladyzhenskaya [7], Babuska [8], and Brezzi [9]. According to

Elman et al., the inf-sup condition is a sufficient condition for the unique solvability of linear systems of equations in the form of Equation (15). Satisfying the inf-sup condition for Equation (15) will guarantee a unique pressure in a closed domain flow.

The use of approximation functions of the same order does not guarantee the inf-sup condition. Therefore, it does not ensure a unique solution for the problem, and the generated results may be non-reliable Hence, finding a system with unique solution requires the use of other methods to approximate the velocity and pressure fields.

Other approximation methods that are recommended for mixed formulation are found in the literature. Elman et al. [10] and Bathe [11] cited the following approximations as satisfactory for the inf-sup condition: Q2-Q1 approximation (bi-quadratic approximation for velocity and bi-linear approximation for pressure), also known as the Taylor-Hood method, Q2-P1 approximation (bi-quadratic approximation for velocity and linear approximation for pressure in each element without guaranteeing continuity), and Q2-P0 approximation (bi-quadratic approximation for velocity and constant for pressure in each element).

However, using same-order shape functions to interpolate pressure and velocity fields has several advantages from a computational perspective. In order to use same-order shape functions for velocity and pressure fields, and to eliminate instability, stabilization techniques can be adopted. Stabilization techniques have the basic idea of relaxing the restriction of incompressibility, ensuring the inf-sup condition. Many authors proposed different methodologies that proved to be very efficient in stabilizing different approximation models.

One of the simplest discretization methodologies is the use of bi-linear form functions for the velocity and pressure domains. This type of approximation is known as Q1-Q1 approximation. The Q1-Q1 approximation, in addition to being easy to implement, produces lower computational cost for solving the system of equations. For this reason, many authors seek to use stabilization techniques to make use of this type of discretization. As an example, we can cite the works of Andreassen and Andreasen [1], Yang et al. [5], Karim et al. [12], Braack and Schieweck [13], Codina and Blasco [14], and Becker and Hansbo [15].

To construct a stable computational model of Q1-Q1 discretization, this work follows the same stabilization adopted by Andreassen and Andreasen [1]. According to Andreassen and Andreasen [1], the stabilization of the Q1-Q1 model can be obtained by adding a new term with a stabilization coefficient in the continuity equation. The weak form of the continuity equation with the stabilization term is defined by

$$\int_\Omega (\mathbf{N}^p)^T \mathbf{B}^u \, d\Omega \bar{\mathbf{u}} - \int_\Omega \tau (\mathbf{N}^p)^T \mathbf{N}^p \, d\Omega \, \bar{\mathbf{p}} = 0, \tag{19}$$

where τ is the pressure stabilization parameter, defined as $\tau = h^2/12$ for quadrilateral elements, where h is the size of the largest diagonal of the finite element.

The integral with the pressure stabilization term results in a matrix that must be added to the equation system of Equation (15), resulting in

$$\begin{bmatrix} \mathbf{K} & \mathbf{G} \\ \mathbf{G}^T & \mathbf{P} \end{bmatrix} \begin{Bmatrix} \bar{\mathbf{u}} \\ \bar{\mathbf{p}} \end{Bmatrix} = \begin{Bmatrix} \mathbf{f} \\ 0 \end{Bmatrix}, \tag{20}$$

where $\mathbf{P}$ represents the pressure stabilization matrix.

$$\mathbf{P} = \int_\Omega \tau (\mathbf{N}^p)^T \mathbf{N}^p \, d\Omega. \tag{21}$$

The introduction of the new term in the continuity equation causes the coefficient matrix to become positively defined. It allows the system to be solved in the direct form.

4. Implementation of Boundary Conditions

As the volume to be analyzed increases, the property measured for that volume approaches the material's effective property. When the increasing analyzed volume reaches the effective property of the material, the volume can be considered representative. Using periodic boundary conditions, the convergence of the volume's effective properties to the REV's effective properties happens much faster (with smaller volumes). This type of boundary condition simulates the volume of interest within a periodic region, which means that the volume of interest repeats in all directions, as shown in Figure 1. Thus, it is possible to obtain homogeneous and periodic results by modeling smaller volumes, thereby reducing computational effort. The computational implementation of periodic boundary conditions can be performed by assigning the same degrees of freedom to correspondent nodes at the opposite boundaries. As a consequence, the opposite boundaries will have the same behavior, i.e., the velocities and pressure values will be the same at the opposite boundary nodes, enforcing that the volume of interest is indeed a periodic medium.

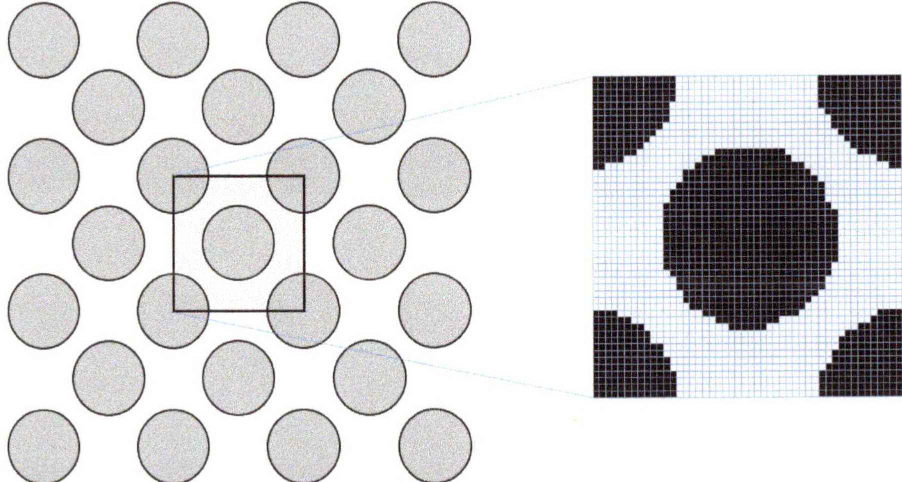

Figure 1. Representation of a periodic medium in an image representing the medium's representative elementary volume (REV).

The simulation of flows in a porous medium poses challenges to the problem modeling. The porous media modeling performed in this work is based on fluid modeling only. Ignoring solid elements represented in the image requires imposing the following condition to the solid-fluid interface: the degrees of freedom in the solid-fluid node interface relative to the velocity components are null.

5. Octave/Matlab Implementation

The code proposed in this paper is designed to receive as input an image (in special micro-CT images) of the problem to be modeled. It means that a pixel-based finite element mesh would be a natural choice of modeling strategy. In this approach, each pixel corresponds to a bi-linear finite element. The benefit of this strategy is that it results in a structured regular mesh (all the elements are squares of the same size) and, therefore, in an extremely low computational cost to generate the mesh. Taking full advantage of the fact that all elements in the mesh are of the same size, we compute the element matrices only once, working solely in the "element domain".

It is important to mention that obtaining accurate values of permeability of real samples through computational simulations relies on contemplating the narrowest constrictions in the porous medium. Covering the material's fine-scale features requires, firstly, images with high resolutions. The resolution has to be high enough to give good representation of the features in the image. On the other hand, the accuracy of the simulation also relies on modeling the image's fine features appropriately. It means that narrowest constrictions such as pore throats and cavities represented by few pixels must be modeled with a finer mesh. Therefore, convergence tests are necessary to compute permeability, taking into account the influence of image's fine-scale features.

The code described in this section is published in Zenodo, an open-source repository. The complete code is available at the following URL: https://doi.org/10.5281/zenodo.3612168.

5.1. Input and Initialization

The developed program performs simulation of Stokes flow in binary images. As initial data, the user should enter a binary image (line 03 of the code) representing the REV (as in Figure 1) and the dimensions (lines 11 and 12 of the code) of that image (model) (see Listing 1).

Listing 1. Implementation of input data and initializations

```
1    %% Determining permeability of porous materials
2    %% Reads the image
3    z = imread('filename.tif');
4    %% Increases the image resolution by n times
5    n = 1;
6    z = repelem(z,n,n);
7    %% Displays the image
8    Figure 1
9    imshow(z)
10   %% Determines the effective permeability
11   lx = 1;% horizontal dimension
12   ly = 1;% vertical dimension
13   KH = compute_permeability(lx,ly,z);
```

The MatLab function "repelem" (code line 06) is used to refine the mesh by a factor n, dividing each image pixel in a group of n × n square finite elements. This is an important procedure to perform convergence tests.

Note that, in the initialization step, the matrix image is transformed into a vector that indicates the nature of each element in the mesh (whether it is solid or fluid).

It is convenient that input data (lines 01 to 13) be stored in their own file, thus decoupling the main program (which calculates the permeability) from the data inputs (model: image and dimensions). This way, we can save the data of several different problems. Therefore, the main program is defined in its own function, which is presented from line 14 of the code.

To determine permeability, we call the function "compute_permeability", which takes as parameters the dimensions of the model and the information of each element whether it is solid or fluid (Listing 2):

Listing 2. Implementation of the main function to compute permeability.

```
14   function KH = compute_permeability(lx,ly,z)
15   %% Initializations
16   [nely, nelx] = size(z);
17   dx = lx/nelx;
18   dy = ly/nely;
19   nel = nelx*nely;
```

It is also necessary to create a matrix containing the element connectivity, which indicates the nodal incidences of each finite element. The connectivity matrix is assembled so that the image is

5.5. Obtaining Permeability

After assembling the vector with a nodal variable, permeability is calculated relating the average velocities to the prescribed pressure gradient. In this step, the mean velocities of each element are calculated integrating the velocity field inside each element. The velocity field of each element is achieved using the nodal velocities, and the shape function matrix **N** is used to approximate the velocity field. This process can be performed using Listing 9.

Listing 9. Alternative implementation for permeability calculation.

```
%% Permeability Calculation
KH = zeros(2);
for elem=1:nelemF
    velem_c1(:,elem) = SN*X(m_dofF(1:8,elem),1);
    velem_c2(:,elem) = SN*X(m_dofF(1:8,elem),2);
end
vel_c1 = sum(velem_c1,2);
vel_c2 = sum(velem_c2,2);
KH(1,1) = vel_c1(1,1);
KH(1,2) = vel_c1(2,1);
KH(2,1) = vel_c2(1,1);
KH(2,2) = vel_c2(2,1);
KH = KH/nel/dx/dy;
end
```

In the above algorithm, the element average velocity is calculated in each element during the for loop. The permeability is determined using the micro and macro relationships stated in Equations (1) and (2). However, for a more elegant and efficient way, taking advantage of the already created matrix of degrees of freedom, we use Listing 10.

Listing 10. Efficient implementation of permeability calculation.

```
124    %% Permeability Calculation
125    % sum of elements' velocities in case 1
126    vel_c1 = SN*sum(reshape(X(m_dofF(1:8,:),1),8,nelemF),2);
127    %sum of elements' velocities in case 2
128    vel_c2 = SN*sum(reshape(X(m_dofF(1:8,:),2),8,nelemF),2);
129    KH = [vel_c1'; vel_c2']/(nel*dx*dy);
130    disp(KH)
131    end
```

5.6. Validation Test

In order to validate the program presented in this work, a Stokes flow simulation was performed in the porous medium represented in Figure 3. The permeability obtained by the code demonstrated in this work, which was based on the code of Andreassen and Andreasen [1], was compared with the analytical solution proposed by Drummond and Tahir [16]. According to Drummond and Tahir [16], the permeability of the medium shown in Figure 3 can be determined using Equation (22).

$$k = r^2(-\ln(c) - 1,476 + 2c - 1,774c^2)/(8c), \tag{22}$$

where permeability (k) is a function of the grain's radius (r) and the volume fraction of solid (c).

Figure 3. Representation of the material's REV to be analyzed.

In Reference [14], the proposed analytical equation is valid for small values of c. Therefore, for program validation, the REV of the medium was modeled with unit dimensions and grains with radius equal to 0.1. For these adopted values, the corresponding c can be registered, and the permeability $k = 0.0281 \times 10^{-3}$ is obtained via the analytical solution. Figure 4 shows the results obtained through numerical modeling using the program proposed in this paper.

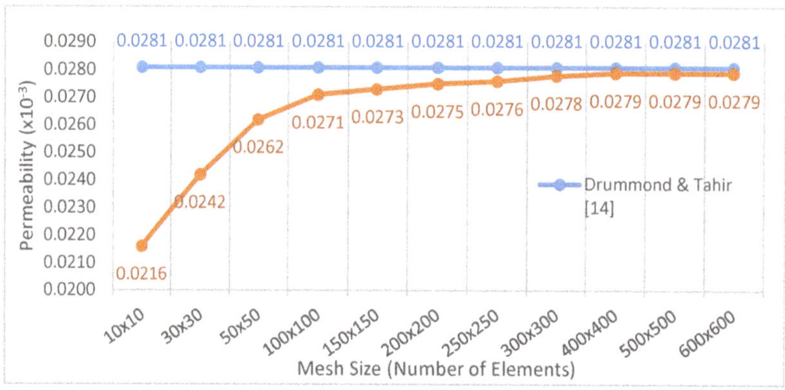

Figure 4. Convergence of permeability obtained by numerical results.

As Figure 4 shows, the convergence of the permeability of the medium displayed in Figure 3 can be obtained from mesh size 400×400. From this mesh size, the difference between the numerical result and analytical result is 0.71%. This means that the program gives good results and fulfills its purpose.

6. Real Case Application: Permeability of Sandstone

6.1. Image Acquisition: Microcomputed Tomography

The convergence of numerical results to experimental data has direct dependence on the fidelity of the representation of the computational model. One way to create reliable computational models is by using X-ray microcomputed tomography (micro-CT). Micro-CT is a technique used for internal three-dimensional visualization of any material on the micrometer scale. This technique consists of generating multiple radiographic projections taken from various angles to produce a 3D image of an object's internal structure. The generation of a micro-CT involves several steps: sample preparation and assembly, projection acquisition, reconstruction, image processing and segmentation, mesh generation, and computer simulations, as illustrated in Figure 5.

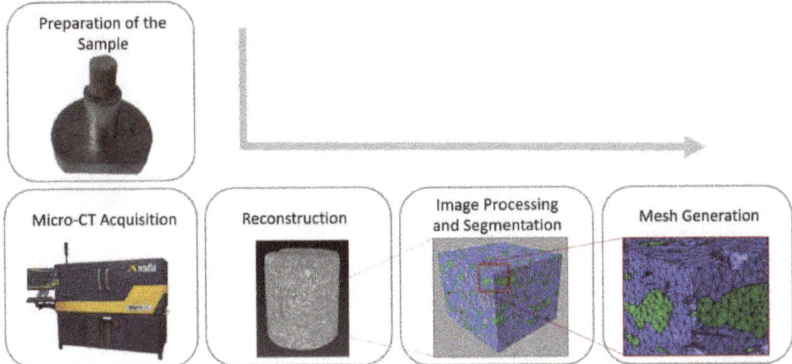

Figure 5. Workflow in digital petrophysics from sample preparation to computer simulation, covering image acquisition, reconstruction of the virtual sample, image processing, segmentation, and mesh generation.

In the micro-CT acquisition process, the sample to be digitalized is placed on a turntable inside the micro-CT scanner that rotates around its own axis while radiographies are taken. The radiographic projections are used by a reconstruction algorithm to create a 3D image of the object. The reconstruction is done by an inverse analysis that determines the level of X-ray absorption of a spatial point inside the material. This inverse analysis is made based on the different absorption levels on each radiographic projection. The result of a micro-CT is a 3D digital image of an object. A digital image is a visual representation of an object formed by small elements called voxels. Each voxel is given a specific value that is responsible for attributing it a color. Figure 6 illustrates the result of a micro-CT performed by the authors of a sandstone sample.

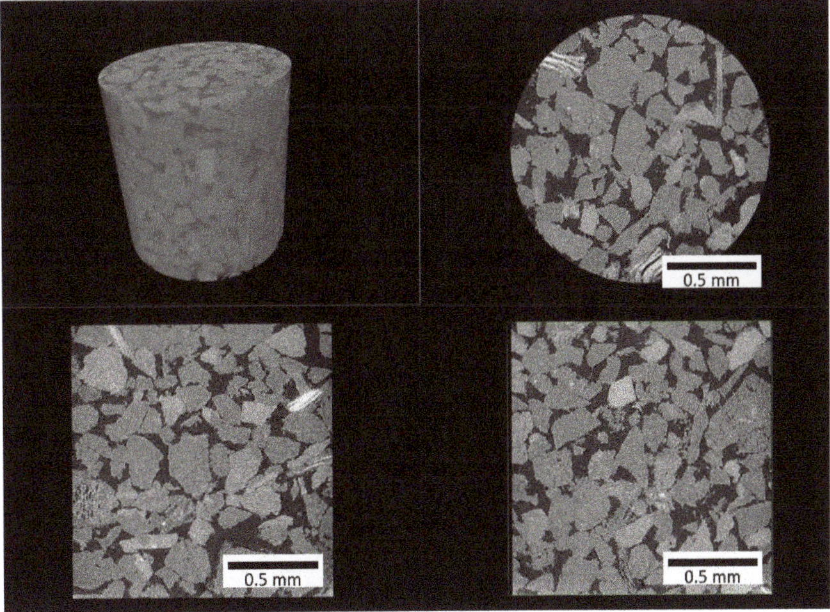

Figure 6. Microcomputed tomography from a sandstone's sample.

6.2. Image Processing and Segmentation

After scanning and reconstruction, a 3D virtual object is generated. The visualization of the material's internal microstructure can be done by cutting planes, as illustrated in Figure 6. However, for quantitative analyzes, such as porosity, permeability, and other effective properties, images must be processed. The process of image processing includes selection of region of interest (ROI), filter application for noise reduction, and image segmentation.

To determine sandstone permeability from 2D models, a slice (2D image) that was considered to be representative was chosen. After that, image filters were used. Filters are tools that reduce interference and artefacts, and that enhance certain features of images. They are based on sophisticated algorithms that facilitate the extraction of specific information from an image by modifying the image's gray histogram.

As a last step, the image was segmented. Since the image obtained by micro-CT is a grayscale image, it is important to determine which regions of the material will receive their respective properties (e.g., which voxels will be defined as solid or pores). Segmentation consists of determining the material phases according to the gray tones of the micro-CT voxels. This process can be performed by using the image histogram. By the use of the image histogram, one can define the range of gray tones that represent the pores and the range of gray tones that represent the material. There are different strategies that can be used in the image segmentation process. This process can take a lot of time, and it demands the user's experience to produce good results. After basic phase separation of the sandstone virtual sample, a tool for grain separation was used to enhance the quality of the segmentation. Using the grain separation tool allows for a richer analysis of fluid flow through the grains, which is essential for permeability determination. The grain separation tool used is based on a distance map function that determine the boundaries of the grains considering the 3D geometry of pores and grains. Since this tool is performed on 3D images, the result of the grain separation in a 2D image slice can present connections to pores (as presented in Figure 7) that seems to artificially enforce pore connection. The use of this tool was necessary to ensure that the fluid simulated in this model would be able to flow inside the material. It is possible to make a comparison between before and after image processing and segmentation of the sandstone in Figure 7. In Figure 7, blue represents the solid region and black represents the porous connected region where the fluid flows.

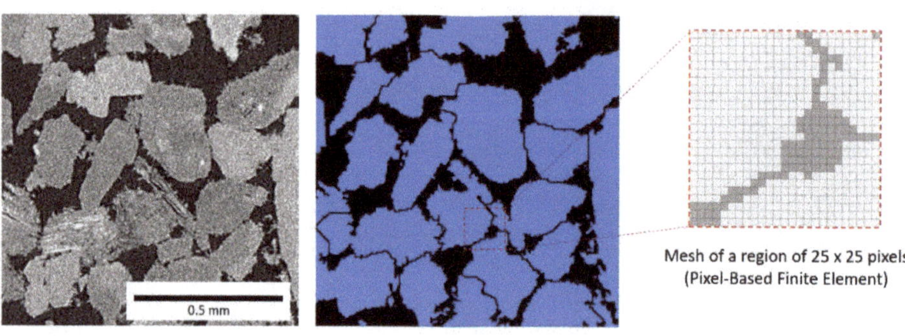

Figure 7. Two-dimensional (2D) slice of a sandstone microtomography, segmentation of the same 2D slice, and representation of the mesh in a region of 25 × 25 pixels.

Note that different users and different tools would eventually generate different segmented images. It is important to reinforce that the segmentation of the 2D slice of sandstone is used here as an example of input for the program to compute absolute permeability. Different results of segmentation of the sandstone sample can be used as input for the program as any other 2D image.

6.3. Sandstone Absolute Permeability

After validation, the program was used to measure the permeability of a sandstone sample represented by a 2D image. The two-dimensional image of the sandstone to be analyzed is exactly the same image presented in Figure 6, which has dimensions of 1 mm × 1 mm. Since the 2D image of the sandstone sample is much smaller than the sample size tested experimentally in the laboratory, periodic boundary conditions must be applied to the sandstone virtual model; thus, no change in the proposed program is necessary. Table 1 presents the obtained values for different mesh sizes (number of elements to represent the image). Figure 8 shows the sandstone permeability convergence as the finite element mesh is refined.

Table 1. Permeability in the x- and y-direction for each mesh size.

Mesh	K_{xx} (μm^2)	K_{yy} (μm^2)
600 × 600	0.2769	0.2952
800 × 800	0.2018	0.2343
1000 × 1000	0.1638	0.2050
1200 × 1200	0.1419	0.1888
1400 × 1400	0.1282	0.1789
1600 × 1600	0.1191	0.1723
1800 × 1800	0.1127	0.1678
2000 × 2000	0.1081	0.1646
2200 × 2200	0.1046	0.1621
2400 × 2400	0.1019	0.1603

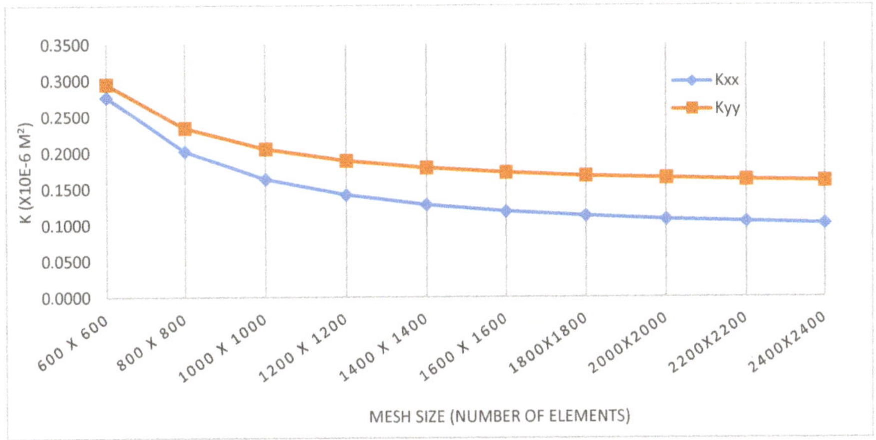

Figure 8. Convergence test of sandstone's permeability.

The permeability obtained for the 2D image of the sandstone sample analyzed with the most refined mesh was $Kxx = 0.1019$ μm^2 and $Kyy = 0.1603$ μm^2. This means that the sample is anisotropic with respect to permeability. The mesh size of 2400 × 2400 was limited by the random-access memory (RAM) of 32 Gb. A computer with 16 Gb RAM would allow refinement up to 1600 × 1600.

Note that the purpose of this paper was to present the methodology to compute permeability through computational tool in an educational way. The computation of permeability of the sample of sandstone here was used as an example of the application of the methodology. In order to obtain more accurate results for this sample through convergence tests with more refined meshes, we suggest the use of more powerful computers or the implementation of the presented methodology in higher-performance languages, such as C.

7. Conclusions

This paper presented an educational program capable of computing the permeability of porous materials by simulating fluid flows governed by the Stokes equation. The necessary theoretical basis for modeling the problem was explained, and the peculiarities of the FEM applied to computational fluid dynamics were discussed. Different ways for assembling the system of equations to be solved were explained, as one of the formulations capable of reducing significantly the processing time.

Because it is part of the digital petrophysical field, the program proposed here uses binary images as data input. This feature allows the permeability of real porous materials to be determined by means of micro-CT, which is capable of revealing the internal microstructure of the material through different voxels to represent both the fluid and the solid phases. In this work, the porous medium was modeled in a way such that only the fluid domain was considered. This methodology is extremely efficient for materials that have low porosities, as it allows the permeability to be obtained using less computational effort compared to solid and fluid domain modeling strategies.

The program described by the authors was validated through a simple problem that has a known analytical solution. Hereafter, the effective permeability of a sandstone sample was computed by a 2D analysis. Although the validation of the code confers credibility to the proposed computational program, the sandstone permeability value calculated in two dimensions may not be representative for a three-dimensional sandstone sample due to the three-dimensional geometric characteristics of its pores. However, the development of the 2D program presented here gives the reader the necessary knowledge to extend the program for 3D cases in a simple and intuitive way.

Author Contributions: Conceptualization, R.S.V. and A.M.B.P.; methodology, R.S.V., A.M.C., and A.M.B.P.; software, R.S.V., A.M.C., and A.M.B.P.; validation, R.S.V., A.M.C., and A.M.B.P.; formal analysis, R.S.V., A.M.C., R.B.V.A., R.L., and A.M.B.P.; investigation, R.S.V., A.M.C., R.L., and A.M.B.P.; writing, review, and editing, R.S.V., A.M.C., R.B.V.A., R.L., and A.M.B.P. All authors read and agreed to the published version of the manuscript.

Funding: This research received no external funding.

Acknowledgments: This research was partially supported by the CNPq (National Council for Scientific and Technological Development) Brazilian agency, grant number 404593/2016-0.

Conflicts of Interest: The authors declare no conflicts of interest.

References

1. Andreassen, E.; Andreasen, C.S. How to Determine Composite Material Properties Using Numerical Homogenization. *Comput. Mater. Sci.* **2014**, *83*, 488–495. [CrossRef]
2. Aarnes, J.E.; Gimse, T.; Lie, K.A. An Introduction to the Numerics of Flow in Porous Media using Matlab. In *Geometric Modelling, Numerical Simulation, and Optimization*; Springer: Berlin/Heidelberg, Germany, 2007.
3. Sheng, X.Y.; Zhi, X.X. A New Numerical Method for the Analysis of the Permeability Anisotropy Ratio. *Chin. Phys.* **2002**, *11*, 1009–1963.
4. Akanji, L.T.; Matthai, S.K. Finite Element-Based Characterization of Pore-Scale Geometry and Its Impact on Fluid Flow. *Transp. Porous Media* **2010**, *81*, 241–259. [CrossRef]
5. Yang, L.; Yang, J.; Boek, E.; Sakai, M.; Pain, C. Image-Based Simulation of Absolute Permeability with Massively Parallel Pseudo-Compressible Stabilized Finite Element Solver. *Comput. Geosci.* **2019**, *23*, 881–893. [CrossRef]
6. Reddy, J.N. *An Introduction to The Finite Element Method, 3rd ed*; McGraw-Hill: New York, NY, USA, 2005.
7. Ladyzhenskaya, O.A. *The Mathematical Theory of Viscous Incompressible Flow*, 2014th ed.; Gordon and Breach: New York, NY, USA, 1969.
8. Babuška, I. Error-Bounds for Finite Element Method. *Numer. Math.* **1971**, *16*, 322–333. [CrossRef]
9. Brezi, F. On the Existence, Uniqueness and Approximation of Saddle-Point Problems Arising from Lagrangian Multipliers. Analyse Numérique. *Rev. Française D'automatique Inform. Rech. Oper.* **1974**, *8*, 129–151.
10. Elman, H.C.D.; Silvester, J.; Wahereen, A.J. *Finite Elements and Fast Iterative Solvers: With Applications in Incompressible Fluid Dynamics*; Oxford University Press: Oxford, UK, 2006.
11. Bathe, K.J. *Finite Element Procedures*, 1st ed.; Prentice Hall: Upper Saddle River, NJ, USA, 1982.

12. Karim, M.R.; Krabbenhoft, K.; Lyamin, A.V. Permeability Determination of Porous Media Using Large-Scale Finite Elements and Iterative Solver. *Int. J. Numer. Anal. Methods Geomech.* **2014**, *38*, 991–1012. [CrossRef]
13. Braack, M.; Schieweck, F. Equal-Order Finite Elements with Local Projection Stabilization for the Darcy–Brinkman Equations. *Comput. Methods Appl. Mech. Eng.* **2011**, *200*, 1126–1136. [CrossRef]
14. Codina, R.; Blasco, J. A Finite Element Formulation for The Stokes Problem Allowing Equal Velocity-Pressure Interpolation. *Comput. Methods Appl. Mech. Eng.* **1997**, *143*, 373–391. [CrossRef]
15. Becker, R.; Hansbo, P. A Simple Pressure Stabilization Method for The Stokes Equation. *Int. J. Numer. Methods Biomed. Eng.* **2008**, *24*, 1421–1430. [CrossRef]
16. Drummond, J.E.; Tahil, M.I. Laminar Viscous Flow Through Regular Arrays of Parallel Solid Cylinders. *Int. J. Multiph. Flow* **1984**, *10*, 515–540. [CrossRef]

 *fluids*

Article

Floodopoly: Enhancing the Learning Experience of Students in Water Engineering Courses

Manousos Valyrakis [1,*], **Gaston Latessa** [1,2], **Eftychia Koursari** [1,3] **and Ming Cheng** [4]

[1]   Water Engineering Lab, School of Engineering, University of Glasgow, Glasgow G12 8QQ, UK;
     Gaston.Latessa@jacobs.com (G.L.); Eftychia.Koursari@amey.co.uk (E.K.)
[2]   Jacobs, Principal Hydraulic Modeler, Water and Environment, Glasgow G2 7HX, UK
[3]   Amey Consulting, Structures, Glasgow ML1 4UR, UK
[4]   Faculty of Education, Edge Hill University, Ormskirk L39 4QP, UK; ChengM@edgehill.ac.uk
*   Correspondence: Manousos.Valyrakis@glasgow.ac.uk

Received: 22 December 2019; Accepted: 6 February 2020; Published: 8 February 2020

Abstract: This study focuses on the utilisation of lab-based activities to enhance the learning experience of engineering students studying water engineering and geosciences courses. Specifically, the use of "floodopoly" as a physical model demonstration in improving the students' understanding of the relevant processes of flooding, infrastructure scour and sediment transport, and improve retention and performance in simulation of these processes in engineering design courses, is discussed. The effectiveness of lab-based demonstration is explored using a survey assessing the weight of various factors that might influence students' performance and satisfaction. It reveals how lab-centred learning, overall course success is linked with student motivation and the students' perception of an inclusive teaching environment. It also explores the effectiveness of the implementation of student-centred and inquiry-guided teaching and various methods of assessment. The analysis and discussion are informed by students' responses to a specifically designed questionnaire, showing an improvement of the satisfaction rates compared to traditional class-based learning modules. For example, more students (85%) reported that they perceived the lab-based environment as an excellent contribution to their learning experience, while less students (about 57%) were as satisfied for a traditional class-based course delivery. Such findings can be used to improve students' learning experience by introducing physical model demonstrations, similar to those offered herein.

Keywords: laboratory demonstration; physical modelling; flood risk assessment; sediment transport; infrastructure scour hazards; outreach; water engineering

---

1. Introduction

Students have an inherent ability in learning and adapting to their continuously changing environment, and instructors have a role not only on offering knowledge to students, but also in developing their ability to learn more efficiently. One learning strategy that may be followed in doing so involves increasing students' desire to learn, by engaging students with questions before arriving to an answer, according to [1,2]. Considering that one of the essential first stages for knowledge discovery is observation, it is useful to enthusiastically engage with the students with lab demonstrations early in the course, in helping to educate the future engineers and geoscientists. Topics such as fluid mechanics and hydraulic and water engineering can highly benefit from such practices, as demonstrations of first principles on which engineering designs or physical processes are based may involve simple bench top to more elaborate flume demonstrations. For example, such phenomena may be mixing of fluids at different flow rates, as Osborne Reynolds did in his famous 1883 experiment [3].

A dynamic learning environment is in alignment with the perspective presented in a quote usually attributed to Socrates: "Education is the kindling of a flame, not the filling of a vessel". This can be best realized if one understands the need to continuously adjust and improve their knowledge, whether it is professional knowledge and experience or soft (e.g., communication) skills, to dynamically adapt to a constantly changing techno-economical and socio-political environment. Physical model demonstrations can allow the students to practice these skills by allowing them to work collaboratively towards inquisitive critical thinking for building their knowledge more effectively. A collaborative environment, where all students freely contributed their own perspectives and are welcoming of each other's views, as well as learning by mistakes, while participating in an open discussion about the best possible methods and tools to be employed towards problem solving, offers an ideal setting. This is because this approach exposes the individual to the collective intelligence of their group and allows them to efficiently achieve a better understanding of the mechanics and processes involved in reaching a solution, thus contributing more to their own learning experience compared to just working on their own.

Physical model demonstrations can also allow for teaching to be more student-centred, fostering active learning. Developing active learning strategies can be realized by implementing a student-centred approach to teaching and using visually rich and engaging lecture presentations, informative lecture notes and tutorials. During the lecture students can be asked questions and offered empirical paradigms relevant to the discussed topic or relevant research to cultivate students' engineering intuition. Asking students questions is very important, as it will improve students' metacognition as well encourage them to take a more active role in their learning [4]. This can also be achieved with the use of more modern technologies involving field work [5] or "virtual laboratories" [6–10], taking advantage of advances in the fields of virtual reality or augmented reality, as an alternative to physical laboratories. However, these technologies are yet not highly accessible for large cohorts of students, are relatively costly and require technical knowledge in using these and developing demonstration applications. Thus, the "floodopoly" setup, as the first author's own invention, is discussed and presented in this research study, as a tool aiming at improving the learning experience of engineering students, as well as engaging pupils in STEM themed subjects.

Student-centred learning has been proved as being effective in encouraging deep approaches to learning and in making students take responsibility for their own learning [11]. Focusing on students, as individual practicing engineers will help them to self-identify and develop their own talents as well as achieve a better understanding of how their unique skill sets and personality will fit in and potentially benefit their peer group's dynamics and performance. Focusing on developing students' strengths is an inherently energizing process for them, and allows the individuals to be in the flow zone of learning (appropriately challenged and motivated), allowing them to feel nurtured into developing as well rounded professionals and future leaders in their fields.

Over the last decade the above concepts have been deployed in shaping the delivery of modules in fluid dynamics, hydraulics and water engineering [12], for diverse cohorts and educational systems, including Virginia Tech (USA) and University of Glasgow (UK), such as the Environmental Fluid Mechanics (CEE3304) and Engineering Hydraulics 3 (ENG3085), respectively. Successful teaching should encourage and develop students' ability for active learning, a belief which, according to the current teaching practice for these courses, is influenced by perspectives early presented by Greek philosophers, such as Socrates, on learning. Specifically, according to the quote usually attributed to Socrates, one "cannot teach anybody anything, but can only make them think". Highly engaging teaching for fluid dynamics-themed courses commonly employ physical model demonstrations and can be even more impactful if their delivery is student-centred, with the students actively encouraged to maintain a sustained engagement and self-motivation towards knowledge and skills development learning, with strong focus on the development of each individual's strengths (which requires a diverse way of assessment). The value of physical model or lab-based demonstrations in introducing a student to the scientific method, the observation of a system or process of interest, the formulation of a research hypothesis and the design of an appropriate framework is indispensable. This can be done by discussing with the students the methods researchers use to build a wealth of scientific information by means of observing physical processes, encouraging them to reproduce experimental results and suggesting how to demonstrate their critical thinking. This approach helps equip students with appropriate tools to carry out research in the future as academics or gain confidence as engineering professionals, as [13] have demonstrated.

Here the use of a physical lab demonstration is presented as one of many student-inspiring demonstrations offered within our UK-wide top degree program in Civil Engineering at the University of Glasgow (ranking top three across the whole UK over the years 2017–2019, and first in 2019, according to the National Student Survey). A distinct feature of this degree is the significant amount of design-centred and practical skills development-oriented courses, many of which involve lab-based or computer demonstrations. In addition, a low staff-to-student ratio, a high entry level for the students and the fact that the University of Glasgow is a research-intensive environment contribute to building an identity as an engineering researcher or innovator; becoming a "world-changer" as the university community prides. Small-group teaching is found to help develop an active learning environment, as the instructor has a better chance for clarifying points made in lectures, helping the students gain a more comprehensive understanding and fostering discussion and communication of ideas [14]. The contained size of the class and lab-based sessions (split into small working groups of four to eight undergraduate students each) works well in helping to encourage discussion of problems and seeking peers' feedback.

Specifically, the use of a novel physical demonstration, namely "floodopoly", as part of a design class, which aims to introduce to the students the principles of flood risk modelling and assessment, is detailed herein. The course in which the "floodopoly" demonstration is introduced involves a problem-based learning environment, and it focuses on developing higher cognitive functions, for example, critical thinking, encouraging problem solving and bringing exciting and tangible lab-based research observations into the course. Such a course design, that links teaching with research, is shown as effective in extending beyond traditional student knowledge, by empowering students and advancing intellectual curiosity [15]. Further, research-based course design has been associated with deep, rather than superficial, student learning [16,17], and is believed to have the potential to improve students' satisfaction and benefit students' professional development. The results from a survey offered to the students are further discussed to demonstrate how physical demonstrations are contributing to an enhanced student learning experience.

2. Physical Model Demonstration

The main goal of the "floodopoly" physical model demonstration is to allow the students to observe the physical mechanisms of flooding, which students are assessing later through numerical modelling. This hands-on, active learning student-centred experience facilitates their grasping of key aspects of flooding and associated hazards, in urban environments.

2.1. Experimental Setup

"Floodopoly" is an interactive demonstration aiming at allowing the students to play out a number of scenarios, simulating the effects of rising hydrograph and flow unsteadiness in a physically relevant and engaging manner, towards knowledge discovery. This allows the students to have a significantly improved understanding and intuition-led critical thinking for the studied dynamical processes, before embarking on using appropriate software (such as HEC-RAS) for flood modelling and flood risk assessment, in the later part of the course.

A number of improved designs have been utilised over the many years, but there are essentially two versions of the "floodopoly" setup; (a) a versatile mobile setup (Figure 1a) and (b) a permanent one (Figure 1b). Both setups involve a fit-for-purpose model city in a sandbox, which is already built for the demonstrations as described in the following. The model city uses building blocks from Lego© along with custom paper-made or 3D-printed physical models of built infrastructure to simulate a typical built environment. Means of transport (such as vehicles, buses and trucks), elements of the natural environment (such as trees) and built infrastructure (houses and skyscrapers), including iconic buildings, such as the Eiffel tower and Parthenon (Figure 1a), or the London Eye and Big-Ben (Figure 1b), are all scaled to fit within the sandbox model city and can be featured for enhanced dramatization during the demonstration.

The mobile "floodopoly" version (Figure 1a) has a 3D-printed topography of the model city and complex river geometry, allowing the floodplains to be covered in water impacting infrastructure in the extreme hydrologic scenarios. The water tank of the setup is made of transparent 7 mm thick acrylic panels and comprises of a tank with external dimensions of 90 cm length, 50 cm width, and 30 cm height. Blue-dyed water is circulated through the model for the demonstrations to improve the visualization of the flooding process. This water is then collected "downstream" so the experiment can be run again. The sandbox containing the 3D-printed model surface replicating the complex topography of the model city, involving a river and its floodplains, is of slightly smaller dimensions, allowing for recirculating water on its surface through an appropriately constructed inlet (aligned with the river's main channel) using a configurable pump introduced at the water tank of the setup. Focus is given on producing a visual effect on the impact of flooding, so some of the built infrastructure and monuments can be purposely placed at or near the floodplains. After the demonstration, water can be drained from a small outlet valve attached at the lower corner of the acrylic box.

The lightweight, modular arrangement (all items such as the built infrastructure and 3D-print surface topography are also lightweight and removable) and small size of this setup allows it to be easily carried to a specific location for engaging with target audiences, beyond the classroom. This versatile setup has been successfully presented to thousands of pupils and students during recruitment events (such as the School of Engineering Open days), as well as outreach events (such as Widening Participation schemes, Glasgow Science Festival, Glasgow Science Museum). It can also be shown in an open-air classroom next to the river Kelvin (within walking distance from the classroom) for the Engineering Hydraulics classes and Civil Design project, for third and fourth year engineering students.

Figure 1. *Cont.*

(**a**)

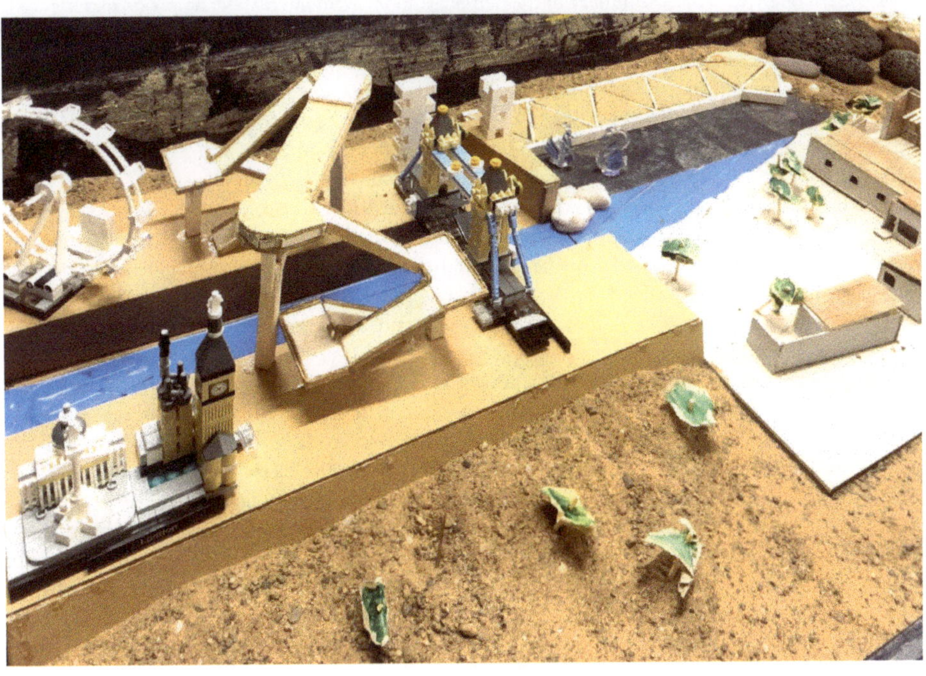

(**b**)

Figure 1. Demonstration of the "floodopoly" demonstration: (**a**) The flooded state of the mobile setup, at the "meet the experts" session (during the Glasgow Science Festival 2016, at the Glasgow Science Centre); and (**b**) the dry state of the permanent (flume-based) setup, at the Water Engineering Lab of the University of Glasgow.

The permanent design (Figure 1b) involves replicating part of the river Thames and the relevant infrastructure along it. The model city is built with Lego$^{©}$ inside a standard Armfield-type demonstration flume repurposed specifically for this project. The flume is substantially longer (7 m long in total) and 60 cm wide, while much bigger flow rates can be run in a recirculating fashion. Both setups have flows running through the river via the use of small pumps of controlled flow rate to simulate a rising or decreasing hydrograph. The students are encouraged to replicate an extreme flash flood scenario by pouring large volumes of water at a fast pace, aiming to demonstrate the dramatic impacts of flooding, resulting in highly-visual damage to built infrastructure, and activation of floodplains in routing the flood. The topography of the models can be sprinkled with sand of different sizes to also demonstrate scour and sediment transport processes and how increased flow or shear flow forces can result in increased rate of transfer of smaller or coarser grains, helping the students understand that coarser grains can generally withstand increased stresses, thus are more suitable for use in hydraulic protections and armouring infrastructure near water.

2.2. Demonstration Protocol

Both of the above setup versions can be used to run the "floodopoly" demonstration in an interactive manner. More complex game mechanics can be further introduced depending on the time allowed for the demonstration, thus the name is chosen to rhyme with the popular game monopoly$^{©}$. The following steps are offered as an indicative guidance to the demonstrator.

Firstly, the demonstration space is introduced and any important health and safety issues are presented, along with a brief summary of the demonstration, as an interactive, hands-on experience for students to investigate the impacts of flooding, and to develop a first understanding of flood defence designs, by means of tangible observation of the involved hydraulic and geomorphic processes.

Secondly, the audience is familiarized with the different model city areas, including the historical areas (model castles and churches), energy related facilities (model coal extraction and oil and gas facilities), industrial plants (behind the model floodwalls), housing areas and residential properties, and hydraulic infrastructure including model bridges, roads/highways and locks.

Thirdly, the impacts of flooding and scouring of infrastructure are presented. Flood risks associated with floods, such as loss of life and infrastructure costs, are discussed. Emphasis is given at the fact that floods have been always happening, but because of increased urbanisation and climate change, societal catastrophic impacts due to floods have continued increasing at an alarming rate. In an inquiry-led session, the students are asked to consider which areas are more flood-prone and what makes them more susceptible to flooding.

The processes of scour [18] and sediment transport are discussed. For example, the demonstrator can introduce some additional fine and coarser sediment, across the setup, and ask the students to observe the locations and the size of sediment in relation to eroding and settlement. The students are allowed to think out loud and discuss with each other their observations, towards identifying the first principles around these physical processes in a collaborative manner. A further inquiry can be around devising ways by which such first principles can be used to design engineering against scour, eventually observing that, in the real world, engineering practitioners use coarser sediment as rip-rap. Since pier shape can also affect the scour geometry [19–21], this installation may be used to demonstrate influence of pier shape on scouring process, by placing piers with different shapes inside the fluvial model.

To further enhance the students' tangible observation-led learning of the physical processes, as well as the impact of flooding on the environment, infrastructure, people and their property, two more highly-interactive role-playing scenarios can be run:

1.   The Scenario of Urbanisation

Individual students from the participating audience are given the same number of 3D-printed miniature houses (similar design to what is found in the housing blocks of Monopoly$^{©}$) and assume the role of citizens and engineers contributing to the development of the city's floodplains. The students are asked to take turns in placing one of their buildings on one of the many predefined plots for development on the floodplain. The demonstrator describes that during this scenario the city gets increasingly urbanised over a period of cycles, as each of the student is occupying a new plot with their building. The students are asked to contemplate and consider the first physical principles discussed earlier towards identifying which of these plots could face a greater risk of flooding. Naturally, most of the students will opt for the higher ground, and, as such, plots become less available, other principles will have to be employed, such as avoiding building next to a segment of the river with low slope, as the flow depth (for the same river geometry and flow rate) will have higher flow depth (thus a higher risk of flooding).

The mechanics of the interactive demonstration can be further gamified, but the focus here is rather on collaborative learning by means of observation. Even though the demonstration is student-centred, the demonstrator is guiding and pacing the process by inquiring, for example, about the students' expectations as the flow rate increases and asking students to identify the areas with greater risks. As the floodplains get flooded, the students get a chance to seriously reflect on their choice, as if their own property had flooded in the real world. This also stimulates their sense of responsibility as an engineer. Last, the students are offered a chance to identify direct (e.g., by placing floodwalls, embankments, channel geometry modification) or indirect protection measures (e.g., having an upstream storage scheme or changing the land use around the catchment, by planting more vegetation to slow the surface runoff before it becomes accumulated stream-flow).

2. The Climate Change Scenario

In this scenario the students are asked to simulate extreme hydrologic events, due to climate change. They are given plastic cups of different sizes (thus varying the volume of water to be dropped), prefilled with blue-dyed water, and are allowed to spill it at the upstream end of the river, running already with the highest base-flow. The students can empty their cup as fast or as slow they desire, observing that the rate of volume offered to the base-flow rather than the volume alone is important in defining the change in the hydrographs. As they observe the peak of the hydrograph upstream propagate downstream, they also have a chance to observe the attenuation of the peak flow rate. This is the most visually impactful part of the demonstration, with many students enjoying the activity of flooding the landscape of the sandbox and impacting infrastructure within it. As the pressure from climate change is rapidly increasing, a whole scenario is purposely devoted to demonstrating its potentially catastrophic impacts. Given that the nature of this physical demonstration is to generate a visually-induced emotional response around the increasing frequency and magnitude of hydrologic events, it is purposefully involving a qualitative rather than a quantitative comparison.

3. Results

3.1. Design Course Background and Assessment

A range of both formative and summative assessment methods are used in the design course where the "floodopoly" demonstration is offered, which proves to be effective in improving students' learning [22]. Specifically, assessment methods range from assigning extra credit quizzes at the end of each lecture to performance assessment quizzes and assignments on a weekly or bi-weekly basis. Assigning customized exams and quizzes as well as a final project report helps them by enhancing their writing and reporting skills and improving their critical thinking capacity, allowing them to progress in more advanced topics. The use of a flexible assessment scheme for the final presentation of their output, where one has the option to prepare a video, a poster or oral presentation, allows a student to capitalize on what they perceive to be their strength and feel greater satisfaction. Based on student feedback, the offered assessment methods are suitable and the diversity of assessment methods and the opportunity to reflect and identify their own strengths is appreciated.

3.2. Student Survey

To assess the effectiveness of the "floodopoly" demonstration as a novel highly visual and tangible way for reinforcing inquiry-led learning, a purpose-designed questionnaire is offered to the students. The use of questionnaires as a data collection method is well established and has often been used for course evaluation [23]. A total of 25 out 30 students completed the questionnaire in full, after the completion of the course. Specifically, the questionnaire covers several aspects that affect student learning, performance and satisfaction, such as students' motivation, factors to effective learning (also assessed by follow-up quizzes) and methods of communication and assessment; in an effort to identify how a physical demonstration may be a more engaging and beneficial learning activity compared to other options, towards better achieving the intended learning objectives of the course within a broad context of the engineering and geosciences curriculum. The presentation of results is given in the following main themes explored below: (a) Students' motivation for attending the water engineering courses, (b) effectiveness of course environment and material offered, (c) suitability of methods of communication, (d) learning and assessment methods and (e) assessment of level of difficulty, performance and overall satisfaction.

3.2.1. Students' Motivation for Attending the Water Engineering Courses

The first set of questions aims to access one of the most important aspects of learning—the level of motivation and enthusiasm for the theme studied. Motivation can affect attendance and focus during the class and can depend on the how each student perceives the course to be useful to their professional development. Student replies are visually offered by means of word clouds showing the most frequently used words by students per question (Figure 2a–d).

Regarding students' perception of their future role as civil engineers (question a1), most replies involved design and construction (building) of infrastructure, as well as problem solving applied to challenging aspects of our society, which are indeed amongst the most fundamental aspects of civil engineering (Figure 2a). Some replies are original and indicative of a high level of inspiration "civil engineer = builder of dreams", while several (at least three) other students chose to identify with known quotes of the role of the civil engineers (e.g., "Harness the power of nature for the greater good of mankind").

For the second question (a2), on what excites them most about the profession they study, answers signifying intrinsic or extrinsic motivation to different levels are observed (Figure 2b). Replies, such as "fulfilment from design to structure completion" and "the opportunity to be part of the design or construction of a landmark, or design to improve the life of people", show a high level of intrinsic satisfaction expected by the practice of this profession. From this question, several highly-motivated students can be identified due to extrinsic factors, such as "being part of a well-known project and opportunity to travel abroad" which reveals focus on external rewards (influence, esteem and travel).

Replies from the next question (a3: "What do you aspire to do in 5 to 10 years from now?") covered virtually all disciplines of civil engineering, namely, structural, water/environmental, geotechnical and transportation engineering (Figure 2c). Most students confidently reported they wish to eventually secure "chartered" status with a professional body, as can also be seen from the word cloud (Figure 2c). More students showed an interest to work in industry/consultancy, than government and NGO's.

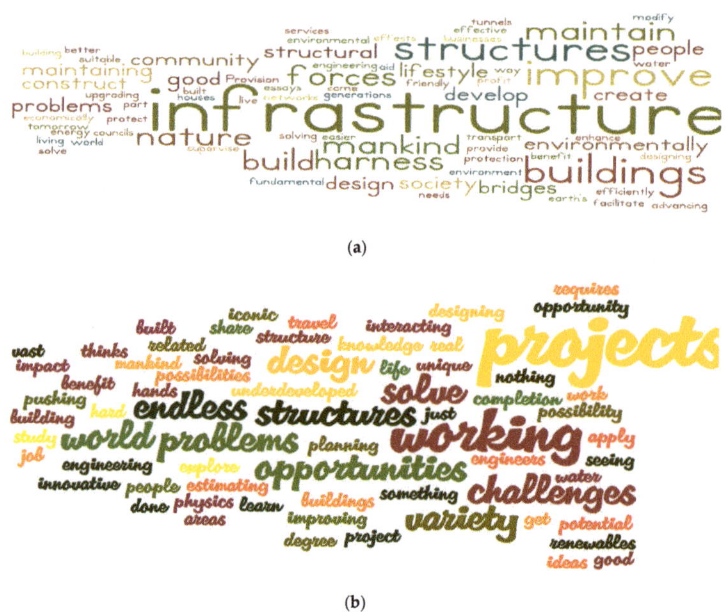

(a)

(b)

Figure 2. *Cont.*

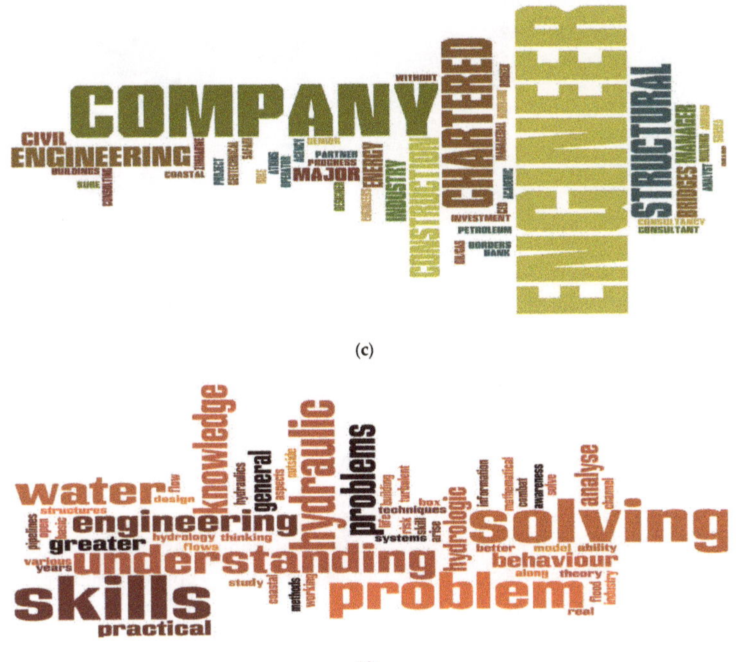

Figure 2. Word cloud representation of collective replies from the first part of the questionnaire assessing the students': (**a**) Role as professionals, (**b**) motivation, (**c**) career aspirations and (**d**) expectations from this course.

With regard to the students' expectations for this course (question a4), the replies were centred around both skills and knowledge to be gained (Figure 2d). The former ranged from group work, report writing and problem solving to the ability to utilise effective modern computational tools and techniques, while the latter involved extending the fundamental theoretical background with reference to practical experience and best practices from real life problems. Indicative is the following student reply, demonstrative of a need to link the taught material with the professional practice, while enriching their interest on the discipline: "Practical knowledge of water engineering and give me something I can actually use in my professional career. I hope to build on my interest and passion for the subject".

3.2.2. Effectiveness of Course Environment and Material Offered

The following set of questions (b1–b3) focus on the perceptions of student's knowledge, aiming to explore the confidence of an individual in thinking about each specific question rather than assessing the true knowledge on this (for which a test/quiz type of question would be required, and which is explored in sequent questions). The specific questions (SQ1–SQ6) are representative of each of the six course objectives and thematic sections that the students were taught in class. As seen from Figure 3a, on average the students feel confident with the overall level of understanding they have acquired during the taught course (average for overall knowledge is 3.4, on a scale from 0 to 5, with 5 being excellent). The students can answer fundamental theoretical questions with more confidence (SQ2, SQ3, SQ5), while more advanced (SQ4) or abstract (SQ6) concepts, requiring critical thinking, were ranked relatively lower. It is interesting to note that while the standard deviation for the first few questions ranges from 0.7 to 0.8, it is above 1.1 for the last one (SQ6), showing that some students felt less equipped in higher cognitive skills compared to others.

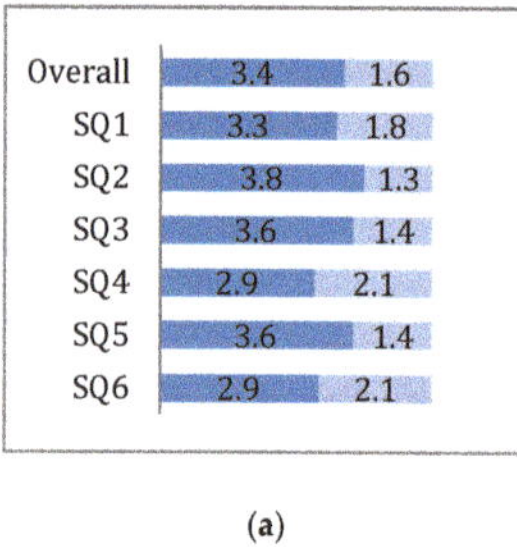

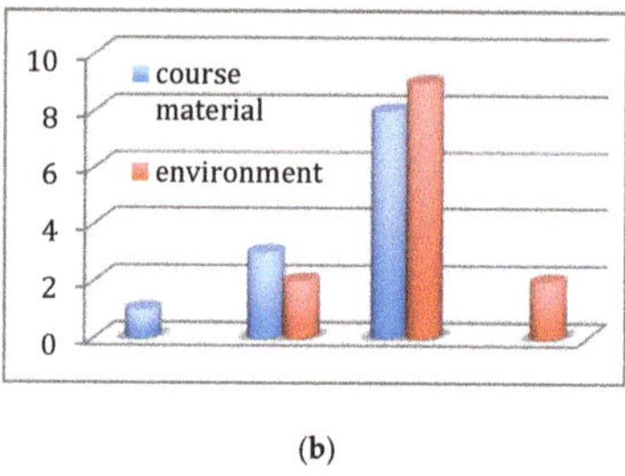

(**a**)

(**b**)

Figure 3. Students' perception of the: (**a**) Knowledge gained during the course (assessed on a scale from 0 to 5). The students are queried in relation to their overall knowledge first and then specific questions about their disciplinary knowledge (SQ1–SQ6) and (**b**) course material and environment (ranging from neutral (left) to excellent (right)).

The next two questions concern student satisfaction with the course material, including notes posted online (e.g., Moodle) or handouts distributed in the class and the lab sessions (b2), as well as the environmental factors during these sessions including the lab-based demonstrations (b3). The feedback offered is summarized in Figure 3b. Even though it is observed that the overwhelming majority of the students who answered indicate their high level of satisfaction with both factors, the shift of the distribution to the left demonstrates clearly that the use of the lab for physical model demonstrations is even more appreciated and valued, compared to only teaching in the classroom. Specifically, about 85% of the students perceived the lab-based environment as a great or excellent contribution to their learning experience, while fewer students (about 57%) were as satisfied for a traditional class-based course delivery.

3.2.3. Suitability of Methods of Communication

Following, the utility of other means of communication and updating students on course content and feedback (in addition to Moodle and e-mail) is inquired (c1–c3). It is interesting to note that many students choose such modern platforms for this goal, many accessed at least on a daily basis or more often. This view is supported by the results illustrated in Figure 4a, where the various online means students frequently use and would be willing to use for the purposes of class communication are shown. Facebook ranks second in popularity (with 16 students, half of whom prefer very frequent communication) compared to the online web pages and blogs (with 20 students). Only a few (six) of the students believe there is a need to establish a Facebook group page for such purposes (Figure 4b). On the contrary there are many who feel this is not required as " … Moodle and email is sufficient for the class needs" or even a few who may feel even stronger and stating an opposition in subscribing to such services "due to the terms of use". Thus if an inclusive method is to be used, Moodle and email suffice for reaching out to the class. The optimal frequency to contact students varies greatly from "every other day" to "only when needed", with more than half of those who answered, believing that it is best to be contacted at least once a week (Figure 4c).

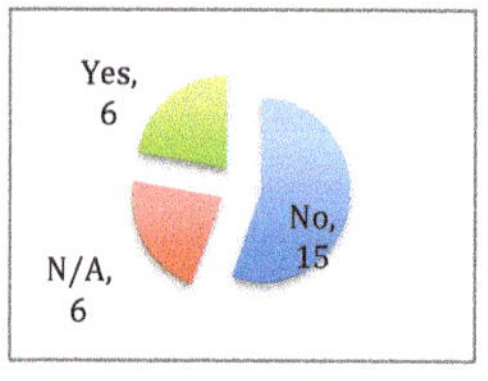

(**a**)

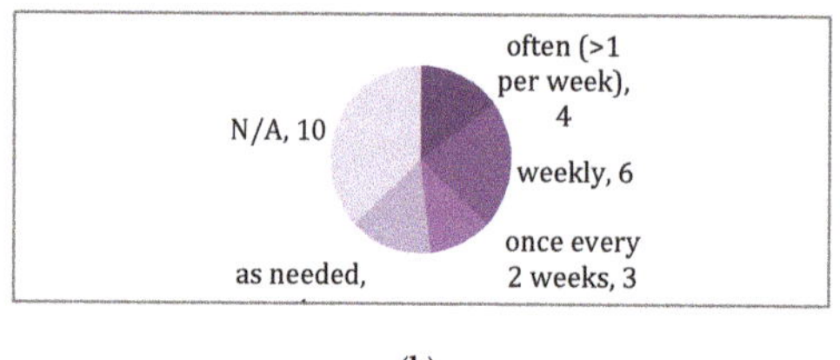

(**b**)

Figure 4. *Cont.*

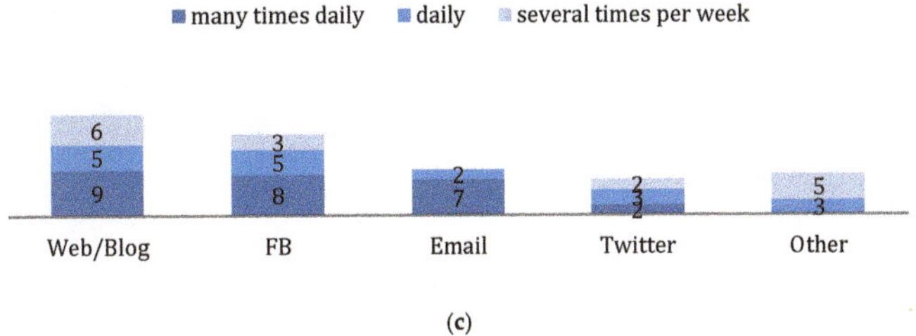

Figure 4. Students' perception of: (**a**) The utility of a Facebook (FB) group page for the course, (**b**) the optimal contact frequency for receiving course updates and feedback, and (**c**) the overall utility of online platforms on class communication and feedback.

3.2.4. Learning and Assessment Methods

Two important questions follow (d1–d2) pertaining the methods of assessment, beyond the traditional exam-based assessment that is employed for this course. Figure 5a shows the results grouped per type of assessment and percentage contribution for each (out of a total of 100%). Students are allowed to choose more than one option, with the most popular ones being lab work, home assignment and technical reporting for about 10%. This is consistent with the students' desire to learn hands on, practical skills at the lab as well as improving their problem solving and technical writing skills, respectively. Fewer preferred the presentation with slides, as unfortunately at this level they may have not gained the experience required to present with confidence, as opposed to later years. It is generally agreed that the course assessment should not be entirely exam-based. The importance of addressing the need to use a versatile scheme that includes different assessment approaches and which allows the individual to reflect on their own skills, putting emphasis on growing their own skills/talents, is also reflected below with only one individual being indifferent and most of those who replied believing it is imperative (Figure 5b).

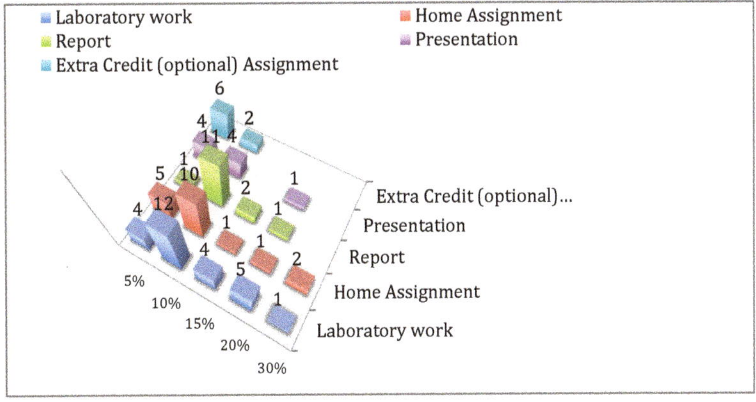

Figure 5. *Cont.*

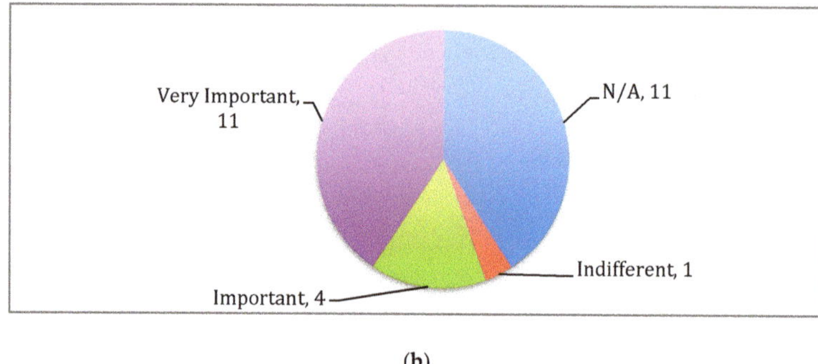

(b)

Figure 5. Students' perception of the: (**a**) Usefulness of a variety of assessment methods and (**b**) importance of adopting a versatile scheme for assessment.

3.2.5. Assessment of Level of Difficulty, Performance and Satisfaction with Individual Performance

The last three questions (e1–e3) relate to the perceived difficulty for the class (a measure relevant to expected individual performance or how challenging the course is, compared to other courses offered), overall performance and degree of satisfaction with the overall grade achieved. More than two thirds of those who replied regarded the class to be generally quite more than challenging and "moderately hard", while only a third perceived it as being at an acceptable "average" level of difficulty (Figure 6a). The reported performance levels vary greatly covering the whole spectrum of grades, but with a good portion of high grades (Figure 6b). This is not unreasonable given that the exams cover material that has been seen by the students in class or tutorials, reducing the uncertainty over the questions that might be asked. Likewise, the satisfaction results (Figure 6c) demonstrate a degree of polarization with an about equal share for those who are satisfied and dissatisfied with their performance.

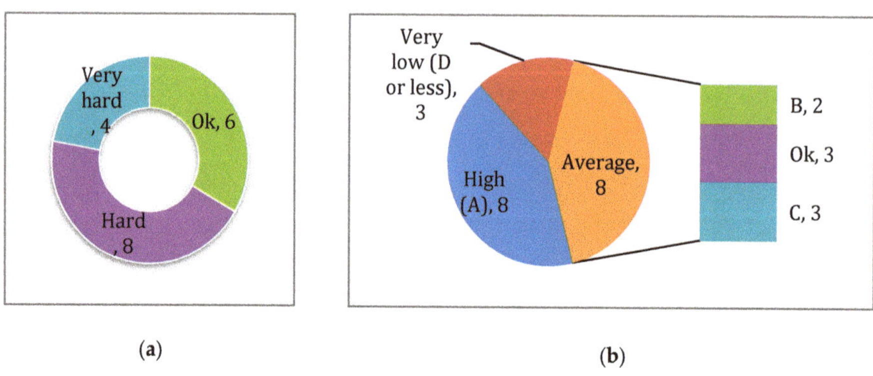

(a)   **(b)**

Figure 6. *Cont.*

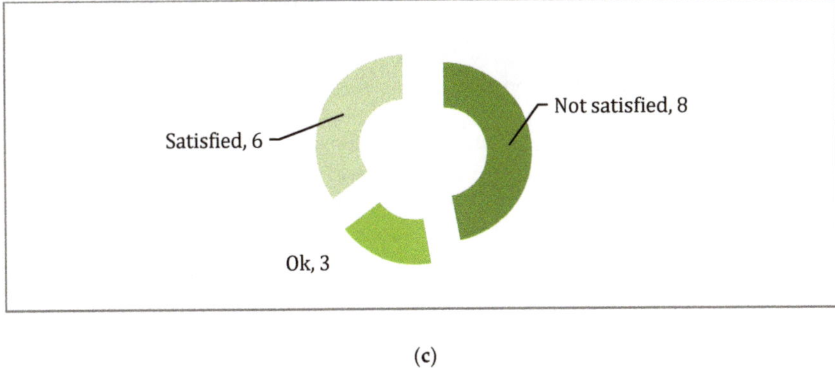

(c)

Figure 6. Demonstration of the students': (**a**) Perception of the difficulty of the course, (**b**) performance achieved overall and (**c**) own satisfaction with their individual performance at this course.

4. Discussion

The role of students' motivation on achieved learning outcomes can be further evaluated. The findings suggest that students' motivation could be either intrinsic or extrinsic. Students demonstrated well how they perceive their role as future professionals and they showed a great sense of duty. Examples of primarily extrinsic drive may be seen in replies to question a3, where some individuals were driven by professional status and income. Further, a good number of students (about 40%) explicitly stated their interest or specific focus in water and environmental engineering in addition to other disciplines (such as structural and geotechnical), which signifies their particular interest for this class (Figure 7). A good correlation holds for those who stated explicitly primarily intrinsic motivation, along with specific aspirations. This demonstrates that motivation is an important factor in positively influencing the learning outcomes, which echoes the argument by Baeten et al. [24].

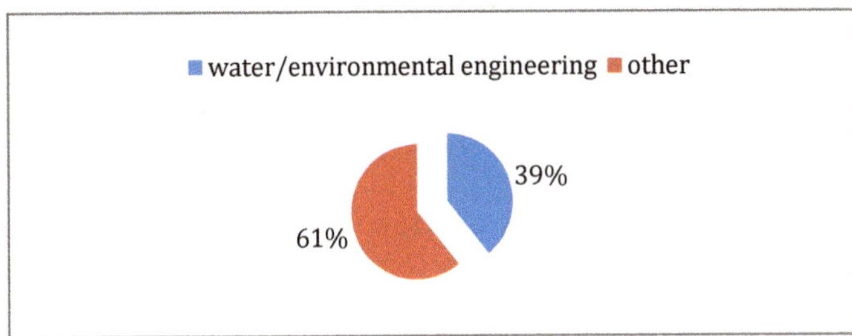

Figure 7. Students' focus on particular engineering disciplines.

Indeed intrinsically motivated students will typically tend to perform better than the rest of the students, as they work harder and more independently, regardless of the learning environment. However, a student-centred environment with highly-visual physical demonstrations, as those described herein, allows for accommodating a wider range of student learning styles, having a greater potential in ensuring that all students do even better, compared to a traditional learning environment. It is the belief of the authors that the time and effort put into the search for active learning strategies is valuable, given the increase in terms of both student satisfaction and experiential learning. Driven by a deep interest for the topic, the instructor can further support the development of a student focus group, consisting of actively engaged students who are motivated in utilising more of their potential while polishing some of their lab work, technical writing, teamwork and presentation skills. The group could be meeting on a regular (e.g., bi-weekly) basis to discuss/present active student projects, opportunities for volunteering or ongoing research in water engineering as well as promoting lab and research activities.

Kabilan et al. [25] argues for the use of Facebook to enrich the learning environment and to reach out to students more efficiently and timely. Even though it seems that some potential exists in the use of social media for effective teaching, the use of physical lab experiments is definitely more defining for promoting active learning, because students can feel present in an inclusive environment, receiving direct feedback and having the chance to interact in a focused manner. For example, if an environment is cultivated aiming at actively engaging with the students, then students will develop a deep learning approach [22,26]. By contrast, students are likely to focus on exams and performance, if the class is perceived as overly difficult with a high workload. Regarding the use of modern online technologies, such as Facebook, in alignment with students' replies, it is worth investigating the potential to assist learning, but only on a complementary basis and in addition to existing platforms currently in use (Moodle and email). Amongst the greatest advantages would be offering direct, timely and personalized course feedback. However, to facilitate learning the tool has to be fully integrated and promote the course objectives (e.g., with live links to video and other resources visible in the timeline) as the course progresses [25].

With regard to assessment methods, even though the exam retains a significant weight of the overall course grade (at least 60%), having more options for assessment rewards the students who are consistent and deep learners, rather than those who simply prepare better for the exams. There have been some who expressed disappointment over the exam being "too short" or "did well in homework but messed up with final exam". So having more assessment components can relieve concerns due to a "hard" exam. Further, having a plethora of assessment methods delivered during the course allows many students to get involved and engaged to learn early on rather than just prepare for the exam, focusing on the grade as a goal to itself. Generally, good performance correlated well with satisfaction for the course, with the exception of one individual who reported being unsatisfied, despite the high grade (A4), as they know "more about the exam than the course". Evidently the students are overall quite confident in knowing well most of the specific thematic questions, long after the class had finished, which was demonstrative of deep learning achieved, as confirmed by Lizzio et al. [26] and Baeten et al. [24].

The students need be able to clearly see that assessment is strongly linked with the intended learning outcomes of the course, and that achieving these takes "time and effort" [27]. As many new students may get easily distracted in our technological era, it is important to encourage explicit interactive sessions that promote a deep learning approach according to Nicol [27]. It is also important to emphasize that extrinsically motivated students may not find value in undertaking assessments that do not contribute towards their final grade. Thus widening the range of offered methods of assessment, by using online quizzes as part of the summative assessment, can also help by offering timely constructive feedback. Well-designed online (Moodle) quizzes offer a great way to offer automated, direct feedback that enables students to act upon improving any mistakes (e.g., by allowing multiple attempts and offering a rationale on every wrong answer), helping to motivate and engage students [27].

To this goal, the role of lab-based demonstrations is of high value in achieving active student learning and offering a supportive, interactive and inclusive environment. The use of *floodopoly* has been demonstrated here for helping reinforce learning and promote the learning objectives of a design course on assessment and modelling of flood risk [28]. The activity described herein is part of a design project, which typically does not involve exams. However, the students in this Design Project as well as in traditional modules which involve other types of assessment (lab reports, assessments and exams) typically explicitly demonstrate a significantly high satisfaction rate for the part of the module involving the physical lab component.

"Floodopoly" is highly versatile and can also be used for outreach purposes, to showcase the detrimental impacts of flooding, for both natural ecosystems and built infrastructure (see relevant posts on Twitter https://twitter.com/WaterEngLab/status/758270564561784832 and Youtube https://youtu.be/H5oThT6QaTc). Highly interactive sessions were used, whereby the students simulate the scenarios of "urbanisation" (by placing more buildings on the floodplains) and "climate change", where more extreme flow rates have to be routed through a greater extent of the floodplains, having even more catastrophic consequences. Using such novel student-centred demonstrations, the students are given a truly unique experience and are encouraged to appreciate engineering principles and design approaches towards installing contemporary flood protections for protecting the model city from extreme hydrologic events. This demonstration can be also combined with targeted research article reading on the processes of flooding and sediment transport [29,30] to break down any access barriers in engineering innovation and promote students' confidence in understanding and eventually performing research. Students who took part in this taught module, and chose a relevant career to become hydraulic engineers or flood modelers, expressed their satisfaction and how much they have been positively impacted by the physical lab demonstrations and research-led component of the course. Indicative is the unsolicited statement of a recent graduate student, currently employed as a Graduate Engineer at a big international engineering consultancy, who described how he has been inspired to possibly pursue engineering innovation or even engineering research in the future: "I know I enjoyed the research aspect, investigating and creating something new ... ultimately I would like to continue ... ".

The demonstration has been very successful as it has been presented to thousands of students in the School's Open days, Widening Participation events, Glasgow Science Festival, Glasgow Science Museum and Engineering Hydraulics classes and Design projects, for more than eight years, while it can be also used for engaging aspiring future female engineers, actively promoting diversity in engineering (e.g., see the International Women in Engineering Day, INWED events, where the authors are also actively engaged). As an example, the Head of the Civil Engineering Discipline has been very supportive of the project and has emphasized its importance for the design courses and outreach events, alike:

"Floodopoly is a successful interactive demonstration tool depicting hydrology design projects carried out by Civil Engineers all over the world. It allows students to immediately see the impact of changes in water supply and flood direction on simulated towns and due to the fact that it is portable and adaptable, it has been used at a variety of outreach activities. It was developed by Dr Valyrakis about a decade ago, is an excellent example of research led teaching and has been utilised in a variety of Hydraulics laboratory sessions, School of Engineering Open Days and Glasgow Science Festival activities."

5. Conclusions

This study presents the use of a novel approach, namely the "floodopoly" demonstration, in facilitating the teaching of fundamental physical principles around the hydraulic processes that shape the Earth's surface and impact our society, while also informing about engineering designs for flood protection. The design of the "floodopoly" setup replicates the topography of an urbanised town, along with its rivers, and demonstrates the impacts of flooding (induced artificially using small water pumps in a recirculating demonstration flume or custom-made portable sand-box, at the Water Engineering Lab of the University of Glasgow).

A purpose-built survey is used at the end of the design course to assess the effectiveness of the environment and lab demonstrations on enhancing the learning experience of the students. The findings suggest that students with a high level of intrinsic motivation are deep learners and are also top performers in a student-centred learning environment. A supportive teaching environment with a plethora of resources and feedback made available over different platforms has the potential to improve student satisfaction and their learning experience. This can be greatly enhanced by interactive lab activities, as offered by the "floodopoly" demonstration. These results have deep implications about student learning and can be used to further improve the design and delivery of water engineering courses in the future. Depending on resources, technical support and expertise available, it is expected that the physical demonstration can become even more engaging if it is combined with modern technological advances, such as augmented reality (AR), to enhance the visual elements of flood impacts and render the activity even more interactive.

Author Contributions: Conceptualization, design, methodology, conduct, formal analysis, visualisation, supervision, project administration, funding acquisition and writing—original draft preparation M.V.; writing—review and editing, M.V., G.L., E.K. and M.C. All authors have read and agreed to the published version of the manuscript.

Funding: Funding from the University of Glasgow's, Chancellor's Fund is gratefully acknowledged in generating improved versions of "floodopoly".

Acknowledgments: The help of the Water Engineering Lab Technician, Tim Montgomery, in constructing the tank and sand-box for "floodopoly", is greatly acknowledged.

Conflicts of Interest: The authors declare no conflicts of interest.

References

1. Bonwell, C.C.; Eison, J.A. Active Learning: Creating Excitement in the Classroom. In *ASHE-ERIC Higher Education Report*; School of Education and Human Development, George Washington University: Washington, DC, USA, 1991; p. 104, ISBN 1-878380-08-7.
2. Bonwell, C.C.; Sutherland, T.E. The active learning continuum: Choosing activities to engage students in the classroom. *New Dir. Teach. Learn.* **1996**. [CrossRef]
3. Reynolds, O. An experimental investigation of the circumstances which determine whether the motion of water shall be direct or sinuous, and of the law of resistance in parallel channels. *Proc. R. Soc. Lond.* **1883**, *35*, 84–89. [CrossRef]
4. King, A. Improving lecture comprehension: Effects of a metacognitive strategy. *Appl. Cogn. Psychol.* **1991**, *4*, 331–346. [CrossRef]
5. Chanson, H. Enhancing Students' Motivation in the Undergraduate Teaching of Hydraulic Engineering: Role of Field Works. *J. Prof. Issues Eng. Educ. Pract.* **2004**, *130*, 259–268. [CrossRef]
6. De Jong, T.; Sotiriou, S.; Gillet, D. Innovations in STEM education: The Go–Lab federation of online labs. *Smart Learn. Environ.* **2014**, *1*, 3. [CrossRef]
7. Koretsky, M.; Amatore, D.; Barnes, C.; Kimura, S. Education of student learning in experimental design using a virtual laboratory. *IEEE Trans. Educ.* **2008**, *51*, 76–85. [CrossRef]
8. Xiang, S.; Wang, L.C. VGLS: A virtual geophysical laboratory system based on C# and viustools and its application for geophysical education. *Comput. Appl. Eng. Educ.* **2017**, *25*, 335–344.
9. Mirauda, D.; Erra, U.; Agatiello, R.; Cerverizzo, M. Mobile augmented reality for flood events management. *Water Study* **2018**, *13*, 418–424. [CrossRef]
10. Pauniaho, L.; Hyvönen, M.; Erkkilä, R.; Vilenius, J.; Koskinen, K.T.; Vilenius, M. Interactive 3D Virtual Hydraulics. In *E–Training Practices for Professional Organizations*; Springer: Boston, MA, USA, 2005; Volume 167, pp. 273–280.
11. O'Neill, G.; McMahon, T. Student-centred learning: What does it mean for students and lecturers. In *Emerging Issues in the Practice of Learning and Teaching*; All Ireland Society for Higher Education (AISHE), Dublin City University: Dublin, Ireland, 2005; p. 173, ISBN -0–9550134–0–2.
12. Valyrakis, M.; Cheng, M. Enhancing learning in geosciences and water engineering via lab activities. *EGU Gen. Assem. Conf. Abstr.* **2016**, *18*, EGU2016-17324-1.
13. Jenkins, A.; Healey, M.; Zetter, R. *Linking Research and Teaching in Disciplines and Departments*; HE Academy: York, UK, 2007; p. 99.
14. Bogaard, A. Small Group Teaching: Perceptions and problems. *Politics* **2005**, *25*, 116–125. [CrossRef]
15. Thomas, C.; Reeves, J.; Laffey, M. Design, Assessment, and Evaluation of a Problem-based Learning Environment in Undergraduate Engineering. *High. Educ. Res. Dev.* **1999**, *18*, 219–232. [CrossRef]
16. Healey, M. Linking Research and Teaching to Benefit Student Learning. *J. Geogr. High. Educ.* **2005**, *29*, 183–201. [CrossRef]
17. Prince, M.J.; Felder, R.M.; Brent, R. Does Faculty Research Improve Undergraduate Teaching? An Analysis of Existing and Potential Synergies. *J. Eng. Educ.* **2007**, *96*, 283–294. [CrossRef]

18. Liu, D.; Valyrakis, M.; Williams, R. Flow hydrodynamics across open channel flows with riparian zones: Implications for riverbank stability. *Water* **2017**, *9*, 720. [CrossRef]

19. Gaudio, R.; Tafarojnoruz, A.; Calomino, F. Combined flow–altering countermeasures against bridge pier scour. *J. Hydraul. Res.* **2012**, *50*, 35–43. [CrossRef]

20. Ferraro, D.; Tafarojnoruz, A.; Gaudio, R.; Cardoso, A.H. Effects of pile cap thickness on the maximum scour depth at a complex pier. *J. Hydraul. Eng.* **2013**, *139*, 482–491. [CrossRef]

21. Valyrakis, M.; Michalis, P.; Zhang, H. A new system for bridge scour monitoring and prediction. In Proceedings of the 36th IAHR World Congress, The Hague, The Netherlands, 28 June–3 July 2015; pp. 1–4.

22. Looney, J.W. Integrating Formative and Summative Assessment: Progress Toward a Seamless System. *OECD Educ. Work. Pap.* **2011**. [CrossRef]

23. Ramsden, P. *Learning to teach in Higher Education*; Routledge: Oxford, UK, 1992. [CrossRef]

24. Baeten, M.; Kyndt, E.; Struyven, K.; Dochy, F. Using student–centred learning environments to stimulate deep approaches to learning: Factors encouraging or discouraging their effectiveness. *Educ. Res. Rev.* **2010**, *5*, 243–260. [CrossRef]

25. Kabilan, M.K.; Ahmad, N.; Abidin, M.J.Z. Facebook: An online environment for learning of English in institutions of higher education? *Internet High. Educ.* **2010**, *13*, 179–187. [CrossRef]

26. Lizzio, A.; Wilson, K.; Simons, R. University students' perceptions of the learning environment and academic outcomes: Implications for theory and practice. *Stud. High. Educ.* **2002**, *27*, 27–52. [CrossRef]

27. Nicol, D. *Transforming Assessment and Feedback: Enhancing Integration and Empowerment in the First Year*; First Year Enhancement Theme Report; QAA: Scotland, UK, 2009.

28. Valyrakis, M.; Solley, M.; Koursari, E. Flood risk modeling of urbanized estuarine areas under uncertainty: A case study for Whitesands, UK. *Br. J. Environ. Clim. Chang.* **2015**, *5*, 147–161. [CrossRef] [PubMed]

29. Valyrakis, M.; Diplas, P.; Dancey, C.L. Entrainment of coarse grains in turbulent flows: An extreme value theory approach. *Water Resour. Res.* **2011**, *47*. [CrossRef]

30. Valyrakis, M.; Diplas, P.; Dancey, C.L. Prediction of coarse particle movement with adaptive neuro–fuzzy inference systems. *Hydrol. Process.* **2011**, *25*, 3513–3524. [CrossRef]

Article

Suite-CFD: An Array of Fluid Solvers Written in MATLAB and Python

Nicholas A. Battista

Department of Mathematics and Statistics, The College of New Jersey, 2000 Pennington Road, Ewing Township, NJ 08628, USA; battistn@tcnj.edu; Tel.: +1-609-771-2445

Received: 1 October 2019; Accepted: 19 February 2020; Published: 25 February 2020

Abstract: Computational Fluid Dynamics (CFD) models are being rapidly integrated into applications across all sciences and engineering. CFD harnesses the power of computers to solve the equations of fluid dynamics, which otherwise cannot be solved analytically except for very particular cases. Numerical solutions can be interpreted through traditional quantitative techniques as well as visually through qualitative snapshots of the flow data. As pictures are worth a thousand words, in many cases such visualizations are invaluable for understanding the fluid system. Unfortunately, vast mathematical knowledge is required to develop one's own CFD software and commercial software options are expensive and thereby may be inaccessible to many potential practitioners. To that extent, CFD materials specifically designed for undergraduate education are limited. Here we provide an open-source repository, which contains numerous popular fluid solvers in 2*D* (projection, spectral, and Lattice Boltzmann), with full implementations in both MATLAB and Python3. All output data is saved in the *.vtk* format, which can be visualized (and analyzed) with open-source visualization tools, such as VisIt or ParaView. Beyond the code, we also provide teaching resources, such as tutorials, flow snapshots, measurements, videos, and slides to streamline use of the software.

Keywords: fluid dynamics education; applied mathematics education; computational fluid dynamics; fluids visualization; open-source CFD; projection method; spectral fluid solver; Lattice Boltzmann Method; cavity flow; circular flow; interacting vortices; flow past cylinder; flow past porous cylinder; MATLAB; Python

---

1. Introduction

Computational Fluid Dynamics (CFD) models are being applied to problems across all sciences and engineering. From designing more aerodynamic sportswear and vehicles, to understanding animal locomotion, to personalized medicine, to disease transmission, and to predicting hurricanes and atmospheric phenomena, it is difficult to find situations where greater knowledge of the underlying fluid dynamics isn't desired or valuable. Due to its immense importance, many scientists have dedicated their careers to the field, whether they study particular fluid phenomena or develop tools, either experimental or numerical, for other scientists and engineers to use. Possibly because of how vital understanding fluid dynamics is to human flourishing, proving existence, uniqueness, and the possible breakdown of solutions to the system of non-linear partial differential equations that govern fluid dynamics is one of the *Millennium Problems*, to which the Clay Institute offers a 1 million dollar prize for solving [1]. On the other hand, CFD allows us to solve these equations, the Navier-Stokes equations, which are the equations that detail the conversation of momentum and mass for a fluid. For an incompressible, viscous fluid, they can be written as follows,

$$\rho\left[\frac{\partial \mathbf{u}}{\partial t}(\mathbf{x},t) + (\mathbf{u}(\mathbf{x},t)\cdot\nabla)\,\mathbf{u}(\mathbf{x},t)\right] = -\nabla p(\mathbf{x},t) + \mu\Delta\mathbf{u}(\mathbf{x},t) \tag{1}$$

$$\nabla\cdot\mathbf{u}(\mathbf{x},t) = 0 \tag{2}$$

where $\mathbf{u}(\mathbf{x}, t)$ and $p(\mathbf{x}, t)$ are the fluid's velocity and pressure, respectively. The physical properties of the fluid are given by ρ and μ, its density and dynamic viscosity, respectively. The independent variables are the time, t, and spatial position, $\mathbf{x}$. Here all variables that pertain to the fluid, e.g., $\mathbf{u}$ and p, are written in an Eulerian framework on a fixed Cartesian mesh, $\mathbf{x}$. Think of a fixed Cartesian mesh as a rectangular grid to which over the course of the simulation physical quantities, like velocity, pressure, etc., are measured at specific lattice points. In an Eulerian framework, specific blobs of fluid are not tracked over time, as though they are billiard balls, but rather, we focus on measuring quantities of interest at specific locations in space, located on a fixed grid. Note that Equations (1) and (2) provide the conservation of momentum and mass, respectively.

Many techniques have been developed to solve the above system of equations, such as projection methods [2–6], spectral methods [7–9], and Lattice Boltzmann methods [10–12]; however, they are typically left out of undergraduate curricula due to the mathematical background required for implementation, while they are commonly found in graduate curricula [13]. Moreover, it is not only mathematical knowledge that is a barrier; CFD requires expertise in at least four knowledge domains [14]: flow physics, numerical methods, computer programming, and validation, either experimental or theoretical. Thankfully, this has not stopped academics from developing and researching methods to integrate CFD into undergraduate courses [13–21]. Some have offered best practices guides and advice for future CFD undergraduate courses [13,21]. Stern et al. [13] has suggested for beginners to focus on the overall CFD process and flow visualizations, in order to solidify their fundamental understanding of fluid physics. Appropriate flow visualizations also help students validate and verify their simulations through careful inspection [21].

As computer literacy is rapidly becoming more of a central focus in undergraduate curricula, CFD could be a natural outlet where students may not only implement and test algorithms, but also practice proper data analysis and data visualization, while building computational experience. Much of the fluid dynamics software that has been used in previous courses is either foreign to students or only commercially available [14,17,19,20]. Commercial software has been recommended by the American Physical Society and American Association of Physics Teachers as a tool to help students prepare for 21st century careers by exposing them to a greater range of experiences and opportunities [22,23]. Unfortunately due to expensive software licenses, some schools, particularly smaller institutions, may not be able to afford such software, especially if they would only be used intermittently to complement the existing course material.

Recently, popular programming languages selected for many undergraduate curricula in science, mathematics, and engineering have been either MATLAB or Python [24–28]. MATLAB [29] and Python [30] are both interpreted languages, making it easier (and more attractive) for students to begin programming [31]. Moreover if students have already been exposed to these programming languages, giving them a platform to refine their skills in these languages while at the same strengthening their intuition in fluid dynamics may prove beneficial.

For this reason, many scientists, engineers, and mathematicians who are passionate about education have begun developing CFD software intended for educational purposes written in these languages [32–36]. Such open source software packages can be integrated into courses in a well-streamlined fashion, with tutorials, activities, exercises, and notes provided, and may even blur the line between classroom activities and contemporary research [32,35–37]. In particular L.A. Barba and G. Forsyth [33] have developed a guide for students to numerically solve the Navier-Stokes equations in 12 scaffolding steps and L.A. Barba and O. Mesnard [34] have developed a similar scaffolding lesson structure for students to study classical aerodynamics using potential flow models. Both are written in Python using Jupyter Notebooks [38]. Pawar and San [39] developed modules for teaching advanced undergraduate and graduate students how to develop their own fluid solvers in the Julia programming language [40].

Rather than develop a collection of modules that teach how to implement particular CFD numerical schemes, we offer the scientific community a variety of popular fluid solvers that solve traditional problems in fluid dynamics, with two independent but equal implementations written in MATLAB and Python. The emphasis is not placed on students having to develop or implement numerical schemes, but to allow them the opportunity to get an accelerated start in CFD, and run, tweak, and analyze simulations of traditionally studied problems in fluids courses, as well as explore data visualization and how to present data that is appropriate for fluid physics interpretation. All of this can be done while testing their own hypotheses in familiar programming environments. In the remainder of this manuscript, we will give a high-level overview of the CFD schemes currently implemented in the software (Section 2), guided tutorials on how to run, visualize, and analyze simulations (Section 3), and provide numerous examples to which students may elect to study or modify as well (Section 4).

Furthermore, we provide an in-depth mathematical description of each numerical scheme in Appendix B. The open-source software repository can be accessed at https://github.com/nickabattista/ Holy_Grail. Any simulation data produced from the software can be visualized and analyzed in open-source software, such as VisIt [41], maintained by Lawrence Livermore National Laboratory, or ParaView [42], developed by Kitware, Inc., Los Alamos National Laboratory, and Sandia National Labs. We also provide teaching resources, such as slides, figures, and movies in the Supplementary Materials.

2. Brief Overview of the Three Fluid Solvers

In this section we will give a brief overview of the three fluid solvers currently implemented in the software—a projection method, a spectral (FFT) method, and the Lattice-Boltzmann method. Our goal is to provide students a high-level overview of the methods and how they compare to one another. Further details regarding their mathematical foundations and implementations are given in Appendix B.

2.1. Projection Method

The projection method was first introduced by Chorin in 1967 [2] and independently by Temam in 1968 [4], to solve the viscous, incompressible Navier-Stokes equations. Projection methods are finite difference based numerical schemes [43], in which the fluid equations are solved in an Eulerian framework, i.e., the computational grid is static and fluid dynamics variables, like velocity, pressure, etc., are measured at particular locations on the grid over time during a simulation. Projection methods have been extensively used throughout the fluid dynamics community for decades, in numerous applications across many fields, while being continually improved for efficiency and accuracy [5,44–46].

In a nutshell, this method decouples the velocity and pressure fields, using operator splitting and a Helmholtz-Hodge decomposition. This makes it possible to explicitly solve Equations (1) and (2) in a few steps [2] while also increasing computational efficiency. Please see Appendix B.1 for further mathematical details.

Furthermore, projection methods allow one to define explicit boundary conditions (BCs) for velocity on edges of the computational domain, whether they are *Dirichlet, Neumann,* or *Robin* boundary conditions [5,47]. In our software, we allow users to impose tangential velocity boundary conditions along the edges of the rectangular domain, while the normal direction boundary conditions for velocity are explicitly set to zero. The choice to set normal BCs to zero was made in order to simplify the code by ensuring that volume conservation would be automatically satisfied, i.e., situations will not arise in which a specified amount of fluid is being pumped into the domain while a different amount is leaving the domain per unit time.

Having defined the fluid velocity as $\mathbf{u} = (u, v)$, one can set the boundary conditions for u on the top and bottom of the domain ($v = 0$ along these edges) or v on the left and right sides ($u = 0$ along these edges). See Figure 1 for an illustration. Later in Section 4, we will showcase two different traditional problems in fluid dynamics with a projection method:

1.   Cavity Flow
2.   Circular Flow in a Square Domain

The examples are different due to the changes in boundary conditions that the user can impose, see Figure 1. In Figure 1 we illustrate the boundary conditions for each example: cavity flow (a) and circular flow (b). This figure also illustrates that the domain could be constructed to be either rectangular or square and that the boundary conditions do not have to be uniform among all sides, nor across the domain.

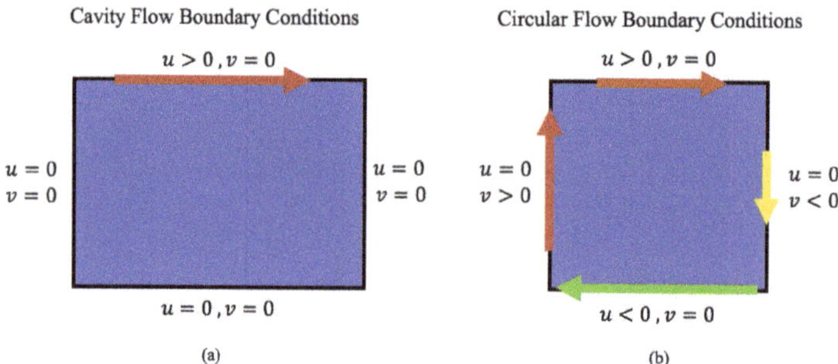

Figure 1. Boundary conditions are depicted pertaining to (**a**) cavity flow and (**b**) circular flow for the projection methods examples.

We will also compare different fluid scales in both examples, using the Reynolds Number, *Re*. The Reynolds Number is given by

$$Re = \frac{\rho L V}{\mu},\tag{3}$$

where ρ and μ are the fluid's density and dynamic viscosity, respectively, while L and V are a characteristic length- and velocity-scale of the system. In essence, *Re* is a ratio of inertial forces to viscous forces in a system. If $Re \gg 1$, inertia dominates, if $Re \ll 1$, viscous forces dominate, and if $Re = 1$, the inertia and viscous forces are balanced. An example of high *Re* flow would be the flow around a marble quickly falling through air. An example of low *Re* flow would be the fluid flow around a marble slowly falling through honey.

2.2. Spectral Method (FFT)

Spectral methods are a different class of differential equation solvers. They are well-known for achieving highly accurate solutions, i.e., spectral accuracy [48–50]. Spectral accuracy occurs when error decreases exponentially with a small change in grid resolution, e.g., there is a linear relationship between the logarithm of error and grid resolution, i.e.,

$$\log(error) \sim N,$$

where N is the number of grid points. These methods approximate solutions using orthogonal expansions, such as Fourier Series or other orthogonal basis of functions, like Chebyshev Polynomials or Legendre Polynomials. In contrast to finite difference based schemes, like the projection method, where solutions are approximated at specific locations in space, in spectral methods a particular orthogonal series expansion is chosen and the earnest is on finding the coefficients of the expansion.

Here we chose to use the basis of Discrete Fourier Transform (DFT). The DFT can be used to express functions that are not periodic, unlike Fourier Series expansions, which are used to represent periodic functions. Moreover, we used the Fast Fourier Transform (FFT), which yields the same numerical values as a DFT, but is considerable more computationally efficient, i.e., fast.

Thus, we implemented a FFT based spectral scheme to solve the viscous, incompressible Navier-Stokes equations in $2D$. We also chose to solve these equations in their vorticity formulation (see Appendix B.2 for mathematical details). The vorticity formulation will lead to a few numerical benefits.

In a nutshell, this will recast our problem into solving for the vorticity, $\boldsymbol{\omega} = \omega\hat{k}$, and the streamfunction, ψ. Velocity can be defined in terms of the curl of the streamfunction, i.e., a vector potential, see below

$$\mathbf{u} = \nabla \times \psi\hat{k}. \tag{4}$$

Hence this allows us to find the velocity field, $\mathbf{u} = (u, v)$, once we find ψ, e.g.,

$$u = \frac{\partial \psi}{\partial y} \quad \text{and} \quad v = -\frac{\partial \psi}{\partial x}.$$

Moreover, the incompressibility condition (Equation (2)) will be automatically satisfied since $\mathbf{u}$ is defined to be the curl of a scalar, and the divergence of the curl of a vector is always identically zero.

Furthermore, working in the vorticity formulation requires us to only solve a Poisson equation for the streamfunction in terms of the vorticity, ω, i.e.,

$$\Delta \psi = -\omega. \tag{5}$$

Notice that Equation (5) is a linear equation. Moreover, upon taking the FFT of the Equation (5), the computation is transformed into frequency (Fourier) space, which offers two advantages—increased speed and accuracy.

In practice, we will solve for the streamfunction at the next time-step based on the previously computed vorticity and then update the vorticity using information from the newly updated streamfunction. We chose to use a Crank-Nicholson time-stepping scheme [51] to update the vorticity to the next time-step. The Crank-Nicholson scheme is well-known for being unconditionally-stable for diffusion problems and as well as for being 2nd order accurate in time and space, as an implicit method [52]. These accuracy and stability properties make the Crank-Nicholson an ideal candidate for numerically solving such equations.

Although using a FFT (spectral method) preserves high accuracy of the spatial information at a particular time-step, numerical error still unfortunately creeps in due to the time-stepping nature of solving an evolution equation for the fluid's momentum (see Equation (A15) in Appendix B.2). That is, evolving the system forward in time is where the brunt of the error takes place, rather than the spatial discretization.

In contrary to the projection method, this FFT-based approach allows us to explore periodic boundary conditions. For example, if fluid flow is moving vertically-upwards through the top of the computational domain, it will re-enter moving vertically-upwards through the bottom of the domain. This has advantages and disadvantages. Some advantages are that explicit boundary conditions do not need to be satisfied nor enforced, which aids in computational speed-up. However, to that extent, one disadvantage is that this makes it more difficult to enforce particular boundary conditions when desired. FFT-based fluid solvers have been used in the fluid-structure interaction community, in particular within immersed boundary methods [32,37,53], when one wants to study the interactions of an object and the fluid to which it's immersed. As the focus is on the object and fluid interactions, it is typically modeled away from the domain boundaries in the middle of the computational domain to avoid boundary artifacts. For a mathematically detailed description of this FFT-based spectral method see Appendix B.2.

For this particular FFT-based fluid solver for the vorticity-formulation of the viscous, incompressible Navier-Stokes equations, we will highlight the following examples:

1. Side-by-Side Vorticity-Region Interactions
2. Evolution of Vorticity from an Initial Velocity Field

Contrary to the examples shown for the projection method in Section 2.1, these examples are not differentiated by different boundary conditions, but rather different initial vorticity configurations in the computational domain. Figure 2 gives the initial vorticity configurations for the two cases considered: side-by-side vorticity region interactions (left) and an initial vorticity field that defined by a velocity vector field (right). In Figure 2a, counterclockwise (*CCW*) and clockwise (*CW*) correspond to regions of uniform vorticity, where vorticity initialized as a positive or negative constant for *CCW* and *CW*, respectively. In all other regions of the computational domain, the vorticity is either initialized to zero (Figure 2a). Although not shown, one could also initialize simulations on a rectangular grid. In Figure 2b, the initial vorticity is defined by computing the vorticity from a simulation's velocity vector field. The velocity vector field was taken from a time-point in a *Pulsing_Heart* simulation [36] in the open-source *IB2d* software [32,37]. In this example, we illustrate that if a snapshot of the velocity field is known, it is possible to evolve the fluid dynamics forward using this spectral (FFT) method, even though it is based on the vorticity formulation of the Navier-Stokes equations.

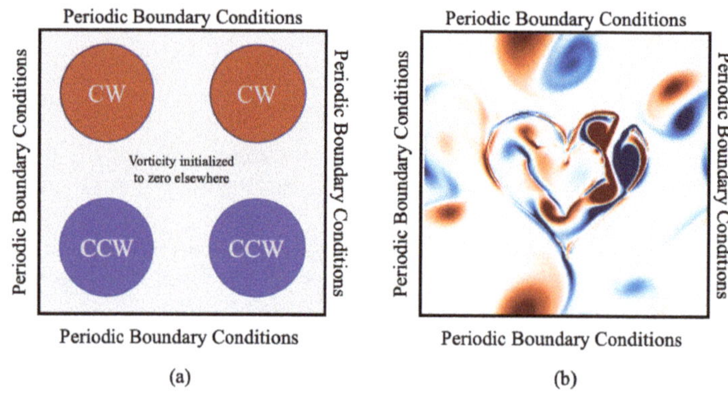

Figure 2. Illustrations of the boundary conditions and vorticity initialization for the cases of (**a**) interacting vorticity regions and (**b**) an initial vorticity field defined from a velocity vector field.

2.3. Lattice Boltzmann Method

The Lattice Boltzmann method (LBM) does not explicitly (or implicitly) solve the incompressible, Navier-Stokes equations, rather it uses discrete Boltzmann equations to model the flow of a fluid [11]. In a nutshell, the LBM tracks fictitious particles of fluid flow, thinking of the problem more as a transport equation, e.g.,

$$\frac{\partial f}{\partial t} + \mathbf{u} \cdot \nabla f = \Omega, \tag{6}$$

where $f(\mathbf{x}, t)$ is the particle distribution function, i.e., a probability density, $\mathbf{u}$ is the fluid particle's velocity, and Ω is what is called the collision operator. However, rather than these particles moving in a Lagrangian framework, the Lattice Boltzmann method simplifies this assumption and restricts the particles to the nodes on a lattice.

In traditional CFD methods, one typically discretizes the fluid equations in an Eulerian framework, as in a projection or spectral method. When you are free of such restriction, as in the LBM, it allows one to more readily deal with complex boundaries, model porous structures and multi-phase flow, as well as incorporate microscopic interactions, while also allowing for massive parallelization of the algorithm, leading to increased computational efficiency [12,54–56].

The LBM involves *collision* and *streaming* steps, where fictional fluid particles consecutively propagate (and collide) over a discretized lattice, and the fluid density is evolved forward. There are multiple ways to handle boundary conditions in the LBM. One such way is to use *bounce-back boundary conditions*, which are executed by masking points in the domain via boolean expressions, thus defining regions where the fluid can and cannot flow [55]. In short, the propagating (streaming) directions are simply reversed when they hit a boundary node, see Figure A3 in Appendix B.3. The bounce-back conditions can be used to enforce the necessary and physical *no-slip* boundary conditions for fluid flow as well as defining solid objects in the interior of the grid (for an example see Figure 33). For a detailed mathematical description of the LBM see Appendix B.3.

To highlight the LBM fluid solver, we showcase the following examples:

1.   Flow past one or more cylinders
2.   Flow past porous cylinder

3. How to Run the Simulations, Visualize, and Analyze

In this section we will give instructions on how to run, visualize, and analyze the simulations produced in this software. We will explain how to change parameters below as well. In particular, we will provide both overviews of what is possible to visualize and analyze as well as a guided walk-through of the steps required, designated by *Guide* in their title. We broke this section into the following subsections:

1.   Running a Simulation
2.   Visualizing the Data in VisIt
3.   *Guide*: Running the Spectral (FFT) Method's 'bubble3' Example
4.   *Guide*: Visualizing the Spectral (FFT) Method's 'bubble3' Data
5.   *Guide*: Analyzing the Spectral (FFT) Method's 'bubble3' Data

3.1. Running a Simulation

Here we will briefly describe how to run each fluid solver's corresponding simulations. To illustrate the software structure and how to run the simulations, we will use the MATLAB (https://www.mathworks.com/products/matlab.html) [29] version. Note that the Python implementation is consistent in its structure as well as naming conventions.

To run the simulation, one would need to do the following:

1.   Either clone the Holy_Grail repository or download the Holy_Grail zip file at https://github.com/nickabattista/Holy_Grail/ to your local machine. Note you can download or clone this repository to any directory on your local machine.

2.   Open MATLAB and go to the appropriate sub-directory for the particular fluid solver example you wish to run, e.g.,

     - `Projection_Method`: for the projection method solver and its examples
     - `FFT_NS_Solver`: for the spectral (FFT) method solver and its examples
     - `Lets_Do_LBM`: for the Lattice Boltzmann method solver and its examples

     and go into the MATLAB subfolder. (If you wanted to run an example in Python, you would change directories until you are in the complementary Python subfolder.)

3.   The main script name for each fluid solver is named accordingly, e.g.,

     - `Projection_Method.m` - main script for the projection method
     - `FFT_NS_Solver.m` - main script for the spectral (FFT) solver
     - `lets_do_LBM.m` - main script for the Lattice Boltzmann solver

     Each of these scripts contains all the functions and modules necessary to run each example. The file `print_vtk_files.m` produces the *.vtk* data files during each simulation. The user should not have to change this file, unless they do not wish to save some of the data for storage reasons. If this is the case, this can be accomplished by commenting out the lines corresponding to whichever data is not to be saved.

4. Depending on the subfolder that you are in, to run an example, you would type its main script name into the MATLAB command window and click `enter`.

5. The simulations are not instantaneous; they may take a few minutes. Upon starting the simulation, a folder named `vtk_data` is produced, which will be used for storing the simulation data files in *.vtk* format. As the simulation runs, it will dump more data into this folder. Note that depending on the simulation you choose and the grid resolution, the simulations may take from a few minutes to $\sim$30 min. Moreover, if you wish to run multiple simulations, note that the default folder name is set to `vtk_data`, and so running additional simulations will overwrite any previously saved data in the previous `vtk_data` folder. Please manually change the name of the folder after each simulation to avoid this.

Digging deeper into each fluid solver script:

1. If one desires to change parameters for a particular simulation, e.g., grid resolution, viscosity (either μ (Projection) or ν (FFT) or τ (LBM), etc.), options are available to the user without digging deep into any of the sub-functions or individual modules beyond the main function itself, see Figure 3.

2. Figure 3 highlights the main choices the user can make for each of the three fluid solvers (a) Projection, (b) Spectral (FFT), and (c) Lattice Boltzmann. The user can identify which of the built-in examples they wish to run by changing the appropriate string flag labeled `choice`.

3. Note that each example will run as described in Section 4 if all the parameters are selected to be those in the parameter Table that provided within each example's section. *Caution*: numerical stability conditions may be violated by choice of other parameters. To test other parameter values, we suggest making incremental variations, e.g., try doubling or halving viscosity, rather than testing values orders of magnitude apart. Numerical instability will appear if solvers cannot converge or begin dumping NaNs (*not-a-number*), which represent unrepresentable numbers to computers, due to their inherent floating-point arithmetic.

4. As the entire fluid-solver is contained within each script, the user can go deeper into the code to modify examples or create their own. This can be done in the following sub-functions:

    -   Projection Method: *please_Give_Me_BCs* (choice)
    -   Spectral (FFT) Solver: *please_Give_Initial_Vorticity_State* (choice, NX, NY)
    -   Lattice Boltzmann Solver: *give_Me_Problem_Geometry* (choice, Nx, Ny, percentPorosity)

```
33   %
34   % GRID PARAMETERS %
35   %
36 - Lx = 1.0;           % Domain Length in x
37 - Ly = 1.0;           % Domain Length in y
38 - Nx = 256;           % Spatial Resolution in x
39 - Ny = 256;           % Spatial Resolution in y
40 - dx = Lx/Nx;         % Spatial Distance Definition in x (NOTE: keep dx = Lx/Nx = Ly/Ny = dy);
41
42   %
43   % SIMULATION PARAMETERS %
44   %
45 - mu = 0.25;          % Fluid DYNAMIC viscosity (kg / m*s)
46   %                       (mu=1000,100,10,1 for Re=4,40,400,4000 respectively for Cavity Flow examples
47   %                       (mu=0.25,1.0,2.5 for Re=4000,1000, and 400, respectively for Circular Flow examples)
48 - rho = 1000;         % Fluid DENSITY(kg/m^3)
49 - nu=mu/rho;          % Fluid KINEMATIC viscosity
50 - numPredCorr = 3;    % Number of Predictor-Corrector Steps
51 - maxIter=200;        % Maximum Iterations for SOR Method to solve Elliptic Pressure Equation
52 - beta=1.25;          % Relaxation Parameter
53
54   %
55   % CHOOSE SIMULATION (gives chosen simulation parameters) %
56   % Possible choices: 'cavity_top', 'whirlwind'
57   %
58 - choice = 'whirlwind';
```

(a)

```
34   %
35   % Simulation Parameters
36   %
37 - nu=1.0e-3;   % kinematic viscosity (dynamic viscosity/density)
38 - NX = 256;    % # of grid points in x
39 - NY = 256;    % # of grid points in y
40 - LX = 1;      % 'Length' of x-Domain
41 - LY = 1;      % 'Length' of y-Domain
42
43   %
44   % Choose initial vorticity state
45   % Choices:  'half', 'qtrs', 'rand', 'bubble3', 'bubbleSplit','bubble1','jets'
46   %
47 - choice='bubble3';
```

(b)

```
35   %
36   % Simulation Parameters
37   %
38 - tau=0.53;                        % tau: relaxation parameter related to viscosity
39 - density=0.01;                    % density to be used for initializing whole grid
40 - w1=4/9; w2=1/9; w3=1/36;         % weights for finding equilibrium distribution
41 - Nx=640; Ny=160;                  % number of grid cells in x and y directions, respectively
42 - Lx = 2; Ly = 0.5;                % Size of computational domain
43 - dx = Lx/Nx; dy = Ly/Ny;          % Grid Resolution in x and y directions, respectively
44
45   %
46   % Chooses which problem to simulate
47   %
48   % Possible Choices: 'cylinder1', 'cylinder2', 'porousCylinder'
49   %
50 - choice = 'porousCylinder';
51 - percentPorosity = 0.625;              % Void fraction (if no porosity, no effect)
52 - [BOUND,Bound2,deltaU,endTime] = give_Me_Problem_Geometry(choice,Nx,Ny,percentPorosity);
53                                          %BOUND: gives geometry, deltaU: incremental increase to inlet velocity
```

(c)

Figure 3. Options for practitioners to change the grid resolution and size, fluid properties (density, viscosity), and other solver attributes within the software for the (**a**) Projection method, (**b**) Fast Fourier Transform (FFT)-based spectral solver, and (**c**) Lattice Boltzmann solver. The possible choices for built-in examples are also listed for each.

3.2. Visualizing the Data in VisIt

In this section we will briefly describe how produce visualizations, such as those Section 4. We will briefly detail the steps to visualize the data using the open-source software VisIt (https://visit.llnl.gov) [41]. Note that each simulation produces data in the *.vtk* format and so one could alternatively use the open-source software ParaView [42] for visualization purposes as well. Later, in Section 3.3 we will guide a user through running an example (the spectral (FFT) method's bubble3 example) followed by a step-by-step guide to visualizing its corresponding data in Section 3.4. We used VisIt version 2.13.3.

To visualize the data, one would need to do the following:

1. Open VisIt [41]
2. Open the desired Eulerian data (magnitude of velocity, vorticity, velocity vector field, etc.) from the `vtk_data` folder.
3. A table of the possible data stored and their corresponding name when saved is found below in Table 1.

Table 1. Data stored to *.vtk* format and its corresponding storage name.

Parameter	Type	'*.vtk*' Name
Magnitude of Velocity	Scalar	*uMag*
Velocity Vectors	Vector	*u*
Horizontal Velocity	Scalar	*uX*
Vertical Velocity	Scalar	*uY*
Pressure (*projection only*)	Scalar	*P*
Vorticity	Scalar	*Omega*
Geometry (*LBM only*)	Mesh	*Bounds*

4. *To Visualize Geometry (LBM only):*

    (a) Click `Open`
    (b) Go to the `vtk_data` data folder that the simulation produced
    (c) Click on the grouping of `Bounds`, click `OK`
    (d) In VisIt, click on `Add` then `Mesh`→`mesh`.
    (e) Then click `Draw`
    (f) You can elect to change the color of boundary or size by double clicking on the Mesh in the VisIt database listing window.

5. *To Visualize Velocity Vectors:*

    (a) Click `Open`
    (b) Go to the `vtk_data` data folder that the simulation produced
    (c) Click on the grouping of `u`, click `OK`
    (d) In VisIt, click on `Add` then `Vector`→`u`
    (e) Then click `Draw`. An error message might pop up during at time-step zero (first data point) if velocity is identically zero
    (f) You can elect to change the color, size, scaling, and number of velocity vectors by double-clicking on *u* in the VisIt database listing window
    (g) To add streamlines of the velocity field, first make sure that your *Active Source* is set to *u* (see Figure 4a), then click `Add`→`Pseudocolor`→`operators`→`IntegralCurve`→`u` (see Figure 4b). Note you have the option to put numerous seed points for streamlines to stem in the domain. Here we have chosen to use a line of points, sampled it at 4 evenly spaced locations within that line, and specified particular integration tolerances, and a maximum number of steps (see Figure 4c). Click `Draw`. You can also adjust the streamline aesthetics to how you desire, e.g., color, thickness, etc.

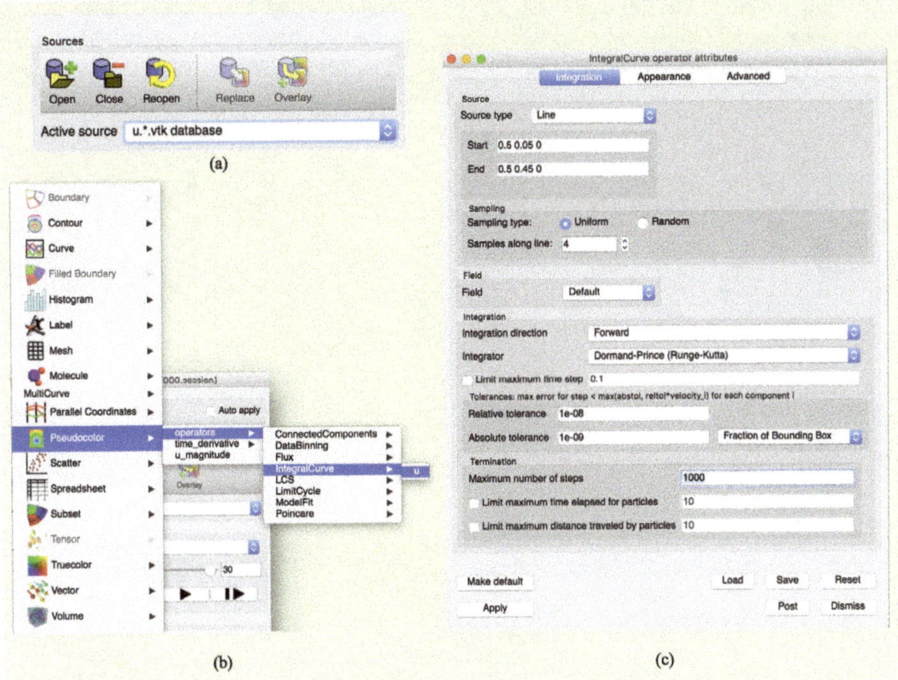

Figure 4. VisIt graphical user interfaces (GUI) interface showing steps to illustrate velocity streamlines, e.g., (**a**) active source must be set on velocity field, (**b**) sequence to plot streamlines, and (**c**) streamline seeding location(s), accuracy tolerances, and solver-type.

6.   *To Visualize the Eulerian scalar data (e.g., Vorticity, Magnitude of Velocity, etc.):*

(a)   Click `Open`

(b)   Go to the `vtk_data` data folder that the simulation produced

(c)   Click on the grouping of the desired Eulerian scalar data, for example, `Omega` (for Vorticity), click `OK`

(d)   In VisIt, click on `Add` then `Pseudocolor`→`Omega`.

(e)   Then click `Draw`

(f)   You can elect to change the colormap and/or colormap scaling by double clicking on Omega in the VisIt database listing window. The following table (Table 2) details the specific colormap used for each scalar variable in the manuscript,

Table 2. Colormaps used for each scalar variable for visualization.

Parameter	Colormap
Magnitude of Velocity	RdYlBu or BrBG (Section 4.1 only)
Horizontal Velocity	RdYlBu
Vertical Velocity	RdYlBu
Pressure	orangehot (Section 4.1 only)
Vorticity	hot_desaturated
FTLE	PuRd

(g)   To visualize the finite-time Lyapunov exponent (FTLE), your *Active Source* in Visit (see Figure 5a) must be on the velocity vectors, u. Then click `Add`→`Pseudocolor`→ `operators`→`LCS`→`u` (see Figure 5b). We used the attributes in Figure 5c for computing the FTLE. In particular, we limited the maximum advection time to 0.05 and maximum number of steps to 10. Once those are changed, click `Draw`.

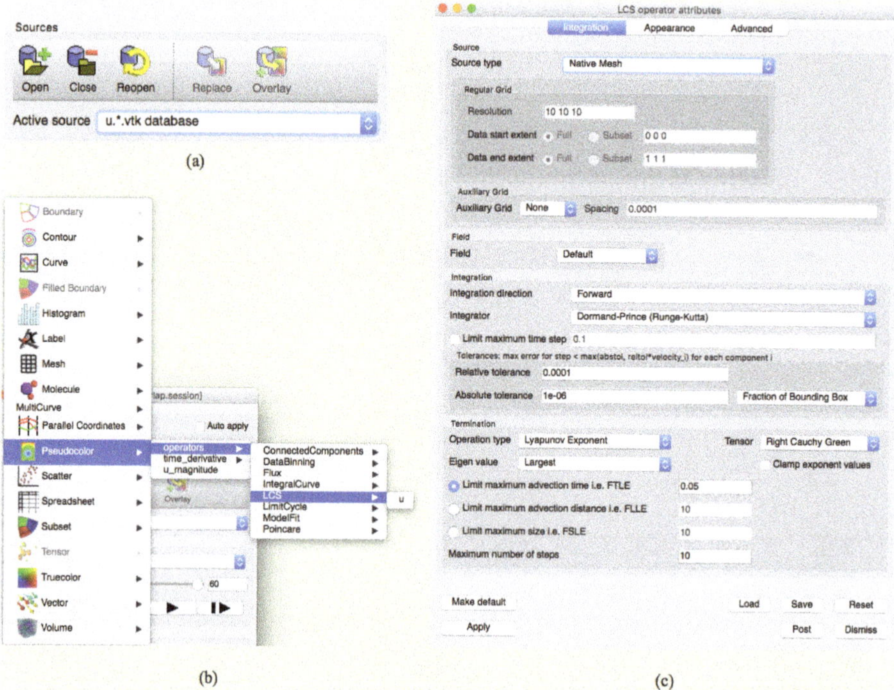

Figure 5. VisIt GUI interface showing steps to illustrate finite-time Lyapunov exponents (FTLE), e.g., (**a**) active source must be set on velocity field, (**b**) sequence to plot FTLE (e.g., Lagrangian Coherent Structures, LCS), and (**c**) attributes for plotting, e.g., accuracy tolerances, solver-type, etc.

(h)   To add contours of the desired scalar variable, first make sure that your *Active Source* (see Figure 5a) is set to the correct variable, then click `Add`→`Contour`→`<desired variable name>`. Note that you have the option to scale the contours separate from the colormap of the variable itself. Moreover, for FTLE your active source must be u and you can modify how FTLE are computed, e.g., Figure 5c; however, we used consistent values for both FTLE computations.

3.3. Guide: Running the Spectral (FFT) Method's 'bubble3' Example

In this guided walk-through, we wish to illustrate how to run a simulation and change its parameters while also observing what information is being printed to the screen. We will use the Spectral (FFT) Method's *bubble3* example and run two different examples where each uses a different fluid kinematic viscosity, ν.

To run the simulation(s), do the following:

1.   First we must make sure that we are inside of MATLAB the corresponding directory for the MATLAB version of the spectral-based solver. To do this click through: `Holy_Grail`→`FFT_NS_Solver`→`MATLAB`, see Figure 6. Note that I have downloaded (or cloned) the software into my `Documents` folder.

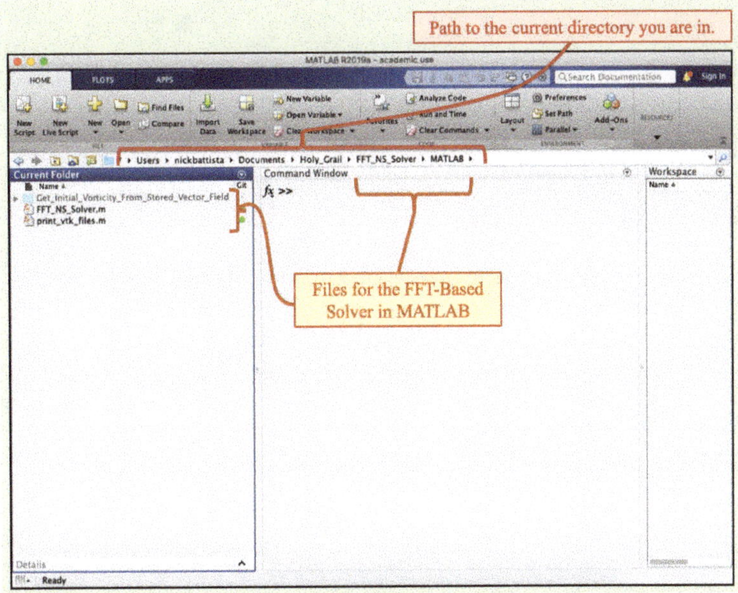

Figure 6. MATLAB GUI interface showing the path to the current directory folder for the MATLAB FFT-Based Spectral Solver as well as what files are contained within it.

2. To run this simulation, type `FFT_NS_Solver` into MATLAB's command window and click `enter`. See Figure 7 for what should print to the command window shortly thereafter. Note that the *bubble3* example is the default built-in example upon downloading the software (see Figure 3b).

Upon starting the simulation, information regarding the simulation solver and simulation is printed to the screen. Moreover, the `current_time` within the simulation gets printed to the screen at the time-points in which the data is stored. The data is stored in the *vtk_data* folder, which is made upon starting the simulation.

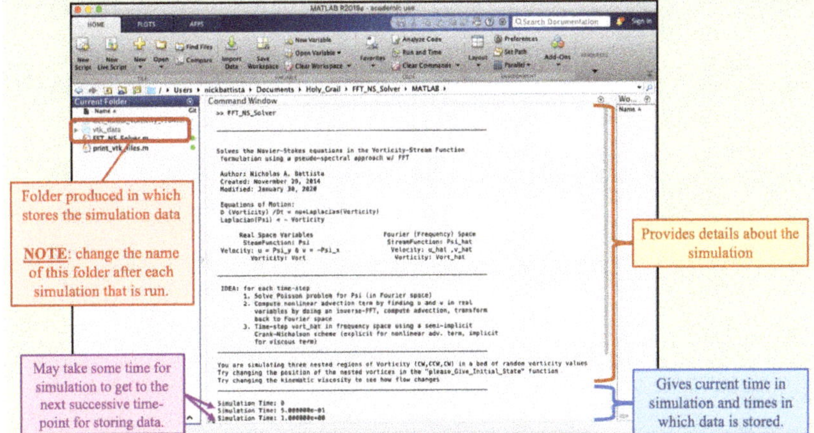

Figure 7. Showing the output on the screen when this example begins running. Note that the *vtk_data* folder is produced, in which stores the simulation data at predetermined time-points.

3. Once the simulation has finished running, we will change the *vtk_folder*'s name to *Simulation_1_Data*. This folder contains all of the data that was stored during the simulation, i.e., the vorticity, fluid velocity field, magnitude of velocity, etc.. It is imperative to change this folder name after each simulation if more simulations are to be run, or else the new data will printed over the previous simulation's data.

4. Once the folder's name has been changed, we will open the FFT_NS_Solver.m script to change one (or more) of the parameters. Here we will only change the fluid's kinematic viscosity, ν; however, there are other options in which you could change, such as the computational domain size and/or grid resolution, see Figure 3b. As an example change nu from $1.0 \times 10^{-3} \rightarrow 1.0 \times 10^{-2}$, see Figure 8. Once you have changed ν, you can close the script.

   Note that if you chose to change the grid resolution and domain size, make sure that the resolution in x and y are equal, i.e., $dx = L_x/N_x = L_y/N_y = dy$. Moreover if you increase the resolution (increase N_x, N_y) the simulations will take longer to run and the resulting data will require more storage.

```
34        %
35        % Simulation Parameters
36        %
37 -      nu=1.0e-2;   % kinematic viscosity (dynamic viscosity/density)
38 -      NX = 256;    % # of grid points in x
39 -      NY = 256;    % # of grid points in y
40 -      LX = 1;      % 'Length' of x-Domain
41 -      LY = 1;      % 'Length' of y-Domain
```

Figure 8. Changing the kinematic viscosity, ν, to a different value.

5. Finally, run this new simulation case by typing FFT_NS_Solver into the command window and pressing enter. Once that simulation has finished running, change its folder name to *Simulation_2_Data*. Now that we have a couple of simulations run, and have their corresponding data saved, we will next focus on visualizing each simulation's dynamics using the open-source software VisIt [41].

3.4. Guide: Visualizing the Spectral (FFT) Method's 'bubble3' Data

Here we will guide the user through visualizing the bubble3 simulation data from Section 3.3. In particular we will visualize the magnitude of velocity data. Note that other Eulerian scalar data, like vorticity, the x- or y-component of velocity, etc., may be visualized analogously, as we are following the protocol from the previous section on data visualization, Section 3.2's *Visualizing the Eulerian scalar data*. To visualize the fluid's velocity vector field, please follow Section 3.2's steps for *To Visualize Velocity Vectors*.

To visualize the uMag data, do the following:

1. First we must open the desired magnitude of velocity data. Recall that the fluid's magnitude of velocity data was saved under the name *uMag*. Click open in the VisIt toolbar and in the dialog box that pops up, locate the desired data directory, here it is *Simulation_Data_1* from earlier, and select the uMag group. Figure 9 is provided as a visual aid for these steps.

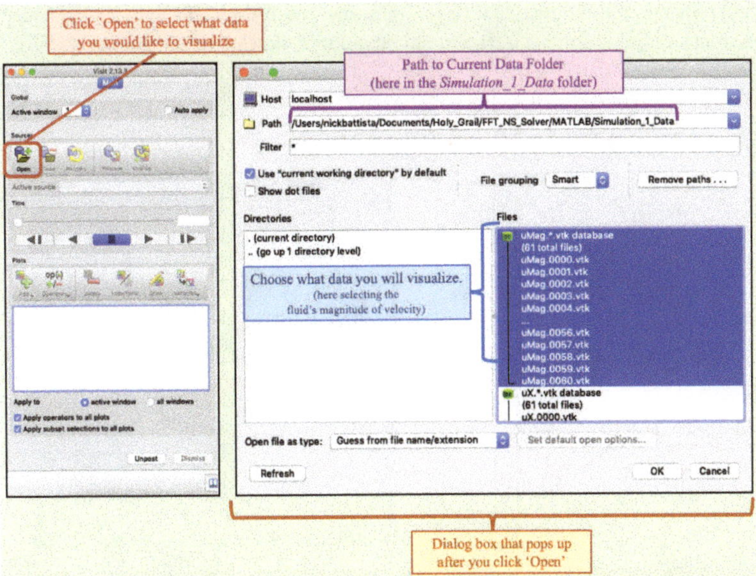

Figure 9. Opening specific simulation data to visualize in VisIt.

2. Although we have opened the data, nothing will appear visualized, yet. However, now uMag is the *Active Source* in the window, see Figure 10. We must now tell VisIt how to visualize the data. We will visualize magnitude of velocity as a background colormap, as it is a scalar quantity. To do this click 'Add' and then select Pseudocolor from the menu that appears, followed by uMag, i.e., Add→Pseudocolor→uMag.

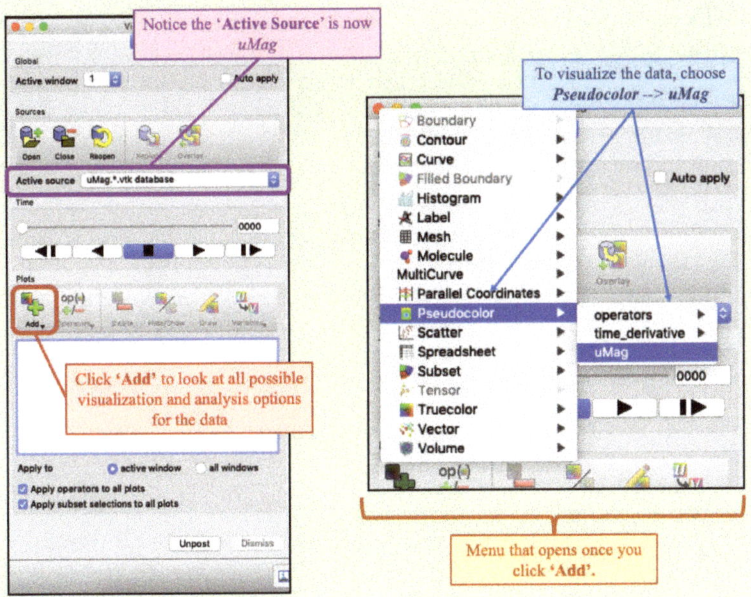

Figure 10. Opening specific simulation data to visualize in VisIt.

3. To finally visualize the data, click `Draw`. Window 1 will then show the fluid's magnitude of velocity data at time-point 0, i.e., the initial magnitude of velocity, see Figure 11.

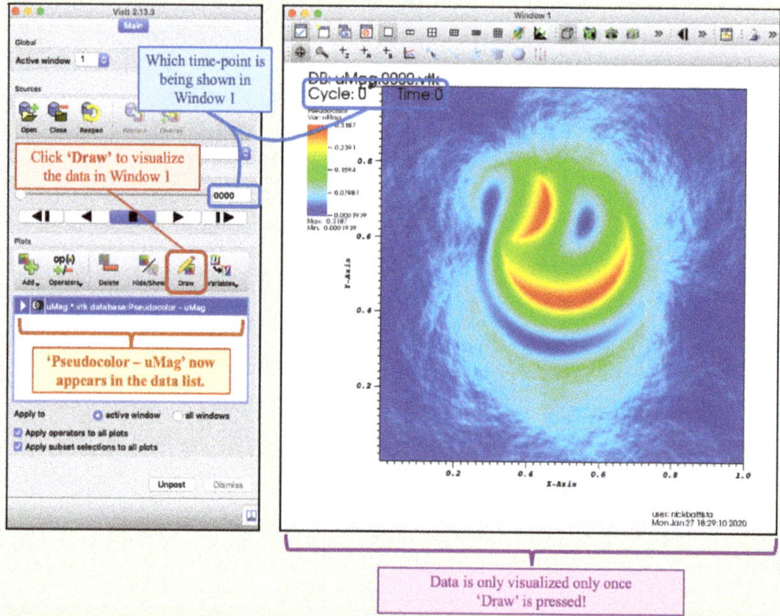

Figure 11. 'Drawing' the data in VisIt. The data shown in Window 1 here corresponds to time-point 0, as given from the *time slider* value.

4. Furthermore, we note that certain choices of colormaps are better for both conveying the data, but also for people who are colorblind. Generally, the default colormap, which was used in Figure 11, is a bad choice, as it is impossible for people who are red-green colorblind to interpret the field [57]. Thus, we will change the colormap to another, e.g., the `RdYlBu` (red-yellow-blue) colormap, by double-clicking on the `uMag.*.vtk database:Pseudocolor uMag` in the database list and appropriately selecting the desired colormap. Other properties may also be changed in this box, such as the colormap's scaling.

    Also, you can visualize the magnitude of velocity at the different time-points that were saved during the simulation by moving the time-slider. Figure 12 shows the data at time-point 30. Moreover, the colormap and its properties may be modified by double-clicking on `uMag.*.vtk database:Pseudocolor uMag` in the database list.

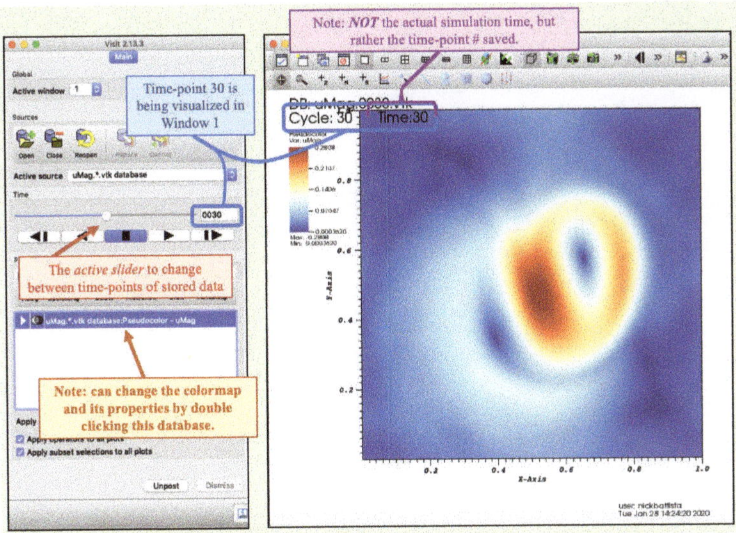

Figure 12. Changing the time-point to the 30th time-point stored.

5. Note that the time in the simulation does not generally correspond to the time-point stored. For example, in this case the simulation modeled 30 s of time and 60 total data time-points were stored. Thus, data was saved every 0.5 s of simulation time. Furthermore, you could repeat Steps 1–4 for data in *Simulation_2_Data* from the second simulation you ran. Instead of comparing the 30th time-point's data, we will compare the 60th, i.e., the last time-point. Repeating this entire process for the other simulation's data yields the comparison as shown in Figure 13, below.

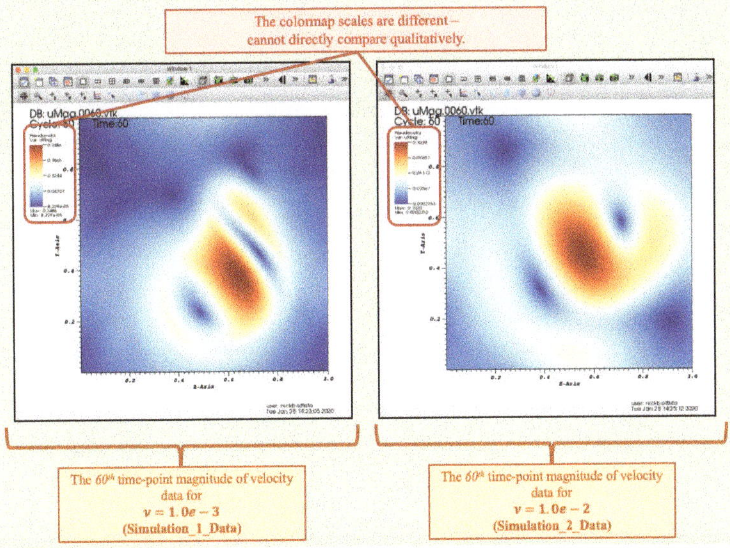

Figure 13. Comparing the data between both simulations' last time-point (60th time-point stored). Note that a direct qualitative comparison is not possible because the scaling of each colormap is different.

However, we cannot qualitatively compare the magnitude of velocity between each of these simulations as they are in Figure 13. Each simulation's colormap has a different scale. In order to compare, we must make both scales equivalent.

6. To change the scale, double-click on each of the databases. This will bring up a dialog window (as described earlier) where you can change the scaling, e.g., minimum and maximum, as well as the colormap itself. Change the minimum and maximum values to 0.0 and 0.2, respectively. Figure 14 illustrates where to change the minimum and maximum for the colormap's scale. Once those values are changed, click `Apply`. Note that you must do this for both of the databases individually.

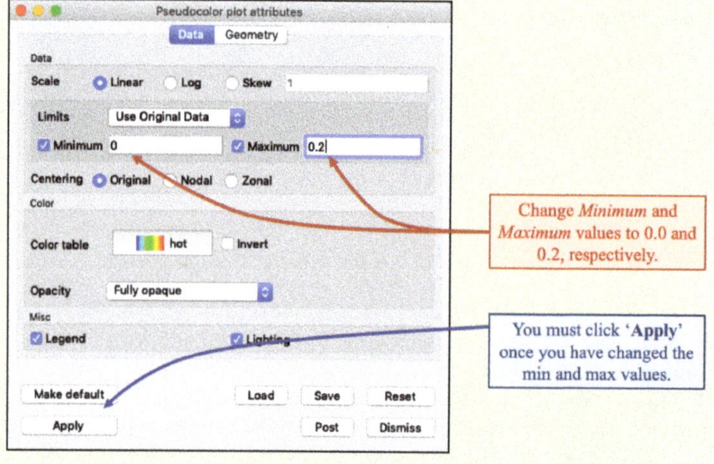

Figure 14. Changing the magnitude of velocity's colormap scaling.

7. After changing the scale, the comparison is significantly different than when first visualized, see Figure 15.

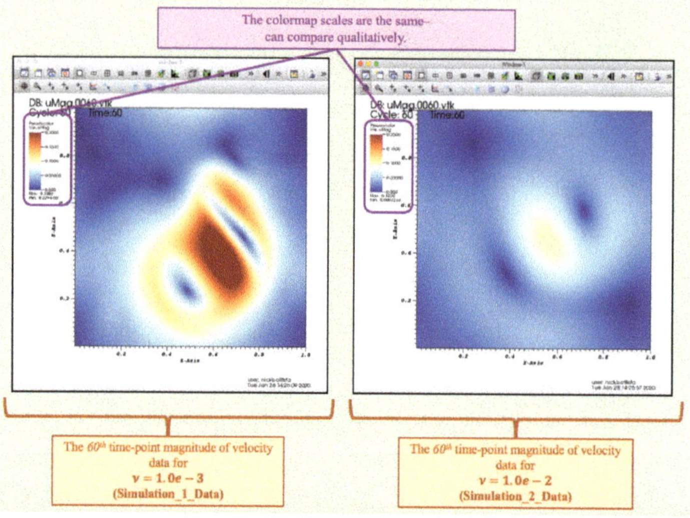

Figure 15. Comparing each simulation's magnitude of velocity data at the 60th time-point. The scaling of each colormap is identical which leads to straight-forward direct qualitative comparison.

Furthermore, we also provide visualizations of other time-points and/or scalar data in Appendix C.

3.5. Guide: Analyzing the Spectral (FFT) Method's 'bubble3' Data

To analyze the data in VisIt, we will provide a walk-through of the steps that necessary to measure a fluid quantity (particular variable) across a line of interest within the computational domain using the Spectral (FFT) Method's bubble3 example from Section 3.3. We will assume that you have visualized the data for the *magnitude of velocity* that corresponds to both of the simulations. In this tutorial we will compare the magnitude of velocity for both simulations across a vertical line through the center of the computational domain at the last time-point stored, i.e., the 60th time-point. See Figure 16 for an idea of where the measurement will take place.

To analyze the uMag data, do the following:

1. Make sure that *Active source* is the uMag data corresponding to the data from *Simulation_1_Data* in VisIt, see Figure 16.

    Once the *Active Source* has been appropriately selected, in the command bar, in the VisIt menu go to Controls→Query (see Figure 16).

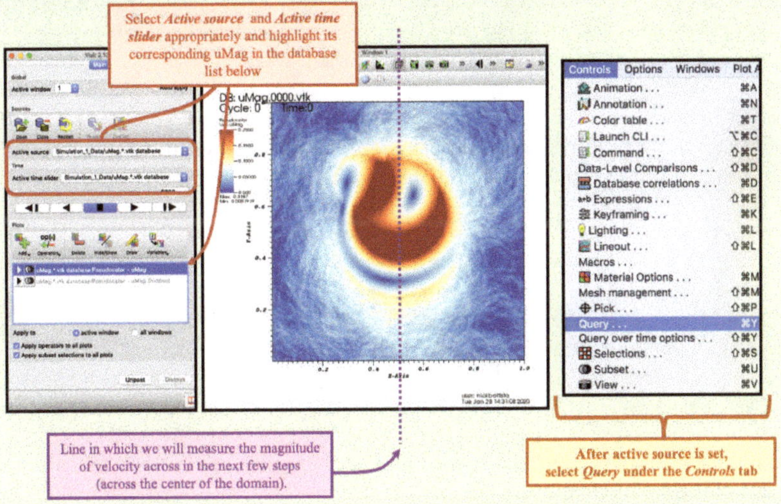

Figure 16. To analyze data in VisIt, the *active source* and *active time slider* must be set to specific data that is desired to be analyzed, e.g., here it is the magnitude of velocity, uMag, for the *Simulation_1_Data*.

2. Next select Lineout under *Queries* and then under *Variables* select Scalars and then uMag, i.e., Variables→Scalars→uMag, see Figure 17.

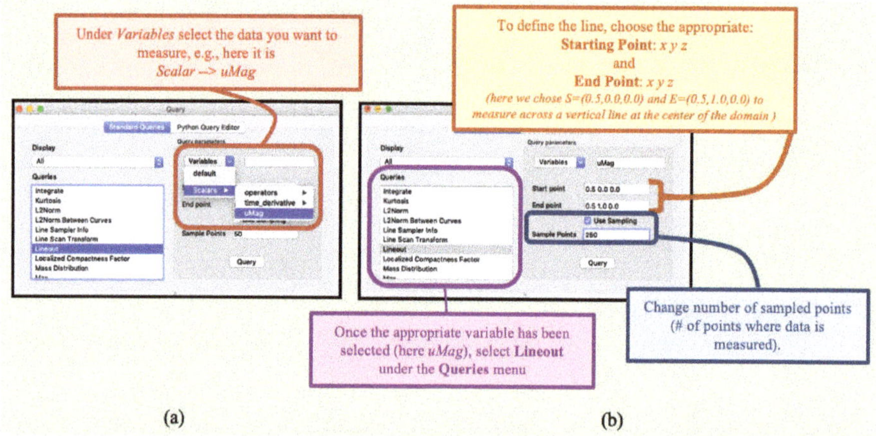

Figure 17. Within the *Query* dialog box, select (**a**) the chosen data to measure, e.g., here it is uMag, and (**b**) the `Lineout` option to measure the data across a line within the domain that is defined by a starting and ending point.

3. Next we will define the line segment in which we intend to measure the data across. To do this the user gets to choose the starting point and ending point of the line segment, see Figure 17b. We will measure across a vertical line through the center of the domain, hence we choose $(x, y, z) = (0.5, 0.0, 0.0)$ and $(x, y, z) = (0.5, 1.0, 0.0)$ as the starting and ending point, respectively. We also will measure the value of uMag at 250 sampled points. Thus, change *Sample Points* to 250, see Figure 17b and select `Use sampling`. VisIt will automatically interpolate the data so that it is measured at number of equally-spaced sample points that is designated by the user.

4. Once you click `Query` a plot should pop up in a new *Active Window* in Visit, i.e., Active window 2, see Figure 18. If you double-click on the *Lineout(uMag)* bar in the VisIt database listing window, you can change various attributes about the line, such as its thickness or color among others. Moreover, you can use the time-slider to observe how the magnitude of velocity across the chosen line varies from time-point to time-points.

5. To get back to the colormap window, select Active Window 1, rather than 2.

6. You can repeat this procedure and overlay multiple uMag measured data on the same plot, each corresponding to a different simulation, e.g., repeat this process for *Simulation_2_Data*. Be sure to choose the correct data for the *Active source*.

7. If you repeat this entire process for the *Simulation_2_Data* and move the time-slider such that the data being shown in *Active window 2* corresponds to the last time-point stored (the 60th), you should see what is give in Figure 19. Also, note that we have changed each line's thickness individually.

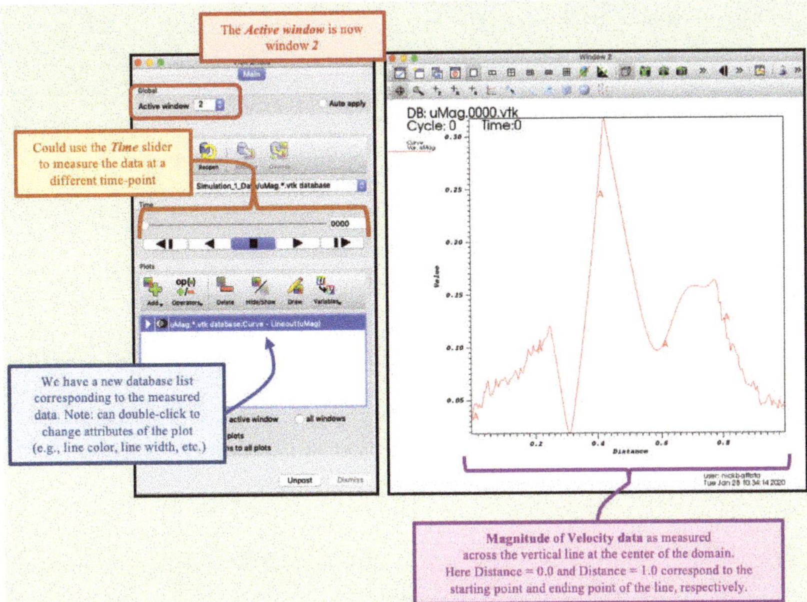

Figure 18. Illustrating the *Active window* change to window 2 as well as a plot of the magnitude of velocity (selected data) as measured across a vertical line at the center of the domain.

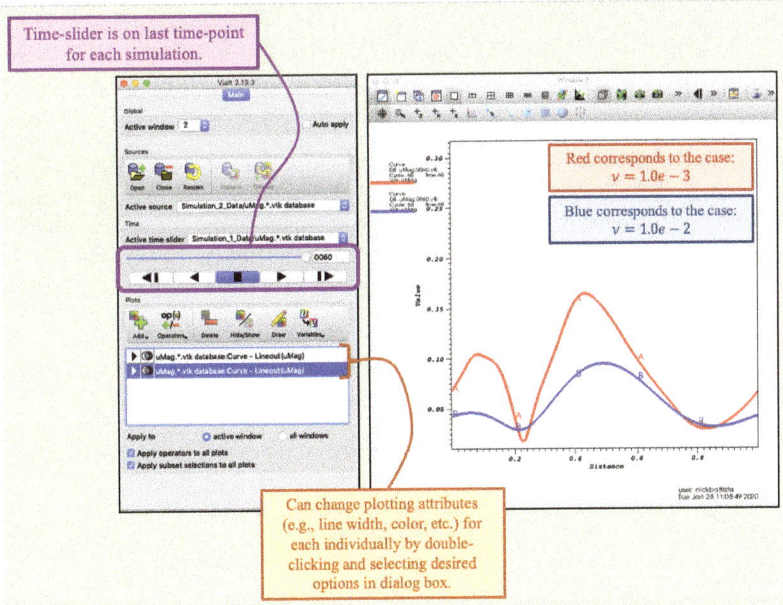

Figure 19. Comparing the measured magnitude of velocity data between both simulations at the last time-point data saved.

4. Built-in Examples (for Each Fluid Solver)

In this section we will highlight a subset of the built-in examples that can be run immediately upon download (or `git clone`) of the software. These examples were selected as they are either traditional problems in fluid dynamics or problems that could naturally lead to fruitful discussion and analysis of the underlying dynamics. The examples we will discuss are:

1. Lid-Driven Cavity Flow via the *Projection Method*
2. Circular Flow in a Square Domain via the *Projection Method*
3. Side-by-side Vorticity Region Interactions via the *Spectral (FFT) Method*
4. Evolution of Vorticity from an Initial Velocity Field via the *Spectral (FFT) Method*
5. Flow Past One or More Cylinders via the *Lattice Boltzmann Method*
6. Flow Past a Porous Cylinders via the *Lattice Boltzmann Method*

For each of these examples, we will provide the necessary simulation details, including a brief background about problem as well as the numerical parameters that were used in each simulation. Furthermore, we will also make observations regarding each simulation's dynamics, investigate some of the data, and suggest future ideas for how students could either modify or explore each example further.

These examples are available within both the MATLAB and Python implementations in the software. Note that Section 3 described how to run, visualize, and analyze the simulations, and thus the visualizations and analysis contained here can be recreated by students and other practitioners.

4.1. Cavity Flow (via Projection Method)

In this example we will use the Projection Method in the software to run a cavity flow problem. Note that this example is contained in the Projection Method folder and may be run by selecting the `cavity_top` option (see line 58 in Figure 3a).

For this lid-driven cavity flow problem, the non-zero horizontal velocity on the top wall, $u = u_T$, is set to $u_T = 4.0$ m/s, while all other tangential (and normal) velocities along the boundaries are set to zero (see Figure 1a). Note that the top wall velocity is not immediately set at u_T, but rather the flow ramps up from 0 to the preset u_T along this wall, to avoid instantaneous acceleration artifacts. For the simulations shown below with $Re = 4000$, use the computational parameters listed in Table 3. The Reynolds Number was set by defining the characteristic length scale to be $L = L_x$, the horizontal length of the domain, and the characteristic velocity to be $V = u_T$, e.g., $Re = \frac{\rho L_x u_T}{\mu}$. To change the Reynolds Number, the dynamic viscosity was varied. For $Re = 4000, 400, 40$ and 4, we used $\mu = 1, 10, 100$, and 1000 kg/(m·s), respectively. Note that for the cases with $Re = 4$ and 40, we decreased the time-step to $dt = 5 \times 10^{-5}$ s to ensure numerical stability of solutions. If we had not, errors would have been magnified every time-step of the simulation and/or the numerical solutions may have blown up. In CFD it is common that one may need to vary the time-step depending on the Re or other parameters of the system. Situations in which very small time-steps need to be taken to ensure numerical stability of solutions are known as *stiff equations*, for more information please see [43].

Table 3. Numerical parameters for the Cavity simulation for $Re = 4000$.

Parameter	Variable	Units	Value
Domain Size	$[L_x, L_y]$	m	$[1, 2]$
Spatial Grid Resolution	$[N_x, N_y]$		$[128, 256]$
Spatial Grid Size	$dx = dy$	m	$L_x/N_x = L_y/N_y$
Time Step Size	dt	s	10^{-3}
Total Simulation Time	T	s	6
Fluid Density	ρ	kg/m^3	1000
Fluid Dynamic Viscosity	μ	kg/(m·s)	1

Upon having run the cavity model for $Re = 4000$, we visualized its corresponding simulation data using VisIt [41], as illustrated by Figure 20, which gives colormaps of vorticity, magnitude of velocity, velocity in the horizontal direction throughout the domain, and pressure, as well as depictions of the velocity field, both as a scaled and non-scaled quantity. For the scaled velocity field snapshot, the scale factor was equal to the largest magnitude of velocity across the entire computational domain. In the non-scaled plot, streamlines are also given. Streamlines show the path that a passive particle would take in the flow at a particular moment in time. The data shown here was taken at $t = 6.0$ s, when the simulation ended.

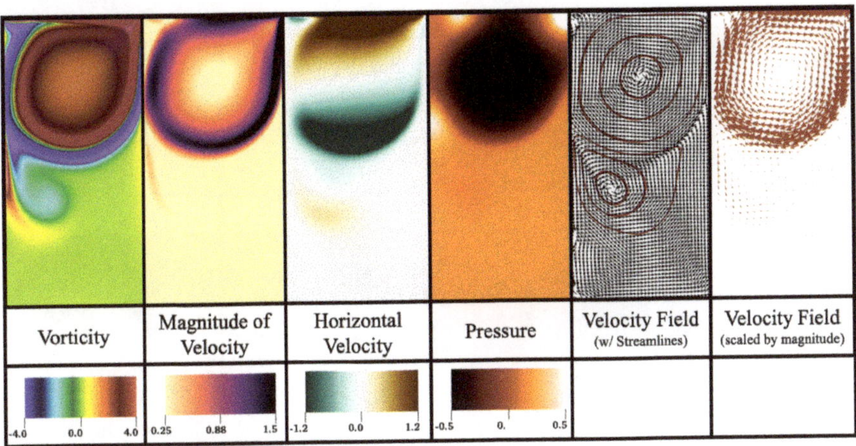

Figure 20. Illustration of all the simulation data produced from the projection method example of cavity flow (at $Re = 4000$). This snapshot was taken at 6.0 s, as the simulation ended.

A fully-formed vortex developed by that time near the top of the cavity, while a smaller vortex appears to be forming below (see Vorticity pane in Figure 20). Snapshots illustrating the formation of such vortices from this case of $Re = 4000$ are presented in Figure 21, which gives a colormap of vorticity and the velocity vector field. As fluid is moving along the top edge of the domain from left-to-right, the fluid nearest to the top begins moving in the same direction. Eventually fluid down in the cavity begins moving in the same direction as well, until it reaches the right boundary, where it must move downwards. It is easier for the fluid to move downward because of the faster flows moving towards the right above it, due to the boundary condition. As the fluid moves downward along the wall, fluid towards the middle of the cavity begins moving downwards as well. In tandem, with the fluid moving left-to-right along the top and top-to-bottom along the right, these fluid patterns initiate the formation of a clockwise-spinning vortex.

Now that there is a vortex spinning clockwise, it causes the fluid below the vortex (moving right-to-left) to begin moving right-to-left until it reaches the wall on the left and an analogous situation occurs to the above, except an oppositely-spinning (counter-clockwise) vortex begins to form. Recall that these vortices can both be traced back to the fluid moving along the top of the domain left-to-right. Within the region that the second vortex forms, much of the energy (velocity) has dissipated away, resulting in a smaller, less strong vortex.

We can explicitly measure the horizontal velocity descending down the cavity. This data is presented in Figure 22, which gives the horizontal velocity vs. depth in the cavity during four different snapshots in the simulation. We also measure the horizontal velocity across three different lines descending into the cavity. Near the surface of the cavity (at depth = 0 m), the velocity is equal to the boundary condition. As one descends into the cavity, there are spikes in both the positive and negative horizontal velocities. These spikes correspond to locations near the edges of the vortices, where there is faster moving fluid.

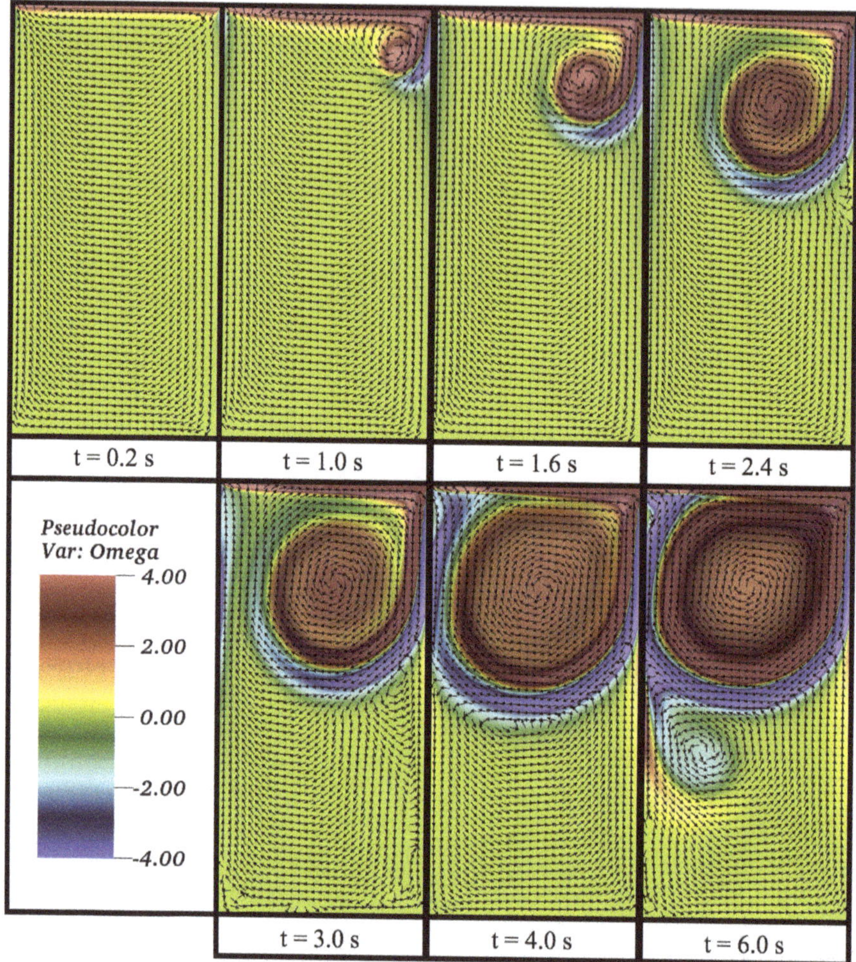

Figure 21. Snapshots showing the evolution of vortical flow patterns during a simulation with $Re = 4000$. The background colormap is of vorticity and the velocity vector field is also given.

Finally, one can change the Re by varying the dynamic viscosity, μ to observe how vortex formation changes within the cavity, as well as, how quickly energy gets dissipated in higher viscosity settings. Recall that if μ is changed to 40 or 4 the time-step, dt, was adjusted for as described above. Figure 23 illustrates how vortex formation changes within the cavity for different Re at three different snapshots during the simulations. Moreover, it gives the horizontal velocity vs. cavity depth, as measured across the middle of the cavity. Note that the $Re = 4$ and $Re = 40$ data virtually overlaps in the figure, and velocity (and system's energy) quickly dissipates to zero going further into the cavity.

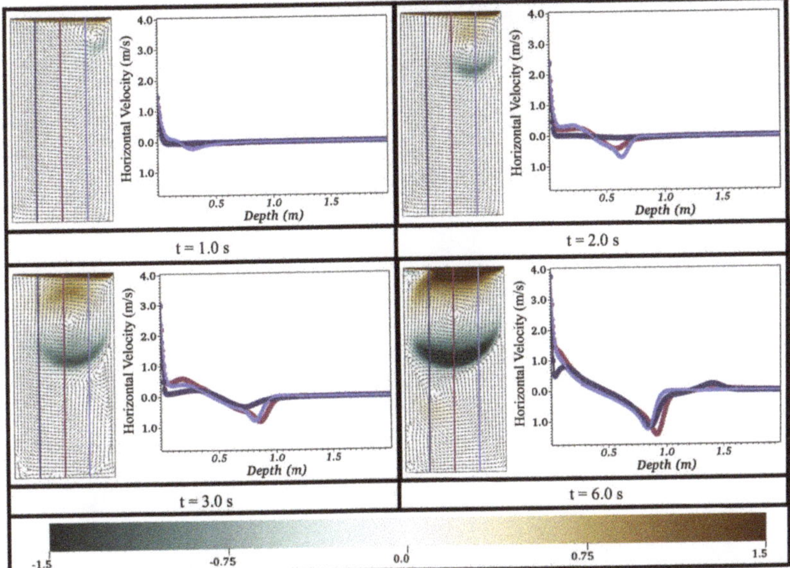

Figure 22. Snapshots of horizontal velocity measurements, when measured down the cavity in three places for $Re = 4000$.

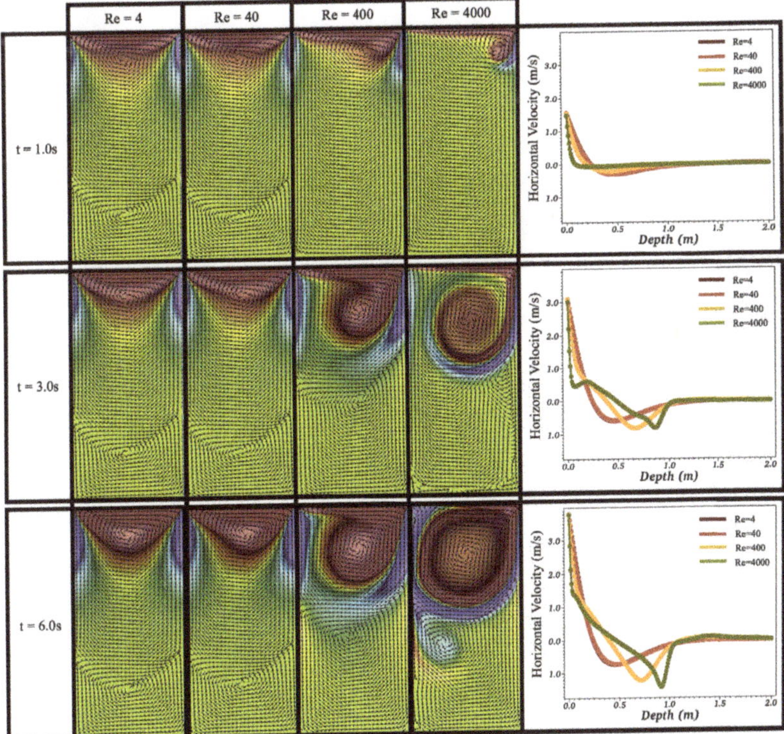

Figure 23. A comparison of vorticity, velocity field, and horizontal velocity measurements down the center of the cavity between cases of $Re = 4, 40, 400,$ and 4000.

Students may elect to try the following:

1. Recreate the above results for cavity flow with the geometric and simulation parameters given in Table 3 for one or more *Re*.
2. Change the domain width (cavity width) to observe differences as the cavity gets wider for a given *Re*.
3. Vary the lid-velocity to observe differences (e.g., varying *Re* for different velocities as opposed to differing viscosities, μ).

4.2. Circular Flow in a Square Domain (via Projection Method)

In this example we will use the Projection Method in the software to run an example involving circular flow in a square domain. Note that this example is in the Projection Method folder and may be run by selecting the 'whirlwind' option (see line 58 in Figure 3a).

In this case, all the tangential boundary conditions along the computational domain were set to be $U = u_{top} = -u_{bot} = v_{left} = -v_{right} = 1.0$ m/s, see Figure 1b. As mentioned in Section 4.1 above, the tangential wall velocities are not immediately set to U, but rather the flow ramps up along each edge from 0 to the preset u_{top}, u_{bot}, v_{left}, or v_{right} along this wall, to avoid instantaneous acceleration artifacts. For the simulation shown below with $Re = 4000$, use the computational parameters listed in Table 4. Note that as $Re = \frac{\rho L U}{\mu}$, with $L = L_x = L_y$ and $U = 1.0$ m/s, one can vary the dynamic viscosity, μ, to change the *Re*. If you decide to make *Re* smaller (e.g., increasing μ), you may also need to decrease the time-step, dt, significantly to ensure numerical stability, as mentioned in Section 4.1. The requirement that very small time-steps are necessary to ensure numerical stability denotes what are called *stiff equations*. For more information on stiff equations, please see [43].

Table 4. Numerical parameters for the whirlwind simulation for $Re = 4000$.

Parameter	Variable	Units	Value
Domain Size	$[L_x, L_y]$	m	$[1, 1]$
Spatial Grid Resolution	$[N_x, N_y]$		$[256, 256]$
Spatial Grid Size	$dx = dy$	m	$L_x/N_x = L_y/N_y$
Time Step Size	dt	s	10^{-3}
Total Simulation Time	T	s	24
Fluid Density	ρ	kg/m^3	1000
Fluid Dynamic Viscosity	μ	kg/(m $\cdot$ s)	0.25

Having run the circular flow simulation for $Re = 4000$, we visualized its corresponding simulation data using VisIt [41], see Figure 24, which gives colormaps of vorticity, magnitude of velocity, velocity in the horizontal and vertical directions, and the finite-time Lyapunov exponent (FTLE). The FTLE can be used to find the rate of separation in the trajectories of two infinitesimally close parcels of fluid. Maxima in the FTLE have been used to determine approximations to the true Lagrangian Coherent Structures (LCSs). LCSs are used to determine distinct flow structures in the fluid [58–61] can be used to divide the fluid's complex dynamics into distinct regions to better understand transport properties of flow [62–64]. True LCSs would require knowledge of the infinite-time Lyapunov exponent (ITLE); however, the FTLE can serve as a proxy to the ITLE. In this paper, we computed the forward-time FTLE field, whose maximal ridges give approximate LCSs corresponding to regions of repelling fluid trajectories and whose low values give rise to regions of attraction [61]. Thus, the FTLE can be used to help find regions of high fluid mixing [58,59,61,65,66].

Figure 24 also provides contours for each quantity. It also includes a depiction of the fluid velocity field, as scaled by the maximum in the entire domain's magnitude of velocity, which also includes its associated streamlines. Recall that streamlines are curves that depict instantaneous tangent lines along direction of the flow velocity. They show the direction that a neutrally-buoyant particle would travel at a particular snapshot in time. These are different than contours (also known as level-sets or isolines), which are curves where a function has a constant value. This snapshot was taken at $t = 24.0\,\text{s}$, when the simulation ended.

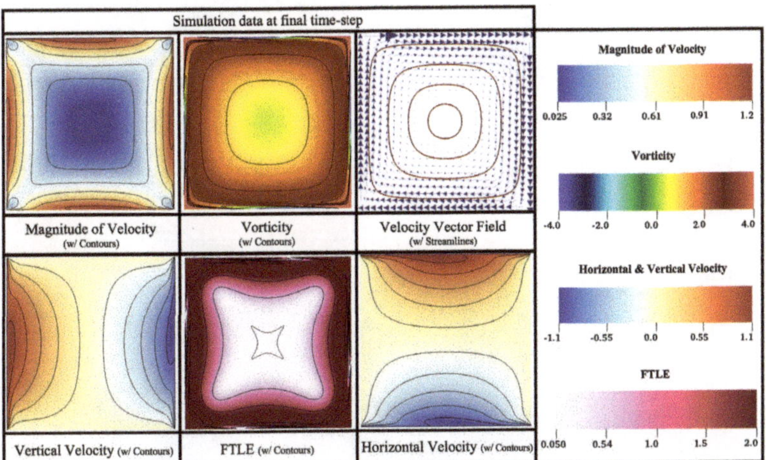

Figure 24. Illustration of some of the data produced for the circular flow example, using the projection method. The data shown is from the final time-step for $Re = 4000$.

As the velocity boundary conditions continually increased, the interior of the domain began to move clockwise, in the same direction as the flow at the boundaries (see Figure 1b). The colormaps and contours in each velocity-related panel reveal that flow speeds decrease towards the interior of the domain. Moreover, as all the boundary conditions are uniform, the vertical and horizontal velocity plots are identical under a 90 degree rotation. Also, the vorticity plot illustrates that vorticity is still low by the ending of the simulation in the center of the domain, even though flow across the whole domain is rotating clockwise. The FTLE plot suggests that the majority of fluid mixing occurs near the edge of the domain, rather than the interior, as higher values of FTLE suggest nearby fluid trajectories move away from each other at higher rates. This is due to larger spatial gradients in the velocity in these regions, where flow is moving at different speeds in different directions.

We also present data comparing simulations for $Re = 400, 1000$, and 4000 (for $\mu = 2.5, 1.0$, and $0.25\,\text{kg}/(\text{m}\cdot\text{s})$, respectively). Figure 25 illustrates colormaps of the magnitude of velocity (with corresponding contours) at various time-points during the simulation. Every colormap uses the same scaling in the figure. Higher velocities appear to creep into the interior of the domain quicker in the lower Re cases. By $t = 24\,\text{s}$ it appears that the $Re = 400$ and 1000 cases look almost identical, the $Re = 4000$ case tells a different story - the interior of the domain is still moving considerable slower than the other two cases. This might seem counter-intuitive at first, as one might generally think that higher Re tends to lead to more fluid motion, especially when we are lowering viscosity to increase Re. However, this is due to differences in time-scales for the diffusion of momentum through the fluid. The viscous diffusion time can be thought of as the time it takes for a fluid parcel to diffuse a particular distance on average. If $\tilde{t}$ is the diffusion time, $\nu = \mu/\rho$ is the kinematic viscosity, and $\tilde{l}$ is the mean-squared distance, then

$$\tilde{t} \sim \frac{\tilde{l}}{\nu} \sim \frac{1}{\nu},$$

if all of the simulation geometries are identical and only μ (and hence ν) is varied. This concept is based off viewing diffusion as a random walk process [67]. Hence for the simulations presented here the time-scales vary between

$$\tilde{t}_{Re=400} \sim \frac{1}{2.5/1000} = 400 \text{ s} \quad \text{and} \quad \tilde{t}_{Re=4000} \sim \frac{1}{0.25/1000} = 4000 \text{ s}.$$

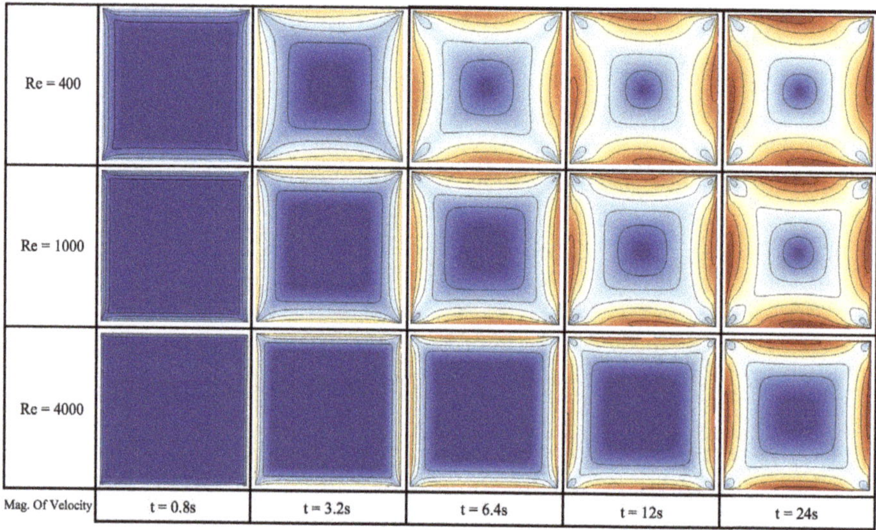

Figure 25. Snapshots showing the evolution of the magnitude of velocity between cases of $Re = 400, 1000,$ and 4000.

Therefore in the $Re = 4000$ case, the diffusion time-scale is $10x$ longer than that of the $Re = 400$ case, and thus the dynamics evolve much slower in the $Re = 4000$ case! Moreover, this phenomena is shown in Figure 26, which illustrates the fluid velocity field along with its associated streamlines at the same snapshots in time.

This is more apparent if we compare flow profiles within the domain. Figure 27 compares the horizontal flow profile for $Re = 400, 1000,$ and 4000 across the vertical mid-line of the domain. As mentioned above the dynamics in the $Re = 4000$ case evolves about $4x$ and $10x$ slower than the $Re = 1000$ and $Re = 400$ cases, respectively. By 24.0 s, the flow profiles in the $Re = 400$ and $Re = 1000$ cases look almost identical, while the flow profile in the $Re = 4000$ resembles that of the the flow profile in the $Re = 1000$ at ~ 6.4 s—approximately a factor of 4 different in time. Furthermore Figure 28 shows the flow profiles for all three cases of Re when the diffusion of momentum has reached the same depth. This occurs at different simulation times due do the variations in viscous diffusion time-scales.

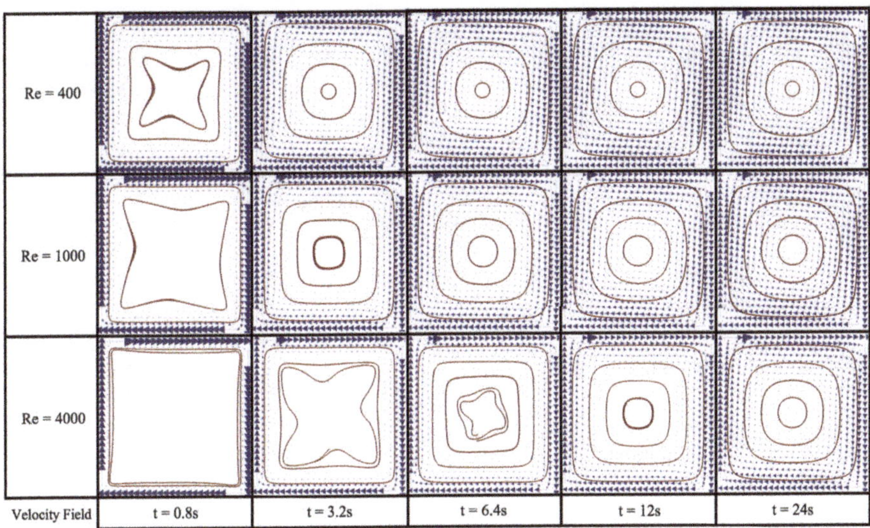

Figure 26. Snapshots showing the velocity field's evolution between cases of $Re = 400, 1000,$ and 4000. Streamlines of the velocity field are also illustrated.

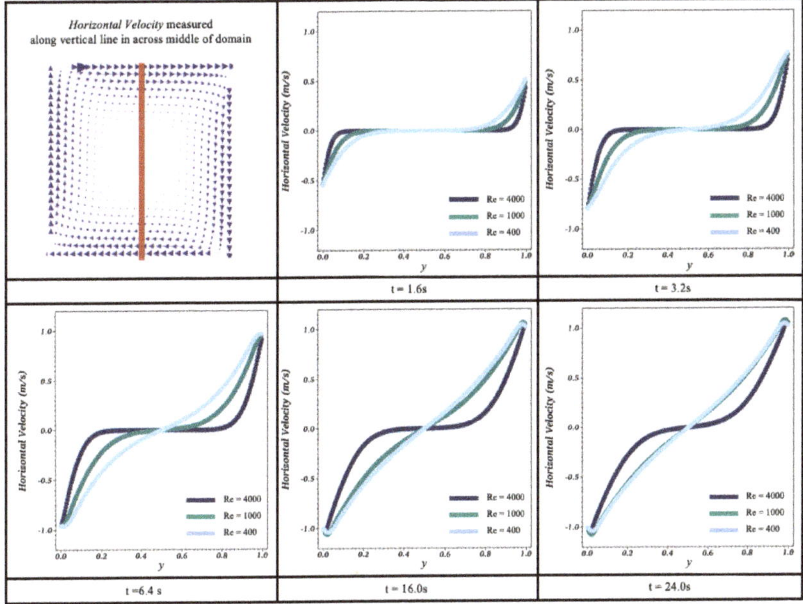

Figure 27. Snapshots showing horizontal velocity measurements across a vertical line centered in the domain among cases for $Re = 400, 1000,$ and 4000.

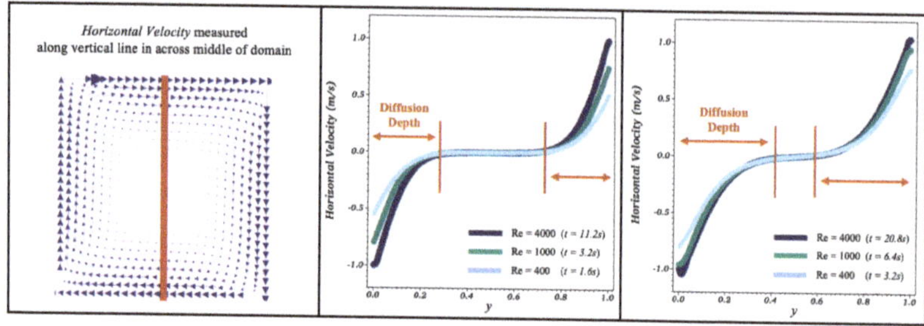

Figure 28. Snapshots showing horizontal velocity measurements across a vertical line centered in the domain among cases for $Re = 400, 1000$, and 4000 in which similar momentum diffusion depths have been reached.

Students may elect to try the following:

1. Recreate the above results with the parameters listed in Table 4
2. Change the computational geometry, e.g., rectangular instead of square
3. Vary the input velocity along each side of the domain to make them non-uniform
4. Vary the Re by changing dynamic viscosity for suggestion (2) or (3)

4.3. Side-by-Side Voritices (via Spectral Method)

In this example we will use a FFT-based fluid solver in the software to run a simulation of multiple regions of vorticity interacting with one another. Note that the first example given here can be run by going into in the *FFT_NS_Solver* script selecting the 'qtrs' option (see line 47 in Figure 3b). The vorticity is initialized as in Figure 2a. We will also highlight the 'half' option in this section as well.

Recall that for this solver, an initial vorticity configuration is needed. For the simulations of four interacting vorticity regions, the CW and CCW vorticity were initialized as $\omega = -1$ and 1 rad/s, respectively. In other cases with only two interacting voricity regions, the strengths of each regions were varied between $-1, -0.5, 0.5$ and 1 rad/s, as appropriate. For the simulations shown below with either 2 or 4 interacting vorticity regions, use the computational parameters listed in Table 5.

Table 5. Numerical parameters for the 4 interacting vortices simulation.

Parameter	Variable	Units	Value
Domain Size (4 case)	$[L_x, L_y]$	m	$[1, 1]$
Spatial Grid Resolution (4 case)	$[N_x, N_y]$		$[512, 512]$
Spatial Grid Size	$dx = dy$	m	$L_x/N_x = L_y/N_y$
Domain Size (2 case)	$[L_x, L_y]$	m	$[1, 0.5]$
Spatial Grid Resolution (4 case)	$[N_x, N_y]$		$[512, 256]$
Spatial Grid Size	$dx = dy$	m	$L_x/N_x = L_y/N_y$
Time Step Size	dt	s	10^{-2}
Total Simulation Time	T	s	5
Fluid Kinematic Viscosity	$\nu = \rho/\mu$	m^2/s	0.001

Figure 29 provides the simulation data for the case with 4 interacting vorticity regions, as described above. It presents colormaps for vorticity, magnitude of velocity, horizontal velocity, vertical velocity, and the finite-time Lyapunov exponent (FTLE), which is used to determine approximations to the true Lagrangian Coherent Structures (LCS) to which can be used to distinguish fluid mixing regions [58,59,61,65,66]. Their corresponding contours are also given. Recall that contours are curves to which the value of a quantity is constant. Figure 29 also gives a snapshot of the velocity vector field. This data is from the last snapshot of the simulation at $t = 5.0$ s.

The top two and bottom two vorticity regions are initialized the same, e.g., clockwise (CW) and counterclockwise (CCW), respectively, by this point in the simulation they are beginning to lean in the direction of rotation. Moreover, it can be seen that fluid is moving quickly horizontally through the top, middle, and bottom of the domain, as illustrated by the magnitude of velocity plot. When this is compared to the horizontal velocity plot, one observes that the fluid along the top and bottom of the domain both are moving left-to-right, while fluid moving horizontally through the middle of the domain is moving right-to-left. These directions align with the direction of the spinning vortices in each of these horizontally-aligned regions—top, middle, and bottom, which can be seen in the velocity vector field depiction.

The high values of FTLE illustrate regions of higher fluid mixing, as higher FTLE suggest faster rates of separation of close fluid blobs. In Figure 29 these are the areas between the top-two and bottom-two vorticity regions. Within these regions there is intense shearing among the vertical flow velocity, see the vertical velocity plot. Red regions in the vertical velocity plot indicate fluid moving upwards while blue corresponds to downward fluid motion. Moreover, from the nature of the FFT-solver, the periodic boundary conditions, illustrate that along vertical walls of the domain on opposite sides there are regions where fluid is moving the the opposite direction. Hence the large FTLE values on the left and ride side of the top-two and bottom-two vortices. There is significantly less mixing in the top, middle, and bottom horizontal strips across the domain because fluid is generally moving unidirectionally in these regions with lower spatial gradients.

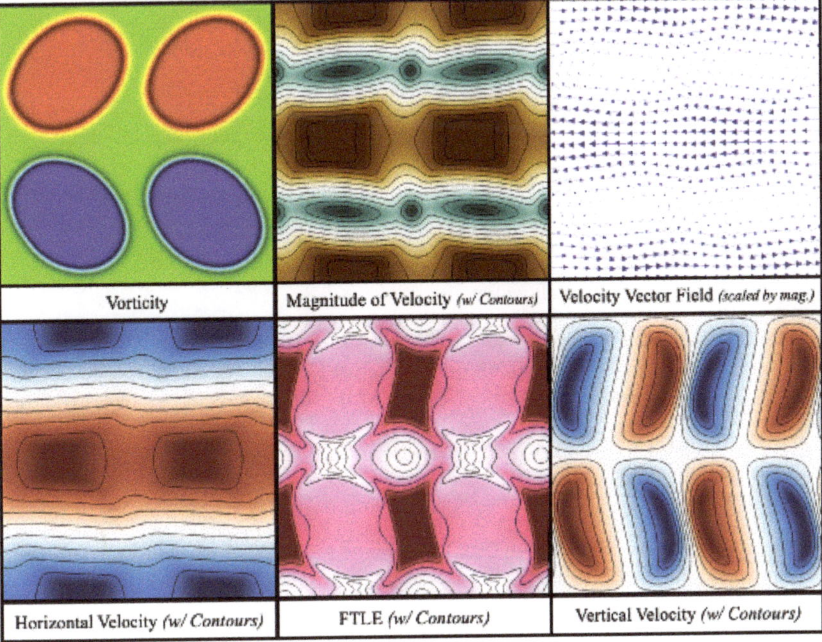

Figure 29. Illustrations of the simulation data produced during the case of 4 interacting vorticity regions. The data shown is from the last time-step.

One could elect to change the size or strength of a subset of these vorticity regions to simulate asymmetric vorticity interactions in this configuration. However, rather than study 4 interacting vorticity regions, we compared 2 interacting regions at a time, and varied the strength (magnitude of vorticity), size, and initial vorticity values (spinning-direction) between them, see Figure 30. Figure 30 provides comparison data of snapshots at $t = 6.0$ s in each simulation for different quantities—either vorticity with velocity vectors, magnitude of velocity with its corresponding contours, and FTLE with its corresponding contours. We use the language that a "strong" vorticity region has a $|\omega| = 1$ rad/s, while a "weaker" region has a vorticity of $|\omega| = 0.5$ rad/s. Each initial size of the vorticity regions had a radius of 0.15 m or 0.3 m.

In general, one can notice the following:

1. When same-spinning and size vorticity regions interact, more fluid mixing occurs directly between them (1st row)
2. When oppositely-spinning and same size vortex regions interact, there is enhanced vertical fluid flow between the vorticity regions, but less mixing between them (2nd row)
3. When oppositely-spinning vorticity regions of different strengths interact, the stronger region (whose magnitude of vorticity is greater), influences fluid flow to move in the same direction further away from it. There is still some enhanced vertical flow through the region between the regions (3rd row)
4. If there are asymmetric region sizes and strengths, where the larger region is weaker (smaller magnitude of vorticity), the region with the larger vorticity magnitude may be able to keep the other region from influencing much of the flow around it, while imposing its flow direction on the weaker region (4th row)
5. Two regions of vorticity with the same sign, but differing strengths, leads to enhanced fluid mixing between the vortices, while flow around the smaller region are significantly less than the other (5th row)
6. If there are asymmetric region sizes but uniform strengths, enhanced vertical flow will be seen between the regions, although the larger vorticity region exerts more influence on flow patterns closer to the smaller region (6th row)
7. If there are asymmetric region sizes and strengths, where the smaller region is weaker (smaller magnitude of vorticity), enhanced vertical flow will be seen between the regions, although the stronger vorticity region exerts more influence closer to the smaller region (7th row)

Students may elect to try the following:

1. Recreate the above results for interacting side-by-side vorticity regions with the geometric and simulation parameters given in Table 5 or for different ν or computational domains.
2. Add more vorticity regions or make the case with four asymmetric in terms of placement and/or size.
3. Change the distance between the vorticity regions to observe how magnitude of velocity or FTLE contours change.

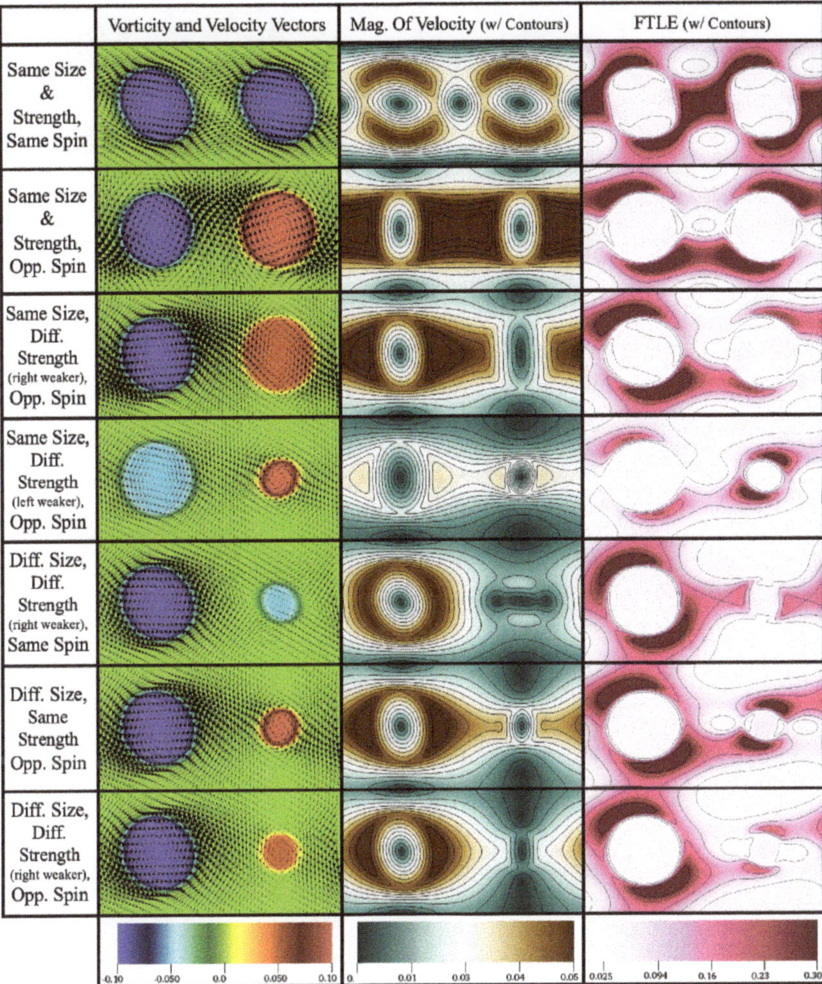

Figure 30. Vorticity and velocity field (**left column**) and magnitude of velocity with its associated contours (**right column**) for cases of two vorticity regions with different sizes, strengths, and vorticity initialization.

4.4. Evolution of Vorticity from an Initial Velocity Field

In this example using a spectral (FFT) solver, we begin with a snapshot of the velocity vector field from an independent CFD simulation, see Figure 31a,b. Note that this snapshot was taken from an open-source fluid-structure interaction example of a pulsing cartoon heart [36]. We used a stored time-point's velocity field data, to which we extracted the horizontal and vertical velocity components, and then computed its associated vorticity at that particular time-step, see Figure 31c. We note that there are numerical errors in computing the vorticity, ω, as a centered finite difference scheme [43] was used to compute each first-order partial derivative in calculating vorticity, i.e., $\omega = \frac{\partial v}{\partial x} - \frac{\partial u}{\partial y}$. Moreover, since this fluid solver uses periodic boundary conditions, we computed the partial derivatives at each boundary using data from the opposite side of the computational domain. Note this example may be run by selecting the 'jets' option in the *FFT_NS_Solver* script (see line 47 in Figure 3b).

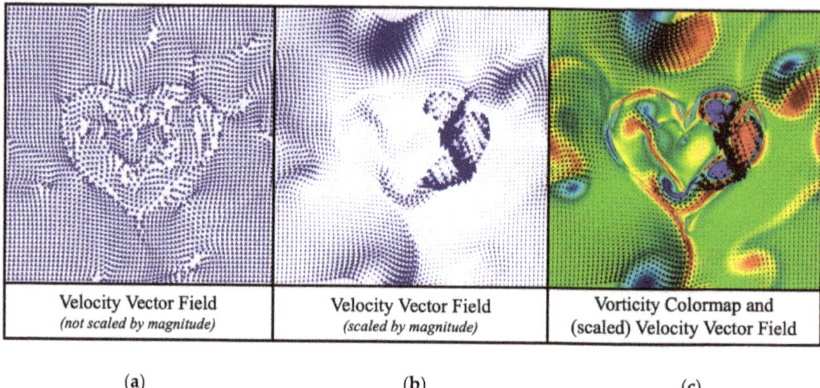

Figure 31. (**a**,**b**) Unscaled and scaled velocity vector field, respectively, that was used to define the initial fluid vorticity (**c**).

After having computed the vorticity from a velocity field, the simulation could begin. This simulation used the computational parameters listed in Table 6. We further note that the velocity field had an original grid resolution of $[N_x, N_y] = [512, 512]$, so we down-sampled it to a $[256, 256]$ grid for the example shown here. Furthermore, as the original velocity field was computed from an independent fluid-structure simulation, we comment that there is no complex, moving boundary within the computational domain; however, at the initial point you can see the remnants of the coinciding heart structure, see Figure 31.

Table 6. Numerical parameters for case of evolving vorticity from an initial velocity field.

Parameter	Variable	Units	Value
Domain Size	$[L_x, L_y]$	m	$[1, 1]$
Spatial Grid Resolution	$[N_x, N_y]$		$[256, 256]$
Spatial Grid Size	$dx = dy$	m	$L_x/N_x = L_y/N_y$
Time Step Size	dt	s	10^{-2}
Total Simulation Time	T	s	50
Fluid Kinematic Viscosity	$\nu = \mu/\rho$	m^2/s	0.001

Figure 32 provides snapshots over the simulation to illustrate how the vorticity and magnitude of velocity evolve. Both Figure 32a,b illustrate the effect of periodic boundary conditions along each edge of the domain, e.g., vortices moving through the top or right side of the domain continue through the bottom or left side of the domain, respectively. Furthermore, in the snapshots provided, the overall vorticity and magnitude of velocity appears to dissipate within the domain. This is because there are no source terms (or explicit boundary conditions) providing additional flow (energy) into the system to induce more flow. Although the original velocity vector field that defined the initial vorticity of the system came from a simulation of pulsing hearts, without the heart moving and thus pushing on the fluid, the fluid will eventually stop moving and reach zero velocity due to the fluid's viscous nature. Therefore initializing the vorticity out of a time-point from another simulation's velocity field, is akin to studying the evolution of the fluid's motion from a singular impulse in the flow.

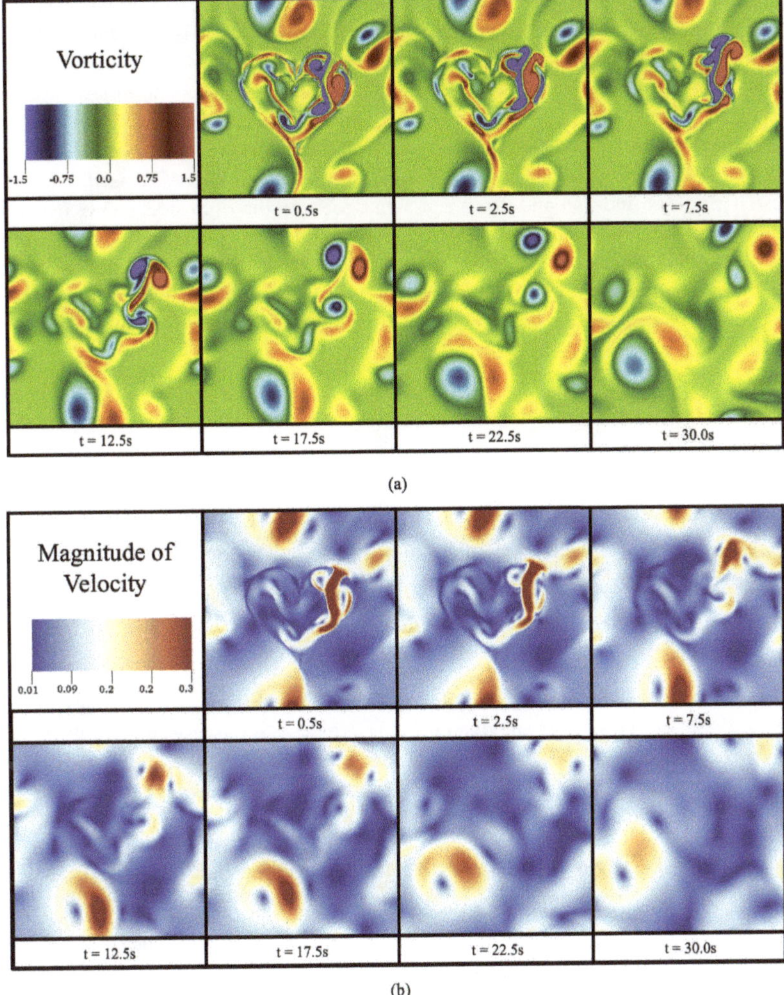

Figure 32. The evolution of (**a**) the fluid vorticity and (**b**) magnitude of velocity during the course of the simulation.

Students may elect to try the following:

1. Recreate the above results with the parameters listed in Table 6 or for different ν.
2. Take a velocity field from one of the Projection method examples, e.g., from Section 4.1 or Section 4.3, and use it to initialize the vorticity for a simulation using this spectral (FFT) method.
3. Find how long it takes the simulation to reach a $1/10$ or $1/100$ of the original maximum magnitude of velocity.
4. Discover how that time varies with changes in viscosity, ν.

4.5. Flow Past One or More Cylinders (via Lattice Boltzmann)

Flow past objects is one of the most heavily studied problems in aerodynamics and thus is a standard example used in many fluid dynamics courses [68–70]. While many texts only consider the problem for inviscid flows and use potential flow methods to solve for exact solutions [69,70], here we include effects of viscosity to illustrate transitions to vortical flow, and upon doing so, a subset of the possible flow separation cases. Furthermore, flow past cylinders is also still an active area of contemporary research [71–77].

We performed simulations for both a single cylinder as well as multiple cylinders. The computational geometry for a single cylinder case is given in Figure 33. Figure 33a gives the boundary conditions considered for the problem for flow past a rigid cylinder contained within a rigid channel. The walls of the channel are rigid and so bounce-back boundary conditions are used. The ends of the channel use periodic boundary conditions, e.g., what goes out the right side, comes in through the left. Note that this is the first example discussed in which uses a mix of explicit boundary conditions and periodic boundary conditions. Moreover, the cylinder itself is modeled as a rigid structure and so bounce-back conditions are used on it. Figure 33b provides visual details how the inflow is modeled in the simulation. Every iteration, the horizontal velocity at the inflow is increased by an amount of ΔU_x to drive fluid flow towards the right.

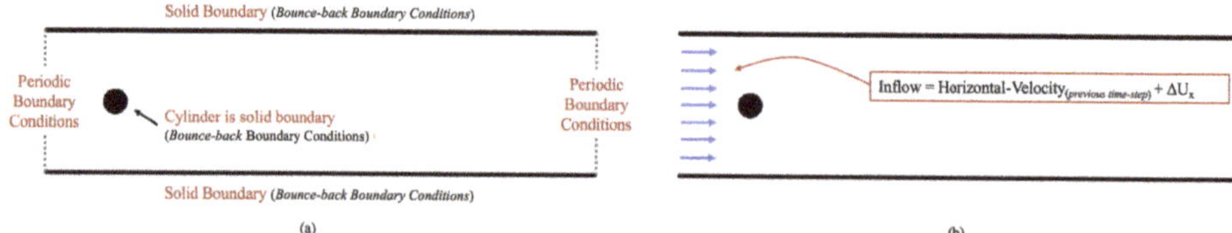

Figure 33. The geometrical setup for flow past cylinders using the Lattice Boltzmann method. (**a**) The boundary conditions specified on all sides of the domain and on the cylinder within the channel and (**b**) the inflow condition.

All simulation parameters (fluid, grid, and geometry) for both cases involving either a single cylinder or multiple rigid cylinders are given in Table 7. Note that the cases with multiple cylinders simulate flow around cylinders with larger radii, use a larger value for ΔU_x, and a different final number of steps and grid resolutions. Furthermore, due to these differences we are not comparing the case of one single cylinder to the case with multiple cylinders.

Table 7. Numerical parameters for flow past one or more cylinders using the Lattice Boltzmann Method.

Parameter	Variable	Units	Value
Domain Size	$[L_x, L_y]$	m	$[2, 0.5]$
Spatial Grid Resolution (1 Cylinder)	$[N_x, N_y]$		$[640, 160]$
Spatial Grid Resolution (Multiple Cylinders)	$[N_x, N_y]$		$[512, 128]$
Spatial Grid Size	$dx = dy$	m	$L_x/N_x = L_y/N_y$
'Density' Initialization	ρ		0.01
Relaxation Parameter	τ		0.53
Total Simulation Steps (1 Cylinder)	N		56000
Total Simulation Steps (Multiple Cylinders)	N		5500
Incremental inflow velocity increase (1 cylinder)	Δu_x		0.00125
Incremental inflow velocity increase (Multiple Cylinders)	ΔU_x		0.01
Radii (1 Cylinder)	r	$\% L_y$	$7.5\% \, L_y$
Radii (Multiple Cylinders)	r	$\% L_y$	$12.5\% \, L_y$

The simulation data obtained at step $n = 49,600$ is given in Figure 34. It presents colormaps (and corresponding contours) for vorticity, magnitude of velocity, horizontal velocity, vertical velocity, and the finite time Lyapunov exponent (FTLE), to which knowledge of the approximate Lagrangian

Coherent Structures (LCS) can be extracted [58,59,61,65,66]. We attribute regions with high FTLE to regions of high fluid mixing, as higher FTLE suggests that nearby fluid parcels separate at a much faster rate than those in lower FTLE regions. From the vorticity panel, significant flow separation is seen as vortices are being shed off the cylinder and flow pushes past it left-to-right. Oppositely spinning vortices are shed off cylinder one after another. Moreover, we observe that there is significantly more horizontal fluid motion than vertical within the channel. Since flow is being pushed left-to-right from the inflow condition, it would require more energy for fluid to move vertically, than to simply...go with the flow. Figure 35 provides snapshots of vorticity (left) and FTLE (right) during the simulation to highlight how smooth flow developed into vortical flow patterns once the horizontal velocity increased past a threshold. Such threshold appears to have occurred around simulation step $\sim 47,000$.

In particular, flow instabilities began to manifest by step $46,000$, as seen by both the vorticity and FTLE snapshots. The majority of fluid mixing occurred in large contours behind the cylinder up to this point. Once flow separation occurs, some mixing occurs in the wake in small patches, but no geometrically long regions of higher fluid mixing are present, as seen by the contours. Note that in the last two panels, for steps $49,200$ and $49,600$, large FTLE-valued regions do not correspond to locations of shed vortices, rather higher FTLE valued regions appear between oppositely spinning vortices.

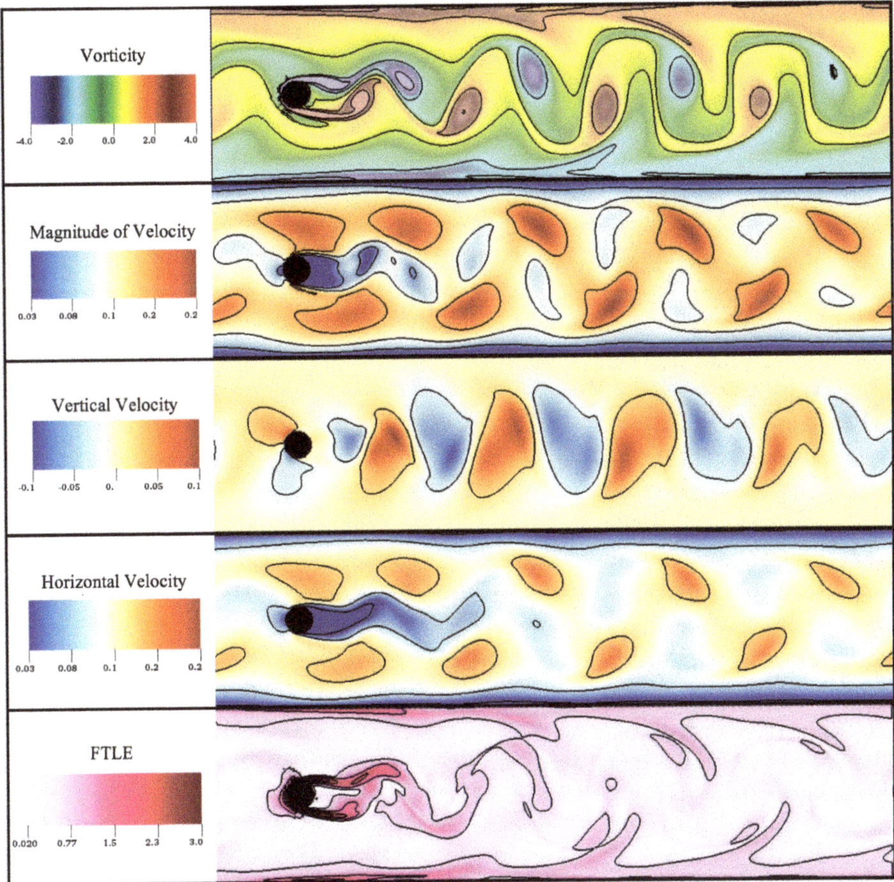

Figure 34. Illustrations of the flow field at the end of a simulation with one cylinder in the channel, showing colormaps of vorticity, magnitude of velocity, vertical and horizontal velocity, and finite-time Lyapunov exponent.

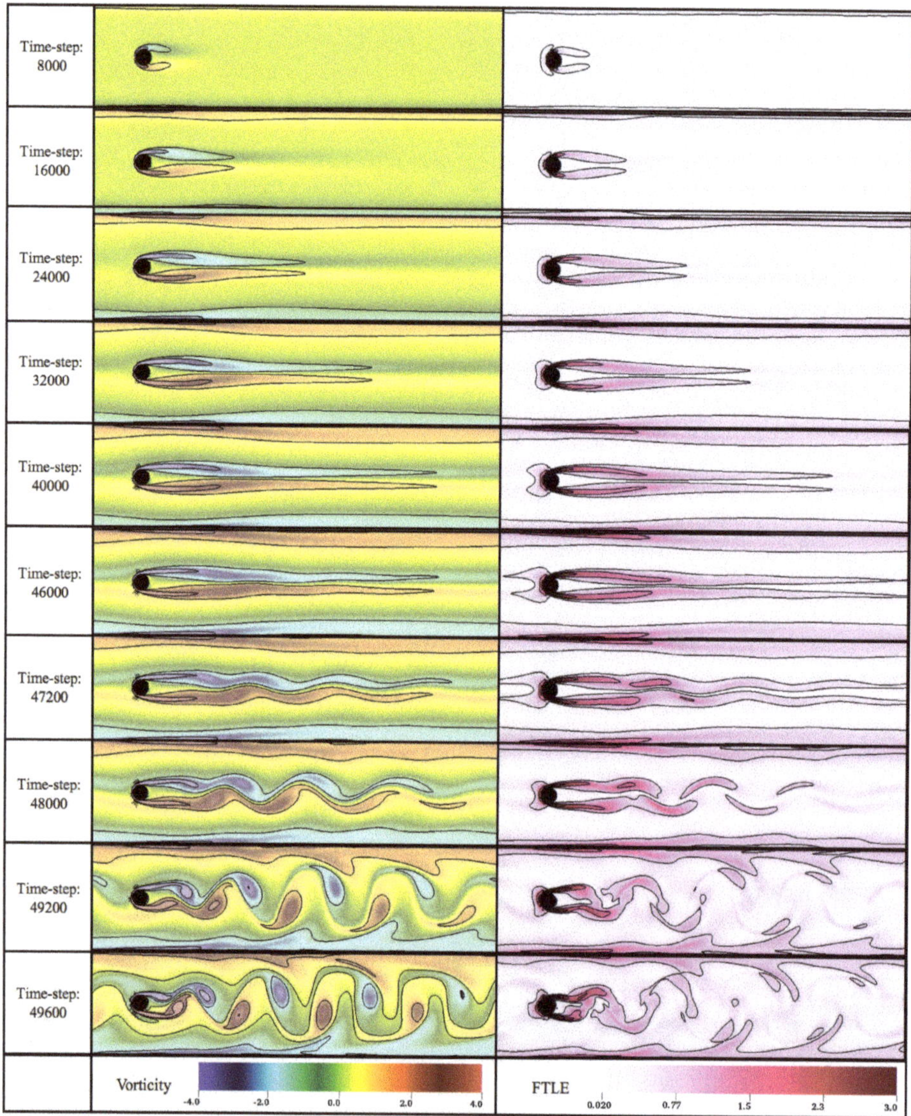

Figure 35. Illustrations depicting the flow's evolution to vortex shedding using vorticity (**left column**) and finite-time Lyapunov exponent (**right column**) with their contours, respectively.

Finally we performed simulations with multiple cylinders in a channel. Each cylinder in these simulations had a radii 66% larger than the single cylinder shown above. We performed three separate simulations with different configurations of the cylinders, each based off a triangle formation. Figure 36 illustrates how flow evolved for each geometric configuration, in particular for (a) vorticity and (b) FTLE.

Vortices are observed shedding off each cylinder, possibly at an enhanced rate due to flow interactions with the other cylinders (see suggested activity below). In the symmetric configurations, e.g., the triangle-formation and vertical-line formation, flow symmetry is preserved, while in the case with only one trailing cylinder, flow asymmetries arise. FTLE plots at the last step shown (5000) illustrate different geometrical configurations not only can lead to different patterns of fluid mixing, but also enhanced mixing, even among cases with only two cylinders. The vertically-aligned cylinder case appears to elicit more downstream mixing in the wake than the case of asymmetrically placed cylinders; however, this may also be an artifact of overall geometry of the system, e.g., the size of the cylinders and the proximity of cylinders to the channel walls.

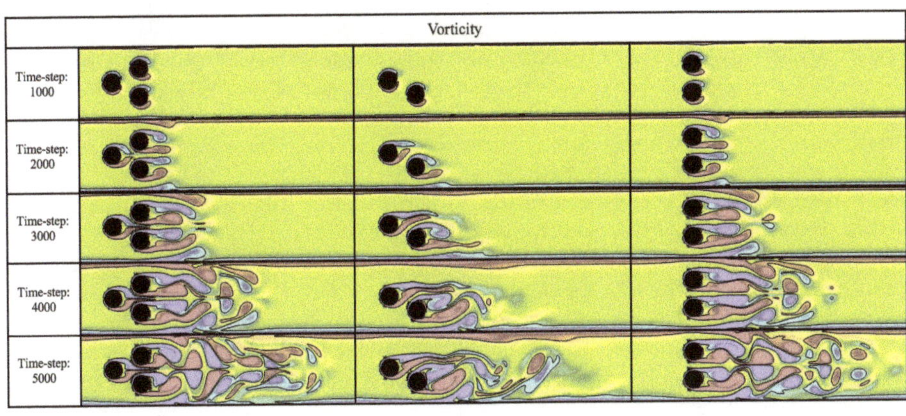

Figure 36. Snapshots from simulations of multiple cylinders in different configurations showing (**a**) vorticity and (**b**) finite-time Lyapunov exponents with their contours, respectively. Note that the colormaps are the same as used in Figure 35.

Students may elect to try the following:

1. Recreate the above results with the parameters listed in Table 7
2. Using the radius for the larger geometry case, run a simulation with only one cylinder present in the middle and compare vorticity and FTLE evolution plots.
3. Vary τ (relaxation parameter relate to viscosity) to test the system's sensitivity to τ
4. Change the number, placement, or size of each cylinder in the domain.
5. Change the shape of the object in the channel, e.g., try an ellipse or airfoil-like shape

4.6. Flow Past a Porous Cylinder (via Lattice Boltzmann)

While flow around solid objects is a popular problem in aerodynamics, flow around porous objects has only recently begun to be investigated within in the past few decades, using either high fidelity numerical simulation or experiments [78–82]. Here, we present an example of flow past a porous cylinder using the LBM. The cylinder geometry, computational setup, and inflow/boundary conditions are identical to the single cylinder case of Section 4.5 (see Table 7). The only difference being that the cylinder geometry itself is porous.

The porosity, or void fraction (VF), of the cylinder was chosen to be $VF \in \{0\%, 10\%, 25\%, 50\%, 62.5\%\}$. To model the void fraction, each grid cell inside the cylinder was given a random number between $[0, 1]$ and then depending on the specific VF, if that grid cell's randomly assigned number was above the VF in decimal form, it was assigned as a solid boundary. For example, for $VF = 25\%$, each grid cell with a randomly assigned number greater than 0.25 was declared a solid boundary point. Figure 37 depicts the porous cylinders considered for each VF case studied below.

Although, the cylinder is no longer uniformly solid, in this example we will not focus our efforts to study flow through the cylinder itself. For example, in Figure 37's $VF = 10\%$ case, there are holes in the interior of the cylinder domain that are not connected to the fluid region outside of the cylinder. Those void regions will not influence the flow outside of the cylinder, as they are blocked from outside fluid penetrating them. However, if you wanted to study the fluid dynamics through a porous structure, you could design a specific void network structure within the cylinder or in another desired geometry. What we will emphasize here is that the porous regions connected to the outside of the cylinder cause asymmetries to develop in the overall flow pattern at an accelerated rate as compared to the non-porous (solid) case. Such asymmetries then lead to quicker vortex shedding.

Figure 37. The porous cylinders considered in this section for different void fractions, $VF = \{0, 10, 25, 50, 62.5\}$. Note that the porous structures were initialized randomly within the cylinder, so there may not be open networks through the cylinder from one side to the other. The slight geometrical perturbations are sufficient enough to initiate quicker transitions to vortex shedding in each case of porous cylinder than the case of $VF = 0$.

Figure 38 shows the evolution of vortex shedding in cases of differing porosities (void fractions) of $\{0\%, 10\%, 25\%, 50\%, 62.5\%\}$. Every case shows that more porosity leads to faster development of vortex shedding. Furthermore, even the case with $VF = 10\%$ has pronounced vortices being shed before the $VF = 0\%$ case even shows any signs of significant flow asymmetry. The increased porosity accelerates flow asymmetries to manifest, leading to vortex shedding occurring earlier on. This example highlights how small perturbations in fluid dynamics problems (here small differences in geometric structure) can lead to the system having significantly different time-scales for flow structures to develop or for a bifurcation in the resulting dynamics to emerge.

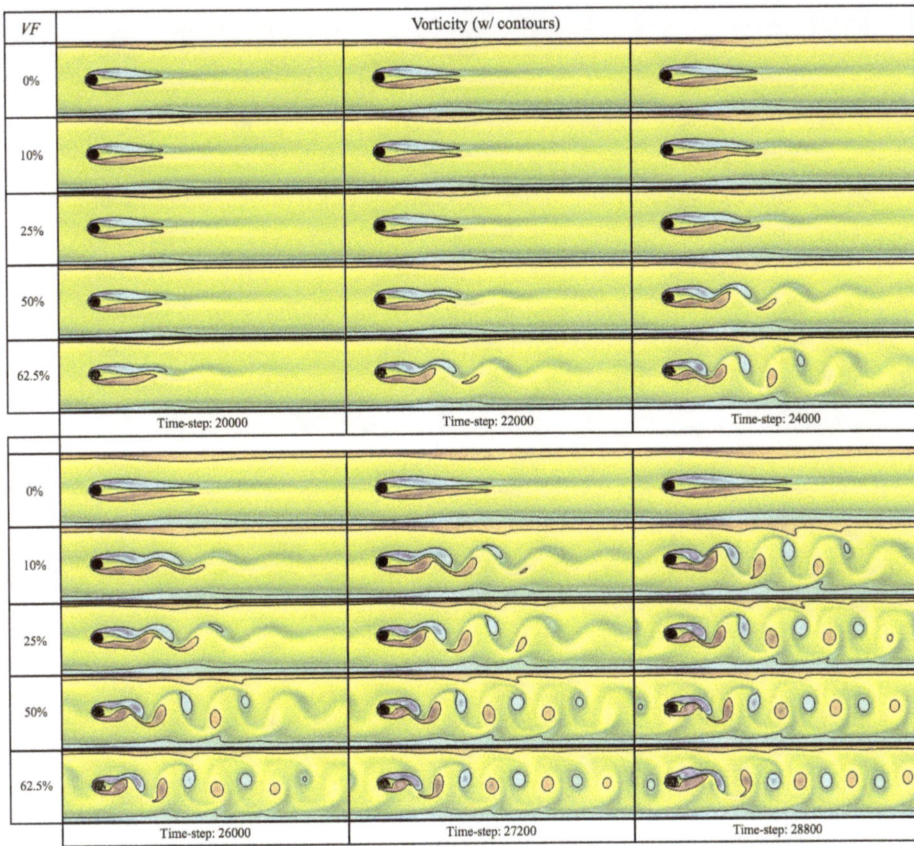

Figure 38. Snapshots illustrating vorticity (with its contours) among cases of differing void fractions, VF.

Students may elect to try the following:

1. Recreate the above results with the parameters listed in Table 7
2. Create the accompanying FTLE plots and discuss how fluid mixing changes due to the porous cylinder
3. Vary τ (relaxation parameter relate to viscosity) to test the system's sensitivity to τ
4. Increase the number of porous cylinder in the domain and vary their placement or size
5. Change the shape of the porous object in the channel, e.g., try an ellipse or airfoil-like shape
6. Create a specific porous network structure through the cylinder and test how flows directly through a porous cylinder may affect vortex shedding

5. Discussion

In this work we provide software that was developed for the specific purpose to make CFD accessible to undergraduate (and graduate) students in order to provide them an opportunity to perform traditional and contemporary CFD simulations as a valuable learning experience. To that end all the fluid solvers contained within were written in both MATLAB and Python, two languages that are commonly familiar to most science and engineering students [24–28,31]. Students also have the opportunity to modify existing examples or create their own examples within the software to test hypothesis or for stimulate further scientific curiosity. To that extent, we provided an overview of how the code is structured (Section 3) and offered interesting variations to each of the examples that we showcased in this paper (Section 4).

Not only can students run CFD simulations, they also have the chance the practice the art of scientific visualization, and use open-source software that is used by many researchers (VisIt [41] or Paraview [42]), to visualize the corresponding data. These software packages are both open-source and have easy-to-use and learn graphical user interfaces (GUIs) that streamline the visualization process. They can also be used to perform data analysis as well, which further reduces the computational learning curve for students to be able probe the data produced. We provided step-by-step guides on how to use VisIt for these tasks in Section 3.

By performing their own CFD experiments, fluid dynamics students have the opportunity to vary parameters (e.g., fluid scale (Re), boundary conditions, geometry, etc.) of a given system and observe the resulting dynamics. More specifically, they are able to decrypt how the dynamics may vary across parameter regimes, both qualitatively and quantitatively. This grants them the ability to challenge their own developing intuition of the resulting flow physics to further accelerate their learning. They also gain more computer experience and witness first-hand some of the advantages that complex computer simulations offer.

If students wish to go a step further and study the underlying mathematical structure of the code, we provide insightful comments everywhere possible within each fluid solver script. To complement this, we provided a detailed mathematical overviews of each fluid solver used here in Appendix B. While it is not our mission here to teach students how implement their own numerical fluid solvers, we believe this work contributes to providing a foundation for capturing the importance (and utility) of different fluid solver schemes. For each solver that was introduced for particular applications in Sections 2.1–2.3, we give a high-level overview of the method. If students wish to continue learning about numerical methods for solving the fluid equations or numerical methods for partial differential equations in greater depth, we suggest Barba and G. Forsyth's *CFD Python: the 12 steps to Navier-Stokes equations* [33], L.A. Barba and O. Mesnard's *Aero Python: classical aerodynamics of potential flow using Python* [34], both of which use Jupyter Notebooks [38], or Pawar and San's [39] modules for developing fluid solvers in the Julia programming language [40].

In this day and age, as the importance of computer literacy increases rapidly [83–86], integrating more hands-on computer-based activities into course structures will be of critical importance. However, students also seeing successful execution of code is paramount [85]. Here we provide gateway software for students to dive into the world of CFD, in familiar programming environments. While proprietary software offers immense advantages to running simulations and analyzing data, students may see a disconnect between the programming knowledge they've acquired and such polished software. We hope to provide them a unique opportunity to see fluid solvers written in familiar, non-intimidating environments that they may be able to tweak and modify comfortably. Our hope is to inspire further ownership of their learning and to stimulate more interest in (lucrative, transferable) computer programming skills.

Supplementary Materials: The following are available at http://www.mdpi.com/2311-5521/5/1/28/s1.

Funding: N.A.B. was funded and supported by the NSF OAC-1828163, the TCNJ Support of Scholarly Activity (SOSA) Grant, the TCNJ Department of Mathematics and Statistics, and the TCNJ School of Science.

Acknowledgments: The author would like to thank Laura Miller for introducing him to the joys of computational fluid dynamics. He would also like to thank Christina Battista, Robert Booth, Christina Hamlet, Alexander Hoover, Matthew Mizuhara, Arvind Santhanakrishnan, Emily Slesinger, and Lindsay Waldrop for comments and discussion. He also would like to sincerely thank the Reviewers, especially Reviewer 2, whose above and beyond effort in providing incredibly thorough, insightful, and constructive feedback led to the manuscript becoming significantly stronger.

Conflicts of Interest: The author declares no conflict of interest.

Abbreviations

The following abbreviations are used in this manuscript:

CFD	Computational Fluid Dynamics
Re	Reynolds Number
BCs	Boundary Conditions
LBM	Lattice Boltzmann Method
VF	Void Fraction (Porosity)
GPU	Graphics Processing Unit
CPU	Central Processing Unit

Appendix A. Instructor Resources

Teaching Resources:

Associated supplemental files contain movies and codes pertaining to all the simulations detailed in this paper. It encompasses the following:

1. **suiteCFD_Supplement.pptx/.pdf**: presentations which may be used in class; slides that tell the story of the paper. Note that the *.pptx* file has embedded movies in *.mp4* format.
2. **Images**: directory containing images (simulation snapshots, data) pertaining to each simulation shown in the manuscript.
3. **Movies**: directory containing movies (*.mp4* format) pertaining to each simulation shown in the manuscript.
4. **suite-CFD-Software-02-06-2019.zip**: zip-file containing all fluid solver codes used in the manuscript (as of 6 February 2020).
5. Note that all codes used in the manuscript can also be found at: https://github.com/nickabattista/Holy_Grail.
6. Visualization software used: VisIt (https://visit.llnl.gov/) (v. 2.12.3).

Appendix B. Select CFD Algorithms

In this appendix we will discuss various numerical algorithms for studying fluid dynamics, namely a projection method [2–5], a spectral methods solver based on the Fast Fourier Transform (FFT) [7–9], and the lattice Boltzmann method [11,12]. Codes are available to test these methods at https://github.com/nickabattista/Holy$_$Grail/. Their corresponding simulation data is saved in the *.vtk* format, as to allow for visualization and analysis using open-source programs, such as VisIt [41] or ParaView [42].

Appendix B.1. Projection Methods

The projection was first introduced by Chorin in 1967 [2] and independently a year later by Temam [4] to solve the incompressible, Navier-Stokes equations [2]. The key feature of this method is that it uses operator splitting and Helmholtz-Hodge decomposition to decouple the velocity and the pressure fields, making it possible to explicitly solve the incompressible, Navier-Stokes equations in only a few steps.

We will begin by introducing some notation. We denote $\mathbf{u}_{ij}^{n} = \mathbf{u}^{n}(x_i, y_j)$ to be the velocity field field at time-step n and spatial location (x_i, y_j), where $\{x_i\}_{i=0}^{N_x}$ and $\{y_j\}_{j=0}^{N_y}$ are the x and y values of the discretized rectangular (Eulerian) computational grid. Furthermore, $\mathbf{u}^{n}$ represents the velocity field at time-step n in its entirety, e.g., not at one specific spatial location.

In the first step, an auxiliary (intermediate) velocity field is computed by ignoring any dependencies on the pressure. This is essentially an operator split. This velocity field found will not be divergence-free, and hence the necessary incompressible condition will not be satisfied (Equation (2)). In discretization terms, this step takes the form of

$$\frac{\mathbf{u}^* - \mathbf{u}^n}{\Delta t} = -\left(\mathbf{u}^n \cdot \nabla\right)\mathbf{u}^n + \nu\Delta\mathbf{u}^n, \tag{A1}$$

where $\mathbf{u}^*$ is an intermediate (auxiliary) velocity field, which is not divergence-free. The second step is known as the projection step, where the pressure gets reintroduced to give a final velocity field, $\mathbf{u}^{n+1}$ that satisfies the incompressibility condition (Equation (2)). This discretized step takes the following form,

$$\frac{\mathbf{u}^{n+1} - \mathbf{u}^*}{\Delta t} = -\frac{1}{\rho}\nabla p^{n+1}, \tag{A2}$$

where p^{n+1} is the pressure field at the next time-step. Because it requires an updated pressure term, we must first find such a pressure. To do this we recall Helmholtz-Hodge Decomposition [87,88], which says any vector field, that is twice continuously differentiable on a bounded domain, say $\mathbf{v}$, can be decomposed into a solenoidal part (divergence-free) and an irrotational part (curl-free), i.e.,

$$\mathbf{v} = \mathbf{v}_{sol} + \mathbf{v}_{irr} = \mathbf{v}_{sol} + \nabla\phi, \tag{A3}$$

where $\mathbf{v}_{sol}$ is the solenoidal part and $\mathbf{v}_{irr}$ is the irrotational part. We note that an irrotational vector field can be written as the gradient of a scalar, e.g., $\mathbf{v}_{irr} = \nabla\phi$, where ϕ is some scalar function (sometimes ϕ is referred to as a *potential*).

Note that if we take the divergence of (A3), we obtain,

$$\nabla \cdot \mathbf{v} = \Delta\phi. \tag{A4}$$

It is then possible to find the divergence-free part of the vector field $\mathbf{v}$ by solving the above Poisson problem in (A4). This motivates the form of the second step for a projection method given in (A2). However, to find the pressure, we take a divergence of (A2) and note that we require that $\mathbf{u}^{n+1}$ be divergence-free, i.e.,

$$\nabla \cdot \mathbf{u}^{n+1} = 0. \tag{A5}$$

Taking the divergence of (A2) and requiring the condition in (A5), we obtain the following Poisson problem for the pressure, p^{n+1}, in terms of the intermediate velocity field $\mathbf{u}^*$,

$$\frac{\Delta t}{\rho}\Delta p^{n+1} = \nabla \cdot \mathbf{u}^*. \tag{A6}$$

Note that this equation can be solved explicitly. Hence once p^{n+1} is found, we can then solve for $\mathbf{u}^{n+1}$ using (A2), e.g.,

$$\mathbf{u}^{n+1} = \mathbf{u}^* - \frac{\Delta t}{\rho}\nabla p^{n+1}. \tag{A7}$$

Appendix B.2. Spectral Methods via Fast Fourier Transform (FFT)

For the spectral (FFT) method fluid solver we choose to work in the vorticity formulation of the viscous, incompressible Navier-Stokes equations. Here we will begin by deriving such equations from the previously written viscous, incompressible Navier-Stokes equations, i.e., Equations (1) and (2),

$$\rho\left[\frac{\partial\mathbf{u}}{\partial t} + (\mathbf{u}\cdot\nabla)\mathbf{u}\right] = -\nabla p + \mu\Delta\mathbf{u}$$

and

$$\nabla \cdot \mathbf{u} = 0.$$

We will use a lot of identities from vector calculus. First, recall the following identity for any vector $\mathbf{F}$,

$$(\mathbf{F} \cdot \nabla)\mathbf{F} = (\nabla \times \mathbf{F}) \times \mathbf{F} + \frac{1}{2}\nabla(\mathbf{F} \cdot \mathbf{F}).$$

Substituting the above identity into the conservation of momentum equation (Equation (1) and dividing by ρ gives us,

$$\frac{\partial \mathbf{u}}{\partial t} + (\nabla \times \mathbf{u}) \times \mathbf{u} + \frac{1}{2}\nabla(\mathbf{u} \cdot \mathbf{u}) = -\frac{1}{\rho}\nabla p + \nu \Delta \mathbf{u}, \tag{A8}$$

where $\nu = \frac{\mu}{\rho}$ is the kinematic viscosity. Note that $\mathbf{u} \cdot \mathbf{u}$ is a scalar quantity, and thus we define it to be $U^2 = \mathbf{u} \cdot \mathbf{u}$. Moreover, we also get to define a new quantity called the vorticity,

$$\omega = \nabla \times \mathbf{u}. \tag{A9}$$

It is tempting to think of ω as describing the global rotation of the fluid, but this is misleading. Although many flows can be characterized by local regions of intense rotation, such as smoke rings, whirlpools, tornadoes, or even the red spot on Jupiter, some flows have no global rotation, but do have vorticity. Vorticity describes the *local* spinning of a fluid near a fixed point in space, as seen by an observer in the Eulerian framework.

Substituting the definitions of vorticity and U^2 into Equation (A8), we obtain

$$\frac{\partial \mathbf{u}}{\partial t} + \omega \times \mathbf{u} + \frac{1}{2}\nabla(U^2) = -\frac{1}{\rho}\nabla p + \nu \Delta \mathbf{u}.$$

Now taking the curl of the above equation we get an equation for the evolution of the vorticity, yields

$$\frac{\partial \omega}{\partial t} + \nabla \times (\omega \times \mathbf{u}) = \nu \Delta \omega, \tag{A10}$$

since $\nabla \times (\Delta \mathbf{u}) = \Delta(\nabla \times \mathbf{u}) = \Delta \omega$. Furthermore we note that the pressure terms drop out as the resulting force from pressure only acts perpendicular to the surface of a fluid blob and not parallel to it, i.e., $\nabla \times (\nabla \Phi) = 0$ for any scalar field Φ.

The viscous, incompressible Navier-Stokes equations can thus be written as follows,

$$\frac{\partial \omega}{\partial t} + \nabla \times (\omega \times \mathbf{u}) = \nu \Delta \omega \tag{A11}$$

$$\nabla \cdot \mathbf{u} = 0 \tag{A12}$$

Shortly we will introduce a vector potential to strive towards introducing a streamfunction, ψ, into our formulation, but first we will use a vector calculus identity to re-write Equation (A11) into a more traditional looking advection-diffusion equation. Using the following identity from vector calculus,

$$\nabla \times (\mathbf{A} \times \mathbf{B}) = (\nabla \cdot \mathbf{B} + \mathbf{B} \cdot \nabla)\mathbf{A} - (\nabla \cdot \mathbf{A} + \mathbf{A} \cdot \nabla)\mathbf{B}, \tag{A13}$$

we can mathematically massage Equation (A11) into the following form,

$$\frac{\partial \omega}{\partial t} + (\mathbf{u} \cdot \nabla)\omega = (\omega \cdot \nabla)\mathbf{u} + \nu \Delta \omega. \tag{A14}$$

Note that the evolution equation for vorticity, Equation (A14), now looks like an advection-diffusion equation, but with an additional extra term, $(\omega \cdot \nabla)\mathbf{u}$. For 2D flows, recall that $\mathbf{u} = (u, v, 0)$ and hence $\boldsymbol{\omega} = (0, 0, \omega)$. Moreover, in 2D flows, all partial derivatives with respect to z are zero; hence the $(\boldsymbol{\omega} \cdot \nabla)\mathbf{u}$ term becomes zero, e.g.,

$$(\boldsymbol{\omega} \cdot \nabla)\mathbf{u} = \left(0\frac{\partial}{\partial x} + 0\frac{\partial}{\partial y} + \omega\frac{\partial}{\partial z} \right) \mathbf{u} = \omega\frac{\partial \mathbf{u}}{\partial z} = 0.$$

Thus, in 2D, we have the following form of the momentum equation in terms of vorticity

$$\frac{\partial \omega}{\partial t} + (\mathbf{u} \cdot \nabla)\omega = \nu\Delta\omega. \tag{A15}$$

Note that the form of Equation (A15) suggests that if $\nu \equiv 0$ and if $\omega = 0$ everywhere at any moment in time, then $\omega = 0$ for all future time. Moreover, since $\omega = \nabla \times \mathbf{u} = 0$, we have irrotational flow. Thus, we would be studying an incompressible *potential flow* problem [70].

Next we introduce the streamfunction, ψ, as part of the vector potential for $\mathbf{u}$,

$$\mathbf{u} = \nabla \times \psi\hat{k}. \tag{A16}$$

Note that if the streamfunction, $\psi = \psi(x, y)$, is known, it is possible to extract the components of the 2D fluid velocity field, $\mathbf{u} = (u, v)$, from it e.g.,

$$u = \frac{\partial \psi}{\partial y} \quad \text{and} \quad v = -\frac{\partial \psi}{\partial x}. \tag{A17}$$

Furthermore, taking the curl of (A16), we are able to get a Poisson problem for ψ in terms of ω,

$$\Delta\psi = -\omega, \tag{A18}$$

where we have used the following vector calculus identity,

$$\nabla \times \nabla \times \mathbf{A} = \nabla(\nabla \cdot \mathbf{A}) - \nabla^2\mathbf{A}, \tag{A19}$$

and the fact that $\frac{\partial \psi}{\partial z} = 0$.

At this point, the idea is that if we are able to solve for the streamfunction, ψ, from the vorticity, ω, we can then get the fluid velocity ,$\mathbf{u}$, and it will automatically satisfy the incompressibility condition by definition of the vector potential, i.e., Equation (A16). In essence this is the algorithm; however, within this algorithm, we will work as much as possible in the Fourier frequency space, granted to us by taking the Fast Fourier Transform (FFT). Before diving into the 4 main steps of this scheme, we will introduce some notation involving Discrete Fourier Transforms (DFT) and hence FFT. Note that the FFT yields the same results as the DFT but does so in a more computationally efficient, i.e., *fast*, manner.

- Taking the Discrete Fourier Transform (DFT) of a set of complex numbers produces another set of complex numbers. The notation $\mathscr{F}\{\mathbf{z}\}$ denotes taking the Discrete Fourier Transform of a set of N-values, $\{z_n\}_{n=0}^{N-1}$, e.g., for $k = 0, 1, \ldots, N-1$, we define

$$\hat{z}_k = \mathscr{F}\{\mathbf{z}\} = \sum_{n=0}^{N-1} z_n\, e^{-2\pi i \frac{kn}{N}}.$$

- Variables with a hat, such as $\hat{z}_k$, denote variables that have been transformed into frequency space via the Discrete Fourier Transform. Moreover, the indices k are known as the DFT's wave-numbers.

- We can also take the Inverse Discrete Fourier Transform (IDFT) to return our quantities of interest from frequency space to real space. We denote the IDFT as

$$z_n = \mathscr{F}^{-1}\{\hat{z}\} = \frac{1}{N}\sum_{k=0}^{N-1} \hat{z}_k\, e^{-2\pi i \frac{kn}{N}},$$

  for all $n = 0, 1, 2, \ldots, N-1$.
- Next we will define the discretized quantities in the algorithm. A quantity such as f_{ij}^n denotes that quantity's value at the nth time-step at spatial location (x_i, y_j) within the rectangular computational grid. Hence we have a set of numbers $\{f_{ij}^n\}_{i=0,j=0}^{N_x-1,N_y-1}$ (or matrix that changes as $n \to n+1$), and can transform it in analogous manner using a DFT.
- We define K_X and K_Y to be matrices of the DFT's wave-numbers, where $K_X, K_Y \in \mathbb{R}^{N_x \times N_y}$. They are defined as:

$$K_X = \begin{bmatrix} 0 & 1 & 2 & \cdots & N_y-1 \\ 0 & 1 & 2 & \cdots & N_y-1 \\ \vdots & \vdots & & \vdots \\ 0 & 1 & 2 & \cdots & N_y-1 \end{bmatrix} \quad \text{and } K_Y = \begin{bmatrix} 0 & 0 & 0 & \cdots & 0 \\ 1 & 1 & 1 & \cdots & 1 \\ 2 & 2 & 2 & \cdots & 2 \\ \vdots & & \vdots & & \vdots \\ N_x-1 & N_x-1 & N_x-1 & \cdots & N_x-1 \end{bmatrix}.$$

  Note that each row of the matrix K_X is a vector that we denote $\mathbf{k}_X$ with components $\mathbf{k}_X = (0, 1, 2, \ldots, N_y-1)^T$. Similarly each column of the matrix K_Y is a vector that we denote $\mathbf{k}_Y = (0, 1, 2, \ldots, N_x-1)$.
- In order to benefit from the FFT we will assume our spatial grid has a resolution of $N_x \times N_y$, where both N_x and N_y are powers of 2 [89].

We will now dive into the the 4 main steps in this spectral (FFT) method's algorithm. They are as follows:

1. *Update the streamfunction to the current time-step, n:*

   From the previous time-step's vorticity, ω^n, we can solve the Poisson problem (Equation (A18)) for the streamfunction at the current time-step, ψ^n, i.e.,

$$\hat{\psi}_{ij}^n = \frac{\hat{\omega}_{ij}^n}{k_{X_i}^2 + k_{Y_j}^2}, \tag{A20}$$

   where k_{X_i} and k_{Y_j} are the Fourier wave-numbers, e.g., the ith and jth components of the vectors $\mathbf{k}_X$ and $\mathbf{k}_Y$, respectively.

2. *Obtain the components of velocity and the gradient of vorticity:*

   With the newly updated $\hat{\psi}^n$ as well as $\hat{\omega}^n$ from the previous time-step, we are able to compute the components of the velocity field at the current time-step, $\mathbf{u}^n = (u^n, v^n)$. To do this, we take derivatives of the streamfunction and vorticity in frequency space. This then gives us the components of velocity (see Equation (A17)) and the gradient of vorticity, in frequency space respectively. We can then use the IDFT to transform these quantities back into in real space, e.g.,

$$u^n = \mathscr{F}^{-1}\left\{2\pi i\, K_Y \circ \hat{\psi}^n\right\} \tag{A21}$$

$$v^n = \mathscr{F}^{-1}\left\{-2\pi i\, K_X \circ \hat{\psi}^n\right\} \tag{A22}$$

$$\omega_x^n = \mathscr{F}^{-1}\left\{2\pi i\, K_X \circ \hat{\omega}^n\right\} \tag{A23}$$

$$\omega_y^n = \mathscr{F}^{-1}\left\{2\pi i\, K_Y \circ \hat{\omega}^n\right\} \tag{A24}$$

where the operation $A \circ B$ between two matrices of equal size is called the Hadamard product. The Hadamard product is element-wise multiplication, e.g., if A and B are matrices of the same size, for all components i, j, $(A \circ B)_{ij} = A_{ij} B_{ij}$.

3. *Compute the advection term in frequency space from Equation (A15):.*

Once you have the velocity field (u^n, v^n) and partial derivatives of vorticity $\omega_x^n = \frac{\partial \omega^n}{\partial x}$ and $\omega_y = \frac{\partial \omega^n}{\partial y}$ at the current time-step, it is now possible to compute the advection term from Equation (A15), i.e., $\mathbf{u} \cdot \nabla \omega$. We define $F^n_{\text{adv}_{ij}}$ to be the above advection term, and hence get that

$$F^n_{\text{adv}_{ij}} = u_{ij}^n \cdot \omega_{x_{ij}}^n + v_{ij}^n \cdot \omega_{y_{ij}}^n. \tag{A25}$$

Furthermore, we can apply the DFT to Equation (A25) to transform it into frequency space, e.g.,

$$\hat{F}^n_{\text{adv}_{ij}} = \mathscr{F}\left\{ F^n_{\text{adv}_{ij}} \right\}. \tag{A26}$$

This will allows us to update vorticity to the $(n+1)^{st}$ time-step using Equation (A15) in frequency space.

4. *Update the vorticity to next time-step:*

Finally we use the Crank-Nicholson scheme to update the vorticity to the next time-step, $\hat{\omega}^{n+1}$,

$$\hat{\omega}_{ij}^{n+1} = \frac{\left[1 + \frac{\nu \Delta t}{2}\left(K_{X_{ij}}^2 + K_{Y_{ij}}^2\right)\right]\hat{\omega}_{ij}^n - \Delta t\, \hat{F}^n_{\text{adv}_{ij}}}{1 - \frac{\nu \Delta t}{2}\left(K_{X_{ij}}^2 + K_{Y_{ij}}^2\right)} \tag{A27}$$

Note that this method is semi-implicit; we explicitly discretize the advection term, while we implicitly discretize the diffusion term. The Crank-Nicholson scheme is second order accurate in time and space [51], and is unconditionally stable for an array of parabolic problems of the type $w_t = a w_{xx}$ [52].

Now that the vorticity has been updated, $\omega^n \to \omega^{n+1}$, you can repeat this process again.

Appendix B.3. Lattice Boltzmann Methods

As mentioned in Section 2.3, the Lattice Boltzmann method (LBM) does not explicitly (or implicitly) solve the viscous, incompressible Navier-Stokes equations, rather it uses discrete Boltzmann equations to model the fluid dynamics. In a nutshell tracks fictitious particles of fluid flow, thinking of the problem more as a transport equation, e.g.,

$$\frac{\partial f}{\partial t} + \mathbf{u} \cdot \nabla f = \Omega, \tag{A28}$$

where $f(\mathbf{x}, t)$ is the particle distribution function, i.e., a probability density, $\mathbf{u}$ is the fluid particle's velocity, and Ω is what is called the collision operator. However, rather than have these particles moving in a Lagrangian framework, the Lattice Boltzmann method simplifies this assumption and restricts the particle movements to nodes of a lattice. While we will only discuss a two dimensional implementation of the LBM, three dimensional implementations follow analogously.

From the assumption restricting the fluid particles to reside on a lattice, there are only 9 possible directions that a particle could potentially stream, or pass information, along to. These directions are either horizontal (left/right) or vertical (up/down) or forward or backward along both diagonal directions, as well as, staying at rest on its current node. These directions are illustrated in Figure A1, and these streaming velocities, $\{\mathbf{e}_i\}$, are called the *microscopic velocities*. The directions illustrated in Figure A1 is commonly called the *D2Q9* Lattice Boltzmann Model.

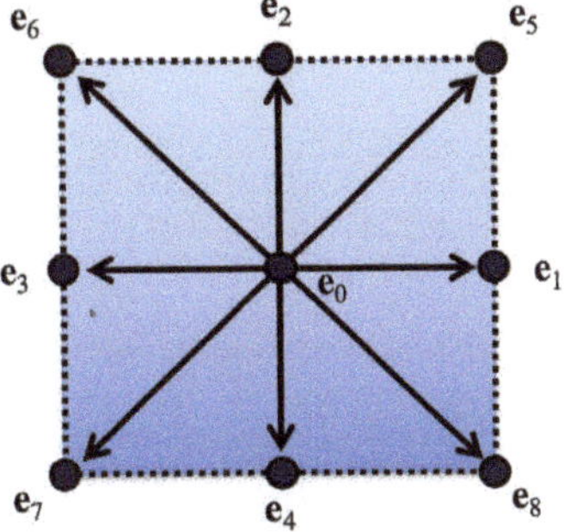

Figure A1. Figure illustrating the possible streaming directions, $\{\mathbf{e}_i\}$ for the D2Q9 Lattice Boltzmann model.

Every point on the lattice has is a probability function, $f(\mathbf{x}, t)$, associated with it. Accounting for the possibility of moving in only 9 directions, we rewrite the probability function as its discretized counterpart, $f_i(\mathbf{x}, t)$, where f_i now gives the probability of streaming in a particular direction $\mathbf{e}_i$. Using this discretization, we can define the macroscopic fluid density to be the sum of all possible f_i, e.g.,

$$\rho(\mathbf{x}, t) = \sum_{i=0}^{8} f_i(\mathbf{x}, t). \tag{A29}$$

Similarly, we can define the *macroscopic velocity* of the fluid as an average of the microscopic velocities in each direction weighted by their associated particle distribution functions f_i using (A29),

$$\mathbf{u}(\mathbf{x}, t) = \frac{1}{\rho} \sum_{i=0}^{8} c f_i(\mathbf{x}, t) \mathbf{e}_i, \tag{A30}$$

where $c = \frac{\Delta x}{\Delta t}$ and is referred to as the lattice speed. The key elements that are left to discuss are exactly what it means to *stream* the particle distributions, f_i, as well as what it meant by the *collision*, Ω. However, they both are encompassed within the steps in the LBM algorithm, so we will explicitly define these procedures while also describing the algorithm. The steps are detailed below:

1. The first step is to stream the particle densities to propagate in each direction. Explicitly you calculate the following intermediate particle density, f_i^*,

$$f_i^*(\mathbf{x} + c\mathbf{e}_i \Delta t, t + \Delta t) = f_i^n(\mathbf{x}, t), \tag{A31}$$

where n is the time-step and where for each direction i, you would in practice compute

$$
\begin{aligned}
f_1^*(x_i, y_j) &= f_1^n(x_{i-1}, y_j), & f_2^*(x_i, y_j) &= f_2^n(x_i, y_{j-1}), & f_3^*(x_i, y_j) &= f_3^n(x_{i+1}, y_j), \\
f_4^*(x_i, y_j) &= f_4^n(x_i, y_{j+1}), & f_5^*(x_i, y_j) &= f_5^n(x_{i-1}, y_{j-1}), & f_6^*(x_i, y_j) &= f_6^n(x_{i+1}, y_{j-1}) \\
f_7^*(x_i, y_j) &= f_7^n(x_{i+1}, y_{j+1}), & f_8^*(x_i, y_j) &= f_8^n(x_{i-1}, y_{j+1}), & f_9^*(x_i, y_j) &= f_9^n(x_i, y_j)
\end{aligned}
\tag{A32}
$$

This idea of streaming is shown in Figure A2.

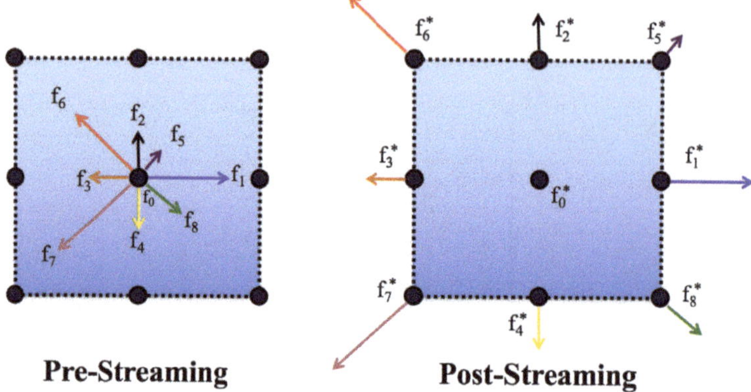

Figure A2. Figure illustrating the idea of streaming by showing color correlated particle probability functions, f_i, before the streaming process and post-streaming, f_i^*.

2. The second step involves finding what is referred to as the equilibrium distribution. This step is a part of the *collision* step, where you want to relax the particle density distributions towards a local equilibrium. The local equilibrium is denoted $f_i^{eq}(\mathbf{x}, t)$. First we must compute macroscopic the properties (density and velocity) from the intermediate particle distributions f_i^* using (A29) and (A30).

Once we have these quantities, we can now define the equilibrium distributions, f_i^{eq}. We note that there are many equilibrium distributions one could use in practice; however, each depends on your specific model and its assumptions. The Lattice Boltzmann method implemented here uses what is called the Bhatnagar-Gross-Krook (BGK) collision mode [90]. The BGK collision model is useful for simulating single phase flows [12] and is most often the classic model to use for solving the incompressible, viscous Navier-Stokes equations, although it can also be useful for simulating compressible flows at low Mach numbers [11]. See [11] for a good review of the BGK model. The BGK model's equilibrium distribution can be written as follows

$$f_i^{eq}(\mathbf{x}, t) = w_i \rho + \rho s_i(\mathbf{u}(\mathbf{x}, t)), \tag{A33}$$

where w_i is a weight and $s_i(\mathbf{u}(\mathbf{x}, t))$ is defined as

$$s_i(\mathbf{u}(\mathbf{x}, t)) = w_i \left[3 \frac{\mathbf{e}_i \cdot \mathbf{u}}{c} + \frac{9}{2} \frac{(\mathbf{e}_i \cdot \mathbf{u})^2}{c^2} - \frac{3}{2} \frac{\mathbf{u} \cdot \mathbf{u}}{c^2} \right]. \tag{A34}$$

The corresponding weights, w_i are given as

$$w_i = \begin{cases} \frac{4}{9} & i = 0 \\ \frac{1}{9} & i \in \{1, 2, 3, 4\} \\ \frac{1}{36} & i \in \{5, 6, 7, 8\} \end{cases}. \tag{A35}$$

3. Finally we compute the collision step associated with the BGK model as follows

$$f_i^{n+1} = f_i^* - \frac{f_i(\mathbf{x}, t) - f_i^{eq}(\text{bf})}{\tau}, \tag{A36}$$

where τ is the relaxation parameter and intuitively is related to the viscosity of the fluid, i.e.,

$$v = \frac{2\tau - 1}{6} \frac{\Delta x^2}{\Delta t}.$$

(A37)

Before we mention how to handle boundary conditions we will briefly discuss some of the advantages of the LBM. One of the biggest advantages of LBM is its implementation lends itself toward massive GPU or CPU parallelization. Due to parallelization it can be an incredibly fast way of solving fluid problems that are coupled with equations that model heat transfer or chemical processes [91]. Moreover, the algorithm also prides itself for the ability to compute flows through complex geometries and porous structures rather easily and efficiently [55]. From the structure of the streaming step, one can easily prescribe boundary conditions and regions in the grid where fluid is not allowed to flow easily. For our considerations here we only will introduce what are referred to as *bounce-back boundary conditions* [55].

The bounce-back boundary conditions are used to enforce no-slip conditions; however, as we will show, they are not only used on the edges of the domain, but can be implemented on the interior to create complex geometries. In a nutshell the incoming streaming directions of the distribution functions are simply reversed when they hit a boundary node. This idea is depicted in Figure A3. In practice, one can simply mask these boundary points on the domain using boolean logic.

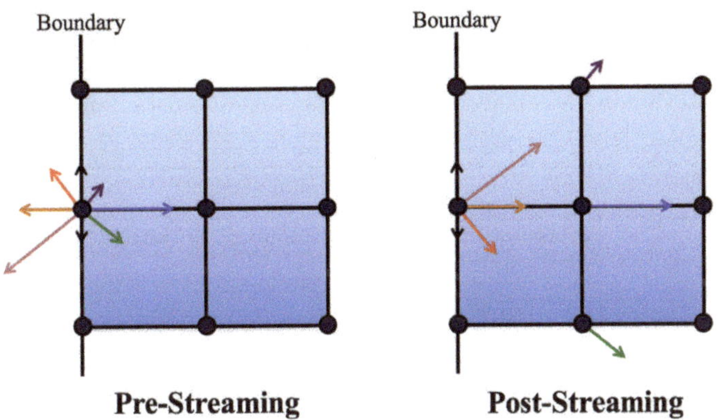

Figure A3. Illustration of bounce-back boundary conditions. During the pre-streaming step there are microscopic velocities set on the boundary and then they are reversed during the streaming step

Appendix C. Extra Visualizations of Data from Section 3: Spectral (FFT) Method's `bubble3` Example

These visualizations are to complement those already presented in the guided tutorials from Section 3. We will also give a bit of background for the simulation as well. This example used the spectral (FFT-based) fluid solver in the software and models multiple regions of vorticity 'overlapping' at the beginning. Note that first example given here can be run by going into in the *FFT_NS_Solver* script selecting the 'bubble3' option. The vorticity is initialized as in Figure A4. Recall that counterclockwise (*CCW*) and clockwise (*CW*) correspond to regions of uniform vorticity, where vorticity initialized as a positive or negative constant for *CCW* and *CW*, respectively.

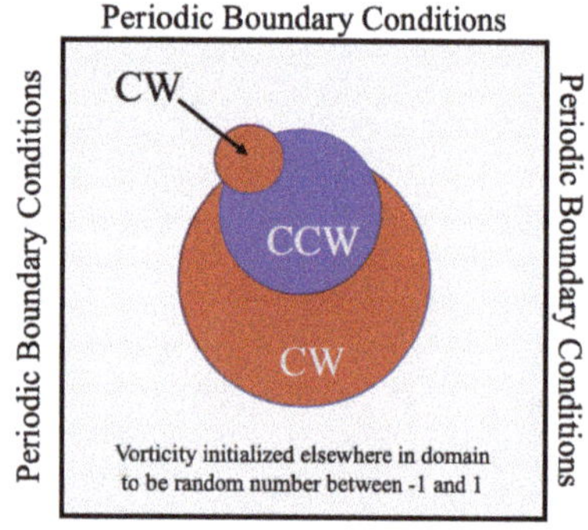

Figure A4. Illustrations of the boundary conditions and vorticity initialization for the cases of the `bubble3` example.

In this example, three circular regions of vorticity are placed, partially on top of one another. The largest region begins with a uniform value of $+0.4$, followed by the smaller regions with -0.5 and $+0.5$, respectively. The remainder of the computational domain is initialized with a random value of vorticity between $[-1, 1]$. We note that initializing a random values of vorticity will generally not satisfy the incompressibility condition (Equation (2)); however, here we do it only to illustrate that the solver is able to handle initial random noise. This simulation used the computational parameters found in Table A1.

Table A1. Numerical parameters for case with 'overlapping' vorticity regions

Parameter	Variable	Units	Value
Domain Size	$[L_x, L_y]$	m	$[1, 1]$
Spatial Grid Resolution	$[N_x, N_y]$		$[256, 256]$
Spatial Grid Size	$dx = dy$	m	$L_x/N_x = L_y/N_y$
Time Step Size	dt	s	10^{-2}
Total Simulation Time	T	s	30
Fluid Kinematic Viscosity	$v = \mu/\rho$	m^2/s	0.001

Figure A5 provides the simulation data at the beginning of the simulation (a) and at the last time-step (b). It presents colormaps (and corresponding contours) for vorticity, magnitude of velocity, horizontal velocity, vertical velocity, and the finite-time Lyapunov exponent (FTLE). It also gives a snapshot of the velocity vector field. The last snapshot of the simulation was from $t = 30.0$ s.

From Figure A5, it is evident that the random vorticity values at the start of the simulation eventually interact and dissipate in the flow, as observed in the vorticity panel. Initially there appears to be a lot of noise in the background vorticity (it was initialized to be random between $[-1,1]$) and by the end it appears virtually averaged out. Moreover, due to the background vorticity noise at the beginning, there are a lot of tiny patches of oppositely moving fluid. This leads to an initial background of high FTLE values, which suggests there is significant fluid mixing occurring. Similarly to vorticity, by the end, the background has significantly less mixing (e.g., smaller FTLE values) overall in areas away from the interacting vortical structures, which started off as overlapping vorticity regions. The area of high mixing in the FTLE panel at the last time-step is in a region where there is a lot of oppositely moving fluid, both horizontally as well as vertically.

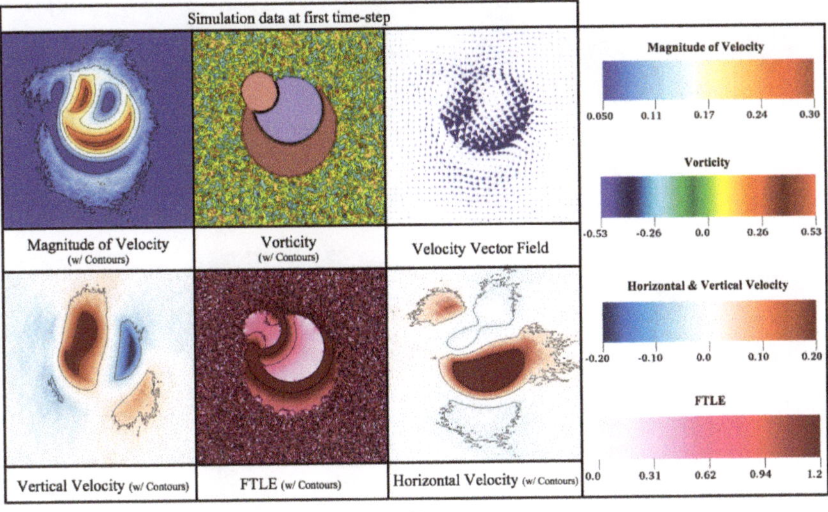

(a)

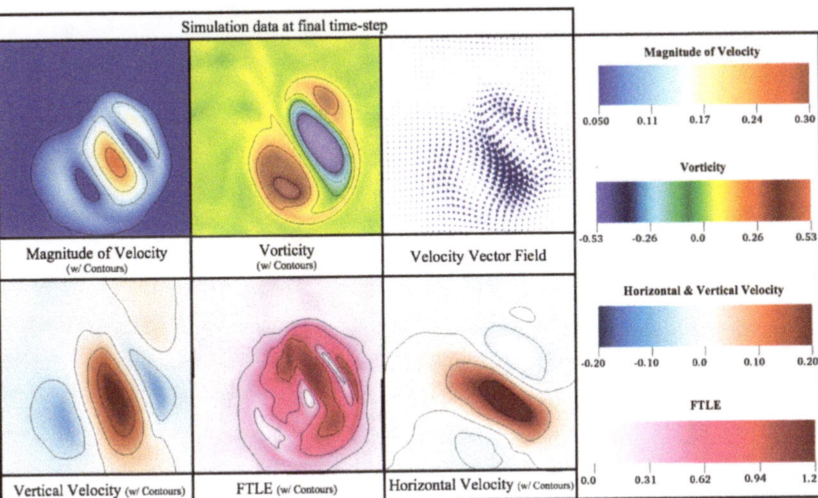

(b)

Figure A5. Simulation data from the case of overlapping vorticity regions during (**a**) the first time-step and (**b**) the final time-step.

Figure A6 provides snapshots over the simulation to illustrate how the magnitude of velocity and vorticity evolved over time. Figure A6a shows that the overlapping regions of vorticity induce the highest flows overall, as quantified by magnitude of velocity. That is, although the background was initialized to random values between [−1,1], which may include values that are higher than initial vorticity of the overlapping regions (+0.4, −0.5, and +0.5 for largest to smallest, respectively), it did not significantly contribute to bulk flow within the domain. Moreover, as suggested earlier, the initial random vorticity configuration quickly dissipates itself out, see Figure A6b.

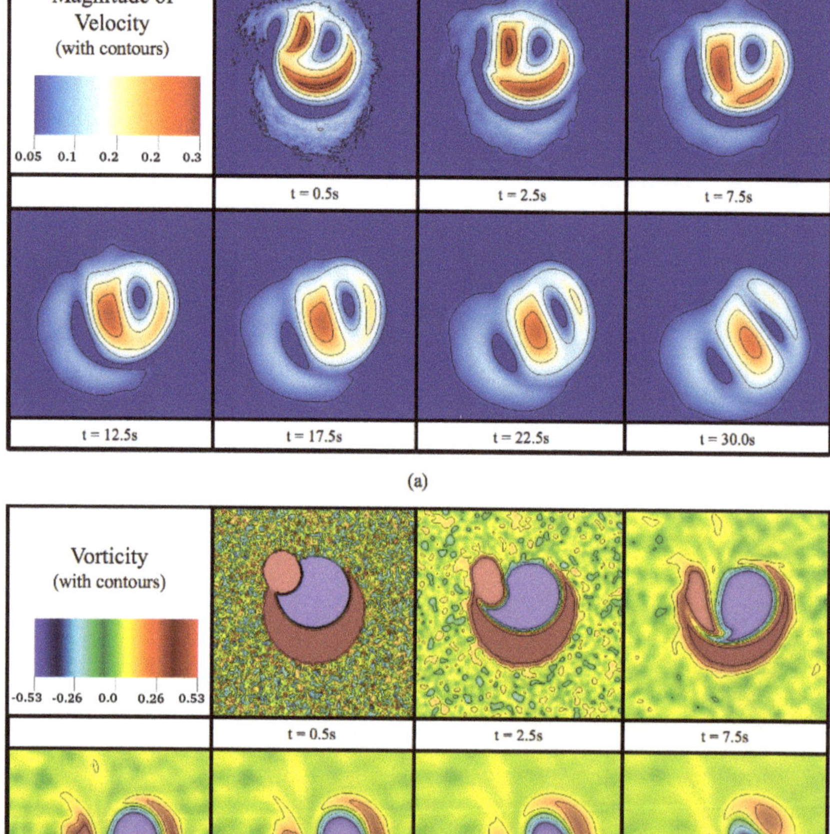

(a)

(b)

Figure A6. The evolution of (**a**) the magnitude of velocity (with its contours) and (**b**) vorticity (with its contours) during the course of the simulation.

Students may elect to try the following:

1. Recreate the above results with the parameters listed in Table A1 or for different ν or computational domains.
2. Change the initial vorticity in each 'overlapping' region.
3. Change the placement of where each vorticity region is.
4. Modify the example to have more/fewer overlapping regions of vorticity.
5. Try initializing a completely random background vorticity without any other vorticity structures.

References

1. Fefferman, C.L. Existence and Smoothness of the Navier-Stokes Equation. In *The Millenium Prize Problems*; Clay Mathematics Institute: Cambridge, MA, USA, 2006; pp. 57–67.
2. Chorin, A.J. The numerical solution of the Navier-Stokes equations for an incompressible fluid. *Bull. Am. Math. Soc.* **1967**, *73*, 928–931. [CrossRef]
3. Chorin, A.J. Numerical Solution of the Navier-Stokes Equations. *Math. Comp.* **1968**, *22*, 745–762. [CrossRef]
4. Temam, R. Une méthode d'approximation de la solution des équations de Navier-Stokes. *Bull. Soc. Math. Franc.* **1968**, *96*, 115–152. [CrossRef]
5. Brown, D.L.; Cortez, R.; Minion, M.L. Accurate Projection Methods for the Incompressible Navier-Stokes Equations. *J. Comp. Phys.* **2001**, *168*, 464–499. [CrossRef]
6. Griffith, B.E. An accurate and efficient method for the incompressible Navier-Stokes equations using the projection method as a preconditioner. *J. Comput. Phys.* **2009**, *228*, 7565–7595. [CrossRef]
7. Costa, B. Spectral Methods for Partial Differential Equations. *CUBO Math. J.* **2004**, *6*, 1–32.
8. Uecker, H. A short ad hoc introduction to spectral methods for parabolic PDE and the Navier-Stokes equations, 2009. In Proceedings of the Lecture given at International Summer School Modern Computational Science, Oldenburg, Germany, 16–28 August 2009.
9. Suzuki, M. Fourier-Spectal Methods For Navier-Stokes Equations in 2D, 2014. Available online: http://www.math.mcgill.ca/gantumur/math595f14/NSMashbat.pdf (accessed on 29 June 2019).
10. Hardy, J.; Pomeau, Y.; de Pazzis, O. Time evolution of a two-dimensional classical lattice system. *Phys. Rev. Lett.* **1973**, *31*, 276–279. [CrossRef]
11. Chen, S.; Doolen, G.D. Lattice Boltzmann method for fluid flows. *Annu. Rev. Fluid Mech.* **1998**, *30*, 282–300. [CrossRef]
12. Succi, S. *The Lattice Boltzmann Equation for Fluid Dynamics and Beyond*; Oxford University Press: Oxford, UK, 2001.
13. Stern, F.; Xing, T.; Yarbrough, D.B.; Rothmayer, A.; Rajagopalan, G.; Prakashotta, S.; Caughey, D.; Bhaskaran, R.; Smith, S.; Hutchings, B.; et al. Hands-On CFD Educational Interface for Engineering Courses and Laboratories. *J. Eng. Edu.* **2006**, *95*, 63–83. [CrossRef]
14. Cummings, R.; Morton, S., Computational Aerodynamics Goes to School: A Course in CFD for Undergraduate Students. In Proceedings of the 43rd AIAA Aerospace Sciences Meeting and Exhibit, Reno, NV, USA, 10–13 January 2005; AIAA: Reston, VA, USA,2005. [CrossRef]
15. Ormiston, S.J. Incorporating CFD into the undergraduate Mechanical Engineering Programme at the University of Manitoba. In Proceedings of the Ninth Annual Conference of the CFD Society of Canada: CFD2001, Waterloo, ON, Canada, 27–29 May 2001; Schneider, G., Ed.; CFD Society of Canada: Waterloo, ON, Canada, 2001; pp. 333–337.
16. Aung, K. Design and Implementation of an Undergraduate Computational Fluid Dynamics (Cfd) Course, 2003. In Proceedings of the 2003 American Society for Engineering Education Annual Conference, Nashville, TN, USA, 22–25 June 2003. Available online: https://peer.asee.org/design-and-implementation-of-an-undergraduate-computational-fluid-dynamics-cfd-course.pdf (accessed on 9 September 2019)
17. Stern, F.; Xing, T.; Yarbrough, D.; Rothmayer, A.; Rajagopalan, G.; Otta, S.P.; Caughey, D.; Bhaskaran, R.; Smith, S.; Hutchings, B.; et al. Development of hands-on CFD educational interface for undergraduate engineering courses and laboratories. In Proceedings of the 2004 American Society for Engineering Education Annual Conference & Exposition, Salt Lake City, UT, USA, 20–23 June 2004; American Society for Engineering Education: Washington, DC, USA, 2004; pp. 1526–1555.
18. Adair, D. Incorporation of Computational Fluid Dynamics into a Fluid Mechanics Curriculum. In *Advances in Modeling of Fluid Dynamics*; Liu, C., Ed.; IntechOpen: London, UK, 2012; Chapter 5, pp. 97–122.
19. Stern, F.; Yoon, H.; Yarbrough, D.; Okcay, M.; Oztekin, B.U.; Roszelle, B. Hands-on integrated CFD educational interface for introductory fluids mechanics. *Int. J. Aerodyn.* **2012**, *2*, 339–371. [CrossRef]
20. Ray, B.; Bhaskaran, R. Integrating Simulation into the Engineering Curriculum: A Case Study. *Int. J. Mech. Eng. Edu.* **2013**, *41*, 269–280. [CrossRef]
21. Eldredge, J.D.; Senocak, I.; Dawson, P.; Canino, J.; Liou, W.; LeBeau, R.; Hitt, D.; Rumpfkeil, M.; Cummings, R. A Best Practices Guide to CFD Education in the Undergraduate Curriculum. *Int. J. Aerodyn.* **2014**, *4*, 200–236. [CrossRef]

22. Heron, P.; McNeill, L. Phys21: Preparing Physics Students for 21st-Century Careers (A Report by the Joint Task Force on Undergraduate Physics Programs), 2016. American Physical Society and the American Association of Physics Teachers. Available online: https://www.compadre.org/JTUPP/report.cfm (accessed on 7 January 2020).

23. Heron, P.; McNeill, L. Preparing Physics Students for 21st-Century Careers. *Phys. Today* **2017**, *70*, 38.

24. Fefferman, C.L. A Comparison of C, MATLAB, and Python as Teaching Languages in Engineering. In *Computational Science—ICCS 2004*; Bubak, M., van Albada, G.D., Sloot, P.M., Dongarra, J., Eds.; Springer: Berlin/Heidelberg, Germany, 2004; pp. 1210–1217.

25. Spencer, R.L. Teaching computational physics as a laboratory sequence. *Am. J. Phys.* **2005**, *73*, 151–153. [CrossRef]

26. Peng, L.; Bao, L.; Huang, M. Application of Matlab/Simulink Software in Physics. In *High Performance Networking, Computing, and Communication Systems*; Wu, Y., Ed.; Springer: Berlin/Heidelberg, Germany, 2011; Chapter 21, pp. 140–146.

27. Sangwin, C.J.; O'Toole, C. Computer programming in the UK undergraduate mathematics curriculum. *Int. J. Math. Edu. Sci. Technol.* **2017**, *48*, 1133–1152. [CrossRef]

28. Wang, Y.; Hill, K.J.; Foley, E.C. Computer programming with Python for industrial and systems engineers: Perspectives from an instructor and students. *Comput. Appl. Eng. Educ.* **2017**, *25*, 800–811. [CrossRef]

29. MATLAB. *version 8.5.0 (R2015a)*; The MathWorks Inc.: Natick, MA, USA, 2015.

30. Van Rossum, G. *Python*; version 3.5. 2015. Available online: https://www.python.org (accessed on 31 August 2019).

31. Carey, M.A.; Papin, J.A. Ten simple rules for biologists learning to program. *PLoS Comput. Biol.* **2018**, *14*, e1005871. [CrossRef]

32. Battista, N.A.; Strickland, W.C.; Miller, L.A. IB2d: A Python and MATLAB implementation of the immersed boundary method. *Bioinspir. Biomim.* **2017**, *12*, 036003. [CrossRef]

33. Barba, L.A.; Forsyth, G. CFD Python: The 12 steps to Navier-Stokes equations. *J. Open Source Edu.* **2018**, *1*, 21. [CrossRef]

34. Barba, L.A.; Mesnard, O. Aero Python: Classical aerodynamics of potential flow using Python. *J. Open Source Edu.* **2019**, *2*, 45. [CrossRef]

35. Battista, N.A.; Mizuhara, M.S. Fluid-Structure Interaction for the Classroom: Speed, Accuracy, Convergence, and Jellyfish! *arXiv* **2019**, arXiv:1902.07615.

36. Battista, N. Fluid-structure Interaction for the Classroom: Interpolation, Hearts, and Swimming! *SIAM Rev.* **2018**, in press.

37. Battista, N.A.; Strickland, W.C.; Barrett, A.; Miller, L.A. IB2d Reloaded: A more powerful Python and MATLAB implementation of the immersed boundary method. *Math. Methods Appl. Sci.* **2018**, *41*, 8455–8480. [CrossRef]

38. Kluyver, T.; Ragan-Kelley, B.; Pérez, F.; Granger, B.; Bussonnier, M.; Frederic, J.; Kelley, K.; Hamrick, J.; Grout, J.; Corlay, S.; et al. Jupyter Notebooks—A publishing format for reproducible computational workflows. In *Positioning and Power in Academic Publishing: Players, Agents and Agendas*; Loizides, F., Schmidt, B., Eds.; IOS Press: Amsterdam, The Netherlands: 2016; pp. 87–90.

39. Pawar, S.; San, O. CFD Julia: A Learning Module Structuring an Introductory Course on Computational Fluid Dynamics. *Fluids* **2019**, *4*, 159. [CrossRef]

40. Bezanson, J.; Edelman, A.; Karpinski, S.; Shah, V.B. Julia: A fresh approach to numerical computing. *SIAM Rev.* **2017**, *59*, 65–98. [CrossRef]

41. Childs, H.; Brugger, E.; Whitlock, B.; Meredith, J.; Ahern, S.; Pugmire, D.; Biagas, K.; Miller, M.; Harrison, C.; Weber, G.H.; et al. VisIt: An End-User Tool For Visualizing and Analyzing Very Large Data. In *High Performance Visualization–Enabling Extreme-Scale Scientific Insight*; Bethel, E.W., Childs, H., Hansen, C., Eds.; Chapman and Hall/CRC: Boca Raton, FL USA, 2012; pp. 357–372.

42. Ahrens, J.; Gerveci, B.; Law, C. *ParaView: An End-User Tool for Large Data Visualizations*; Elsevier: Atlanta, GA, USA, 2005.

43. Burden, R.L.; Faires, J.D. *Numerical Analysis*, 5th ed.; Prindle, Weber and Schmidt: Boston, MA USA, 1993.

44. Minion, M.L. Higher-Order Semi-Implicit Projection Methods. In *Numerical Simulations of Incompressible Flows*; Hafez, M.M., Ed.; World Scientific Publishing Company: Waterloo, ON, Canada, 2003; pp. 126–140.

45. Guermond, J.L.; Minev, P.; Shen, J. An overview of projection methods for incompressible flows. *Comp. Methods Appl. Mech. Eng.* **2006**, *195*, 6011–6045. [CrossRef]
46. Almgren, A.S.; Aspden, A.J.; Bell, J.B.; Minion, M.L. On the Use of Higher-Order Projection Methods for Incompressible Turbulent Flow. *SIAM J. Sci. Comput.* **2013**, *35*, B25–B42. [CrossRef]
47. Bell, J.B.; Colella, P.; Glaz, H.M. A second order projection method for the incompressible Navier-Stokes equations. *J. Comput. Phys.* **1989**, *85*, 257–283. [CrossRef]
48. Atkinson, K.E. *An Introduction to Numerical Analysis*, 2nd ed.; John Wiley & Sons: Hoboken, NJ, USA, 1989.
49. Trefethen, L.N. *Spectral Methods in MATLAB*; SIAM: Philadelphia, PA, USA, 2001.
50. Battista, N.A. Spectrally Accurate Initial Data in Numerical Relativity. Master's Thesis, Rochester Institute of Technology, Rochester, NY, USA, 2010.
51. Crank, J.; Nicholson, P. A practical method for numerical evaluation of solutions of partial differential equations of the heat conduction type. *Proc. Camb. Phil. Soc.* **1947**, *43*, 50–67. [CrossRef]
52. Thomas, J.W. *Numerical Partial Differential Equations: Finite Difference Methods*; Springer: New York, NY, USA, 1995.
53. Battista, N.A.; Baird, A.J.; Miller, L.A. A Mathematical Model and MATLAB Code for Muscle-Fluid-Structure Simulations. *Integr. Comp. Biol.* **2015**, *55*, 901–911. [CrossRef]
54. Zhang, J. Lattice Boltzmann method for microfluidics: Models and applications. *Microfluid. Nanofluidics* **2011**, *10*, 1–28. [CrossRef]
55. Bao, Y.B.; Meskas, J. Lattice Boltzmann Method for Fluid Simulations, 2011. Available online: http://www.cims.nyu.edu/~billbao/report930.pdf (accessed on 19 September 2019).
56. Tu, J.; Yeoh, G.H.; Liu, C. *Computational Fluid Dynamics*, 3rd ed.; Butterworth-Heinemann: Oxford, UK, 2018.
57. Ishihara, I. Tests for color blindness. *Am. J. Ophthal.* **1918**, *1*, 457. [CrossRef]
58. Shadden, S.C.; Lekien, F.; Marsden, J.E. Definition and properties of Lagrangian coherent structures from finite-time Lyapunov exponents in two-dimensional aperiodic flows. *Physica D* **2005**, *212*, 271–304. [CrossRef]
59. Shadden, S.C. Lagrangian Coherent Structures: Analysis of Time Dependent Dynamical Systems Using Finite-Time Lyapunov Exponent, 2005. Available online: https://shaddenlab.berkeley.edu/uploads/LCS-tutorial/index.html (accessed on 19 September 2019).
60. Shadden, S.C.; Katija, K.; Rosenfeld, M.; Marsden, J.E.; Dabiri, J.O. Transport and stirring induced by vortex formation. *J. Fluid Mech.* **2007**, *593*, 315–331. [CrossRef]
61. Haller, G.; Sapsis, T. Lagrangian coherent structures and the smallest finite-time Lyapunov exponent. *Chaos* **2011**, *21*, 023115. [CrossRef]
62. Shadden, S.C.; Dabiri, J.O.; Marsden, J.E. Lagrangian analysis of fluid transport in empirical vortex ring flows. *Phys. Fluids* **2006**, *18*, 047105. [CrossRef]
63. Lukens, S.; Yang, X.; Fauci, L. Using Lagrangian coherent structures to analyze fluid mixing by cilia. *Chaos* **2010**, *20*, 017511. [CrossRef]
64. Cheryl, S.; Glatzmaier, G.A. Lagrangian coherent structures in the California Current System—Sensitivities and limitations. *Geophys. Astrophys. Fluid Dyn.* **2012**, *106*, 22–44.
65. Haller, G. Finding finite-time invariant manifolds in two-dimensional velocity fields. *Chaos* **2000**, *10*, 99–108. [CrossRef]
66. Haller, G. Lagrangian Coherent Structures. *Annu. Rev. Fluid Mech.* **2015**, *47*, 137–162. [CrossRef]
67. Truskey, G.A.; Yuan, F.; Katz, D.F. *Transport Phenomena in Biological Systems*; Pearson Prentice Hall Bioengineering: Upper Saddle River, NJ, USA, 2004.
68. Rayleigh, L. On the flow of compressible fluid past an obstacle. *Lond. Edinb. Dublin Philos. Mag. J. Sci.* **1916**, *32*, 1–6. [CrossRef]
69. Acheson, D.J. *Elementary Fluid Dynamics*; Oxford University Press: Oxford, UK, 1990.
70. Batchelor, G.K. *An Introduction to Fluid Dynamics*; Cambridge University Press: Cambridge, UK, 2000.
71. Morton, C.; Yarusevych, S. Vortex shedding in the wake of a step cylinder. *Phys. Fluids* **2010**, *22*, 083602. [CrossRef]
72. Bao, Y.; Wu, Q.; Zhou, D. Numerical investigation of flow around an inline square cylinder array with different spacing ratios. *Comput. Fluids* **2012**, *55*, 118–131. [CrossRef]
73. Carini, M.; Gianetti, F.; Auteri, F. On the origin of the flip-flop instability of two side-by-side cylinder wakes. *J. Fluid Mech.* **2014**, *742*, 552–576. [CrossRef]

74. Younis, M.Y.; Alam, M.M.; Zhou, Y. Flow around two non-parallel tandem cylinders. *Phys. Fluids* **2016**, *28*, 125106. [CrossRef]

75. Gao, Y.; Chen, W.; Wang, B.; Wang, L. Numerical simulation of the flow past six-circular cylinders in rectangular configurations. *J. Mar. Sci. Technol.* **2019**, 1–25. [CrossRef]

76. Ji, C.; Cui, Y.; Xu, D.; Yang, X.; Srinil, N. Vortex-induced vibrations of dual-step cylinders with different diameter ratios in laminar flows. *Phys. Fluids* **2019**, *31*, 073602.

77. Ji, C.; Yang, X.; Yu, Y.; Cui, Y.; Srinil, N. Numerical simulations of flows around a dual step cylinder with different diameter ratios at low Reynolds number. *Eur. J. Mech. B/Fluids* **2019**, in press. [CrossRef]

78. Fransson, J.H.; Konieczny, P.; Alfredsson, P.H. Flow around a porous cylinder subject to continuous suction or blowing. *J. Fluids Stuct.* **2004**, *19*, 1031–1048. [CrossRef]

79. Chen, X.; Yu, P.; Winoto, S.; Low, H. Numerical analysis for the flow past a porous square cylinder based on the stress-jump interfacial-conditions. *Int. J. Num. Meth. Heat Fluid Flow* **2008**, *18*, 635–655. [CrossRef]

80. Naito, H.; Fukagata, K. Numerical simulation of flow around a circular cylinder having porous surface. *Phys. Fluids* **2012**, *24*, 117102. [CrossRef]

81. Shahsavari, S.; Wardle, B.L.; McKinley, G.H. Interception efficiency in two-dimensional flow past confined porous cylinders. *Chem. Eng. Sci.* **2014**, *116*, 752–762. [CrossRef]

82. Ledda, P.G.; Siconolfi, L.; Viola, F.; Gallaire, F.; Camarri, S. Suppression of von Kármán vortex streets past porous rectangular cylinders. *Phys. Rev. Fluids* **2007**, *3*, 103901. [CrossRef]

83. Gupta, G.K. Computer literacy: Essential in today's computer-centric world. *ACM SIGCSE Bull.* **2005**, *38*, 115–119. [CrossRef]

84. Shein, E. Should everybody learn to code? *Commun. ACM* **2014**, *57*, 16–18.

85. Sterling, L. Coding in the curriculum: Fad or foundational? *ACER Res. Conf.* **2016**, *4*, 72–84.

86. Baker, M. Scientific computing: Code alert. *Nature* **2017**, *541*, 563–565. [CrossRef]

87. Helmholtz, H. Uber Integrale der hydrodynamischen Gleichungen, welcher der Wirbelbewegungen entsprechen. *J. Reine Angew. Math.* **1858**, *55*, 25–50.

88. Bladel, J. *On Helmholtz's Theorem in Finite Regions*; Midwestern Universities Research Association: Madison, WI, USA, 1958.

89. Cooley, J.W.; Tukey, J.W. An algorithm for the machine calculation of complex Fourier series. *Math. Comput.* **1965**, *19*, 297–301. [CrossRef]

90. Bhatnagar, P.L.; Gross, E.P.; Krook, M. A Model for Collision Processes in Gases. I. Small Amplitude Processes in Charged and Neutral One-Component Systems. *Phys. Rev.* **1954**, *94*, 511–525. [CrossRef]

91. Asinari, P. Multi-Scale Analysis of Heat and Mass Transfer in Mini/Micro Structures. Ph.D. Thesis, Energy Engineering, Politecnico di Torino, Turin, Italy, 2005.

Article

Understanding Fluid Dynamics from Langevin and Fokker–Planck Equations

Andrei Medved [1], **Riley Davis** [2] **and Paula A. Vasquez** [1,*]

[1] Department of Mathematics, University of South Carolina, Columbia, SC 29208, USA; amedved@email.sc.edu

[2] Department of Mechanical and Aerospace Engineering, Case Western Reserve University, Cleveland, OH 44106, USA; rileywd2000@gmail.com

* Correspondence: paula@math.sc.edu

Received: 17 February 2020; Accepted: 15 March 2020; Published: 23 March 2020

Abstract: The Langevin equations (LE) and the Fokker–Planck (FP) equations are widely used to describe fluid behavior based on coarse-grained approximations of microstructure evolution. In this manuscript, we describe the relation between LE and FP as related to particle motion within a fluid. The manuscript introduces undergraduate students to two LEs, their corresponding FP equations, and their solutions and physical interpretation.

Keywords: Langevin; Fokker–Planck; microrheology; Stokes–Einstein relation; mobility; fluctuation dissipation; Matlab GUI

---

1. Introduction

This review focuses on two idealized scenarios involving microscopic particles embedded in a fluid. In the first one, we consider the uncoupled motion of individual Brownian probes, while in the second one, we consider the dynamics of an ensemble of such probes. These two cases allow us to explore the relation between two well-known families of equations in fluids dynamics: the Langevin equations (LE) and Fokker–Planck (FP) equations. By no means is this meant to be a comprehensive review of either of these equations, but rather a bird's-eye view of their relationship and how they can be used to better understand fluid dynamics at the microscale. The article is written for undergraduate students and highlights different concepts from undergraduate courses in calculus and differential equations and their applications to fluid dynamics problems. In addition, whenever pertinent, the reader will be referred to more specialized publications for a more in-depth treatment of the different subjects.

To elucidate the relation between these two types of approaches, Figure 1 shows the relation between a LE and a FP description of particles moving as a result of simple Brownian motion in two dimensions. This process describes the random migration of small particles arising from their motion due to thermal energy. The term *Brownian motion* was coined after the botanist Robert Brown, who was the first to describe this phenomenon in 1828 during his investigation of the movements of fine particles, like pollen, dust, and soot, on a water surface. In 1905, Albert Einstein explained Brownian motion in terms of random thermal motions of fluid molecules bombarding the microscopic particle and causing it to undergo a random walk [1]. Nonetheless, the range of applications of Brownian motion goes beyond the study of microscopic particles and includes modeling of thermal noise, stock prices, and random perturbations in many physical, biological, and economic systems [2,3]. From an observational point of view, the Langevin equation is easier to understand than the Fokker–Planck equation. The LE approach directly uses the concept of time evolution of the random variable describing the process; in the case of Figure 1, this corresponds to the individual particle's position. In contrast, the FP approach follows the time evolution of the underlying probability distribution. That is, instead of describing a particle position, it describes the likelihood of finding a particle at a given position.

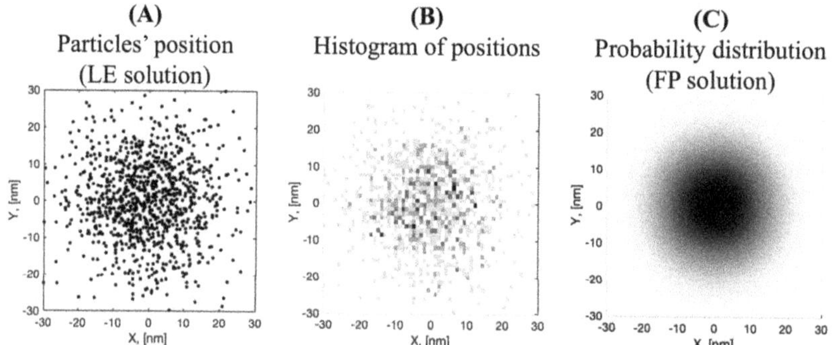

Figure 1. Relation between the Langevin equations (LE) and Fokker–Planck (FP) solutions. (**A**) LE solutions give particle positions as functions of time, each point represents a particle position at some time t. All particles have initial position $(0, 0)$ and follow different trajectories dictated by their respective noise history. (**B**) The distribution of particle positions can be summarized using a histogram. (**C**) FP solutions shows the probability of finding a particle at a given position as a function of time; darker color indicates a higher probability of finding a particle at that position.

In this paper, we briefly describe the basics of LE equations and investigate their relation with their corresponding FP equations. The special cases discussed in the following sections are aimed at understanding how information is represented under these two different descriptions and illustrating how one can gather data under one approach and be able to infer behavior under the other.

2. Langevin Approach

To understand the Langevin description, we start by considering a particle immersed in a fluid. The particle "feels" a force arising from the collisions with the fluid's molecules. This force consists of two parts: (a) a deterministic hydrodynamic drag, which resists motion; and (b) a fluctuating stochastic force, caused by thermal fluctuations. Newton's second law gives the evolution equation governing the dynamics of the particle as

$$
\begin{aligned}
m\mathbf{a} &= \sum \mathbf{F} \\
&= \text{drag force} + \text{random force}.
\end{aligned}
$$

Assuming a linear drag force (force = $-$drag coefficient $\times$ velocity) and a white noise, the resulting equation is known as the Langevin equation:

$$
m\mathbf{a} = m\dot{\mathbf{v}} = -\zeta\mathbf{v} + \mathbf{f}. \tag{1}
$$

White noise describes a random term that assumes no correlation on the fluctuating forces; this is captured by drawing a random number from a Gaussian distribution with mean and variance given by

$$
\begin{aligned}
\langle \mathbf{f}(t) \rangle &= 0, \\
\langle f_i(t) f_j(s) \rangle &= \Gamma \delta_{ij} \delta(t - s),
\end{aligned}
$$

where Γ represents the variance of the distribution or strength of the noise; i, j indicate vector components; δ_{ij} is the Kronecker delta; and $\delta(t)$ is a Dirac delta function. Note that both the Kronecker delta and Dirac delta function capture the zero-correlation of the forces both spatial, $\delta_{ij} = 0$ for $i \neq j$, and temporal, $\delta(t - s) = 0$ for $t \neq s$. Moreover, for any interval $[a, b]$ contained in interval $[c, d]$, we have the following rule:

$$
\int_a^b d\tau \int_c^d f(\tau, s) \delta(\tau - s) \, ds = \int_a^b f(\tau, \tau) \, d\tau \tag{2}
$$

Since $\mathbf{f}$ represents a stochastic term, Equation (1) is part of a broad class of differential equations known as stochastic differential equations (SDEs) [3,4].

Assuming white noise, one can solve Equation (1) formally using basic solution techniques for ordinary differential equations (ODEs). In particular, Equation (1) can be treated as a first order, non-homogeneous differential equation of the form

$$
\frac{dy(t)}{dt} + p(t) y(t) = q(t),
$$

with integrating factor and solution given by

$$
v(t) = \int p(t) \, dt, \qquad\qquad y(t) = e^{-v(t)} \int e^{v(t)} q(t) \, dt.
$$

For Equation (1), $y(t) = \mathbf{v}$, $p(t) = \zeta/m$ and $q(t) = \mathbf{f}$, so that its formal solution is given by

$$
\mathbf{v}(t) = \mathbf{v}(0) e^{-\zeta t/m} + \frac{1}{m} \int_0^t e^{-\zeta(t-\tau)/m} \, \mathbf{f}(\tau) \, d\tau = \mathbf{v}(0) e^{-t/\tau_B} + \frac{1}{m} \int_0^t e^{-(t-\tau)/\tau_B} \, \mathbf{f}(\tau) \, d\tau. \tag{3}
$$

The quantity $\tau_B = m/\zeta$ has units of time and is usually referred to as the Brownian relaxation time of the particle velocity.

Note that, in the absence of random noise, $\mathbf{f}(t) = 0$, Equation (3) gives $\mathbf{v}(t) = \mathbf{v}(0) e^{-t/\tau_B}$, which implies that $\mathbf{v} \to 0$ as $t \to \infty$. However, according to the equipartition theorem, the velocity should satisfy

$$
\lim_{t \to \infty} \left\langle \frac{1}{2} m v^2(t) \right\rangle = \frac{d}{2} k_B T,
$$

where the brackets $\langle \cdot \rangle$ represent averages; k_B is the Boltzmann's constant; T is the temperature; $v^2(t) = \mathbf{v}(t) \cdot \mathbf{v}(t)$; and d represents the degrees of freedom, or dimensionality, $d = 1, 2$ or 3. The fact that the equipartition theorem states that the velocity cannot approach zero as time goes to infinity implies that the random force is necessary to obtain the correct equilibrium condition. Furthermore, the strength of the noise, Γ, should be such that equipartition theorem is satisfied.

To determine the strength of the random force, Γ, we take the average of $v^2(t)$ using Equation (3) as

$$
\begin{aligned}
\langle \mathbf{v}(t) \cdot \mathbf{v}(t) \rangle &= \mathbf{v}(0) \cdot \mathbf{v}(0) e^{-2t/\tau_B} + \frac{2}{m} \int_0^t e^{-(2t-\tau)/\tau_B} \left[\mathbf{v}(0) \cdot \langle \mathbf{f}(t) \rangle \right] d\tau + \\
&\quad \frac{1}{m^2} \int_0^t \int_0^t e^{-(2t-\tau-s)/\tau_B} \langle \mathbf{f}(\tau) \cdot \mathbf{f}(s) \rangle \, d\tau \, ds, \\
\left\langle v^2(t) \right\rangle &= \mathbf{v}(0) \cdot \mathbf{v}(0) e^{-2t/\tau_B} + \frac{2}{m} \int_0^t e^{-(2t-\tau)/\tau_B} \left[0 \right] d\tau \\
&\quad + \frac{d}{m^2} \int_0^t \int_0^t e^{-(2t-\tau-s)/\tau_B} \left[\Gamma \delta(\tau - s) \right] d\tau \, ds, \\
&= v^2(0) \, e^{-2t/\tau_B} + \frac{d\,\Gamma}{2\zeta\, m} \left[1 - e^{-2t/\tau_B} \right].
\end{aligned}
\tag{4}
$$

In the last step of Equation (4), we used the property of the Dirac delta function given in Equation (2). Taking the limit as $t \to \infty$ in Equation (4), and comparing it to the condition given by the equipartition theorem, gives

$$
\begin{aligned}
\lim_{t \to \infty} \left\langle \frac{1}{2} m v^2(t) \right\rangle &= \frac{d}{2} k_B T \qquad &&\text{Equipartition} \\
&= \frac{1}{2} m \left[\frac{d\,\Gamma}{2\zeta\, m} \right] = \frac{d\Gamma}{4\zeta} \qquad &&\text{Equation (4).}
\end{aligned}
$$

Therefore, the strength of the noise should satisfy

$$
\Gamma = 2\zeta\, k_B T.
\tag{5}
$$

This relation between the strength of the fluctuations of the stochastic forces (Γ) and the dissipative term given by the drag force (ζ) is a special case of a more general result known as the *fluctuation–dissipation theorem* [5].

The fluctuation–dissipation theorem states that equilibrium is brought about by a dissipation force, in our case drag, between the particle and the medium, and whatever the mechanism of the dissipation, it has to be the same process that produces random, fluctuating forces on the particle. In other words, both the frictional force and the random force must be related since they have the same origin: fluid molecules "bombarding" the particle and inducing mobility.

Finally, after solving Equation (1), the particle position can be obtained as

$$
\mathbf{x}(t) - \mathbf{x}(0) = \int_0^t \mathbf{v}(\tau) \, d\tau.
\tag{6}
$$

For a given SDE such as Equation (1), in order to make inferences based on its solution, it is necessary to find the average over many realizations. To illustrate this, consider the 2D version of Equation (1) solved three different times using the same initial position but subject to different random noises. The resulting trajectories are shown in Figure 2.

The fact that each trajectory is very different from the others implies that we cannot infer any behavior from the system by just considering a handful of solutions. That is, just as one would not be able to determine whether a coin is fair by just a couple of tosses, to be able to infer behavior based on Equation (1) one needs to look at many realizations of particle trajectories. This can be done numerically by solving the equation many times and then finding the average of such solutions or can be done analytically by using time-correlation functions, as discussed next.

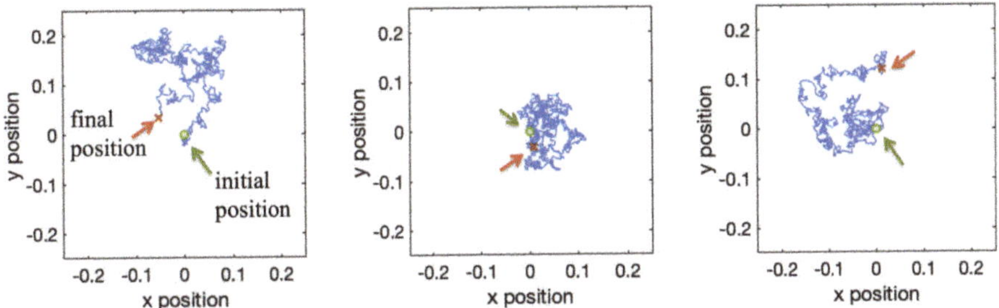

Figure 2. Evolution of 2D particle position for three different solutions of Equation (1). All three trajectories start from an initial position $(0, 0)$. In all three figures, $\Delta t = 0.001$ and the simulation ran for 1000 time steps.

2.1. Moments of a Stochastic Process

In mathematical statistics, the sth moment of a set of stochastic observations $\{X_i\}_{i=0}^{n}$ is defined as

$$M_s\left(\mathbf{X}\right) = \frac{1}{n} \sum_{i=1}^{n} \mathbf{X}_i^s.$$

Note that in the present application, our variables denote displacement of the particle with respect to the zero-time position, and the different moments measure deviations of the observations from the mean of the values. Within the context of fluid dynamics and Langevin equations, we are interested in the first ($s = 1$) and second ($s = 2$) moments of the particle positions. That is, the mean value of the stochastic variable and its spread or variance with respect to the mean.

- *First moment*
  For the velocity in Equation (3), taking into account that $\langle \mathbf{f} \rangle = 0$, the mean is given by

$$M_1(\mathbf{v}) = \langle \mathbf{v}(t) \rangle = \mathbf{v}(0)e^{-t/\tau_B},$$

  with a long-time or equilibrium value given by

$$M_1(\mathbf{v}) \to 0, \quad \text{as} \quad t \to \infty.$$

  For the particle position,

$$\begin{aligned}
M_1(\mathbf{x}) = \langle \mathbf{x} - \mathbf{x}(0) \rangle &= \int_0^t \langle \mathbf{v}(s) \rangle \, ds = \int_0^t \mathbf{v}(0)e^{-s/\tau_B} \, ds, \\
&= \left. -\mathbf{v}(0)\, \tau_B e^{-s/\tau_B} \right|_0^t \\
&= \mathbf{v}(0)\, \tau_B \left[1 - e^{-t/\tau_B} \right].
\end{aligned}$$

  In addition, its long-time behavior is

$$M_1(\mathbf{x}) \to \mathbf{v}(0)\, \tau_B, \quad \text{as} \quad t \to \infty.$$

- *Second moment*
  From Equation (4), the second moment for the velocity is

$$M_2(\mathbf{v}) = \left\langle v^2(t) \right\rangle = v^2(0)\, e^{-2t/\tau_B} + \frac{dk_B T}{m}\left[1 - e^{-2t/\tau_B}\right].$$

For the particle position, we have

$$
\begin{aligned}
M_2(\mathbf{x}) = \left\langle |\mathbf{x}(t) - \mathbf{x}(0)|^2 \right\rangle &= \\
\left\langle x^2(t) \right\rangle &= \left\langle \left[\int_0^t v(\tau)\, d\tau\right]^2 \right\rangle \\
&= \left\langle \int_0^t \mathbf{v}(s)\, ds \cdot \int_0^t \mathbf{v}(\tau)\, d\tau \right\rangle \\
&= \int_0^t \int_0^t \langle \mathbf{v}(s) \cdot \mathbf{v}(\tau) \rangle \, d\tau\, ds \\
&= \int_0^t \int_0^t \left(v^2(0)\, e^{-\frac{s+\tau}{\tau_B}} + \frac{dk_B T}{m}\left[e^{-\frac{|s-\tau|}{\tau_B}} - e^{-\frac{s+\tau}{\tau_B}}\right] \right) d\tau\, ds \\
&= v^2(0) I_1 + \frac{dk_B T}{m} I_2 - \frac{dk_B T}{m} I_1.
\end{aligned}
$$

In this derivation, we have again used Equation (2) and the properties of the Dirac-delta function. For clarity, we solve each integral separately:

$$I_1 = \int_0^t \int_0^t e^{-(s+\tau)/\tau_B}\, d\tau\, ds = \tau_B^2 \left(1 - e^{-t/\tau_B}\right)^2$$

$$
\begin{aligned}
I_2 = \int_0^t \int_0^t e^{-|s-\tau|/\tau_B}\, d\tau\, ds &= 2\int_0^t \int_0^s e^{-|s-\tau|/\tau_B}\, d\tau\, ds \\
&= 2\tau_B t - 2\tau_B^2 \left(1 - e^{-t/\tau_B}\right).
\end{aligned}
$$

Substituting I_1 and I_2 into the equation for $M_2(\mathbf{x})$ gives

$$
\begin{aligned}
\left\langle x^2(t) \right\rangle &= \left[v^2(0) - \frac{dk_B T}{m}\right]\left[\tau_B^2 \left(1 - e^{-t/\tau_B}\right)^2\right] + \frac{dk_B T}{m}\left[2\tau_B t - 2\tau_B^2 \left(1 - e^{-t/\tau_B}\right)\right] \\
&= v^2(0)\tau_B^2 \left(1 - e^{-t/\tau_B}\right)^2 + \frac{dk_B T}{m}\tau_B^2\left[-\left(1 - e^{-t/\tau_B}\right)^2 + \frac{2}{\tau_B}t - 2\left(1 - e^{-t/\tau_B}\right)\right] \\
&= v^2(0)\tau_B^2 \left(1 - e^{-t/\tau_B}\right)^2 + \frac{dk_B T}{m}\tau_B^2\left[\frac{2}{\tau_B}t - 3 + 4e^{-t/\tau_B} - e^{-2t/\tau_B}\right].
\end{aligned}
\tag{7}
$$

The quantity $\left\langle x^2(t) \right\rangle$ is called a mean squared displacement (MSD) and represents the square of the mean distance a particle has traveled in a given time interval. In practice, the MSD is one of the most commonly used experimental measures to determine material properties, as discussed in the next section.

2.2. Applications of the MSD

Performing a Taylor expansion of the MSD about $t = 0$ gives

$$\left\langle x^2(t) \right\rangle \approx v^2(0)\, t^2 + O(t^3),$$

that is, at short times, the MSD grows quadratically in time. Similarly, at large times, we obtain

$$\left\langle x^2(t) \right\rangle \to \frac{2dk_BT}{m} \tau_B\, t, \quad \text{as} \quad t \to \infty.$$

Using the definition of τ_B,

$$\left\langle x^2(t) \right\rangle \to \frac{2dk_BT}{m} \frac{m}{\zeta} t = \frac{2dk_BT}{\zeta} t = 2dD\, t \quad \text{as} \quad t \to \infty, \tag{8}$$

where we have introduced the *diffusion coefficient*, $D = k_BT/\zeta$.

The result in Equation (8) constitute a powerful tool in the characterization of fluids. The diffusion coefficient characterizes the mobility of particles of a given size in a given medium at a given temperature. For example, for spherical particles the drag coefficient is given by $\zeta = 6\pi\eta a$, where a is the particle radius and η is the viscosity of the fluid. By embedding spherical particles in a fluid of unknown properties, one can estimate the viscosity of the fluid based on the particle trajectories.

Assume we had tracked the trajectories of n spherical particles diffusing in a Newtonian fluid, $\{\mathbf{x}_i(t)\}_{i=1}^{n}$, where $\mathbf{x}(t) = [x_i(t),; y_i(t); z_i(t)]$. We can calculate the 1D, 2D, and 3D MSD as follows:

$$
\begin{aligned}
MSD_{1D}(\Delta t) &= \left\langle [x(t+\Delta t) - x(t)]^2 \right\rangle, \\
MSD_{2D}(\Delta t) &= \left\langle [x(t+\Delta t) - x(t)]^2 + [y(t+\Delta t) - y(t)]^2 \right\rangle, \\
MSD_{3D}(\Delta t) &= \left\langle [x(t+\Delta t) - x(t)]^2 + [y(t+\Delta t) - y(t)]^2 + [z(t+\Delta t) - z(t)]^2 \right\rangle.
\end{aligned}
$$

Examples of these MSDs are shown in Figure 3 for different Δt's and the Matlab code used to generate them can be found in Appendix A. Note that in Appendix A we have used the zero-mass limit of the LE equation, see Section 2.4 for details of this limiting case.

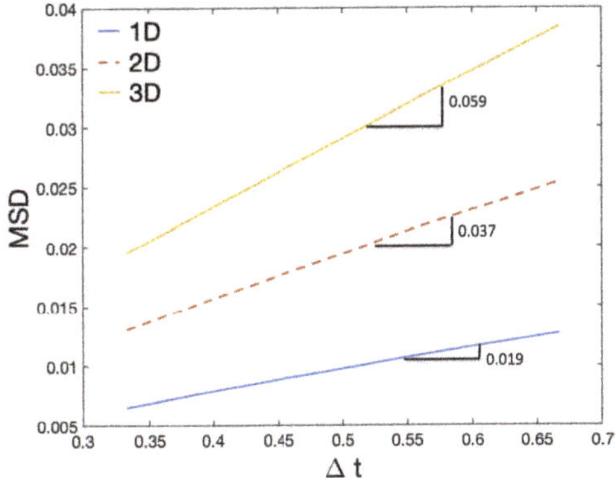

Figure 3. Mean squared displacement (MSD) in 1D, 2D, and 3D. Each line has been fitted to a function of the form $y = m\, x$, the best fitting value for the slope m is shown for each line. The slope for 1D is $2D = 0.019$, giving $D = 0.0095$, the slope for 2D gives $4D = 0.037 \Rightarrow D = 0.0093$, and for 3D it is $6D = 0.059 \Rightarrow D = 0.0098$. As a reference, the diffusion coefficient used to generate the particle trajectories is $D = 0.01$.

Once the diffusion coefficient is found from the particle trajectories and the MSD, the fluid viscosity can be determined by

$$D = \frac{k_B T}{\zeta} = \frac{k_B T}{6\pi\eta a} \quad \Rightarrow \quad \eta = \frac{k_B T}{6\pi a D}. \tag{9}$$

This type of inference can also be used with more complex fluids and/or different types of particles. For instance, the Einstein–Smoluchowski–Sutherland relation states that [6]

$$D = \mu k_B T,$$

where μ is the particle's mobility. This mobility is given by Stokes' law in terms of the particle hydrodynamic radius, a_H,

$$\mu = \frac{1}{c\,\pi\,\eta\,a_H},$$

where both the constant c and a_H depend on the particle size and shape. Note that, for spherical particles, we return to the so called Stokes–Einstein relation,

$$D = \frac{k_B T}{6\pi\eta a},$$

however, the relation in Equation (9) stills holds for non-spherical particles.

In addition, complex fluids, such as viscoelastic materials, exhibit MSDs that do not depend linearly on time [7,8]. For example, the long-time MSD of some fluids obeys

$$MSD \sim t^\alpha.$$

This type of behavior is known as anomalous diffusion and the power law exponent, α, indicates the type of diffusion: for $\alpha < 1$, it is called subdiffusion, for $\alpha = 1$, regular diffusion, and for $\alpha > 1$, superdiffusion [9]. Although Equation (9) no longer holds in this case, material properties can still be inferred from the MSD of these fluids as discussed in [7,8].

2.3. Generalized Langevin Equations (GLE)

The GLE, as its name implies, is a generalization of Equation (1) and it can be similarly derived from Newton's second law assuming that the forces acting over the particle are a stochastic force, a drag force, and some external conservative force [3]:

$$m\ddot{\mathbf{x}}(t) = \mathbf{F}^R + \mathbf{F}^D + \mathbf{F}^E.$$

In the GLE approach, the drag coefficient is considered dynamic, so that the drag force is given by [5]

$$F^D = -\int_0^t K\left(t - \tau\right)\dot{\mathbf{x}}\left(\tau\right) d\tau,$$

where $K(t)$ is a memory kernel.

Since the external force is considered conservative, from Vector Calculus we know that this implies it arises from some potential field $V(\mathbf{x})$, such that

$$\mathbf{F}^E = -\nabla V(\mathbf{x}).$$

The resulting equation of motion is

$$m\ddot{\mathbf{x}}(t) = \mathbf{f}(t) - \int_0^t K(t-\tau)\dot{\mathbf{x}}(\tau)\,d\tau - \nabla V(\mathbf{x}). \tag{10}$$

The power of the GLE is that it is able to coarse-grain several degrees of freedom by describing: *(i)* explicitly the dynamics of variables of interests, which in this case corresponds to the position $\mathbf{x}(t)$ of a particle of mass m; and *(ii)* implicitly the remaining degrees of freedom through a memory kernel $K(t)$, a random noise $\mathbf{f}(t)$, and an external potential $V(\mathbf{x})$. For free diffusion, particle mobility is in response *only* to stochastic thermal forces, i.e., $\nabla V(\mathbf{x}) \equiv \mathbf{0}$, but in more complex systems external forces also play a role, $\nabla V(\mathbf{x}) \neq \mathbf{0}$.

The memory kernel, $K(t)$, represents a retarded effect of the frictional force, and to generate the correct equilibrium statistics, the random noise has to be related to this kernel in order to obey the fluctuation–dissipation theorem:

$$\langle f_i(t)f_j(s)\rangle = 2k_B T\,\delta_{ij}\,K(t-s).$$

Physically, the kernel $K(t)$ represents the fact that the medium requires a finite time to respond to any fluctuations in the motion of the particle; this in turn affects how the medium acts back on the particle. Thus, the force that the medium exerts on the particle at a given time depends on what the particle did in the past.

For simple fluids and large Brownian particles, the medium is capable of responding infinitely quickly to changes in the particle position, i.e., it has no memory. In this case the memory kernel is a delta-function and Equation (10) reduces to the Langevin equation previously discussed [3,5]:

$$\begin{aligned}
K(t-\tau) &= \zeta\,\delta(t-\tau) & (11)\\
m\ddot{\mathbf{x}}(t) &= \mathbf{f}(t) - \zeta\dot{\mathbf{x}}(t)\,d\tau - \nabla V(\mathbf{x}). & (12)
\end{aligned}$$

Another extreme is a very sluggish medium that responds slowly to changes in the particle position. In this case, one can assume $K(t) \approx K(0) = K_0$, so that the GLE becomes

$$\begin{aligned}
m\ddot{\mathbf{x}}(t) &= \mathbf{f}(t) - K_0\int_0^t \dot{\mathbf{x}}(\tau)\,d\tau - \nabla V(\mathbf{x})\\
&= \mathbf{f}(t) - K_0\,(\mathbf{x}(t) - \mathbf{x}(0)) - \nabla V(\mathbf{x})\\
&= \mathbf{f}(t) - \nabla\left(V(\mathbf{x}) + \frac{K_0}{2}\|\mathbf{x} - \mathbf{x}(0)\|^2\right), & (13)
\end{aligned}$$

thus adding an extra harmonic term to the potential. Such a term has the effect of trapping the system of particles in certain regions of its configuration space, an effect known as dynamic caging [10].

2.4. Zero-Mass Limit of the Langevin Equation

We finish this section on Langevin-type equations with a simplification. If we assume $K(t) = \zeta\,\delta(t)$, and that the particles are so small that their mass is negligible, we obtain the so-called zero-mass limit:

$$0 \;=\; \mathbf{f}(t) - \zeta \dot{\mathbf{x}}(t) - \nabla V(\mathbf{x}).$$

This limit is also known as the *overdamped* or *inertialess limit* since it assumes that the inertial forces, $m\mathbf{a}$, are negligible compared to the other forces acting on the particle. Rearranging terms and recalling that $\langle f_i(t) f_j(s) \rangle = 2\,\zeta\,k_B T\,\delta_{ij}\delta(t-s)$ gives

$$
\begin{aligned}
\zeta \frac{d\mathbf{x}}{dt} &= -\nabla V(\mathbf{x}) + \mathbf{f}(t), \\
&= -\nabla V(\mathbf{x}) + \sqrt{2\zeta k_B T}\,\mathbf{W}(t), \\
\frac{d\mathbf{x}}{dt} &= -\frac{1}{\zeta}\nabla V(\mathbf{x}) + \frac{1}{\zeta}\sqrt{2\zeta k_B T}\,\mathbf{W}(t), \\
&= -\frac{1}{\zeta}\nabla V(\mathbf{x}) + \sqrt{\frac{2 k_B T}{\zeta}}\,\mathbf{W}(t), \\
\frac{d\mathbf{x}}{dt} &= -\frac{1}{\zeta}\nabla V(\mathbf{x}) + \sqrt{2 D}\,\mathbf{W}(t),
\end{aligned}
\tag{14}
$$

where $\mathbf{W}$ is a normally distributed random noise with $\langle \mathbf{W} \rangle = 0$ and $\langle W_i(t)W_j(s) \rangle = \delta_{ij}\,\delta(t-s)$. For simplicity, in the following sections, we only consider the relation between equations of the form (14) and their respective Fokker–Planck equations.

3. Fokker–Planck Equation

As discussed in the previous section, when a system is described by an LE, a complete description of the macroscopic system will require the solution and averaging of *many* SDEs. An equivalent approach is to describe the system by macroscopic variables which fluctuate as a result of stochasticity, instead of describing the individual evolution of stochastic probes [11]. An excellent explanation of the different representations and their characteristics can be found in Risken's book [11], which we summarize in Figure 4.

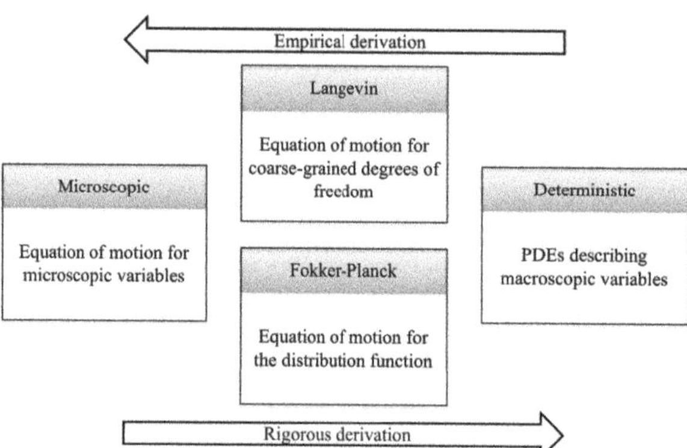

Figure 4. Levels of description of a system, adapted from [11].

A Fokker–Planck (FP) equation is a partial differential equation that describes the evolution of the probability density function (PDF) of a stochastic variable. For Langevin-type equations of the form given by Equation (14), the stochastic variable is a particle's position as a function of time, $x(t)$. The corresponding PDF is the function that gives the probability of a particle being in the position x at time t as $P(x,t)\,dx$. The reader is referred to Appendix B for a brief introduction to PDFs.

The LE equation given by Equation (14) can be written as

$$\begin{aligned}
\frac{dx}{dt} &= -\frac{1}{\zeta}\nabla V(x) + \sqrt{2D}\,W(t), \\
&= g_1(x,t) + g_2(x,t)\,W(t),
\end{aligned}$$

and the corresponding FP is given by [11],

$$\frac{\partial P}{\partial t} = -\nabla \cdot [g_1(x,t)\,P(x,t)] + \frac{1}{2}\nabla^2 \left[(g_2(x,t))^2\,P(x,t) \right]. \tag{15}$$

Note that taking a *deterministic* perspective is equivalent to ignoring the random noise term in the LE, $g_2(x,t) = 0$, which results in the absence of the diffusion term in the FP equation, $\nabla^2 [(\cdot)\,P(x,t)]$. This simple statement helps us identify the relation between fluctuations at the microscale and diffusion at the macroscale. That is, the observed diffusion at the macroscale is the result of fluctuations arising from the fluid's molecules bombarding the probe at the microscale.

3.1. One-Dimensional Examples

In the following examples, we assume that all initial positions of particles are located at zero, that is,

$$\begin{aligned}
x(0) &= 0 \\
P(x,0) &= \delta(0).
\end{aligned}$$

- *Brownian motion without external field for small particles* ($m \to 0$)

$$\frac{dx}{dt} = \sqrt{2D}W(t). \tag{16}$$

In this case, the corresponding FP equation is the well-known diffusion equation, which has the same mathematical form of the heat equation in the context of heat transfer under temperature gradients:

$$\begin{aligned}
\frac{\partial P}{\partial t} &= -\frac{\partial}{\partial x}[0 \cdot P] + \frac{1}{2}\frac{\partial^2}{\partial x^2}\left[\left(\sqrt{2D}\right)^2 P \right] \\
\frac{\partial P}{\partial t} &= D\frac{\partial^2 P}{\partial x^2}. \tag{17}
\end{aligned}$$

We can find the general solution of this equation using similarity solutions with the transformation

$$P(x,t) = t^{-p}\,F(\mu), \tag{18}$$

where t^{-p} represents temporal decay and $\mu = x^2/(4Dt)$ is a shape factor used to reduce the partial differential equation (PDE) to an ordinary differential equation (ODE). For details on how this particular form is obtained, the reader is referred to [12].

We can calculate the derivatives of $P(x,t)$ using Equation (18) as

$$\frac{\partial P}{\partial t} = -pt^{-p-1}F(\mu) + t^{-p}\frac{dF}{d\mu}\frac{\partial \mu}{\partial t} = -pt^{-p-1}F(\mu) - \mu\, t^{-p-1}\frac{dF}{d\mu}$$

$$\frac{\partial P}{\partial x} = t^{-p}\frac{dF}{d\mu}\frac{\partial \mu}{\partial x} = \frac{x\,t^{-p-1}}{2D}\frac{dF}{d\mu}$$

$$\frac{\partial^2 P}{\partial x^2} = \frac{t^{-p-1}}{2D}\frac{dF}{d\mu} + \frac{x\,t^{-p-1}}{2D}\frac{d^2F}{d\mu^2}\frac{\partial \mu}{\partial x} = \frac{t^{-p-1}}{2D}\frac{dF}{d\mu} + \frac{\mu\, t^{-p-1}}{D}\frac{d^2F}{d\mu^2}$$

Substituting in Equation (17) gives

$$-pt^{-p-1}F(\mu) - \mu\, t^{-p-1}\frac{dF}{d\mu} = D\left[\frac{t^{-p-1}}{2D}\frac{dF}{d\mu} + \frac{\mu\, t^{-p-1}}{D}\frac{d^2F}{d\mu^2}\right]$$

$$\mu\frac{d}{d\mu}\left(\frac{dF}{d\mu} + F\right) + \frac{1}{2}\left(\frac{dF}{d\mu} + 2pF\right) = 0.$$

Since we have yet to define a value for p, we conveniently choose it to be $p = 1/2$, so that the two quantities in the parenthesis are the same. Finally, a solution for the resulting differential equation will satisfy

$$\frac{dF}{d\mu} + F = 0,$$

with the general solution

$$F(\mu) = C_0\, e^{-\mu},$$

which gives the solution for $P(x,t)$,

$$P(x,t) = C_0 t^{-1/2}\exp\left(-\frac{x^2}{4Dt}\right).$$

To find the value of the constant of integration C_0, we consider the fact that

$$\int_{-\infty}^{\infty} P(x,t)\, dx = 1,$$

that is, all possible realizations are included, see Appendix B.

To solve the integral, we introduce the error function $\mathrm{erf}(y)$

$$\mathrm{erf}(y) = \frac{2}{\sqrt{\pi}}\int_0^y e^{-s^2}\, ds$$

which has values $\mathrm{erf}(-\infty) = -1$ and $\mathrm{erf}(\infty) = 1$.

$$
\begin{aligned}
\int_{-\infty}^{\infty} P(x,t)\, dx = \int_{-\infty}^{\infty} C_0 t^{-1/2} \exp\left(-\frac{x^2}{4Dt}\right) dx &= C_0 t^{-1/2} \int_{-\infty}^{\infty} \exp\left(-\frac{x^2}{4Dt}\right) dx \\
&= C_0 t^{-1/2} \sqrt{\pi D t}\, \mathrm{erf}\left(\frac{x}{2\sqrt{Dt}}\right)\Big|_{-\infty}^{\infty} \\
&= C_0 t^{-1/2} \sqrt{\pi D t}\,(2) = 2 C_0 \sqrt{\pi D},
\end{aligned}
$$

which gives

$$
C_0 = \frac{1}{2\sqrt{\pi D}} = \frac{1}{\sqrt{4\pi D}}.
$$

Therefore, the solution of the FP equation is

$$
P(x,t) = \frac{1}{\sqrt{4\pi D t}} \exp\left(-\frac{x^2}{4Dt}\right), \tag{19}
$$

which is a one-dimensional Gaussian function centered at zero: $M_1(x) = 0$ and with variance $M_2(x) = 4Dt$.

To compare these results with those obtained in the LE section, we consider the first and second moments of $P(x,t)$.

The first moment is given by

$$
\begin{aligned}
M_1(x) &= \int_{-\infty}^{\infty} x\, P(x,t)\, dx \\
&= \int_{-\infty}^{\infty} x\, \frac{1}{\sqrt{4\pi D t}} \exp\left(-\frac{x^2}{4Dt}\right) dx = \frac{1}{\sqrt{4\pi D t}} \int_{-\infty}^{\infty} x \exp\left(-\frac{x^2}{4Dt}\right) dx.
\end{aligned}
$$

We use integration by substitution ($u = x^2$) to obtain

$$
\int x \exp\left(-\frac{x^2}{4Dt}\right) dx = \int \frac{1}{2} \exp\left(-\frac{u}{4Dt}\right) du = -2Dt \exp\left(-\frac{u}{4Dt}\right) + C = -2Dt \exp\left(-\frac{x^2}{4Dt}\right) + C,
$$

and

$$
M_1(x) = \frac{1}{\sqrt{4\pi D t}} \left[-2Dt \exp\left(-\frac{x^2}{4Dt}\right)\right]_{-\infty}^{\infty} = 0.
$$

For the second moment, we have

$$
\begin{aligned}
M_2(x) &= \int_{-\infty}^{\infty} x^2\, P(x,t)\, dx \\[4pt]
&= \int_{-\infty}^{\infty} x^2\, \frac{1}{\sqrt{4\pi D t}}\, \exp\left(-\frac{x^2}{4Dt}\right) dx \\[4pt]
&= \frac{1}{\sqrt{4\pi D t}} \int_{-\infty}^{\infty} x^2\, \exp\left(-\frac{x^2}{4Dt}\right) dx \\[4pt]
&= \frac{1}{\sqrt{4\pi D t}} \left[-2x\, Dt\, \exp\left(-\frac{x^2}{4Dt}\right) \Big|_{-\infty}^{\infty} + \frac{(4Dt)^{3/2}\sqrt{\pi}}{4}\, \mathrm{erf}\left(\frac{x}{\sqrt{4Dt}}\right) \Big|_{-\infty}^{\infty} \right] \\[4pt]
&= \frac{1}{\sqrt{4\pi D t}} \left[0 + \frac{(4Dt)^{3/2}\sqrt{\pi}}{4}(2) \right] = \frac{1}{\sqrt{4\pi D t}} \left[\frac{(4Dt)^{3/2}\sqrt{\pi}}{2}\sqrt{\pi} \right] \\[4pt]
&= 2Dt,
\end{aligned}
$$

which is the same result we obtained in the limit $t \to \infty$ for Equation (8), when the dimensionality is $d = 1$.

Solutions at different times are shown in Figure 5, together with the normalized histograms obtained from LE data. For details of the histogram normalization see Appendix B.

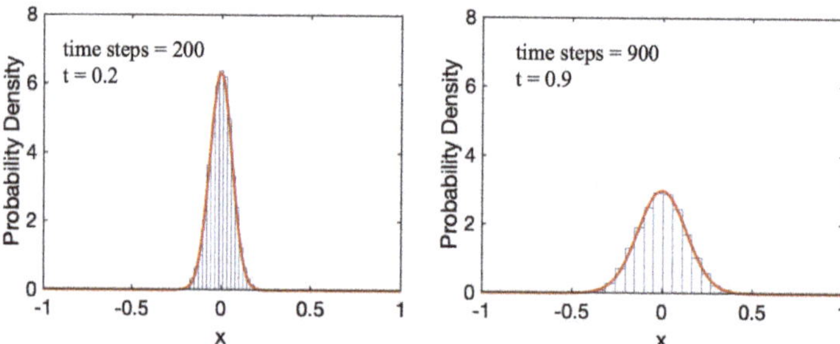

Figure 5. Probability density function (PDF) for position of particles diffusing via one-dimensional Brownian motion. Histograms correspond to LE data and solid lines corresponds to FP solutions. In this figure, $D = 10^{-2}\ \mu\mathrm{m}^2/\mathrm{s}$.

- *Brownian motion with external field for small particles ($m \to 0$)*

A common example of an external field is a background velocity, u, which imposes a drift on the particles:

$$
\frac{dx}{dt} = u + \sqrt{2D}\, W(t). \tag{20}
$$

The corresponding FP equation is the advection–diffusion equation

$$
\begin{aligned}
\frac{\partial P}{\partial t} &= -\frac{\partial}{\partial x}[u \cdot P] + \frac{1}{2}\frac{\partial^2}{\partial x^2}\left[\left(\sqrt{2D}\right)^2 P \right] \\[6pt]
\frac{\partial P}{\partial t} &= -u\frac{\partial P}{\partial x} + D\frac{\partial^2 P}{\partial x^2}
\end{aligned}
\tag{21}
$$

The solution of this PDE is [13]

$$P(x,t) = \frac{1}{\sqrt{4\pi DT}} \exp\left(-\frac{(x-ut)^2}{4Dt}\right).$$

(22)

A comparison between Equations (19) and (22) shows that the only difference between these two solutions is in a "shift" of x by ut. That is, the effect of drift is to move the mean of the Gaussian distribution from zero.

As before we can calculate the first and second moment of the distribution function as [13]

$$
\begin{aligned}
M_1(x) &= ut, \\
M_2(x) &= u^2 t^2 + 2Dt.
\end{aligned}
$$

That is, the mean position of the particle is displaced by the background velocity over a distance ut. In addition, note that the MSD at long times becomes $\sim t^2$ due to the additional *linear* flow in the fluid.

In the case of simple diffusion, the dispersion of the particles can be attained from the MSD or equivalently the variance of the Gaussian. In the case where drift is present, dispersion is superimposed by the background flow, for this reason a more accurate measure of the dispersion is given by the metric [13]

$$\sigma^2 = \int_{-\infty}^{\infty} (x - M_1(x))^2 \, P(x,t) \, dx = 2Dt,$$

which gives back the linear behavior characteristic of standard diffusive processes. Note that in this equation, $M_1(x)$ is a central moment as opposed to the general definition given at the beginning of Section 2.1. These two moments will coincide if the mean is zero.

Plots of Equation (22) are shown in Figure 6, where the solutions with drift are compared to solutions without drift.

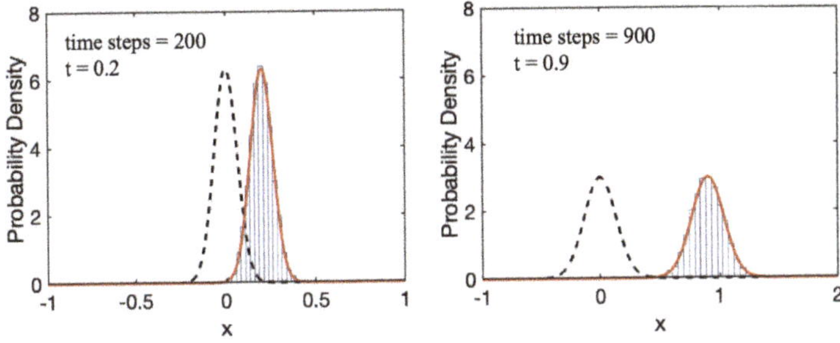

Figure 6. PDF for position of particles diffusing via one-dimensional Brownian motion plus drift. Histograms correspond to generalized Langevin equations (GLE) data and solid lines corresponds to FP solutions. For comparison, FP solutions without drift are shown in dashed lines. In this figure, $D = 10^{-2} \mu m^2/s$, $u = 0.2\mu/s$.

3.2. Two-Dimensional Examples

In this section, we present the 2D equations corresponding to four different cases and their numerical solutions.

- *No external field*

    - Langevin equations

$$\frac{dx}{dt} = \sqrt{2D}W_x(t),$$
$$\frac{dy}{dt} = \sqrt{2D}W_y(t).$$

    In this case, both W_x and W_y are statistically independent white noises; the subscripts are used to denote that the noise histories are for x and y.

    - Fokker–Planck Equation

$$\frac{\partial P}{\partial t} = D\left[\frac{\partial^2 P}{\partial x^2} + \frac{\partial^2 P}{\partial y^2}\right].$$

Solutions for LE and FP representations for this case of no external field are provided in Figure 7.

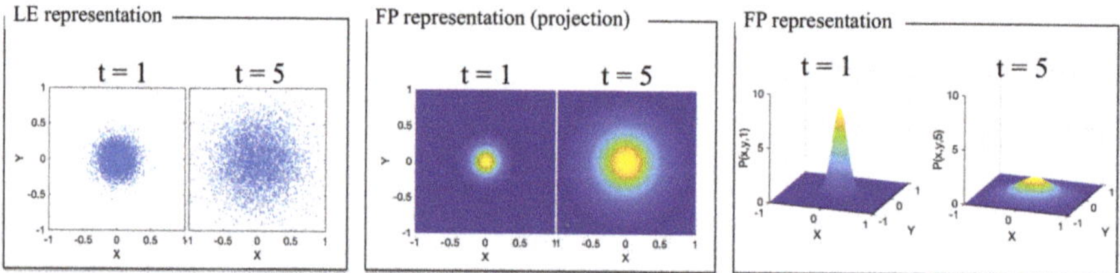

Figure 7. LE and FP solutions for two-dimensional Brownian motion without external field. For both representations, $D = 10^{-2}$ μm^2/s and time is in seconds. For the LE representation, the total number of particles is 5×10^3.

Note that, in two dimensions, we obtain two LEs, but still one FP equation. In general, as the degrees of freedom of the system increase, the choice between LE and FP representations is analogous to the choice between solving many SDEs and solving a single, high-dimensional PDE.

- *Constant drift in the x-direction, $V(x,y) = u\,x$*

    - Langevin equations

$$\frac{dx}{dt} = u + \sqrt{2D}W_x(t),$$
$$\frac{dy}{dt} = \sqrt{2D}W_y(t).$$

–   Fokker–Planck Equation

$$\frac{\partial P}{\partial t} = -u\frac{\partial P}{\partial x} + D\left[\frac{\partial^2 P}{\partial x^2} + \frac{\partial^2 P}{\partial y^2}\right].$$

Solutions for LE and FP representations for this case of constant drift in the x-direction are provided in Figure 8.

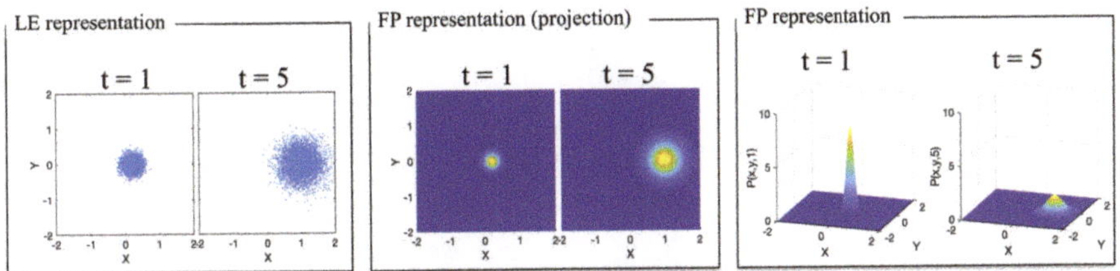

Figure 8. LE and FP solutions for two-dimensional Brownian motion with constant drift. For both representations, $D = 10^{-2}$ µm^2/s, $u = 0.2$ µ/s and time is in seconds. For the LE representation, the total number of particles is 5×10^3.

- *Background flow field,* $\mathbf{u} = \mathbf{u}_0 + \left[u_x(x,y), u_y(x,y)\right]^T$

For any background field of the form where u_0 and v_0 are constants,

$$\mathbf{u} = \left[\begin{array}{c} u_0 + u_x(x,y) \\ v_0 + u_y(x,y) \end{array}\right],$$

the Langevin equations are given by

$$\frac{dx}{dt} = u_0 + \left(\frac{\partial u_x}{\partial x}\right)x + \left(\frac{\partial u_x}{\partial y}\right)y + \sqrt{2D}W_x(t),$$

$$\frac{dy}{dt} = v_0 + \left(\frac{\partial u_y}{\partial x}\right)x + \left(\frac{\partial u_y}{\partial y}\right)y + \sqrt{2D}W_y(t).$$

This equation can be written in vector form as

$$\frac{d\mathbf{x}}{dt} = \mathbf{u}_0 + \kappa \cdot \mathbf{x} + \sqrt{2D}\mathbf{W},$$

where $\kappa_{ij} = \partial u_i / \partial x_j$ is the strain-rate tensor. The corresponding FP equation is

$$\frac{\partial P}{\partial t} = -\nabla \cdot \left[(\mathbf{u}_0 + \kappa \cdot \mathbf{x})\,P\right] + D\left[\frac{\partial^2 P}{\partial x^2} + \frac{\partial^2 P}{\partial y^2}\right].$$

Note that when $u_x = u_y = 0$, the equations reduce to those for constant drift, as discussed above.

- Example: simple shear, $\mathbf{u} = [U\,y, 0]^T$

$$\kappa = \begin{bmatrix} \frac{\partial u_x}{\partial x} & \frac{\partial u_x}{\partial y} \\ \frac{\partial u_y}{\partial x} & \frac{\partial u_y}{\partial y} \end{bmatrix} = \begin{bmatrix} 0 & U \\ 0 & 0 \end{bmatrix}.$$

$$\kappa \cdot \mathbf{x} = \begin{bmatrix} 0 & U \\ 0 & 0 \end{bmatrix} \cdot \begin{bmatrix} x \\ y \end{bmatrix} = \begin{bmatrix} U\,y \\ 0 \end{bmatrix}.$$

$$\nabla \cdot [(\kappa \cdot \mathbf{x})\,P] = \begin{bmatrix} \frac{\partial}{\partial x} \\ \frac{\partial}{\partial y} \end{bmatrix} \cdot \begin{bmatrix} U\,y\,P \\ 0 \end{bmatrix} = \begin{bmatrix} U\,y\frac{\partial P}{\partial x} \\ 0 \end{bmatrix}.$$

  * Langevin equations

$$\begin{aligned} \frac{dx}{dt} &= U y + \sqrt{2D}W_x(t), \\ \frac{dy}{dt} &= \sqrt{2D}W_y(t). \end{aligned}$$

  * Fokker–Planck Equation

$$\frac{\partial P}{\partial t} = -U y \frac{\partial P}{\partial x} + D\left[\frac{\partial^2 P}{\partial x^2} + \frac{\partial^2 P}{\partial y^2}\right].$$

Solutions for LE and FP representations for this case of simple shear flow are provided in Figure 9.

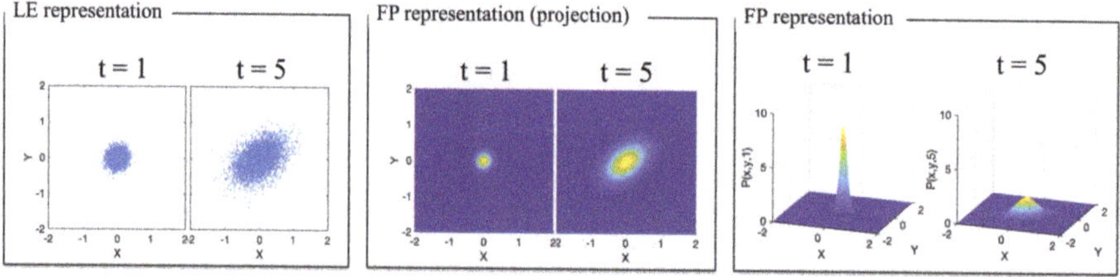

Figure 9. LE and FP solutions for two-dimensional Brownian motion with simple shear flow. For both representations, $D = 10^{-2}\ \mu\mathrm{m}^2/\mathrm{s}$, $U = 0.2\ \mu/\mathrm{s}$ and time is in seconds. For the LE representation, the total number of particles is 5×10^3.

4. Conclusions

In this paper, we analyzed the relation between Langevin (LE) and Fokker–Planck (FP) representations of particles moving in a fluid due to Brownian motion with and without external fields. Although each description offers a different perspective of the underlying fluid dynamics, there is a direct connection between these two approaches, which were explored through simple examples in both one and two dimensions. In studying these two families of equations, we employed subjects from calculus, such as

Taylor expansions and conservative vector fields; ordinary differential equations, such as integrating factors; and partial differential equations, such as similarity solutions.

The LE representation involves stochastic differential equations (SDEs) and offers the advantage of allowing the microscopic process to be easily incorporated into the equations. One of the drawbacks of this representation is that it is necessary to have as many SDEs as degrees of freedom in the system and each SDE needs to be solved many times to reduce statistical noise.

The FP approach involves a partial differential equation (PDE) describing the evolution of the probability density function (PDF) of the stochastic variable. Unlike numerical solutions for SDEs, which are noisy, solutions for the FP are deterministic and as such can truly avoid noise. One of its drawbacks is that the dimensionality of the PDF increases with the number of degrees of freedom of the system, which leads to high algorithmic complexity, and as a result, the corresponding numerical schemes have a high computational cost.

Experimentally, LE equations can be informed by techniques that capture the movement of probes at the microscale such as passive microrheology [8]. On the other hand, FP equations can be informed by experimental techniques that capture the behavior of the ensemble of probes such as light scattering experiments [14].

Author Contributions: Conceptualization, A.M. and P.A.V.; methodology, A.M. and P.A.V.; software, A.M., R.D. and P.A.V.; validation, A.M., R.D. and P.A.V.; formal analysis, A.M. and P.A.V.; investigation, A.M. and P.A.V.; resources, P.A.V.; data curation, P.A.V.; writing—original draft preparation, A.M., R.D. and P.A.V.; writing—review and editing, A.M. and P.A.V.; visualization, A.M., R.D. and P.A.V.; supervision, P.A.V.; project administration, P.A.V.; funding acquisition, P.A.V. All authors have read and agreed to the published version of the manuscript.

Funding: This research was funded by National Science Foundation, grant number NSF-DMS 1751339 and NASA, grant number 80NSSC19K0159 P00004.

Appendix A

```matlab
%% Constants
    D  = 1e-2;  % diffusion coefficient
    dt = 1e-3;  % time step
    Np = 5e2;   % number of particles
    NT = 1e3;   % number of time steps


%% Generate particle positions
    x = zeros(Np,NT);        y = zeros(Np,NT);        z = zeros(Np,NT);
    for k=2:NT
        x(:,k) = x(:,k-1) + sqrt(2*D*dt)*randn(Np,1);
        y(:,k) = y(:,k-1) + sqrt(2*D*dt)*randn(Np,1);
        z(:,k) = z(:,k-1) + sqrt(2*D*dt)*randn(Np,1);
    end


%%  Calculate MSD
    lag = round((1/3)*NT):round((2/3)*NT);
    for k=1:length(lag)
        dx = x(:,1+lag(k):end)-x(:,1:end-lag(k));
        dy = y(:,1+lag(k):end)-y(:,1:end-lag(k));
        dz = z(:,1+lag(k):end)-z(:,1:end-lag(k));
```

```matlab
        msd1(k,:) = mean(dx'.^2);
        msd2(k,:) = mean(dx'.^2+dy'.^2);
        msd3(k,:) = mean(dx'.^2+dy'.^2+dz'.^2);
    end

%% Plot MSDs
    plot(lag*dt,mean(msd1'),'linewidth',3)
    hold on
    plot(lag*dt,mean(msd2'),'—','linewidth',3)
    plot(lag*dt,mean(msd3'),'—.','linewidth',3)

%% Recover D
        fmsd1 = fit(lag'*dt,mean(msd1')','a*x','start',1);
        fmsd2 = fit(lag'*dt,mean(msd2')','a*x','start',1);
        fmsd3 = fit(lag'*dt,mean(msd3')','a*x','start',1);

        fmsd1.a/2
        fmsd2.a/4
        fmsd3.a/6
```

Appendix B

Recall that in Figure 2, the trajectory of three different particles were plotted. Assume we are interested in determining the x-coordinate of the final position of a particle. From the figure, we would have three different answers: $x = 0.0140$, $x = -0.0526$, and $x = 0.0081$. It is clear that the fact that each plot gives a different answer is a result of the randomness of the process. This simple observation implies that the correct question is not *what is the position of the particle?* but rather *what is the most likely position of the particle?* To answer this, we collect the final position of 1000 particles and summarize them using histograms, like the ones shown in Figure A1.

To construct histograms, we divide the data into groups or 'bins' and count how many realizations fall within each bin. The next question is how to extract probabilities out of these histograms. A first attempt to calculate probabilities is to divide the number of realizations within each bin by the total number of realizations. An inspection of Figure A1 shows why this is not the correct approach; although both figures (**B**) and (**D**) correspond to the same data, the answer is different depending on the number of bins used.

In order to *transform* the histogram data into probabilities, it is necessary to eliminate the effect of the bin size. This is done by normalizing the histogram using the area of each bin, rather than the counts in each bin. This normalization is shown in Figure A2.

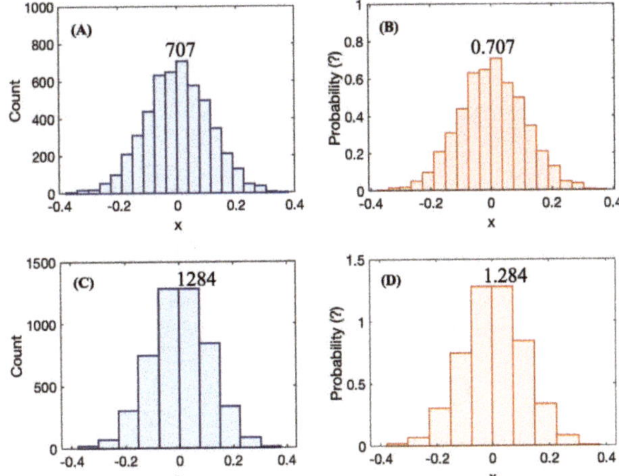

Figure A1. Histograms of particle positions for 1000 particles. (**A**) counts using 20 bins; (**B**) counts from (**A**) divided by total number of particles; (**C**) counts using 10 bins; (**D**) counts from (**C**) divided by total number of particles.

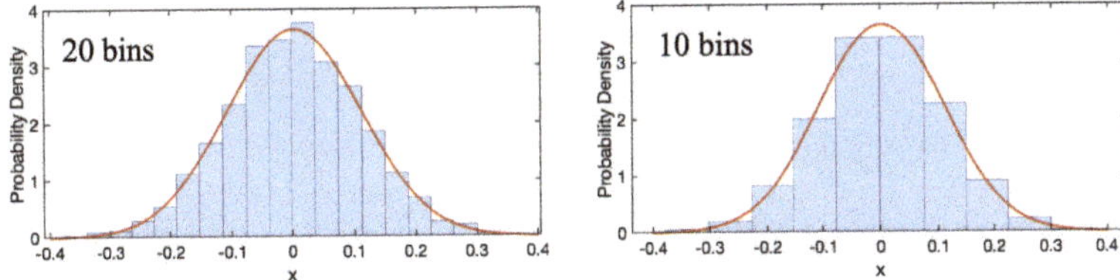

Figure A2. Normalized histogram of particle positions. The code used to generate these plots can be found in Appendix B.

Since the size of the bins does not affect the resulting plot, we could make the size of the bins as small as we wish, to the point of being able to trace a continuous probability density function, represented by the continuous line in Figure A2, which is unique to the data independently of the number of bins used.

Just as its name indicates, a PDF is not exactly a probability. However, just as one can find the mass of an object by multiplying its density by its volume, we can find a probability by multiplying is probability *density* (PDF) by a range. In this sense, the quantity

$$P(x, t)\, dx$$

represents the probability that a particle's position is in the range $[x, x + dx]$ at time t.

In addition, working with the continuous form of the PDF, $P(x, t)$, allows us to use integrals to find different probabilities as areas under the curve, as the example given in Figure A3, or to provide a definition of statistical moments based on PDFs:

$$M_n(x) \;=\; \int_{-\infty}^{\infty} x^n\, P(x, t)\, dx$$

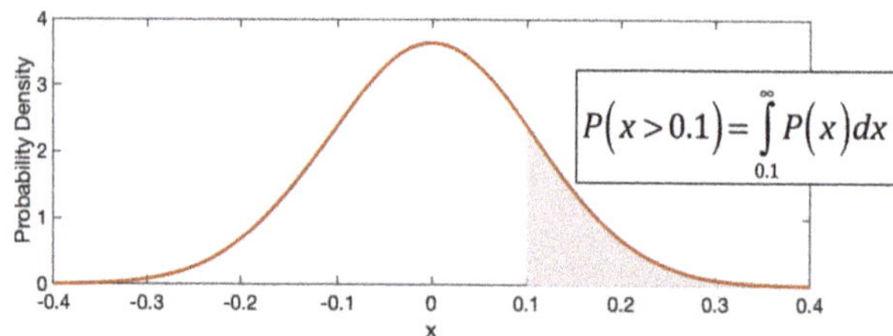

Figure A3. Calculating probabilities using PDFs.

Note that since the area under the PDF should represent all possible realizations, we have the condition that the area under the whole curve should be equal to one:

$$\int_{-\infty}^{\infty} P(x,t)\,dx = 1. \tag{A1}$$

As a final remark, note that not all PDFs are bell-shaped like a Gaussian. The only condition that remains true for all PDFs is that the area under the curve is equal to one as defined in Equation (A1). In addition, not all PDFs have to define a probability density for every location of the sample space. For a more in-depth review of different probability distribution functions, the reader is referred to [15].

```matlab
%% Computing and Normalizing histogram
   [c,n] = hist(x(:,600),20);
   c1 = c./trapz(n,c);

%% Fitting to normal distribution
   [mu,sigma,muci,sigmaci] = normfit(x(:,600));
   Y = normpdf(linspace(-0.4,0.4,1e3),mu,sigma);

%% Plots
   bar(n,c1)
   hold on
   plot(linspace(-0.4,0.4,1e3),Y)
```

References

1. Einstein, A. Uber die von der molekularkinetischen Theorie der Warme geforderte Bewegung von in ruhenden Flussigkeiten suspendierten Teilchen. *Annalen der Physik* **1905**, *322*, 49–560. [CrossRef]
2. Karatzas, I.; Shreve, S.E. *Brownian Motion and Stochastic Calculus*; Springer: New York, NY, USA, 1998.
3. Coffey, W.; Kalmykov, Y.P. *The Langevin Equation: With Applications to Stochastic Problems in Physics, Chemistry and Electrical Engineering*; World Scientific: Singapore, 2012; Volume 27.
4. Arnold, L. *Stochastic Differential Equations: Theory and Applications*; John Wiley and Sons: New York, NY, USA, 1974.
5. Kubo, R. The fluctuation-dissipation theorem. *Rep. Prog. Phys.* **1966**, *29*, 255. [CrossRef]
6. March, N.H.; Tosi, M.P. *Introduction to Liquid State Physics*; World Scientific: Singapore, 2002.
7. Mason, T.G.; Ganesan, K.; Van Zanten, J.H.; Wirtz, D.; Kuo, S.C. Particle tracking microrheology of complex fluids. *Phys. Rev. Lett.* **1997**, *79*, 3282. [CrossRef]
8. Furst, E.M.; Squires, T.M. *Microrheology*; Oxford University Press: Oxford, UK, 2017.

9. Klafter, J.; Blumen, A.; Shlesinger, M.F. Stochastic pathway to anomalous diffusion. *Phys. Rev. A* **1987**, *35*, 3081. [CrossRef] [PubMed]
10. Tuckerman, M. *Statistical Mechanics: Theory and Molecular Simulation*; Oxford University Press: Oxford, UK, 2010.
11. Risken, H. *Fokker-Planck Equation*; Springer: Berlin/Heidelberg, Germany, 1996.
12. Renardy, M.; Rogers, R.C. *An Introduction to Partial Differential Equations*; Springer Science & Business Media: Berlin/Heidelberg, Germany, 2006.
13. Codling, E.A.; Plank, M.J., and Benhamou, S. Random walk models in biology. *J. R. Soc. Interface* **2008**, *25*, 813–834. [CrossRef] [PubMed]
14. Sarmiento-Gomez, E.; Santamaria-Holek, I.; Castillo, R. Mean-square displacement of particles in slightly interconnected polymer networks. *J. Phys. Chem. B* **2014**, *118*, 1146–1158. [CrossRef] [PubMed]
15. Evans, M.J.; Rosenthal, J.S. *Probability and Statistics: The Science of Uncertainty*; Macmillan: New York, NY, USA, 2004

MDPI

St. Alban-Anlage 66

4052 Basel

Switzerland

Tel. +41 61 683 77 34

Fax +41 61 302 89 18

www.mdpi.com

Fluids Editorial Office

E-mail: fluids@mdpi.com

www.mdpi.com/journal/fluids